Radio Receiver Design

Radio Receiver Design

Kevin McClaning
Tom Vito

Noble Publishing Corporation
Atlanta, GA

Library of Congress Cataloging-in-Publication Data

McClaning, Kevin, 1959-
    Radio receiver design / Kevin McClaning, Tom Vito.
    p. cm.
    Includes bibliographical references and index.
    ISBN 1-884932-07-X
    1. Radio--Receivers and reception--Design and construction. I. Vito, Tom, 1953-II.
Title.

TK6563 .M38 2000
621.384'18--dc21

00-061271

NOBLE
PUBLISHING

Copyright 2000 by Noble Publishing Corporation.

All rights reserved. No part of this book may be reproduced in any form or by any means without prior written permission of the publisher.

Printed in the Unites States of America

ISBN 1-884932-07-X

Contents

Preface   **ix**
    References x

Introduction   **1**
    1.1   Nomenclature 1
    1.2   Rules of Thumb 2
    1.3   Energy, Power, Voltage and Current 2
    1.4   Decibels 2
    1.5   Signal Standards 6
    1.6   Other Standards 10
    1.7   dB Math 12
    1.8   Sanity Checking 18
    1.9   Approximations 20
    1.10   Frequency, Wavelength and Propagation Velocity 21
    1.11   Transmission Lines 25
    1.12   Two Ports 62
    1.13   Matching and Maximum Power Transfer 72
    1.14   Matching 75
    1.15   RF Amplifiers 78
    1.16   Signals, Noise and Modulation 86
    1.17   Introduction Data Summary 133
    1.18   References 140

Filters   **143**
    2.1   Introduction 143
    2.2   Linear Systems Review 144
    2.3   Evaluating Pole/Zero Plots 160
    2.4   Filters and Systems 163
    2.5   Filter Types and Terminology 165
    2.6   Generic Filter Responses 168
    2.7   Classes of Low-Pass Filters 179
    2.8   Low-Pass Filter Comparison 195
    2.9   Filter Input and Output Impedances 199
    2.10   Transient Response of Filters 207
    2.11   Band-Pass Filters 215
    2.12   Other Filters 225
    2.13   Noise Bandwidth 229

2.14	Butterworth Filters in Detail	236
2.15	Filter Technologies and Realizations	250
2.16	Miscellaneous Items	253
2.17	Filter Design Summary	255
2.18	References	258

Mixers   261

3.1	Introduction	261
3.2	Frequency Translation Mechanisms	263
3.3	Nomenclature	267
3.4	Block vs. Channelized Systems	284
3.5	Conversion Scheme Design	285
3.6	Frequency Inversion	292
3.7	Image Frequencies	296
3.8	Other Mixer Products	301
3.9	Spurious Calculations	307
3.10	Mixer Realizations	313
3.11	Single-Ended Mixers (SEM)	313
3.12	Single-Balanced Mixers (SBM)	320
3.13	Double-Balanced Mixers (DBM)	329
3.14	Further Mixer Characteristics	334
3.15	Other Uses of Mixers	337
3.16	Mixer Design Summary	340
3.17	References	343

Oscillators   345

4.1	Introduction	345
4.2	Phase Noise	352
4.3	Phase Modulation Review: Sinusoidal Modulation	364
4.4	Relating $\phi(f_m)$ to $\mathcal{L}(f_m)$	375
4.5	$S_\phi(f_m)$	378
4.6	When the Small β Conditions are Valid	378
4.7	Phase Noise and Multipliers	379
4.8	Phase Noise and Dividers	381
4.9	Incidental Phase Modulation	382
4.10	Incidental Frequency Modulation	389
4.11	Comparison of IPM and IFM	393
4.12	Other Phase Noise Specifications	394
4.13	Spurious Considerations	395
4.14	Frequency Accuracy	399
4.15	Other Considerations	407
4.16	Oscillator Realizations	408
4.17	Numerically Controlled Oscillator (NCO)	417

4.18	Crystal Reference Oscillators	433
4.19	Oscillator Design Summary	435
4.20	References	440

Amplifiers and Noise 443

5.1	Introduction	443
5.2	Equivalent Model for a RF Device	443
5.3	Noise Fundamentals	446
5.4	Thermal Noise Properties	454
5.5	Noise Power	465
5.6	One Noisy Resistor	465
5.7	System Model — Two Noisy Resistors	469
5.8	Internally Generated System Noise Model	479
5.9	Signal-to-Noise Ratio (SNR)	484
5.10	Noise Factor and Noise Figure	487
5.11	Cascade Performance	497
5.12	Examining the Cascade Equations	507
5.13	Minimum Detectable Signal (MDS)	508
5.14	Noise Temperature Measurement	509
5.15	Lossy Devices	519
5.16	Amplifier and Noise Summary Data	536
5.17	References	539

Linearity 541

6.1	Introduction	541
6.2	Linear and Nonlinear Systems	542
6.3	Amplifier Transfer Curve	543
6.4	Polynomial Approximations	550
6.5	Single-Tone Analysis	552
6.6	Two-Tone Analysis	556
6.7	Distortion Summary	565
6.8	Preselection	567
6.9	Second-Order Distortion	568
6.10	Third-Order Distortion	572
6.11	Narrowband and Wideband Systems	575
6.12	Higher-Order Effects	577
6.13	Second-Order Intercept Point	578
6.14	Third-Order Intercept Point	587
6.15	Measuring Amplifier Nonlinearity	595
6.16	Gain Compression and Output Saturation	599
6.17	Comparison of Nonlinear Specifications	603
6.18	Nonlinearities in Cascade	605
6.19	Distortion Notes	622

6.20	Nonlinearities and Modulated Signals	626
6.21	Linearity Design Summary	633
6.22	References	638

Cascade I — Gain Distribution 639

7.1	Introduction	639
7.2	Minimum Detectable Signal (MDS)	641
7.3	Dynamic Range	643
7.4	Spur-Free Dynamic Range	644
7.5	Dynamic Range Notes	652
7.6	Gain Distribution	654
7.7	Gain and Noise	661
7.8	Gain and Linearity	672
7.9	System Nonlinearities	685
7.10	TOI Tone Placement	687
7.11	Automatic Gain Control (AGC)	691
7.12	Cascade Gain Distribution Rules	694
7.13	Cascades, Bandwidth and Cable Runs	697
7.14	Cascade I Design Summary	700
7.15	References	702

Cascade II — IF Selection 703

8.1	Introduction	703
8.2	Review	704
8.3	Image Noise	709
8.4	Upconversion vs. Downconversion	714
8.5	LOs, Tuning Range and Phase Noise	721
8.6	IF Selection Guidelines	724
8.7	Practical Design Considerations	726
8.8	A Typical System	733
8.9	Design Examples	736
8.10	Cascade II Design Summary	750
8.11	References	751
9.1	Gaussian or Normal Statistics	753

Appendix 753

9.2	Statistics and Noise	760
9.3	Cancellation and Balance of Cosine Waves	762
9.4	References	766

Index 767

Preface

> The theory is beautiful — but what can you do with it?
> —Anonymous
>
> My way is to begin with the beginning.
> —Don Juan

We wrote this book as engineers, with a focus on useful and proven concepts and not so much on technology. Some of the most useful engineering books we own were written in the 1940's by Frederick Terman, John Kraus and Mischa Schwartz [1, 2, 3]. Admittedly, their books are full of information on electron tubes and contain little on modern filter design, statistical decision theory and quadrature modulation, but they are clear and well-written. They were written by engineers for engineers and as such they remain useful. Our hope is that, in twenty years, people will say, the McClaning/Vito book is a little dated but it is clear and well-written.

We decided early to follow one cardinal rule: to be clear. If we have succeeded, errors should be easy to detect. If we hear about errors in this book (and I hope we do), I consider it an advantage. It means we have been clear enough to bring doubts into the reader's mind. It also means the reader is thinking about the material and understands it enough to find inconsistencies.

We would like to thank our normally noisy children, Chris and Jenny (McClaning) and Mandy, Nick, Steve and James (Vito) for being quiet while we worked on this book. We would also like to thank our wives, Kitty and Terri, for putting up with us (and not just while we were writing this book).

References

1. Kraus, John D. *Antennas*. New York: McGraw-Hill, 1950.

2. Schwartz, Mischa. *Information Transmission, Modulation, and Noise*. New York: McGraw-Hill, 1959.

3. Terman, Frederick E. *Radio Engineering*. 3rd ed. New York: McGraw-Hill, 1947.

1

Introduction

> Begin at the beginning, and go on 'til you come to the end; then stop.
> —Lewis Carroll, *Alice in Wonderland*
>
> The journey of a thousand miles begins with a single step.
> —Lao Tsu

Radio Receiver Design focuses on understanding and designing radio receivers and systems. Chapter 1 introduces many of the basic concepts that will be expanded on in the following chapters. For a more in-depths treatment of any of the topics under discussion, consult the material listed in the reference section at the end of the chapter.

1.1 Nomenclature

In this book, we work with quantities commonly expressed in both linear and logarithmic (or dB) formats. Any time a quantity is expressed in dB, it will have a dB subscript. For example, *noise figure* expressed in decibels will be written F_{db}. When noise figure is expressed as a linear number, it will not have a subscript (F). However, occasionally a linear quantity may be written as F_{lin}. One of the major sources of mistakes in receiver design results from using a quantity expressed in decibels when the equation requires the linear number.

1.2 Rules of Thumb

Rules of thumb serve as sanity checks and provide valuable insight into problems. They should, however, be used with caution because they are only approximations and cannot be binding. In each chapter, we will examine the appropriate places to apply these rules and point to their limitations.

1.3 Energy, Power, Voltage and Current

A transmitter produces a given amount of energy per unit time. We are interested in collecting this signal energy and making the best use of it. However, the signal energy competes with the noise energy also present in the signal path. As receiver designers, we are interested in received signal energy and noise energy, and we are not interested in voltage, current or power, except as they apply to received energy.

Jupiter Probe

Consider a space probe circling Jupiter. In order for a receiving system on earth to receive data from the probe, the probe must send out enough signal energy to overcome the noise energy we will also receive. Since

$$Energy = (Power)(Time) \qquad 1.1$$

the probe can send out either a large amount of power for a short period of time or it can transmit a smaller amount of power for a longer period of time (for a fixed amount of energy).

Many common signal quality measurements, such as signal-to-noise ratio and bit error rates, are measurements of signal energy versus noise energy that was received over a period of time. But since the signal and noise are processed over the same period of time, the time in Equation 1.1 becomes irrelevant, and we can meaningfully speak of *received signal power* and *received noise power*.

1.4 Decibels

Decibels were originally conceived to make measurements of human hearing more understandable. Human hearing covers a 1:10,000,000,000,000 (1:10^{13}) range from the threshold of hearing to the threshold of pain. This enormous range makes it difficult to plot and analyze data. Decibels make data comparisons easier and graphical data more manageable.

INTRODUCTION | 3

Figure 1-1 shows the range of powers that exist on a typical INMARSAT satellite link. The power levels fall over a 10^{26} range. Since Figure 1-1 uses decibels and a logarithmic scale, the data is very readable. Other examples of the large ranges system designers must accommodate include

- A typical communications receiver can easily process signals whose input powers vary by a factor of 10^{12} or 1 trillion.

- On its journey to a geosynchronous satellite a signal can be attenuated by a factor of 10^{20}.

- Filters routinely provide attenuation factors of 10^6 or 1 million.

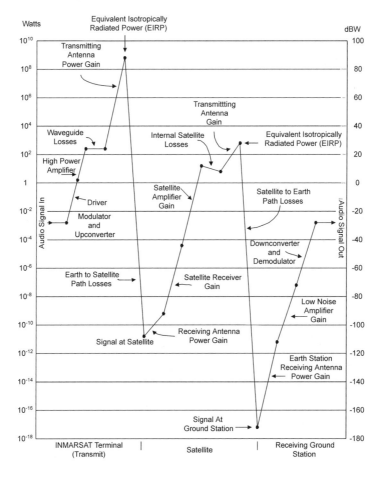

Figure 1-1 *Power budget of an INMARSAT satellite link.*

4 | RADIO RECEIVER DESIGN

To make these measurements more manageable, RF engineers have adapted the decibel to their own use. The decibel is one-tenth of a Bel, a logarithmic power ratio named after Alexander Graham Bell. His studies of human hearing showed that the ear responds to sound in a logarithmic rather than a linear fashion. For example, humans perceive a 10 times increase in sound power as a doubling of sound level.

Definitions

The Bel is defined as

$$\text{Bel} = B = \log\left(\frac{P_2}{P_1}\right) \quad 1.2$$

The decibel (or dB) is one-tenth of a Bel or

$$\text{dB} = 10\log\left(\frac{P_2}{P_1}\right) \quad 1.3$$

where P_2 and P_1 are signal powers. All other decibel relationships are derived from these two definitions. A decibel is always a dimensionless number.

Ratios

Both the Bel and decibel are ratios. It is meaningless to speak of something as "6 dB" unless it is clear which two items are being compared.

Example 1.1 — dB Power Conversions

A RF amplifier accepts $3.16 \cdot 10^{-6}$ watts from a signal source and supplies its load resistor with $2.2 \cdot 10^{-3}$ watts. What is the power gain of the RF amplifier? Answer in dB.

Solution —

Using Equation 1.3, we set P_2 equal to the amplifier's output power and P_1 equal to the amplifier's input power. $P_1 = 3.16 \cdot 10^{-6}$ watts and $P_2 = 2.2 \cdot 10^{-3}$ watts and

$$\begin{aligned} Gain_{dB} &= 10\log\left(\frac{2.2 \cdot 10^{-3}}{3.16 \cdot 10^{-6}}\right) \\ &= 10\log(696.2) \\ &= 28.4 \text{ dB} \end{aligned} \quad 1.4$$

The power gain, or simply the gain, of the amplifier is 28.4 dB.

Decibels, Current and Voltage

Figure 1-2 shows a simplified schematic diagram of an amplifier. We apply a voltage (V_{in}) to the amplifier.

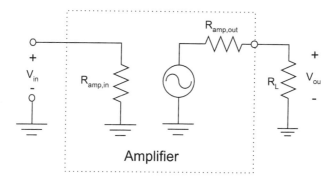

Figure 1-2 *Simplified schematic of an RF amplifier.*

The power being dissipated in the input resistor $R_{amp,in}$ is

$$P_{in} = \frac{V_{in}^2}{R_{amp,in}} \qquad 1.5$$

Similarly, the power delivered by the amplifier to the load resistor R_L is

$$P_{out} = \frac{V_{out}^2}{R_L} \qquad 1.6$$

Using Equation 1.3, the gain of the amplifier in decibels is

$$\begin{aligned} Gain_{dB} &= 10\log\left(\frac{P_{out}}{P_{in}}\right) \\ &= 10\log\left(\frac{V_{out}^2/R_L}{V_{in}^2/R_{amp,in}}\right) \qquad 1.7 \\ &= 20\log\left(\frac{V_{out}}{V_{in}}\right) + 10\log\left(\frac{R_{amp,in}}{R_L}\right) \end{aligned}$$

Equation 1.7 is correct. The last term of Equation 1.7 involving $R_{amp,in}$ and R_L is often ignored even in systems with wildly differing impedance levels. However, most RF systems have identical input and output impedances. In Figure 1-2, for example, $R_{amp,in}$, $R_{amp,out}$ and R_L are the same value (usually 50 or 75 ohms). If $R_{amp,in} = R_L$, then

$$10\log\left(\frac{R_L}{R_{amp,in}}\right) = 0 \quad \left(\text{when } R_L = R_{amp,in}\right) \qquad 1.8$$

and

$$G_{dB} = 20\log\left(\frac{V_{out}}{V_{in}}\right) \quad \text{when } R_L = R_{amp,in} \qquad 1.9$$

Similarly, we can show

$$G_{dB} = 20\log\left(\frac{I_{out}}{I_{in}}\right) \quad \text{when } R_L = R_{amp,in} \qquad 1.10$$

where I_{in} and I_{out} are the signal currents flowing in $R_{amp,in}$ and R_L of Figure 1-2. Equations 1.9 and 1.10 are true only if the source and load resistors are the same.

1.5 Signal Standards

The decibel format is commonly used to express absolute power or voltage levels. To do this, we replace P_1 in Equation 1.3 with a standard value. For example, if we are dealing with a system where milliwatt power levels are common, we might choose P_1 to be 1 mW and Equation 1.3 would become

$$\begin{aligned} P_{\text{Referenced to 1 mW}} &= 10\log\left(\frac{P_{\text{Measured,mW}}}{1 \text{ mW}}\right) \\ &= 10\log\left(P_{\text{Measured,mW}}\right) \\ &= 10\log\left(\frac{P_{\text{Measured,W}}}{0.001}\right) \end{aligned} \qquad 1.11$$

where
$P_{\text{Measured,mW}}$ = the power level measured in mW,
$P_{\text{Measured,W}}$ = the same power level measured in watts.

Note that Equation 1.11 defines the dBm standard.

dBm

dBm is the oldest standard way of expressing measurements. It was originally devised by a telephone company to measure signal levels on their lines and has been widely adapted. The reference power for dBm is 1 milliwatt. (The m in dBm stands for power referenced to 1 mW.) A quantity expressed in dBm represents an absolute power level, for example, P_{out} = 20 dBm.

$$0 \text{ dBm} = 1 \text{ mW} \tag{1.12}$$

$$\begin{aligned} P_{dBm} &= 10\log\left(\frac{P_{Watts}}{0.001}\right) \\ &= 10\log(P_{mW}) \end{aligned} \tag{1.13}$$

The impedance of a telephone line is about 600 ohms; hence 1mW (or 0 dBm) measures a little less than 0.775 V_{RMS}. In a 50-ohm system, a 1 mW or 0 dBm sine wave has a RMS voltage of 0.224 V_{RMS}.

dBW

dBW is a power measurement referenced to 1 watt. dBW is an absolute unit for expressing *power level* and is widely used in the world of transmitting equipment.

$$0 \text{ dBW} = 1 \text{ W} \tag{1.14}$$

$$P_{dBW} = 10\log(P_{Watts}) \tag{1.15}$$

Example 1.2 — dBm and dBW
Find an expression relating dBm and dBW.

Solution —
Solving Equation 1.13 for P_{Watts} produces

$$P_{Watts} = (0.001)\left(10^{P_{dBm}/10}\right) \tag{1.16}$$

Equating 1.16 to Equation 1.15 produces

$$\begin{aligned} P_{dBW} &= 10\log\left[0.001\left(10^{P_{dBm}/10}\right)\right] \\ &= 10\log(0.001) + 10\log\left(10^{P_{dBm}/10}\right) \\ \Rightarrow P_{dBw} &= P_{dBm} - 30 \end{aligned} \tag{1.17}$$

8 | RADIO RECEIVER DESIGN

To convert dBm to dBW, we add 30. For example, 0 dBW is +30 dBm, and both are equal to 1 watt.

Volume Units

This standard is meant for measuring speech or music. Like dBm, VU represents an absolute power level. The reference power for VU is 1 mW.

$$0 \text{ VU} = 0 \text{ dBm} = 1 \text{ mW} \qquad 1.18$$

$$P_{VU} = 10 \log\left(\frac{P_{Watts}}{0.001}\right)$$
$$= 10 \log(P_{mW}) \qquad 1.19$$

The major difference between measurements specified in dBm and VU is that the VU measurements are made with special meters. The meter dynamics are carefully controlled so they will respond appropriately to speech and music signals. The meters are specified to have a 300 msec rise time with 1.5% maximum overshoot and are designed to ignore short bursts.

dBf

This standard is used mostly to specify the sensitivity of consumer receiving equipment. The reference power is $1 \, fW = 1$ femtowatt $= 10^{-15}$ watts. dBf is an absolute power level.

$$0 \text{ dBf} = 1 \text{ fW} = 1 \text{ femtowatt} = 10^{-15} \text{ W} \qquad 1.20$$

$$P_{dBf} = 10 \log(P_{fW}) \qquad 1.21$$

The smaller the number, the more sensitive the receiver.

dBV

This standard is dB referenced to 1 volt$_{RMS}$. This is not a power reference unless some impedance is specified or assumed from the syntax.

$$0 \text{ dBV} = 1 \text{ Volt}_{RMS} \qquad 1.22$$

$$V_{dBV} = 20 \log(V_{Volts,RMS}) \qquad 1.23$$

Note the multiplier of 20 which arises because the voltage is squared to produce power and the missing

$$10\log\left(\frac{R_L}{R_{amp,in}}\right) \qquad 1.24$$

term. A quantity expressed in dBV represents an absolute voltage level in a system (for example, 2 V_{RMS} = 3 dBV). It can represent an absolute power level if the system impedance is specified or understood.

dBmV

dBmV is decibels referenced to 1 millivolt$_{RMS}$. This voltage standard is common in the video and cable TV industries.

$$0 \text{ dBmV} = 1 \text{ mV}_{RMS} \qquad 1.25$$

$$V_{dBmV} = 20\log\left(\frac{V_{Volts,RMS}}{0.001}\right) \qquad 1.26$$
$$= 20\log(V_{mVolts,RMS})$$

dBmV represents an exact voltage measurement (1 V_{RMS} = 30 dBmV) and can express power if the system impedance is known.

Example 1.3 — Power Measurements
Express 24 mW in dBm, dBW, dBf, dBV and dBmV. Assume a 50-ohm system.

Solution —

- dBm. Equation 1.16 produces

$$P_{dBm} = 10\log(P_{Measured,mW})$$
$$= 10\log(24) \qquad 1.27$$
$$= 13.8 \text{ dBm}$$

- dBW. Using Equation 1.15, P_{dBW} = 10log($P_{Measured,Watts}$). Since 24 mW = 24·10^{-3} W then P_{dBW} = 10log(24·10^{-3}) = –16.2 dBW.

- dBf. Using Equations 1.16 and 1.24, we know
$$24 \text{ mW} = 24 \cdot 10^{-3} \text{ Watts}$$
$$= 24 \cdot 10^{12} \text{ fW} \qquad 1.28$$
$$= 133.8 \text{ dBf}$$

To find the *dBV* and *dBmV* solutions, we need to determine how much voltage comes across a 50-ohm resistor that is dissipating 24 mW. Since $P = V_{RMS}^2/R$ with $P = 24 \cdot 10^{-3}$ W and $R = 50\ \Omega$, solving for $V_{RMS} = 1.1$ and $V_{RMS} = 1100$ mV$_{RMS}$.

- *dBV*: Using Equation 1.26, we know

$$V_{dBV} = 20\log(1.1)$$
$$= 0.83\ dBV \qquad 1.29$$

- *dBm*: Using Equation 1.29, we know

$$V_{dBmW} = 20\log\left(\frac{1.10}{0.001}\right)$$
$$= 60.8\ dBmV \qquad 1.30$$

Example 1.4 — Power Measurements

A Yamaha RX-700U Stereo receiver has a "usable sensitivity" specification of 9.3 dBf or 0.8 mV$_{RMS}$ at the antenna terminals. What is the input impedance of the stereo receiver?

Solution —

Using Equation 1.24, we know P_{dBf} = 9.3 dBf which implies

$$9.3\ dBf = 10\log(P_{dBf})$$
$$P_{dBf} = 8.51\ fW \qquad 1.31$$
$$= 8.51 \cdot 10^{-15}\ W$$

Since

$$P_{Watts} = \frac{V_{in,RMS}^2}{R_{Rcvr,in}} \qquad 1.32$$

and $V^2_{in,RMS} = 0.80 \cdot 10^{-6}$ V, then

$$8.51 \cdot 10^{-6}\ W = \frac{(0.8 \cdot 10^{-6}\ V_{RMS})^2}{R_{Rcvr,in}} \qquad 1.33$$
$$R_{Rcvr,in} = 75\ \Omega$$

1.6 Other Standards

Several other quantities are also expressed in decibels. Although they do not fit the exact definition of a decibel because they are not power ratios, they are still useful for convenient representation of large ratios.

dBK

dBK is a temperature measurement. Its reference is 1K.

$$0 \text{ dBK} = 1 \text{ Kelvin} \qquad 1.34$$

$$\text{dBK} = 10\log(\text{Temperature in K}) \qquad 1.35$$

We will use dBK when we discuss amplifier noise, satellite communications and link budgets.

Example 1.5 —dBK
Convert 430 K to dBK.

Solution —
Using Equation 1.35,

$$\text{dBK} = 10\log(430)$$
$$= 26.3 \text{ dBK} \qquad 1.36$$

dBHz

This is a measure of bandwidth. The reference is 1 hertz.

$$0 \text{ dBHz} = 1 \text{ hertz} \qquad 1.37$$

$$\text{dBHz} = 10\log(\text{Bandwidth in hertz}) \qquad 1.38$$

We will use dBHz when we discuss noise bandwidth and receiver sensitivity.

Example 1.6 — dBHz
Convert 30 kHz to dBHz.

Solution—
Using Equation 1.4,

$$\text{dBHz} = 10\log(30{,}000)$$
$$= 44.8 \text{ dBHz} \qquad 1.39$$

1.7 dB Math

Decibels enable the designer to calculate signal levels quickly. For example,

- A 2-times increase in power.

$$10\log\left(\frac{2P_{in}}{P_{in}}\right) = 3.01\,\text{dB}$$
$$\approx 3\,\text{dB}$$

(1.40)

A doubling of power is a 3 dB increase.

- A 2-times decrease in power.

$$10\log\left(\frac{P_{in}}{2P_{in}}\right) = -3.01\,\text{dB}$$
$$\approx -3\,\text{dB}$$

(1.41)

Halving the power means a 3 dB decrease.

- A 4-times increase in power.

$$10\log\left(\frac{4P_{in}}{P_{in}}\right) = 6.02\,\text{dB}$$
$$\approx 6\,\text{dB}$$

(1.42)

A 4-times power increase means doubling the power twice and 3 dB + 3 dB = 6 dB.

- A 4-times decrease in power.

$$10\log\left(\frac{P_{in}}{4P_{in}}\right) = -6.02\,\text{dB}$$
$$\approx -6\,\text{dB}$$

(1.43)

A 4-times power decrease means halving the power twice and −3 dB + −3 dB = −6 dB.

- A 10-times increase in power.

$$10\log\left(\frac{10P_{in}}{P_{in}}\right) = 10\,\text{dB}$$

(1.44)

Table 1-1 summarizes these results.

Table 1-1 Power ratios and their decibel equivalents.

Power Ratio	Decibels
10^{-6}	-60 dB
0.001	-30 dB
0.01	-20 dB
0.1	-10 dB
0.5	-3 dB
1.0	0 dB
2	3 dB
3	4.77 dB
4	6 dB
5	7 dB
8	9 dB
10	10 dB
100	20 dB
1000	30 dB
10^6	60 dB

Orders of Magnitude

Table 1-2 indicates that every 10 dB increase in a quantity represents a factor of 10 increase in that quantity. In other words, we gain an order of magnitude for every 10 dB increase. Also note that a decrease of 10 dB represents multiplying the quantity by 1/10, or we lose an order of magnitude.

Table 1-2 Power levels in dBm and linear formats.

Power Level in dBm	Linear Power Level
-30 dBm	0.001 mW = 1 µW
-20 dBm	0.01 mW = 10 µW
-10 dBm	0.1 mW = 100 µW
0 dBm	1.0 mW = 1000 µW
10 dBm	10 mW
20 dBm	100 mW
30 dBm	1000 mW = 1 W
40 dBm	10 W
50 dBm	20 W

When a number is given in the format of

$$ABCD.EF\ dB$$

the numbers represented by the *ABC* indicate the order of magnitude of the number, and the *D.EF* refer to the position in that order of magnitude.

> ***Example 1.7 — Power Amplifiers and Decibels***
> a. An amplifier is generating an output signal of −30 dBm. If we increase the output power by 3 dB, how much more power (in linear terms) does the amplifier produce?
> b. A power amplifier is generating an output signal of +40 dBm. If we increase the output power by 3 dB, how much more power (in linear terms) does the amplifier produce?
>
> ***Solution —***
> a. Equation 1.13 relates the amplifier's output power in dBm to the power in mW.
>
> $$-30\,\text{dBm} = 10\log(P_{out,mW}) \qquad 1.45$$
> $$\Rightarrow P_{out,mW} = 1 \cdot 10^{-3}\,\text{mW}$$
>
> Increasing this power by 3 dB is equivalent to doubling the power, i.e., the increase is 10^{-3} mW (a small number).
>
> b. Using Equation 1.13 produces
>
> $$+40\,\text{dBm} = 10\log(P_{out,mW})$$
> $$\Rightarrow P_{out,mW} = 1 \cdot 10^{4}\,\text{mW} \qquad 1.46$$
> $$= 10\,\text{W}$$
>
> Doubling the output power represents a 10-watt increase in the amplifier's output power. In summary, a 3 dB power increase is easy to accomplish at relatively small power levels because it does not represent much power. At high output power levels, unit decibel increases can amount to large increases in output power.

Amplifiers, Attenuators and Decibels

Decibel math allows designers to calculate quickly the effects of amplifiers and attenuators on signal strength. When we multiply in the linear domain, we add in the decibel (or logarithmic) domain. In mathematical terms, the relationship

$$C = A \cdot B$$
$$\Rightarrow C_{dB} = A_{dB} + B_{dB} \qquad 1.47$$

is always true. A, B and C are quantities expressed in linear terms, and A_{dB}, B_{dB} and C_{dB} are the same quantities expressed in decibels. For example, if a signal with −82 dBm of power passes through an amplifier with 15 dB of power gain, the output signal is

$$P_{out,dBm} = P_{in,dBm} + G_{p,dB}$$
$$= -82 + 15 \qquad 1.48$$
$$= -67 \ dBm$$

Using only linear expressions, we first convert the input signal power and gain expressions from decibel to linear terms.

$$-82 \, dBm = 10\log(P_{in,mW}) \quad 15 dB = 10\log(G_P) \qquad 1.49$$
$$\Rightarrow P_{in,mW} = 6.31 \cdot 10^{-9} \ mW \quad \Rightarrow G_p = 31.6$$

We then multiply and convert to decibels.

$$P_{out} = P_{in} \cdot G_P$$
$$P_{out,mW} = (6.31 \cdot 10^{-9} \ mW)(31.6)$$
$$= 199.5 \cdot 10^{-9} \, mW \qquad 1.50$$
$$P_{out,dBm} = 10\log(199.5 \cdot 10^{-9})$$
$$= -67.0 \ dBm$$

Gains and Losses

The terms *gain* and *loss* are best illustrated when used in an example. When a device has a 6 dB gain, it means it will accept a signal, multiply its power by 4 times (or add 6 dB) and present that power to the outside world. If the input signal power is P_{in}, the output signal power is

$$P_{out} = 4P_{in} \qquad 1.51$$

or, in decibels,

$$P_{out,dBm} = P_{in,dBm} + 6 \, dB \qquad 1.52$$

16 | RADIO RECEIVER DESIGN

When a device is specified with a 6 dB loss, it means that the device accepts a signal, divides the signal power by 4 (or subtracts 6 dB from the input signal power), then presents that power to the outside world. Let the input signal power be P_{in}. The output signal power will be

$$P_{out} = \frac{P_{in}}{4} \qquad 1.53$$

or

$$P_{out,dBm} = P_{in,dBm} - 6\,\text{dB} \qquad 1.54$$

Hence, a loss of x dB is equivalent to a gain of –x dB (a negative gain is equivalent to a positive loss). For example, if a device causes a signal to be attenuated by 7 dB, we can say it has a gain of –7 dB or a loss of 7 dB. In equation form, gains and losses are related by

$$Gain_{dB} = -Loss_{dB} \qquad 1.55$$

and

$$Gain_{Lin} = \frac{1}{Loss_{Lin}} \qquad 1.56$$

Example 1.8 — *Gains and Losses in Cascades*

Figure 1-3 shows a cascade made up of blocks with gain and losses. Find the cascade gain.

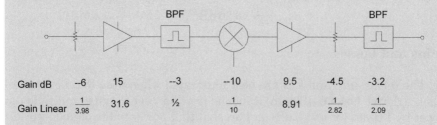

Gain dB	–6	15	–3	–10	9.5	–4.5	–3.2
Gain Linear	$\frac{1}{3.98}$	31.6	½	$\frac{1}{10}$	8.91	$\frac{1}{2.82}$	$\frac{1}{2.09}$

Figure 1-3 Cascade gain example.

Solution —

The cascade gain is

$$\begin{aligned}G_{p,cas,dB} &= (-6) + 15 + (-3) + (-10) + 9.5 + (-4.5) + (-3.2) \\ &= -2.2\,\text{dB}\end{aligned} \qquad 1.57$$

The cascade has a gain of –2.2 dB or a loss of 2.2 dB. The cascade gain can be found using only the linear terms.

$$G_{p,cas} = \left(\frac{1}{3.98}\right)(31.6)\left(\frac{1}{2}\right)\left(\frac{1}{10}\right)(8.91)\left(\frac{1}{2.82}\right)\left(\frac{1}{2.09}\right) \qquad 1.58$$
$$= 0.60$$
$$= -2.2 \text{ dB}$$

Measurements, Significant Figures and Decibels

Suppose we make a power measurement and the meter reads 0.000000 dBm. Is this number of decimal places reasonable? If not, how many are reasonable? Let us assume that the power level is 0.000000 dBm (or 1 mW) exactly. If the power level increases by 1 dB to +1 dBm, then

$$1 \text{ dBm} = 10\log(P_{\text{Measured,mW}}) \qquad 1.59$$
$$\Rightarrow P_{\text{Measured,mW}} = 1.2589 \text{ mW}$$

which is a 26% change. Any reasonable meter would have a good chance of measuring this amount of change accurately. We can conclude that the ones digit of a number expressed in dB is probably meaningful in light of measurement error.

Let us look at the one-tenths (or 0.x) digit. A 0.1 dB increase in a measured quantity results in

$$0.1 \text{ dB} = 10\log(P_{\text{Measured,mW}}) \qquad 1.60$$
$$\Rightarrow P_{\text{Measured,mW}} = 1.0233 \text{ mW}$$

which is approximately a 2% change. This is measurable with care. We conclude that the tenth dB digit is probably meaningful in a carefully controlled environment. Using the same technique as above for the one-hundredth (or 0.0x) dB digit, we find

$$0.01 \text{ dB} = 10\log(P_{\text{Measured,mW}}) \qquad 1.61$$
$$\Rightarrow P_{\text{Measured,mW}} = 1.00233 \text{ mW}$$

which is a 0.2% change. Under normal engineering circumstances, this amount of accuracy is unusual.

When dealing with significant figures the following problems must be addressed:

- Can we reasonably measure the quantity we are expressing to the number of digits we are reporting? This is a complex question involving the accuracy of the equipment used to make the measurement, the experiment itself and the calculations performed with the data.

- Do we need to know the least significant digits? For example, often we need "enough gain to overcome the system noise" or "enough noise figure to see a −125 dBm signal." The absolute number is not important.

- Given the specifications on the components the author has used to design the system, can we justify the number of significant digits that are being used? For example, the gain and noise figure specifications for a typical RF amplifier might be

    Power Gain: 11.0 dB (± 0.5 dB)
    Noise Figure: 2.5 dB (± 0.5 dB)

- In light of these specifications and along with the topology of the system, it might be difficult to justify stating that the power gain of the system is 24.22 dB when the gain variation of just one of the amplifiers is ± 0.5 dB.

1.8 Sanity Checking

The late Nobel prize winning physicist Richard Feynman once related this story about his childhood.

> We had the *Encyclopedia Britannica* at home. When I was a small boy, [my father] used to sit me on his lap and read to me from the *Britannica*. We would be reading, say, about dinosaurs. It would be talking about the Tyrannosaurus rex, and it would say something like, "This dinosaur is twenty-five feet high and its head is six feet across." My father would stop reading and say, "Now let's see what that means. That would mean that if he stood in our front yard, he would be tall enough to put his head through our window up here." (We were on the second floor.) "But his head would be too wide to fit in the window." Everything he read to me he would translate as best he could into some reality.
> (Feynman, *What do you care what other people think?)*

Often we encounter specifications that do not make sense until we put the information into familiar terms. Following are some everyday examples.

Bit Error Rate

A contractor once told the authors his system would produce a bit error rate (BER) of 10^{-20}. Is this reasonable? A bit error rate of 10^{-20} means we will experience one bit error for every 10^{20} bits processed. Let us assume we can measure bit error rates at 10 MHz. In other words, we can test 10 million bits per second. How long would it take to complete the test?

$$\frac{10^{20} \text{ bits}}{10^7 \text{ bits/second}} = 10^{13} \text{ seconds} \qquad 1.62$$
$$= 317{,}000 \text{ years}$$

It would take about 317,000 years on the average to run the test in order to get just *one* bit error. Our conclusion is that this BER is unverifiable.

Space Shuttle Failure Rate

Before the Rogers commission convened on the Challenger accident, NASA maintained that the probability of a catastrophic failure of a shuttle was 1 in 100,000. Is this reasonable? Let us assume we had the resources to perform one launch per day. How long might we have to wait for a failure?

$$\frac{1 \text{ failure}}{100{,}000 \text{ launches}} \Rightarrow 1 \text{ failure every } 100{,}000 \text{ days} \qquad 1.63$$
$$100{,}000 \text{ days} = 274 \text{ years}$$

Could you get into your car every day for 274 years and expect it to start? This figure is probably not reasonable.

Hamburgers

A noted hamburger establishment advertises that they have served over 75 billion hamburgers. Is this a reasonable claim? Let us assume the chain (which has restaurants worldwide) can sell 100 burgers/second for 24 hours/day. How long would it take to sell 75 billion?

$$\left(\frac{75\text{E}9 \text{ burgers}}{100 \text{ burgers/second}}\right)\left(\frac{1 \text{ hour}}{3600 \text{ seconds}}\right)\left(\frac{1 \text{ year}}{8760 \text{ hours}}\right) \qquad 1.64$$
$$= 23.8 \text{ years}$$

Since this restaurant chain has been in business for over 25 years, this estimate seems reasonable.

Gigawatts

A recent motion picture seemed to indicate that 1.21 gigawatts is a lot of power. How much is 1.21 gigawatts? Let us make some assumptions about the average house. For example, let us say we are running the air conditioning which consumes about 5000 watts. We are also running an electric clothes dryer which consumes about 4000 watts, an electric stove (1000 watts) and ten 100-watt light bulbs (for 1000 watts). Our rather busy electric house requires about 11 kW to operate. How many houses can we power with 1.21 gigawatts?

$$\frac{1.21\text{E}9 \text{ W}}{11000 \text{ W}} = 110{,}000 \text{ houses} \qquad 1.65$$

We can light 110,000 average houses with 1.21 gigawatts.

Geostationary Flux Density

The flux density from a typical geosynchronous satellite downlink with a 10-watt power amplifier is about $80.0 \cdot 10^{-15}$ watts/meter². How small is this? A typical two-cell flashlight is a 3-volt system that requires about 150 mA to operate, which comes to 0.45 watts of power. In 1 second, the flashlight will use 0.45 joules (or wattseconds) of energy. Assuming we had a receive antenna with a 1-meter aperture, we can collect $80.0 \cdot 10^{-15}$ watts per second from the satellite.

$$\frac{0.450 \text{ W}}{80.0 \cdot 10^{-15} \text{ W / second}} = 5.63 \cdot 10^{12} \text{ seconds} \qquad 1.66$$
$$= 178{,}000 \text{ years}$$

We would have to collect this energy for 178,000 years to power a flashlight for 1 second.

As these five examples show, it is always a good idea to frame an unfamiliar number into some recognizable perspective where it is easier to understand the meaning of the figures used.

1.9 Approximations

In one of the first atomic bomb tests, a major question concerned the magnitude of the explosion. The exact answer required complex calculations involving parameters such as air temperature, air pressure and the pressure gradient at various points from the bomb. Scientists installed sensors and monitoring equipment around the bomb to obtain the numbers they needed to answer the yield question.

At the moment of the blast, one of the physicists threw a few scraps of paper in the air and noted how fast the wind blew the paper. Using a slide rule and a handful of approximations, he calculated the approximate yield of the bomb. A few weeks later, when the yield of the bomb was calculated with more exact methods, it was found that the two answers differed by only 50%.

The reason this type of calculation works is that the errors in simplifying assumptions are often independent of each other. In other words, if we guess a little high on one number and a little low on another number, the errors will tend to average out. Exact answers to questions such as What are the sidelobe levels of this antenna?, or Will my system work in this crowded spectral environment? frequently require expensive measurements or a precise knowledge of quantities that are difficult to measure. We can often provide answers which are approximate by making some simplifying assumptions and quick calculations. Approximations can also provide an understanding of what is reasonable.

1.10 Frequency, Wavelength and Propagation Velocity

Frequency, wavelength and propagation velocity are related by

$$\lambda f = v \qquad \qquad 1.67$$

where
  λ = the wavelength of the signal,
  v = the velocity of propagation,
  f = the frequency of the signal.

Equation 1.67 is the most general equation form that relates wavelength to propagation velocity. The period T of a wave is given by

$$T = \frac{1}{f} \qquad \qquad 1.68$$

Free Space

When a wave is travelling through empty space or through air, the propagation velocity (v in Equation 1.67) is the speed of light commonly labeled as c. We can write

$$c = 2.9979 \cdot 10^8 \frac{\text{meters}}{\text{second}}$$
$$\approx 3 \cdot 10^8 \frac{\text{meters}}{\text{second}} \qquad 1.69$$

and Equation 1.67 becomes

$$\lambda_0 f = c \qquad 1.70$$

where λ_0 = the wavelength in free space.

Example 1.9 — Wavelength and Frequency
Find the wavelengths of the following frequencies in the atmosphere: 1 MHz (lower HF band), 20 MHz (upper HF band), 100 MHz (middle VHF band), 150 MHz (the two-meter band), 300 MHz, 600 MHz, 1 GHz, 3 GHz, 10 GHz, 30 GHz and 100 GHz.

Solution —
Since we are propagating through the atmosphere, we can use Equation 1.73, which assumes the wave is moving at the speed of light. Liberally applying Equation 1.73, we find

Table 1-3 *Frequency and wavelength relationships.*

Frequency (MHz)	Wavelength			
	meters	yards	feet	inches
1	300.0	328.0	984.0	11,800.0
20	15.0	16.4	49.2	591.0
100	3.0	3.28	9.84	118.0
150	2.0	2.19	6.56	78.7
300	1.0	1.10	3.28	39.4
600	0.50	0.547	1.64	19.7
1,000	0.30	0.328	0.948	11.8
3,000	0.10	0.109	0.328	3.94
10,000	0.03	0.0328	0.0948	1.18
30,000	0.01	0.0109	0.0328	0.394
100,000	0.003	0.00328	0.00948	0.118

When given a frequency, it helps to remember that 300 MHz equals one meter of wavelength and then work out the wavelength using the frequency ratio.

The Speed of Light

Since the speed of light is $2.998 \cdot 10^8$ meters/second, and there are 39.37 inches in a meter,

$$c = \left(2.998 \cdot 10^8 \frac{\text{meters}}{\text{second}}\right)\left(39.37 \frac{\text{inches}}{\text{meter}}\right) \qquad 1.71$$

$$= 11.80 \frac{\text{inches}}{\text{nanosecond}}$$

or

$$c \approx 1 \frac{\text{ft}}{\text{nanosecond}} \qquad 1.72$$

With less than a 2% error, we can say that light travels about one foot in one nanosecond. Conversely, it takes approximately one nanosecond for light to travel one foot.

Example 1.10 — Satellite Delay Time

A geostationary satellite is $35.863 \cdot 10^6$ meters (= 22,284 miles or $117.66 \cdot 10^6$ feet) above the earth. How long does it take for a signal to travel from the surface of the earth to the satellite and back to the earth again?

Solution —

Since the speed of light is $c = 2.9979 \cdot 10^8$ meters/second, it takes

$$\frac{(2)(35.863 \cdot 10^6) \text{ meters}}{2.9979 \cdot 10^8 \text{ meters/second}} = 0.239 \text{ seconds} \qquad 1.73$$

to travel the distance. If we use the 1 nsec/foot rule of thumb, we find

$$(2)(35.863 \cdot 10^6 \text{ meters})\left(\frac{1 \text{ foot}}{0.304 \text{ meters}}\right)\left(\frac{1 \text{ nsec}}{\text{foot}}\right) = 0.236 \text{ sec} \qquad 1.74$$

Physical Size

If we examine the electrical effects of physical structures, it soon becomes apparent that the physical size of a system or device expressed in wavelengths is an important quantity. Consider the following limitations.

- Wires begin to act like antennas when their physical sizes approach wavelength dimensions (typically when their physical dimensions become larger than $\lambda/10$).

- Wires begin to appear similar to transmission lines when the physical length of the wire path is of the same order of magnitude as a wavelength ($\lambda/15$ is the rule of thumb).

- If two wires carrying a signal are physically separated by $\lambda/5$, they begin to act like an antenna and will radiate energy into space.

- When the manufacturing tolerances of a device approach 1/10th of a wavelength, the device will begin to misbehave. Connectors will exhibit excessive losses and VSWR problems and antennas will exhibit gain and sidelobe variations and other disturbances.

We will use these observations to determine how large an antenna which performs at a particular frequency will be, to answer some *electromagnetic interference* (EMI) problems, and to decide when transmission line effects are important and when they are not. Table 1-4 relates frequency to wavelength and physical size.

Table 1-4 Transmission line/antenna effects vs. physical dimensions.

F	λ	$\lambda/2\pi$	$\lambda/20$
10 Hz	30,000 km	4,800 km	1,500 km
60 Hz	5,000 km	800 km	250 km
100 Hz	3,000 km	480 km	150 km
400 Hz	750 km	120 km	38 km
1 kHz	300 km	48 km	15 km
10 kHz	30 km	4.8 km	1.5 km
100 kHz	3 km	480 m	150 m
1 MHz	300 m	48 m	15 m
10 MHz	30 m	4.8 m	1.5 m
100 MHz	3.0 m	0.48 m	15 cm
1 GHz	30 cm	4.8 cm	1.5 cm
10 GHz	3.0 cm	4.8 mm	1.5 mm

where
 F = frequency,
 λ = wavelength,
 $\lambda/2\pi$ = the boundary between antenna near and far fields,
 $\lambda/20$ = antenna effects begin to occur in wires and slots.

1.11 Transmission Lines

Propagation Time

Figure 1-4 illustrates the time it takes for a signal to travel from one part of the circuit to another.

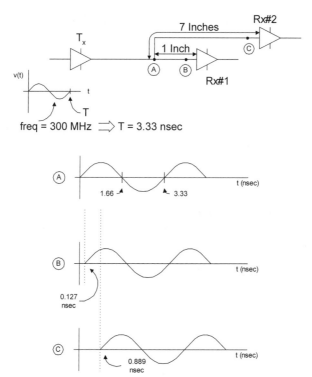

Figure 1-4 Clock skew in digital logic.

For example, we want to drive a 300 MHz signal to two different receivers: one located an inch away from the driver, the other located 7 inches away on another printed circuit card. Since the wave is travelling in a transmission line, it will propagate slower than the speed of light. Let us assume that the signal travels at $2 \cdot 10^8$ meters/second and it will take 0.127 nanoseconds (or $0.127 10^{-9}$ seconds) for the signal to travel one inch. The signal at point B arrives 0.127 nanoseconds after being sent by the driver. The signal at point C arrives 0.889 nanoseconds after it was sent. The difference in propagation time between A-B and A-C is 0.762 nanoseconds, which is an appreciable fraction of a period. If we were using the 300 MHz as a digital clock, the clock edges would be severely misaligned at the two receivers in Figure 1-4. This propagation delay problem, called *clock skew*, is crucial in high-speed computers.

Lumped Elements

One of the implicit assumptions of lumped-element circuit analysis is that it does not take any time for a signal to travel from one point to another. This is clearly not the case with transmission lines. Since the propagation time of the signal is a significant portion of the signal's period, lumped-element circuit analysis procedures are not appropriate.

Parasitic Components

As the frequency increases, the effects of *parasitic components* become important. Every item in the universe has some amount of stray capacitance to ground. Every length of wire, no matter how short, contains a small amount of series parasitic inductance.

Example 1.11 — Parasitic Components
Figure 1-5 (a) shows an amplifier driving another amplifier through a 1/2" piece of wire. Figure 1-5(b) shows an equivalent circuit with the parasitic inductance and capacitance included.

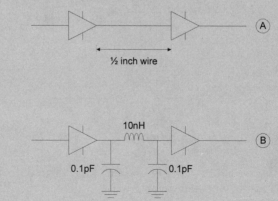

Figure 1-5 *Equivalent circuit of ½-inch of straight wire.*

Figure 1-6 shows the frequency response of the circuit in Figure 1-5(b). At low frequencies, the stray components have almost no effect on the circuit's operation. However, the insertion loss increases as the frequency goes up until we are losing almost 3 dB at 1500 MHz.

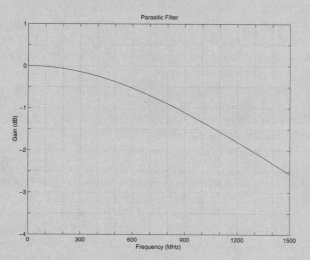

Figure 1-6 Signal loss due to ½-inch of straight wire.

Reflections

Waves travel down transmission lines at finite speeds. Similar to a voice echoing off of a cliff or a radar pulse bouncing off of an airplane, the wave will bounce off of any discontinuity and travel back in the direction it came. This finite propagation time coupled with signal reflection effects will cause radical changes to a signal.

Identifying a Transmission Line

When we are analyzing a system, sometimes it is complicated to decide if transmission line effects will affect the system. Consider these rules.

- *Interconnection length.* If the length of the interconnection is greater than 1/15th of the wavelength of the highest frequency signal it carries, the interconnection is in fact a transmission line.

- *Rise and fall times.* If the time required for the signal to travel the length of the interconnection is greater than 1/8th of the signal rise time, the interconnection is a transmission line.

28 | RADIO RECEIVER DESIGN

Rise time typically refers to the smaller of either the rise or fall time as the shortest duration will have the greatest effect on circuit behavior. It corresponds to the time it takes for the digital signal to travel from its 0% value to its 100% value. Semiconductor manufacturers typically publish the 10% to 90% or the 20% to 80% rise time figures. These figures have to be multiplied by 1.25 or 1.67 respectively to find the linear ramp time from 0% to 100%. The *interconnection length* and rise and fall times rules are equivalent. They both compare the wavelength of the highest frequency present in the system with the physical size of the interconnection.

Example 1.12 — Transmission Lines
A 10 MHz video signal must travel 3 feet to its destination. Are transmission line effects important here?

Solution —
Since the highest frequency present in the signal is 10 MHz, the smallest wavelength, using Equation 1.73, is 30 meters = 98 feet. Since 1/15th of 98 feet is about 6.6 feet, we can conclude that transmission line effects do not matter in this case.

Example 1.13 — Transmission Lines
A digital signal with a 3 nsec rise time and a 2 nsec fall time must travel 5 inches to its destination. Are transmission line effects important here?

Solution —
We are interested in the smallest of the rise or fall time, so we will use the 2 nsec fall time for the necessary computations. 1/8th of the 2 nsec fall time is 0.25 nsec. Using Equation 1.75, we calculate that it takes 0.42 nsec for the signal to travel the 5 inches. Since the 0.42 nsec travel time is greater than 1/8th of the 2 nsec rise time (i.e., the 0.25 nsec figure above), we conclude that transmission line effects are important here.

Bandwidth of Digital Signals
The approximate cutoff frequency for a digital signal is

$$f_{digital} = \frac{1}{\pi t_r} \qquad 1.75$$

where
t_r = the rise or fall time of the digital signal (whichever is smaller).

War Story — Bandwidth of Digital Signals

The bandwidth of a digital signal can be widely different from its clocking rate. For example, a 4800 bit/sec data line can have a huge bandwidth if the signals have very sharp rise and fall times.

The author was working with another engineer on a 4800 bit-per-second signal. They were feeding the signal and its clock into a rising-edge triggered device but the logic was reading bad data. After some work, the author and his colleague found that they were clocking bits into the device on both the rising and falling edges of the clock. They then noticed that the signal had very sharp rise and fall times. Also, their test setup included several meters of cable, several meters of ribbon cable and several EZ clips for good measure. Both realized that transmission line effects were causing reflections on the rising as well as falling edges of the clock. These reflections caused the input logic to see rising clock edges on both the rising and falling edges of the 4800-hertz clock.

The author and his colleague solved the problem by applying a low-pass filter to the 4800-hertz clock line, which limited the bandwidth of the clock signal and reduced the transmission line effects. Part of the problem was that the clock input logic was fast enough to trigger on the reflections present in the setup. If lower-speed logic had been used, the problem might have never been noticed.

Rise Times, Frequency and Length

Table 1-5 relates signal rise times with frequency and physical lengths.

Table 1-5 *Signal rise times and frequencies related to physical lengths.*

t_r	$f_{digital}$	L_{cross}	$L_{cross}/2$	L_{term}
1 nsec	318 MHz	1.0 ft	0.5 ft	3 in
3 nsec	95 MHz	3.0 ft	1.5 ft	9 in
10 nsec	32 MHz	10 ft	5.0 ft	2.5 ft
30 nsec	9.5 MHz	30 ft	15 ft	7.5 ft
100 nsec	3.2 MHz	100 ft	50 ft	25 ft
300 nsec	950 kHz	300 ft	150 ft	75 ft
1 μsec	320 kHz	1000 ft	500 ft	250 ft

where
   t_r is the rise or fall time,
   $f_{digital}$ is the cutoff frequency of the rise time given by Equation 1.75,
   L_{cross} is the length of one rise time in free space,
   $L_{cross}/2$ is the typical length of a rise time on a cable or printed circuit board,
   L_{term} is the maximum allowable unterminated length on a cable or printed circuit board.

Transmission Line Models

Figure 1-7 shows a useful electrical model for transmission line analysis. The combination of V_s and R_s could be an antenna, signal generator, digital logic gate or any other signal source. The resistor (R_L) represents the signal load that could be a receiver, amplifier, and so on.

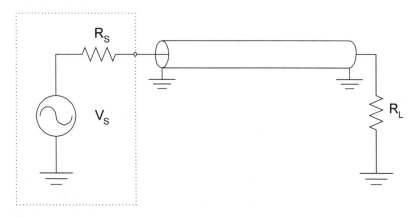

Signal Generator or Antenna

Figure 1-7 *Transmission line analysis model.*

There are always two elements in a transmission line. In Figure 1-7, the center conductor carries the signal, and the second outer conductor serves as both a signal ground and a shielding mechanism. Other physical arrangements are possible.

Figure 1-8(a) shows an equivalent circuit of a transmission line. Each marked-off section in Figure 1-8(a) is an infinitesimal piece of the line. The shunt capacitors (ΔC) represent the capacitance present between the signal conductor and ground. The shunt resistors (ΔB) represent signal loss through the insulation separating the two wires (dielectric loss). The series inductance (ΔL) and the series resistance (ΔR) model the parasitic inductance and resistance of the center conductor. Any transmission line can be modeled as an infinite number of the marked-off equivalent circuits in series. As the number of sections increases, the values of ΔC, ΔB, ΔL and ΔR approach zero.

INTRODUCTION | 31

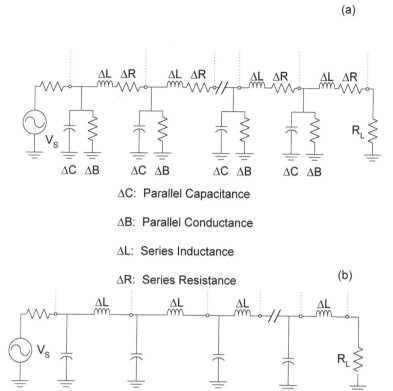

Figure 1-8 *Transmission line equivalent circuits (a) lossy (b) lossless.*

Physical cables always exhibit signal loss if they are not superconducting. In Figure 1-8(a), the signal loss has been modeled as series (ΔR) and shunt resistors (ΔB). Removing the resistors makes the line lossless [see Figure 1-8(b)].

Characteristic Impedance

One of the main properties of a transmission line is its characteristic impedance (Z_0). When we terminate a transmission line with a resistance whose value equals the cable's characteristic impedance, signals propagating down the cable are completely absorbed by the load resistor. When the load value of the load resistor does not equal the value of the cable's characteristic impedance, signals propagating down will reflect off of the load resistor and travel back up the cable, in the direction of the source (see Figure 1-9).

32 | RADIO RECEIVER DESIGN

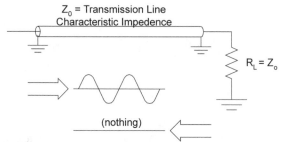

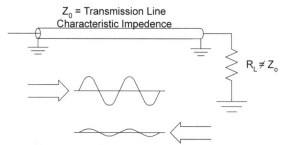

Figure 1-9 *Signals propagating on a transmission line under matched and mismatched conditions.*

These signal reflections have profound effects on the performance of the system as will be shown. A transmission line's characteristic impedance is a function of the physical geometry of the cable and the materials used to build it. According to Figure 1-8(b), the characteristic impedance of the cable is

$$Z_0 = \sqrt{\frac{\Delta L}{\Delta C}} = \sqrt{\frac{L/\text{unit length}}{C/\text{unit length}}} \qquad 1.76$$

Notice that the unit length quantity of Equation 1.76 drops out of the relation, which allows us to measure the inductance and capacitance of any random length of cable and use those numbers directly in Equation 1.76.

Example 1.14 — Characteristic Impedance

We selected a random length of RG-223 cable and performed the following measurements.

- We open-circuited the far end of the cable [as in Figure 1-10(a)] and measured 62.5 pF of capacitance at the near end.

- We short-circuited the far end of the cable [as in Figure 1-10(b)] and measured 177 nH of inductance at the near end.

What is the characteristic impedance of the cable?

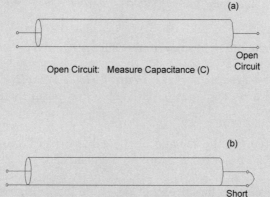

Figure 1-10 *Measurement of unknown characteristic impedance.*

Solution —
Assume the cable is unit length and, knowing the capacitance/unit length = 62.5 pF and the inductance/unit length = 177 nH, we find

$$Z_0 = \sqrt{\frac{177 \cdot 10^{-9}}{62.5 \cdot 10^{-12}}}$$
$$= 53 \, \Omega$$

1.77

Propagation Velocity

Waves travel slower in a transmission line than in free space. The propagation velocity is v. Referring to Figure 1-8(b), we can show that the propagation velocity in a low-loss transmission line is

$$v = \frac{1}{\sqrt{\Delta L \Delta C}}$$

$$= \frac{1}{\sqrt{(L/\text{unit length})(C/\text{unit length})}} \qquad 1.78$$

Unlike Equation 1.79, the cable length does not cancel out in this expression. If we want the propagation velocity to be in meters per second, we have to express ΔL in henries/meter and ΔC in farads/meter.

$$v = \frac{1}{\sqrt{(\text{henries/meter})(\text{farads/meter})}} \left(\frac{\text{meters}}{\text{second}}\right) \qquad 1.79$$

The following example shows how to measure ΔL and ΔC (see Figure 1-10).

Example 1.15 — Measuring the Propagation Velocity

We selected a 6-foot long RG-223 cable and performed the following measurements:

- We open-circuited the far end of the cable [as in Figure 1-10(a)] and measured 62.5 pF of capacitance at the near end.

- We short-circuited the far end of the cable [as in Figure 1-10(b)] and measured 177 nH of inductance at the near end.

What is the velocity factor of the cable?

Solution —

Equation 1.82 demands that the electrical measurements be expressed as a per meter quantity. We know ΔC = 62.5 pF/6 foot = 34.2 pF/meter. Likewise, we know ΔL = 177 nH/6 foot = 96.8 nH/meter. Equation 1.82 produces

$$v = \frac{1}{\sqrt{(96.8 \cdot 10^{-9})(34.2 \cdot 10^{-12})}} \qquad 1.80$$

$$= 550 \cdot 10^6 \, \frac{\text{meters}}{\text{second}}$$

The velocity factor is the ratio of the propagation velocity in the cable to the propagation velocity in free space, so the velocity factor of the transmission line is

$$\text{Velocity Factor} = \frac{v}{c}$$
$$= \frac{550 \cdot 10^6}{3 \cdot 10^8} \quad\quad 1.81$$
$$= 0.545$$

Velocity Factor

Waves usually move slower in a transmission line than in free space. The velocity factor is the ratio of the propagation velocity of a wave in a transmission line to the wave's velocity in free space or

$$\text{Velocity Factor} = v_f = \frac{v}{c} \leq 1 \quad\quad 1.82$$

where
  v_f = the velocity factor of the transmission line ($0 \leq v_f \leq 1$),
  v = the propagation velocity in the transmission line,
  c = the speed of light.

Example 1-16 — Measuring Velocity Factor
We launched an 8 nsec long pulse into a 6-foot long piece of RG-223 coaxial cable. We left the far end of the cable open-circuited and the return pulse came back 19.1 nsec after we launched the first pulse into the cable. What is the velocity factor of this piece of cable?

Solution —
The 19.1 nsec between pulses is the time it takes for the pulse to travel down the cable and return. The one-way travel time is 19.1/2 = 9.55 nsec. Our pulse traveled 6 feet in 9.55 nsec. Using Equation 1.72, the pulse would travel the 6 feet in 6 nsec if it were travelling in free space. The velocity factor of the cable is

$$\text{Velocity Factor} = v_f = \frac{6 \text{ nsec}}{9.55 \text{ nsec}} \quad\quad 1.83$$
$$= 0.63$$

Since waves move slower than the speed of light inside the transmission line, the wavelength inside the transmission line (the *guide wavelength*), will be shorter than the wavelength in free space. The wavelength inside of a transmission line is related to the wavelength in free space through the velocity factor.

$$\text{Velocity Factor} = v_f = \frac{\lambda_g}{\lambda_0} \leq 1 \qquad 1.84$$

where

v_f = the velocity factor of the transmission line ($0 \leq v_f \leq 1$),
λ_g = the wavelength in the transmission line (g = guided wavelength),
λ_0 = the wavelength in free space.

Equation 1.73 relates λ_0 to the frequency.

Transmission Lines and Pulsed Input Signals

Figure 1-11 shows a transmission line experiment. A lossless transmission line (whose characteristic impedance is Z_0) is being fed with a very narrow pulse. We will plot V_{in}, the voltage at the input of the transmission line, over time. We will plot data for various values of the load resistor R_L.

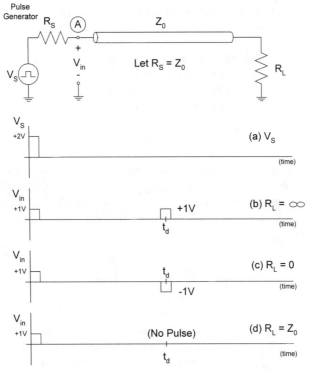

Figure 1-11 *Transmission line under pulsed input conditions with various loads.*

Open-Circuit Line

Note that we have set R_S, the resistor in series with V_s, equal to the characteristic impedance of the cable. The first plot [Figure 1-11(a)] shows the source voltage V_s. We are generating a 2-volt pulse and observe the same waveform regardless of the value of R_L.

Figure 1-11(b) shows V_{in} when the load is an open circuit ($R_L = \infty$). The magnitude of the pulse has changed: it is now only 1 volt — half of what the generator produced. We will also observe a second pulse at V_{in}. The second pulse has the same magnitude as the first pulse but appears at a later time (t_d) from the initial pulse. We can explain the second pulse by considering the transmission line as a simple time delay. The first pulse was generated by the source. The pulse travels through the cable until it encounters the open circuit at the load end. Similar to a voice echoing off of a cliff or a radar pulse bouncing off of an airplane, the pulse reflects off of the discontinuity and travels back in the direction it came. The second pulse at V_{in} is travelling back from the load end of the cable toward the generator. The time (t_d) is the time it takes the pulse to travel down the cable and back again. This is one technique we can use in the lab to determine the cable's propagation velocity.

When the pulse first enters the transmission line and *before* the return pulse has had time to return from the open circuit, the transmission line is a resistor whose value is its characteristic impedance. The change in the magnitude of the pulse when it enters the cable can be easily explained. Since we set the source resistor (R_S) equal to the cable's characteristic impedance Z_0, we have a voltage divider whose output voltage is half the input voltage.

Short-Circuit Line

Figure 1-11(c) shows V_{in} if we short-circuit the load end of the transmission line. Everything is the same except for one minor detail — the return pulse has the same magnitude, but it is opposite in polarity. The short circuit at the end of the cable causes the pulse to come back upside-down or inverted.

Z_0 Terminated Line

Figure 1-11(d) shows a terminated cable with a resistor whose value equals the characteristic impedance of the cable. No pulse returns because the matched load resistor absorbs all of the energy in the pulse. This is usually a desirable situation and one of the reasons to keep system impedances matched.

Reflection Coefficient

Open and short circuits are extreme conditions. What happens if we place a variable resistor at the load end of the cable? In general, the magnitude of the reflected pulse will be

$$V_{Reflected} = \rho V_{Incident} \qquad 1.85$$

where
    ρ is the *reflection coefficient* (ρ can be a complex number if the load impedance is complex).
    $V_{Incident}$ is the voltage incident on the load. In other words, the magnitude of the pulse we sent into the transmission line. $V_{Incident}$ travels from the source to the load.
    $V_{Reflected}$ is the magnitude of the voltage reflected off of the load and back into the transmission line. $V_{Reflected}$ travels from the load to the source.

We can show

$$\rho = \frac{R_L - Z_0}{R_L + Z_0} \qquad -1 \le \rho \le +1 \qquad 1.86$$

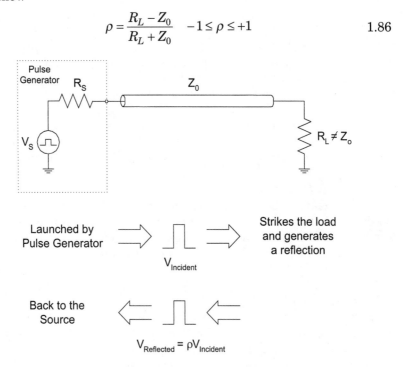

Figure 1-12 *Transmission line reflection coefficient.*

    Figure 1-13 shows a graph of the load resistance vs. the reflection coefficient for a 50-ohm cable. At low values of R_L, the reflection coefficient approaches −1. This means that most of the voltage we send down the cable comes back (although its magnitude is reversed). When R_L is very large, ρ

approaches +1. Note that everything that is sent down into the cable is reflected but the magnitude is not affected.

When the value of the load resistor equals the transmission line's characteristic impedance, the reflection coefficient equals zero. None of the energy we send into the cable returns; all of it is dissipated in the load resistor.

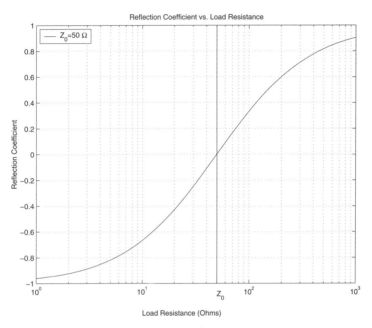

Figure 1-13 Transmission line reflection coefficient vs. load resistance (50-ohm system).

Transmitting Systems

In a transmitting system, a high-power amplifier feeds a cable. The cable, in turn, feeds an antenna. We are interested in sending power into an antenna and emitting that power into free space. If the antenna is not matched to the cable or the cable is not matched to the power amplifier, some of the energy that is being sent into the cable will bounce off the antenna, return and not radiate into space.

Receiving Systems

If we are interested in receiving signals, we normally place an antenna feeding a cable which then feeds a receiver. Any mismatch between the antenna and the cable or between the cable and the receiver will cause energy to reflect off of the mismatch. Signal energy will be lost and the receiving system will not be as sensitive as it could be.

Example 1.17 — Reflection Coefficient
Find the reflection coefficient for the following resistors:
a. 10 ohms,
b. 25 ohms,
c. 50 ohms,
d. 100 ohms,
e. 250 ohms.
Assume a 50-ohm system.

Solution —
Liberally applying Equation 1.86 produces
a.
$$\rho = \frac{10-50}{10+50} = -0.667 \qquad 1.87$$

b.
$$\rho = \frac{25-50}{25+50} = -0.333 \qquad 1.88$$

c.
$$\rho = \frac{50-50}{50+50} = 0 \qquad 1.89$$

d.
$$\rho = \frac{100-50}{100+50} = 0.333 \qquad 1.90$$

e.
$$\rho = \frac{250-50}{250+50} = 0.667 \qquad 1.91$$

Example 1.18 — Mismatched Transmission Line
Figure 1-14 shows a signal source driving a transmission line. The signal source is a step function,

$$V_S = 0 \text{ volts for } t \leq 0$$
$$V_S = 1 \text{ volt for } t > 0 \qquad 1.92$$

and $Z_0 = 50\,\Omega$, $R_S = 15\,\Omega$ and $R_L = 125\,\Omega$. If it takes τ seconds for a wave to propagate from one end of the transmission line to the other (i.e., τ is the one-way travel time), plot the voltage vs. distance over the length of the cable for $t = 0.3\tau$, 1.3τ, 2.3τ, and so on.

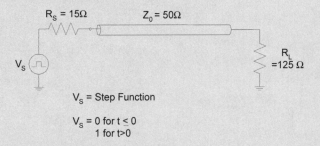

Figure 1-14 *Mismatched transmission line example.*

Solution —

First, we use Equation 1.86 to obtain the reflection coefficients of the source and load. The reflection coefficient of the source is

$$\begin{aligned}\rho_S &= \frac{R_L - Z_0}{R_L + Z_0} \\ &= \frac{15 - 50}{15 + 50} \\ &= -0.538\end{aligned} \qquad 1.93$$

The reflection coefficient associated with the load is

$$\begin{aligned}\rho_L &= \frac{R_L - Z_0}{R_L + Z_0} \\ &= \frac{125 - 50}{125 + 50} \\ &= 0.429\end{aligned} \qquad 1.94$$

Initially, before any reflections begin to propagate down the cable and back the transmission line appears as a resistive element with a value of Z_0. When the generator "turns on" at $t = 0$, the sources recognizes the transmission line as a 50-ohm resistor. Using voltage division, the magnitude of the incident wave propagating down the transmission line is

$$V_{\text{Incident}} = V_S \frac{Z_0}{R_S + Z_0}$$
$$= (1\text{ V})\frac{50}{15+50}$$
$$= 0.770\text{ V}$$

1.95

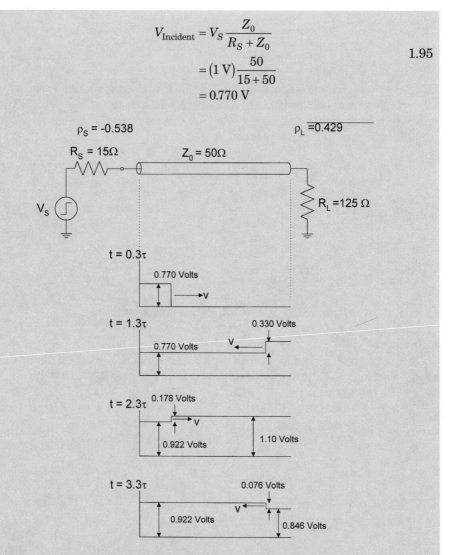

Figure 1-15 *Snapshots of voltages present on a mismatched transmission line at t = 0.3τ, t = 1.3τ, t = 2.3τ and t = 3.3τ.*

The graph labeled "t = 0.3τ" in Figure 1-15 shows the initial pulse travelling from the source to the load. At t = 1.0τ, the incident pulse encounters the load resistor. The load resistor absorbs some of the incident pulse's energy and reflects the rest. Equation 1-85 relates the incident and reflected voltage. The magnitude of the incident voltage is 0.770, and the load reflection coefficient is 0.429. The reflected voltage is

$$V_{\text{Reflected}} = \rho_L V_{\text{Incident}}$$
$$= (0.429)(0.770)$$
$$= 0.330 \text{ V}$$
(1.96)

Figure 1-15 shows the state of the transmission line at time $t = 1\tau$. At $t = 2\tau$ the 0.330-volt wave travelling from the load encounters the source. At this discontinuity, the incident voltage is 0.330 volts and the reflection coefficient is −0.538. Equation 1.88 shows the reflected voltage.

$$V_{\text{Reflected}} = \rho_S V_{\text{Incident}}$$
$$= (-0.538)(0.330)$$
$$= -0.178 \text{ V}$$
(1.97)

Figure 1-15 shows the state of the transmission line at time $t = 2.3\tau$. At $t = 3\tau$, the −0.178-volt waveform transient is incident on the load. The discontinuity produces a reflected wave. The value of the reflected wave is

$$V_{\text{Reflected}} = \rho_L V_{\text{Incident}}$$
$$= (0.429)(-0.178)$$
$$= -0.076 \text{ V}$$
(1.98)

Figure 1-15 shows the state of the transmission line at time $t = 3.3\tau$. If we continue this process, we will find that the voltage on the transmission line eventually settles down to what we would expect from a simple DC analysis. In other words, after all the transients have receded, the final voltage on the transmission line will be

$$V_{\text{Final}} = V_{DC} = V_S \frac{R_L}{R_L + R_S}$$
$$= (1 \text{ V})\left(\frac{125}{15 + 125}\right)$$
$$= 0.893 \text{ V}$$
(1.99)

Note that we can arrive at this result either through DC analysis or by summing up the reflections on the transmission line until they disappear. The results should be identical.

Complex Reflection Coefficient

So far we have examined resistive cable terminations and the astute reader may wonder what happens if we decided to terminate a cable with an inductor or capacitor. The answer is that everything we have discussed will still apply with the exception that the reflection coefficient will be a complex number and have a magnitude and a phase angle. With a resistive load, the reflection coefficient is always real. With a complex load, the time-domain examples are harder to visualize because of the frequency-dependent nature of the load.

For a complex load, $Z_L = R_L + jX_L$, the reflection coefficient is

$$\rho = |\rho|\angle\theta_\rho$$
$$= \frac{Z_L - Z_0}{Z_L + Z_0} \qquad 1.100$$

where
$Z_L = R_L + jX_L$ = the complex load (resistance and reactance),
ρ = the complex reflection coefficient,
$|\rho|$ = the magnitude of the complex reflection coefficient,
θ_ρ = the angle of the complex reflection coefficient.

A complex ρ means that the incident wave suffers a phase change as well as a magnitude change when it encounters the mismatch. All the equations concerning return loss will also apply for the complex reflection coefficient. The magnitude is specified wherever it was required. If it is not specified, we use the complex reflection coefficient, which yields a complex result.

Example 1.19 — Reflection Coefficient and Reactive Terminations
Figure 1-16 shows a transmission line with a reactive termination. Find the reflection coefficient, return loss and VSWR for this load. Assume a 50-ohm system and that the frequency of operation is 4 GHz.

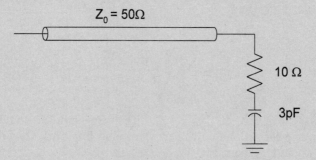

Figure 1-16 *Transmission line with a complex load.*

Solution —
The reactance of this load at 4 GHz is $10 - j13.3\ \Omega$ or $16.6 \angle 53°$. Using Equation 1.100 with $Z_L = 10 - j13.3\ \Omega$ and $Z_0 = 50\ \Omega$ produces

$$\rho = \frac{(10 - j13.3) - 50}{(10 - j13.3) + 50}$$

$$= \frac{-40 - j13.3}{60 - j13.3}$$

$$= -0.589 - j0.352$$

$$= 0.686 \angle -149°$$

(1.101)

Note that the reflection coefficient is complex because of the complex load. Since the load impedance changes with frequency, the complex reflection coefficient changes with frequency.

$$VSWR = \frac{1 + |\rho_L|}{1 - |\rho_L|}$$

$$= \frac{1 + 0.686}{1 - 0.686}$$

$$= 5.4$$

(1.102)

and

$$\text{Return Loss}_{dB} = 20 \log(|\rho|)$$

$$= 20 \log(|0.686|)$$

$$= -3.2 \text{ dB}$$

(1.103)

Mismatch Loss

The voltage reflecting off of a mismatched load at the end of a transmission line represents power loss. If the load were perfectly matched to the transmission line, we would transfer all of the available incident signal power to the load resistor. But since the load is not matched to the transmission line's characteristic impedance, all the available power is not transferred to the load. This *mismatch loss* is defined as

$$\text{Mismatch Loss} = \frac{P_{\text{Available}} - P_{\text{Reflected}}}{P_{\text{Available}}}$$

$$= \frac{P_{\text{Delivered}}}{P_{\text{Available}}}$$

(1.104)

where
 $P_{\text{Available}}$ = the power that is delivered to the matched load,
 $P_{\text{Delivered}}$ = the power that is delivered to the unmatched load,
 $P_{\text{Reflected}}$ = the power that is reflected off of the unmatched load.

We know

$$P_{\text{Available}} = \frac{V_{\text{Incident}}^2}{Z_0} \qquad 1.105$$

and

$$P_{\text{Reflected}} = \frac{V_{\text{Reflected}}^2}{Z_0} \qquad 1.106$$

Combining Equations 1.104 through 1.106 with Equation 1.85 produces

$$\begin{aligned}
\text{Mismatch Loss}_{\text{Load}} &= \frac{V_{\text{Incident}}^2/Z_0 - V_{\text{Reflected}}^2/Z_0}{V_{\text{Incident}}^2/Z_0} \\
&= \frac{V_{\text{Incident}}^2 - V_{\text{Reflected}}^2}{V_{\text{Incident}}^2} \qquad 1.107 \\
&= \frac{V_{\text{Incident}}^2 - \left(|\rho_L| V_{\text{Incident}}\right)^2}{V_{\text{Incident}}^2} \\
&= 1 - |\rho_L|^2
\end{aligned}$$

Mismatch loss is yet another way of describing the relationship between the characteristic impedance of a transmission line and its load. Mismatch loss is specified in decibels,

$$\text{Mismatch Loss}_{\text{Load,dB}} = 10 \log\left(1 - |\rho_L|^2\right) \qquad 1.108$$

Mismatch loss is only relevant in connection with an antenna. In a receiving situation, the mismatch loss of the antenna represents signal loss that affects the system noise floor and minimum detectable signal. In the transmitting case, the mismatch loss represents signal energy that is not delivered to the antenna and so cannot be radiated into free space.

Source and Load Mismatches

We can experience mismatches at both the source and load ends of a transmission line. In a source mismatch, i.e., when the source impedance (R_S) is not equal to the characteristic impedance of its accompanying transmission line, the maximum amount of power is not transferred into the line. The mismatch loss due to source mismatch is

$$\text{Mismatch Loss}_{\text{Source,dB}} = 1 - |\rho_L|^2 \qquad 1.109$$

or

$$\text{Mismatch Loss}_{\text{Source,dB}} = 10\log\left(1 - |\rho_S|^2\right) \qquad 1.110$$

Example 1.20 — Mismatch Loss

Find the mismatch loss (in decibels) for the following resistors in a 50-ohm system.
a. 0 ohms ($\rho = -0.667$),
b. 25 ohms ($\rho = -0.333$),
c. 50 ohms ($\rho = 0$),
d. 100 ohms ($\rho = 0.333$),
e. 250 ohms ($\rho = 0.667$).
f. 20 $-j100$ ohms ($\rho = 0.530 - j0.671 = 0.855 \angle -51.7°$).

Solution —
Liberally applying Equation 1-110 produces
a.

$$\begin{aligned}\text{Mismatch Loss}_{\text{dB}} &= 10\log\left(1 - |\rho|^2\right) \\ &= 10\log\left(1 - |-0.667|^2\right) \\ &= -2.6 \text{ dB}\end{aligned} \qquad 1.111$$

b.

$$\begin{aligned}\text{Mismatch Loss}_{\text{dB}} &= 10\log\left(1 - |\rho|^2\right) \\ &= 10\log\left(1 - |-0.333|^2\right) \\ &= -0.5 \text{ dB}\end{aligned} \qquad 1.112$$

c.

$$\begin{aligned}\text{Mismatch Loss}_{\text{dB}} &= 10\log\left(1 - |\rho|^2\right) \\ &= 10\log\left(1 - |0|^2\right) \\ &= 0.0 \text{ dB}\end{aligned} \qquad 1.113$$

d.
$$\text{Mismatch Loss}_{dB} = 10\log(1-|\rho|^2)$$
$$= 10\log(1-|0.333|^2)$$
$$= -0.5 \text{ dB}$$
1.114

e.
$$\text{Mismatch Loss}_{dB} = 10\log(1-|\rho|^2)$$
$$= 10\log(1-|0.667|^2)$$
$$= -2.6 \text{ dB}$$
1.115

f.
$$\text{Mismatch Loss}_{dB} = 10\log(1-|\rho|^2)$$
$$= 10\log(1-|0.855|^2)$$
$$= -5.7 \text{ dB}$$
1.116

The sign of the reflection coefficient is irrelevant as far as the mismatch loss is concerned.

Transmission Lines and Sine-Wave Input Signals

We have discussed the behavior of pulses as they move up and down the transmission line and when they encounter impedance discontinuities. Since a pulse consists of a series of sine waves, we argue that the same effects occur when we launch a continuous RF carrier (i.e., a sine wave) down a transmission line. Reflections still occur as the wave encounters mismatched impedances. However, because of the continuous nature of the sine wave, the effects are not as intuitive as in the pulsed case.

Figure 1-17 shows an example where the voltage generator V_g is a sine- wave source that turns on at time t = 0. The system is completely mismatched (i.e., $R_S \neq Z_0 \neq R_L$). When we turn the signal generator on at t = 0, the sine wave first encounters the mismatch between the signal generator's source impedance and the impedance of the transmission line. This effect acts to reduce the voltage incident on the line. After the sine wave enters the transmission line, it travels down the line until it encounters the load resistor. The load absorbs some of the signal energy and reflects the rest according to the equation

$$V_{\text{Reflected}} = \rho V_{\text{Incident}} \qquad \qquad 1.117$$

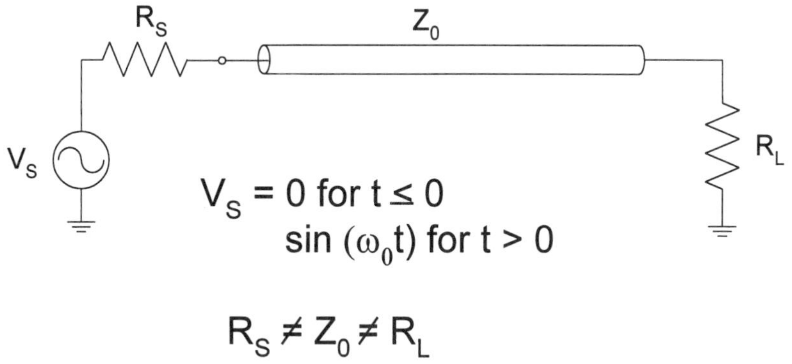

Figure 1-17 *Mismatched transmission line with sinusoidal input voltage.*

The energy that reflects off of the load travels back along the line toward the source. If ρ is complex, the sine wave experiences a phase shift as well as a magnitude change. At the source end of the line, the signal from the load splits into two pieces. Some of the energy is absorbed by the source while the rest is reflected and sent back again toward the load. This process of absorption and reflection at each end of the transmission line continues until eventually the transients recede and the line reaches a steady-state condition. The voltage present at any particular point on the transmission line is the sum of the initial incident wave and all of the reflections.

The mathematical expression describing this process involves an infinite series. To simplify the analysis, we will examine a transmission line with an arbitrary load resistance but with a matched source resistance (see Figure 1-18). When $R_S = Z_0$, the reflections caused by a mismatch at the source end of the cable are eliminated. The voltage at any physical point on the line is due to the incident wave and one reflection from the load. Any signals travelling from the load to the source are absorbed entirely by the source, because, by matching the source to the line, we have set the source reflection coefficient to zero.

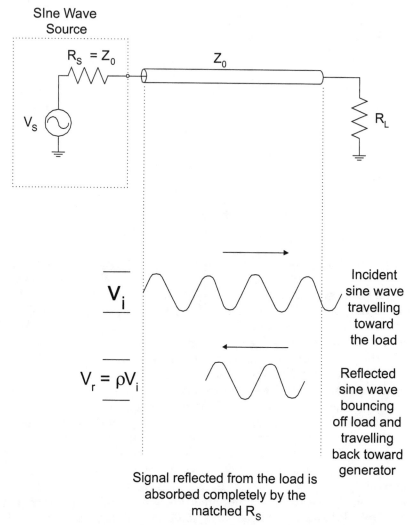

Figure 1-18 *Mismatched transmission line showing incident and reflected voltages.*

Voltage Minimums and Maximums

Figures 1-19 and 1-20 show the magnitudes of the sine waves that can be observed along a length of transmission lines for several load mismatch conditions. Figure 1-19 describes loads that are greater than Z_0 (or 50 ohms, in this case); Figure 1-20 describes the situation when the loads are less than 50 ohms.

INTRODUCTION | 51

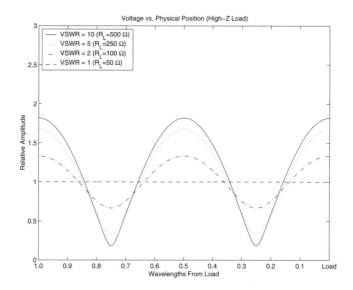

Figure 1-19 *Voltage standing wave ratio (VSWR). The magnitude of the voltages present on a mismatched transmission line when $Z_L > Z_0$.*

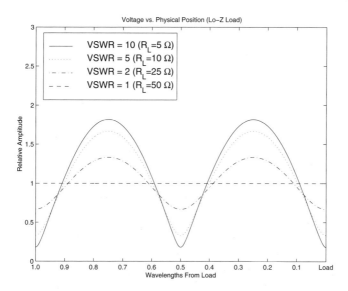

Figure 1-20 *Voltage standing wave ratio (VSWR). The magnitude of the voltages present on a mismatched transmission line when $Z_L < Z_0$.*

Example 1.21 — Transmission Line Voltages (High Z-Load)

Assume we are operating a 50-ohm transmission line with a 250-ohm load resistance. What are the time-domain voltages we would observe at the following distances from the load?
a. $0.10\lambda_g$,
b. $0.25\lambda_g$,
c. $0.40\lambda_g$,
d. $0.50\lambda_g$,
e. $0.60\lambda_g$,
f. $0.65\lambda_g$.

Solution —
Reading from the $R_L = 250\ \Omega$ (or VSWR = 5) curve of Figure 1-19, we find
a. $1.37\cos(\omega t + \phi_1)$,
b. $0.37\cos(\omega t + \phi_2)$,
c. $1.37\cos(\omega t + \phi_3)$,
d. $1.65\cos(\omega t + \phi_4)$,
e. $1.37\cos(\omega t + \phi_5)$,
f. $1.00\cos(\omega t + \phi_6)$.

where
$\omega = 2\pi f$ = the frequency of operation,
ϕ_n = a phase constant depending upon the frequency and the distance from the source to the load.

Example 1.22 — Transmission Line Voltages (Low Z-Load)

Assume we are operating a 50-ohm transmission line with a 10-ohm load resistance (which is equivalent to a VSWR of 5). What are the time-domain voltages we would observe at the following distances from the load?
a. $0.10\lambda_g$,
b. $0.25\lambda_g$,
c. $0.40\lambda_g$,
d. $0.50\lambda_g$,
e. $0.60\lambda_g$,
f. $0.65\lambda_g$.

Solution —
Reading from the $R_L = 10\ \Omega$ (or VSWR = 5) curve of Figure 1-20, we find
a $0.90\cos(\omega t + \phi_1)$,
b. $1.35\cos(\omega t + \phi_2)$,
c. $0.90\cos(\omega t + \phi_3)$,
d. $0.65\cos(\omega t + \phi_4)$,

e. $0.90\cos(\omega t + \phi_5)$,
f. $1.12\cos(\omega t + \phi_6)$.

where

$\omega = 2\pi f$ = the frequency of operation,
ϕ_n = a phase constant depending upon the frequency and the distance from the source to load.

Given a line with a mismatched load, we will define the highest voltage present on the line as V_{max} and the lowest voltage present on the line as V_{min} or,

$$\text{Largest Voltage on the T-Line} \Rightarrow V_{max}\cos(\omega t + \phi_1)$$
$$\text{Smallest Voltage on the T-Line} \Rightarrow V_{min}\cos(\omega t + \phi_2) \qquad 1.118$$

Note that V_{max} and V_{min} change with the value of the load resistor. When the line is matched ($R_L = Z_0$), then $V_{max} = V_{min}$. When the line is not matched, $V_{max} \neq V_{min}$. As the load's reflection coefficient increases (i.e., as the match gets worse), V_{max} and V_{min} become increasingly different.

Voltage Standing Wave Ratio (VSWR)

The characteristic of voltage minimums and voltage maximums along a mismatched transmission line is referred to as the *voltage standing wave ratio* (VSWR) of the line. VSWR is defined as

$$VSWR = \frac{V_{max}}{V_{min}} \geq 1 \qquad 1.119$$

VSWR is a common measure of impedance, similar to the reflection coefficient (ρ). The concept of VSWR was developed from the observations of physical transmission lines and from T-line theory. In the past, VSWR was a very common unit of impedance measurement because it was relatively easy to measure with simple equipment and without disturbing the system under test. RF engineers rarely measure VSWR directly anymore but many vendors still specify it as a measure of how well their systems are matched.

VSWR Relationships

A little algebra with V_{max}, V_{min}, R_L, Z_0 and ρ_L reveals the following relationships:

$$VSWR = \frac{1+|\rho_L|}{1-|\rho_L|} \qquad 1 \leq VSWR \leq \infty \qquad 1.120$$

and

$$|\rho| = \frac{VSWR - 1}{VSWR + 1} \qquad 1.121$$

The absolute value in these two equations contains two separate values of reflection coefficient, i.e., two values of load resistance, which can produce the same value for VSWR. Two values of R_L will produce the same VSWR on a transmission line if the two values satisfy the following equation:

$$R_L = Z_0(VSWR) \quad \text{or} \quad R_L = \frac{Z_0}{VSWR} \qquad 1.122$$

Figure 1-21 shows a plot of VSWR vs. load resistance for a 50-ohm system. When the match is perfect ($R_L = 50\ \Omega$), the VSWR equals 1. The closer the VSWR is to 1, the better the match. VSWR ranges from 1 (a perfect match) to infinity (a short or open circuit).

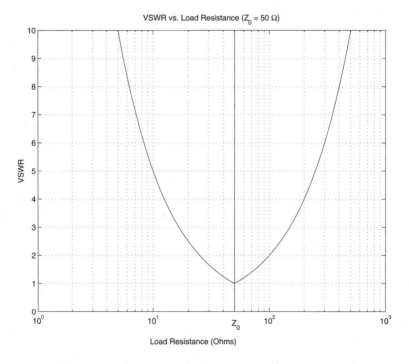

Figure 1-21 *Voltage standing wave ratio (VSWR) vs. load resistance for $Z_0 = 50\ \Omega$.*

War Story — VSWR and Microwave Ovens

If we stretch our imagination, we can think of a microwave oven as a transmission line system. The energy source is the magnetron tube inside the microwave. The transmission line is the cavity where we place the food to be heated; the load is the water in the food.

As long as we have a load on the transmission line, i.e., as long as we have food in the oven, energy transfers from the magnetron tube through the waveguide and into the food. We have a low VSWR inside of the microwave oven and all is well. But if we operate the oven without an adequate load, the oven runs under high VSWR conditions. There will be areas of "high voltages" (large electric fields) inside the cavity that will stress the magnetron tube. Also, since the energy from the magnetron meets a load with a high VSWR (or equivalently, a high reflection coefficient), all of the energy sent out by the magnetron will bounce off of the poor load and travel back into the tube. The magnetron now has to dissipate a large amount of heat and which may cause it to fail.

Even when we operate a microwave oven with food in the cavity, the load is still poorly characterized. In other words, if we think of a microwave oven and its load as a transmission line with a source and a load, the load is not exactly Z_0. After all, what is the impedance of a bologna sandwich? The system still exhibits a nonunity VSWR. From a practical point of view, this effect causes "hot" and "cold" spots in the oven. The "hot" spots are areas of high field energy; the "cold" spots are areas of low field energy. Food that lies in a hot spot heats quickly, and food in a cold spot heats slowly or not at all.

To combat this problem, manufacturers often arrange to rotate the food in the oven. Since the rotating food alternately passes through hot and cold spots in the cavity, the food gets heated more evenly. Another solution is to install stirrers in the cavities. A stirrer is a metal propeller-like structure that rotates slowly in the cavity. It reflects the energy from the magnetron in different directions as it moves. This changes the positions of the hot and a cold spot in the oven and results in more evenly heated food.

War Story — VSWR and Screen Room Testing

Screen rooms are commonly used to test antennas or to measure the electromagnetic radiation emanating from a component. Such tests are necessary to make sure pieces of equipment will not interfere with one another. Frequently, sensitive electronic tests are performed here. A screen room usually takes the form of a large metal room that shields the inside from the ambient electromagnetic fields on the outside of the room. It has metal walls, doors and ceilings with the appropriate fixtures to supply power and air into the room.

After constructing a screen room we need to determine how much shielding the room supplies from outside interference. There are many such tests to verify the performance. In order to test a screen room, the operator often places a transmitter inside the room, closes all the doors, then checks for radiation from the transmitter on the outside of the room.

As it was the case in the microwave oven example, we have to be concerned with the load that the transmitter experiences and with the VSWR inside the room. In an empty metal room, the transmitter effectively has no load, and we will experience a high VSWR condition inside the screen room. High VSWR causes large electromagnetic fields in some areas of the room and small electromagnetic fields in other areas of the room. If an area of small electromagnetic field happens to fall in an area where the screen room is leaky, then we may not detect the leak. If the area of strong electromagnetic field falls in the neighborhood of a good joint, we may detect that the joint is leaky due to the excess field placed across the joint by the high VSWR condition in the room.

Commercial gear avoids this problem by sweeping the transmitter slightly in frequency. The positions of the voltage maximums and minimums on a transmission line depend upon the operating frequency. When we change the frequency, we change the position of the voltage maxima and minima to obtain a more realistic measurement of the attenuation of the screen room.

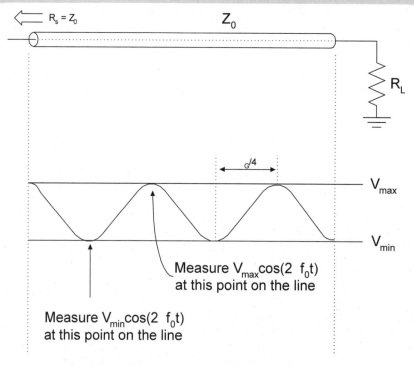

Figure 1-22 *Measuring VSWR on a transmission line.*

Figures 1-19 and 1-20 show another feature of the voltage peaks and valleys. As Figure 1-22 indicates, the distance between two adjacent voltage maximums or two adjacent voltage minimum is $\lambda_G/2$ where λ_G is the wavelength in the transmission line or the guide wavelength. Also, the distance between a voltage minimum and its closest voltage maximum is $\lambda_G/4$.

Return Loss

Return loss is a measure of how well a device is matched to a transmission line. We have mentioned launching pulses down a transmission line. The load absorbed some of the pulse energy and some of the energy was reflected back toward the source. Return loss characterizes this energy loss in a very direct manner (see Figure 1-23).

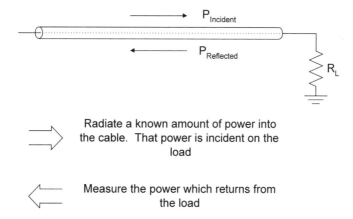

Figure 1-23 *Measuring return loss.*

When a known voltage is incident on a load ($V_{Incident}$), we can imply that a known amount of power is incident on the load ($P_{Incident}$). Similarly, when voltage is reflected from a load ($V_{Reflected}$), that voltage represents power that is reflected off of the load ($P_{Reflected}$). This applies with continuous signals as well as pulses. Return loss is defined as

$$\text{Return Loss}_{dB} = 10 \log\left(\frac{P_{Reflected}}{P_{Incident}}\right) \leq 0 \, dB \qquad 1.123$$

Return loss is measured directly by launching a known amount of power into a transmission line ($P_{Incident}$) and measuring the amount of power that reflects off of a load ($P_{Reflected}$). Return loss measures the power lost when a signal is launched into a transmission line.

$$\begin{aligned}\text{Return Loss}_{\text{dB}} &= 20\log\left|\frac{V_{\text{Reflected}}}{V_{\text{Incident}}}\right| \\ &= 20\log|\rho| \\ &= 20\log\left|\frac{Z_L - Z_0}{Z_L + Z_0}\right|\end{aligned} \qquad 1.124$$

Figure 1-24 shows a plot of return loss vs. load resistance for a 50- ohm system. When the load resistance is very small or very large, most of the power we launch into the transmission line reflects off of the load. Since $P_{\text{Reflected}} \cong P_{\text{Incident}}$, the return loss is near 0 dB. This is a poorly matched condition.

When the load resistance is approximately equal to the system characteristic impedance ($Z_0 = 50\ \Omega$ in this case), most of the power we launch into the line is absorbed by the load and very little returns. The return loss is a large negative number, which indicates a good match.

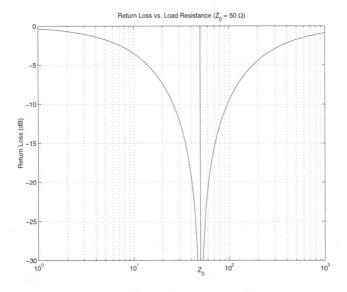

Figure 1-24 *Return loss vs. load resistance for a 50-ohm system.*

Transmission Line Summary

VSWR, reflection coefficient and return loss all measure one parameter: how well the load is matched to the system's characteristic impedance. In the examples, we have used 50 ohms as Z_0, but other common characteristic impedances are 23 ohms, 75 ohms, 93 to 125 ohms and 300 ohms.

INTRODUCTION | 59

War Story — Reflection Coefficients and Speech Synthesis

The use of reflection coefficients is not limited to transmission lines as the following example from speech synthesis and processing shows. Figure 1-25 shows a simplified model of the human vocal tract. The tubes represent the throat, nasal cavities, mouth, and so on. The signal source is noise (when we are making an unvoiced sound such as the "sh" in "should") or an impulse train (when we are making the long "a" sound as in "lake," for example). The dimensions of the throat, mouth and nasal cavities change as different sounds are made. The position of the tongue has a radical effect on the "diameter" of some of the tubes. Each configuration represents the shape of the vocal and nasal tracts as a particular sound is produced.

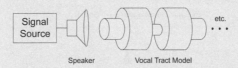

Figure 1-25 *Reflection coefficient model of a human vocal tract.*

When an audio wave travels from the source into the network of tubes, the characteristics of the sound changes as it passes through the tubes. As the wave encounters a transition between two tubes, the interface affects the sound. Some sound energy passes through the interface while some reflects off of the interface. Note that these characteristics are similar to the effects that are linked with signals in transmission lines.

One mathematical model of the vocal tract uses the reflection coefficients between the tubes to describe how the tube structure changes the sound applied to it. Each tube has an equivalent characteristic impedance which depends upon the physical size of the tube.

Transmission Line Input Impedance

Figure 1-26 shows a transmission line terminated in a complex load (a resistance R_L in series with some load reactance jX_L). We are interested in the impedance looking into the cable a physical distance l from the cable's load end.

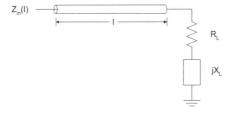

Figure 1-26 *Transmission line input impedance under complex load conditions.*

In the general case,

$$Z_{in}(l) = Z_0 \frac{Z_L + jZ_0 \tan(\beta l)}{Z_0 + jZ_L \tan(\beta l)} \qquad 1.125$$

If Z_L is an open circuit (i.e. $Z_L = \infty$) then

$$Z_{in,OC}(l) = -jZ_0 \cot(\beta l) \qquad 1.126$$

Figure 1-27 shows the input impedance of an open-circuited transmission line. Again, the input impedance is purely reactive and varies from $-j\infty$ to $+j\infty$.

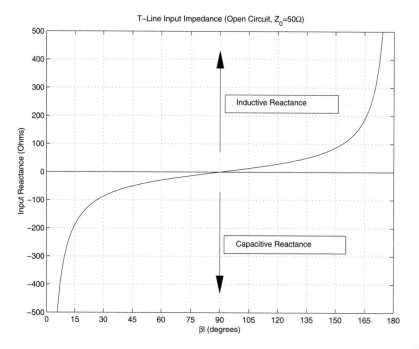

Figure 1-27 *Input impedance of an open-circuited 50-ohm transmission line.*

If Z_L is a short circuit (i.e., $Z_L = 0$) then

$$Z_{in,SC}(l) = jZ_0 \tan(\beta l) \qquad 1.127$$

Figure 1-28 shows the input impedance for an open-circuited transmission line. Note that the input impedance is purely reactive and varies from $-j\infty$ to $+j\infty$.

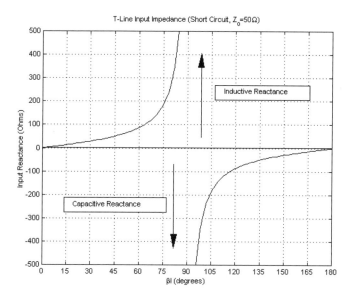

Figure 1-28 *Input impedance of a short-circuited, 50-ohm transmission line.*

For Equations 1.125, 1.127 and 1.126,
l = the physical length from the load to the near end of the cable,
Z_0 = the characteristic impedance of the cable,
Z_L = the load impedance expressed as a complex number,
$\beta = 2\pi/\lambda_g$,
λ_g = the guide wavelength.

Lossy Transmission Lines

Actual transmission lines commonly exhibit some power loss. Any time a signal propagates down a cable whether it is a pulsed input or sine-wave input transmission line, it will experience loss. In the pulsed input case, for example, the pulses applied to the source end of the cable will suffer loss as they travel toward the load. This means that the power and voltage incident on the load will not be as large as we would expect them to be in the lossless case. The reflected signal will also experience signal loss as it travels from the load to the source.

If we were to measure the power of the returning pulse at the sending end of the transmission line, it would be less than it would have been had the cable been lossless. In other words, both $P_{\text{Reflected}}$ and $V_{\text{Reflected}}$ will be smaller because of the line loss. Since

$$\rho = \frac{V_{\text{Reflected}}}{V_{\text{Incident}}} \qquad 1.128$$

the loss in the cable makes the reflection coefficient smaller and the load appears to be closer to Z_0 than it really is. A lossy cable tends to make the loads appear better matches than they really are.

This insight also applies to return loss measurements (see Equation 1.124). We know how much power we initially launched into the source end of the cable (P_{Incident}) and can measure how much returns from the load ($P_{\text{Reflected}}$). However, we do not know how much power was absorbed by the load and how much power was absorbed by the lossy line. Since the reflected power is less than we would see if we had a lossless cable, the load will appear to be more closely matched than if the cable had been lossless.

This is not necessarily to our advantage. For example, in a receiving situation, the cable loss represents signal loss, directly reducing the system's sensitivity. In a transmitting situation, cable loss represents power that has been used up in the cable between the power amplifier and the transmitting antenna. For the continuous sine-wave input model we discussed, we have to keep in mind that, as the signal moves along the cable, it experiences attenuation. Then we mentally examine the sine-wave equations and see if the results agree with the time-domain models.

1.12 Two Ports

Figure 1-29 shows a two-port device with the voltage and current assignments we will assume. We will feed voltage, current or power into one port of the device and we attach a known impedance to the other port. Then we will observe what happens to the voltages and currents on the output terminals of the two-port. A two-port device can be characterized in various ways but, in this book, we focus on the following:

- z-parameters, which describe the two-port in terms of impedance-like or voltage/current quantities,

- y-parameters, which describe the two-port in terms of admittance-like or current/voltage parameters,

- h-parameters, which are useful for transistor amplifier design,

- s-parameters, which are useful for RF and microwave systems.

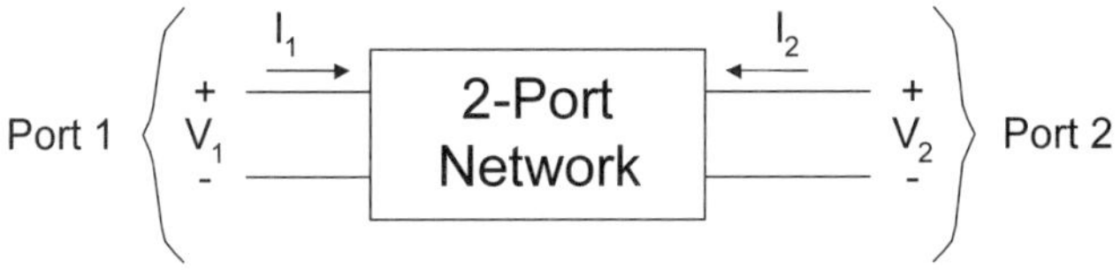

INTRODUCTION | 63

Figure 1-29 Voltage and current definitions for a general two-port device.

Z-Parameters

The z-parameters define a two-port as follows:

$$V_1 = z_{11}I_1 + z_{12}I_2 \quad or \quad \begin{bmatrix} V_1 \\ V_2 \end{bmatrix} = \begin{bmatrix} z_{11} & z_{12} \\ z_{21} & z_{22} \end{bmatrix} \begin{bmatrix} I_1 \\ I_2 \end{bmatrix} \qquad 1.129$$

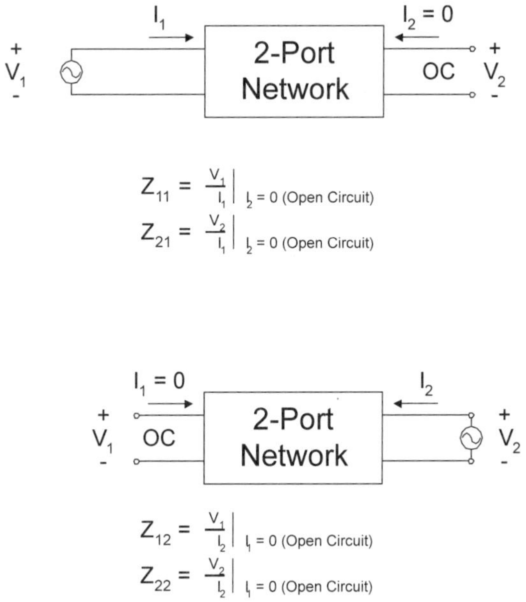

$$Z_{11} = \left.\frac{V_1}{I_1}\right|_{I_2=0} \quad I_2 = 0 \text{ (Open Circuit)}$$

$$Z_{21} = \left.\frac{V_2}{I_1}\right|_{I_2=0} \quad I_2 = 0 \text{ (Open Circuit)}$$

$$Z_{12} = \left.\frac{V_1}{I_2}\right|_{I_1=0} \quad I_1 = 0 \text{ (Open Circuit)}$$

$$Z_{22} = \left.\frac{V_2}{I_2}\right|_{I_1=0} \quad I_1 = 0 \text{ (Open Circuit)}$$

Figure 1-30 Measurement of two of the four z-parameters of a general two-port device. Requires a wideband open circuit.

Figure 1-30 shows the measurement of a network's Z-parameters. We apply a voltage source to port 1 and leave port 2 open-circuited. This forces $I_2 = 0$. We know the value of V_1 and can measure or calculate the values of I_1 and V_2. Combining this insight with Equation 1.129, we can find two of the z-parameters.

$$z_{11} = \left.\frac{V_1}{I_1}\right|_{I_2=0} \quad \text{and} \quad z_{21} = \left.\frac{V_2}{I_1}\right|_{I_2=0} \qquad 1.130$$

To determine the values of the other two z-parameters, we apply a voltage source at the same test frequency to port 2, i.e., we set the value of V_2 and leave port 1 as an open circuit and set $I_1 = 0$. Then, we measure or calculate V_1 and I_2. The other two z-parameters are

$$z_{12} = \left.\frac{V_1}{I_2}\right|_{I_1=0} \quad \text{and} \quad z_{22} = \left.\frac{V_2}{I_2}\right|_{I_1=0} \qquad 1.131$$

Knowing the values of z_{11}, z_{12}, z_{21} and z_{22}, allows us to characterize the two-port at the test frequency. Given an input voltage and load impedance, all of the corresponding voltages and currents can be found. Note that the values of the z-parameters will likely change with frequency. Usually, the parameters at all the frequencies under consideration have to be measured.

In summary, we can measure or calculate the z-parameters of a network using

$$\begin{aligned} z_{11} &= \left.\frac{V_1}{I_1}\right|_{I_2=0} & z_{12} &= \left.\frac{V_1}{I_2}\right|_{I_1=0} \\ z_{21} &= \left.\frac{V_2}{I_1}\right|_{I_2=0} & z_{22} &= \left.\frac{V_2}{I_2}\right|_{I_1=0} \end{aligned} \qquad 1.132$$

Z_{11} is the input impedance of the two-port (when the output port is in the open-circuit condition) and z_{22} is the output impedance of the two-port (when the input port is in the open-circuit condition).

Y-Parameters

Like z-parameters, y-parameters completely define a two-port device.

$$\begin{aligned} I_1 &= y_{11}V_1 + y_{12}V_2 \\ I_2 &= y_{21}V_1 + y_{22}V_2 \end{aligned} \quad \text{or} \quad \begin{bmatrix} I_1 \\ I_2 \end{bmatrix} = \begin{bmatrix} y_{11} & y_{12} \\ y_{21} & y_{22} \end{bmatrix} \begin{bmatrix} V_1 \\ V_2 \end{bmatrix} \qquad 1.133$$

To measure or calculate the y-parameters, we force a current into one port while short-circuiting the opposite port (see Figure 1-31). We calculate

$$y_{11} = \frac{I_1}{V_1}\bigg|_{V_2=0} \quad y_{12} = \frac{I_1}{V_2}\bigg|_{V_1=0}$$
$$y_{21} = \frac{I_2}{V_1}\bigg|_{V_2=0} \quad y_{22} = \frac{I_2}{V_2}\bigg|_{V_1=0}$$

1.134

y_{11} is the input admittance of the two-port (when the output port is in the short circuit condition) and y_{22} is the output admittance of the two-port (when the input port is in the short circuit state).

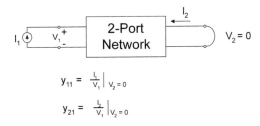

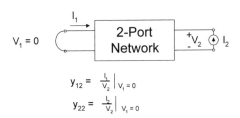

Figure 1-31 *Measurement of the four y-parameters of a general two-port device. Requires a wideband short circuit.*

H-Parameters

H- (for hybrid) parameters are well suited to bipolar transistor design. The defining equations for the h-parameters are

$$V_1 = h_{11}I_1 + h_{12}V_2$$
$$I_2 = h_{21}I_1 + h_{22}V_2 \quad \text{or} \quad \begin{bmatrix} V_1 \\ I_2 \end{bmatrix} = \begin{bmatrix} h_{11} & h_{12} \\ h_{21} & h_{22} \end{bmatrix} \begin{bmatrix} I_1 \\ V_2 \end{bmatrix}$$

1.135

To measure the h-parameters, we again apply voltages or currents to the various ports leaving other ports in the open- or short-circuit condition (see Figure 1-32). We then calculate

$$h_{11} = \frac{V_1}{I_1}\bigg|_{V_2=0} \quad h_{12} = \frac{V_1}{V_2}\bigg|_{I_1=0}$$

$$h_{21} = \frac{I_2}{I_1}\bigg|_{V_2=0} \quad h_{22} = \frac{I_2}{V_2}\bigg|_{I_1=0}$$

1.136

h_{11} is the input impedance of the device (when the output is shorted), h_{22} is the output admittance (when the input is in the open-circuit state), h_{21} is the current gain and h_{12} is the reverse voltage feedthrough.

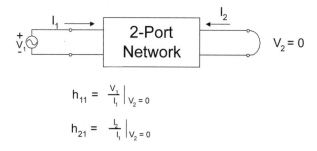

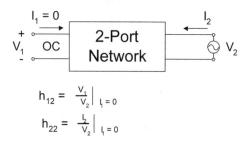

Figure 1-32 *Measurement of the four h-parameters of a general two-port device. Requires wideband open- and short-circuits.*

S-Parameters

Measuring the z-, y- and h-parameters of active devices, such as transistors and FET's at high frequencies, presents several problems.

- At high frequencies, it is difficult to build equipment to measure the current and voltage directly at the ports of a network. However, power can be measured fairly easily.

- Due to stray capacitance and inductance, it is very difficult to realize wideband, accurate short- and open-circuit loads.

- Active devices are often not short- or open-circuit stable. In other words, transistor amplifiers may oscillate if they are subjected to short or open circuits. This completely distorts any measurements planned.

Course of Thought

In the radio receiver arena, we will use most of the circuits we build in systems with some characteristic impedance. In other words, when circuits are operating normally, they are in some environment where they will always see a Z_0 source resistance and a Z_0 load resistance. Systems are best characterized under these conditions.

We can transfer knowledge from transmission line theory. When we analyzed transmission lines, we launched a voltage down the line and observed the incident voltage reflected off of the load. We can use this technique to characterize the devices in a Z_0 environment.

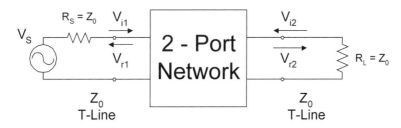

Figure 1-33 *Measurement of the four s-parameters of a general two-port device. Requires only matched loads and sources.*

Figure 1-33 shows the measurement environment we would like to use. First, we terminate both the input and output of the device with a Z_0 load in all of the measurements. Then, we launch an incident wave (V_{i1}) at the two-port. The incident wave will encounter the two-port device and some of the voltage will reflect off of port 1 (is V_{r1}). Some of the incident signal will enter the two-port and find its way to port 2 where it appears as V_{r2} and V_{i2}.

Since we have terminated the device on both ports with a matched load, we know that any signals that occur at the termination will be completely absorbed. In Figure 1-33, for example, the matched load placed on the output port of the device will force the voltage V_{i2} to be zero. By measuring the s-parameters of a network, we are radiating energy at the device and observing how the energy scatters off. The s-parameters of a network are the *scattering parameters*.

Definition

S-parameters are defined as

$$V_{r1} = s_{11}V_{i1} + s_{12}V_{i2}$$
$$V_{r2} = s_{21}V_{i1} + s_{22}V_{i2}$$
or
$$\begin{bmatrix} V_{r1} \\ V_{r2} \end{bmatrix} = \begin{bmatrix} s_{11} & s_{12} \\ s_{21} & s_{22} \end{bmatrix} \begin{bmatrix} V_{i1} \\ V_{i2} \end{bmatrix} \qquad 1.137$$

To calculate the various s-parameters from measured data, we perform

$$s_{11} = \left.\frac{V_{r1}}{V_{i1}}\right|_{V_{i2}=0} \quad s_{12} = \left.\frac{V_{r1}}{V_{i2}}\right|_{V_{i1}=0} \qquad 1.138$$

$$s_{21} = \left.\frac{V_{r2}}{V_{i1}}\right|_{V_{i2}=0} \quad s_{22} = \left.\frac{V_{r2}}{V_{i2}}\right|_{V_{i1}=0}$$

Measurement Technique

To measure the s-parameters of a network, one port is derived with a matched source (a source whose series impedance is Z_0) and the other port will be terminated with a Z_0 load (see Figure 1-34).

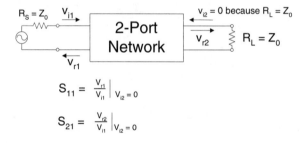

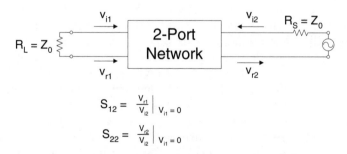

Figure 1-34 *Detailed measurement of the s-parameters of a two-port device indicating incident and reflected voltages.*

The source launches energy at the two-port, which is represented by the voltage V_{i1}. Some of that energy reflects off of the two-port's input, travels back toward the source and becomes V_{r1}. Some of the energy which entered the input port will travel through the two-port and find its way to the right-hand side of the two-port. Some of this energy will exit the right-hand side of the two-port becoming V_{r2}. Since we set the load resistor R_L to equal Z_0, R_L absorbs all of the energy and reflects none. The same effects occur in the bottom half of Figure 1-34 when we reverse the source and load sides.

It is difficult to measure voltage directly at high frequencies, but it is easy to measure power. It is also easy to measure the phase difference between two voltages. To accommodate for easy measurements and for not so easy ones, the defining equations have to be changed.

$$\begin{matrix} b_1 = s_{11}a_1 + s_{12}a_2 \\ b_2 = s_{21}a_1 + s_{22}a_2 \end{matrix} \quad or \quad \begin{bmatrix} b_1 \\ b_2 \end{bmatrix} = \begin{bmatrix} s_{11} & s_{12} \\ s_{21} & s_{22} \end{bmatrix} \begin{bmatrix} a_1 \\ a_2 \end{bmatrix} \qquad 1.139$$

where

$$b_1 = \frac{V_{r1}}{\sqrt{Z_0}} \quad a_1 = \frac{V_{i1}}{\sqrt{Z_0}}$$
$$b_2 = \frac{V_{r2}}{\sqrt{Z_0}} \quad a_2 = \frac{V_{i2}}{\sqrt{Z_0}} \qquad 1.140$$

These equations are identical to Equation 1.137 except that everything was divided by $\sqrt{Z_0}$. Note that

- $|a_1|^2$ = power incident on the input of the network
  = power available from a source of impedance Z_0,

- $|a_2|^2$ = power incident on the output of the network
  = power reflected from the load,

- $|b_1|^2$ = power reflected from the input port of the network
  = power available from a Z_0 source minus the power delivered to the input of the network,

- $|b_2|^2$ = power reflected or emanating from the output of the network
  = power incident on the load,
  = power that would be delivered to a Z_0 load.

Since we can easily measure power and phase angles, we can determine the quantities above and obtain the s-parameters of a given network.

S-Parameter Relationships

The s-parameters are related to power gain and mismatch loss.

$$|s_{11}|^2 = \frac{\text{Power reflected from the network input}}{\text{Power incident on the network input}}$$

$$|s_{22}|^2 = \frac{\text{Power reflected from the network output}}{\text{Power incident on the network output}}$$

$$|s_{21}|^2 = \frac{\text{Power delivered to a } Z_0 \text{ load}}{\text{Power available from a } Z_0 \text{ load}}$$

$\qquad$ = Transducer power gain with Z_0 load and source

$|s_{12}|^2$ = Reverse transducer power gain with Z_0 load and source
$\qquad$ = Reverse isolation

(1.141)

These equations are valid only when the network is driven by a source with a Z_0 characteristic impedance and when the network is terminated with a Z_0 load. The s-parameters are related to the reflection coefficient and the return loss of the two-port.

$$s_{11} = \text{reflection coefficient}$$
$$= \rho$$
$$\text{return loss} = 20\log(|\rho|)$$
$$= 20\log(|s_{11}|)$$

(1.142)

Relationships Among Two-Port Parameters

Each of these sets of parameters *exactly and absolutely* define the two-port network. The network parameters mentioned contain the same information. Since all of these parameters contain the same information in different forms, we can convert from one two-port parameter to any other. The Tables 1.143 through 1.145 show the relationships between two-port parameters.

S and Z

$$s_{11} = \frac{(z_{11}-1)(z_{22}+1)-z_{12}z_{21}}{(z_{11}+1)(z_{22}+1)-z_{12}z_{21}} \qquad z_{11} = \frac{(1+s_{11})(1-s_{22})+s_{12}s_{21}}{(1-s_{11})(1-s_{22})-s_{12}s_{21}}$$

$$s_{12} = \frac{2z_{12}}{(z_{11}+1)(z_{22}+1)-z_{12}z_{21}} \qquad z_{12} = \frac{2s_{12}}{(1-s_{11})(1-s_{22})-s_{12}s_{21}}$$

$$s_{21} = \frac{2z_{21}}{(z_{11}+1)(z_{22}+1)-z_{12}z_{21}} \qquad z_{12} = \frac{2s_{21}}{(1-s_{11})(1-s_{22})-s_{12}s_{21}} \qquad 1.143$$

$$s_{11} = \frac{(z_{11}+1)(z_{22}-1)-z_{12}z_{21}}{(z_{11}+1)(z_{22}-1)-z_{12}z_{21}} \qquad z_{22} = \frac{(1-s_{11})(1+s_{22})+s_{12}s_{21}}{(1-s_{11})(1-s_{22})-s_{12}s_{21}}$$

S and Y

$$s_{11} = \frac{(1-y_{11})(1+y_{22})+y_{12}y_{21}}{(1+y_{11})(1+y_{22})-y_{12}y_{21}} \qquad y_{11} = \frac{(1-s_{11})(1+s_{22})+s_{12}s_{21}}{(1+s_{11})(1+s_{22})-s_{12}s_{21}}$$

$$s_{12} = \frac{-2y_{12}}{(1+y_{11})(1+y_{22})-y_{12}y_{21}} \qquad y_{12} = \frac{-2s_{12}}{(1+s_{11})(1+s_{22})-s_{12}s_{21}}$$

$$s_{21} = \frac{-2y_{21}}{(1+y_{11})(1+y_{22})-y_{12}y_{21}} \qquad y_{12} = \frac{-2s_{21}}{(1+s_{11})(1+s_{22})-s_{12}s_{21}} \qquad 1.144$$

$$s_{11} = \frac{(1+y_{11})(1-y_{22})+y_{12}y_{21}}{(1+y_{11})(1+y_{22})-y_{12}y_{21}} \qquad y_{22} = \frac{(1+s_{11})(1-s_{22})+s_{12}s_{21}}{(1+s_{11})(1+s_{22})-s_{12}s_{21}}$$

S and H

$$s_{11} = \frac{(h_{11}-1)(h_{22}+1)-h_{12}h_{21}}{(h_{11}+1)(h_{22}+1)-h_{12}h_{21}} \qquad h_{11} = \frac{(1+s_{11})(1+s_{22})-s_{12}s_{21}}{(1-s_{11})(1+s_{22})+s_{12}s_{21}}$$

$$s_{12} = \frac{2h_{12}}{(h_{11}+1)(h_{22}+1)-h_{12}h_{21}} \qquad h_{12} = \frac{2s_{12}}{(1-s_{11})(1+s_{22})+s_{12}s_{21}}$$

$$s_{21} = \frac{-2h_{21}}{(h_{11}+1)(h_{22}+1)-h_{12}h_{21}} \qquad h_{12} = \frac{-2s_{21}}{(1-s_{11})(1+s_{22})+s_{12}s_{21}} \qquad 1.145$$

$$s_{11} = \frac{(1+h_{11})(1-h_{22})+h_{12}h_{21}}{(h_{11}+1)(h_{22}+1)-h_{12}h_{21}} \qquad h_{22} = \frac{(1-s_{11})(1-s_{22})-s_{12}s_{21}}{(1-s_{11})(1+s_{22})+s_{12}s_{21}}$$

Normalized Components

The z-, y- h- and s-parameters shown in Equations 1.143 through 1.145 are normalized, i.e., the characteristic impedance is unity. If the characteristic impedance is z_0, convert from z_0 using

$$z_{11} = \frac{z_{11,Z0}}{Z_0} \quad z_{12} = \frac{z_{12,Z0}}{Z_0} \quad z_{21} = \frac{z_{21,Z0}}{Z_0} \quad z_{22} = \frac{z_{22,Z0}}{Z_0} \qquad 1.146$$

$$y_{11} = y_{11,Z0}Z_0 \quad y_{12} = y_{12,Z0}Z_0 \quad y_{21} = y_{21,Z0}Z_0 \quad y_{22} = y_{22,Z0}Z_0 \qquad 1.147$$

$$h_{11} = \frac{h_{11,Z0}}{Z_0} \quad h_{12} = h_{12,Z0} \quad h_{21} = h_{21,Z0} \quad h_{22} = z_{22,Z0}Z_0 \qquad 1.148$$

1.13 Matching and Maximum Power Transfer

We can model any practical signal source as Figure 1-35 indicates. The combination of V_S and R_S can be a signal generated, an antenna, or a RF amplifier. The value of R_S can be small, but it will never be zero. For a various reasons, R_S is set equal to Z_0, the system's characteristic impedance.

Let us assume the source in Figure 1-35 is an antenna. Then V_S represents the signal energy the antenna receives and R_S represents the radiation resistance of the antenna. If the load resistor represents a receiving system, the receiver should be able to take the maximum amount of signal power from the antenna into the load resistor R_L.

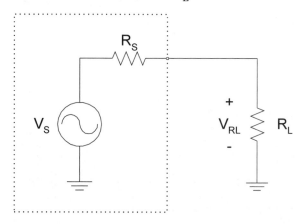

o Signal Generator
o Antenna
o RF Amplifier

Figure 1-35 *Model used to derive maximum power transfer conditions.*

Resistive Loads

Given that the antenna has some nonzero series impedance, what value of R_L will produce the maximum power transfer from V_S to R_L? The power dissipated in R_L is

$$P_{RL} = \frac{V_{RL}^2}{R_L} \qquad 1.149$$

We know

$$V_{RL} = V_S \frac{R_L}{R_L + R_S} \qquad 1.150$$

Hence

$$P_{RL} = V_S^2 \frac{R_L}{(R_L + R_S)^2} \qquad 1.151$$

The maximum power transfer will occur when

$$\frac{\partial P_{RL}}{\partial R_L} = 0 \qquad 1.152$$

$$\Rightarrow V_S^2 \frac{(R_L + R_S)^2 - 2R_L(R_L + R_S)}{(R_L + R_S)^2} = 0$$

which can be simplified to

$$R_S = R_L \qquad 1.153$$

The receiver will pull the maximum amount of power from the antenna if we set the input impedance of the receiver equal to the output impedance of the antenna (the "matched" condition). Figure 1-36 shows a graph of P_{RL} versus R_L with $R_S = 100\,\Omega$ and $V_S = 1$ V. The curve peaks when $R_S = 100\,\Omega = R_L$.

Let us assume we are operating in a 50-ohm system and the combination of V_S and R_S is an antenna. In order to get the maximum amount of signal energy from the antenna, we have to set $R_S = 50\,\Omega$. The farther R_L is from 50 ohms, the less power we will be able to draw out of the antenna.

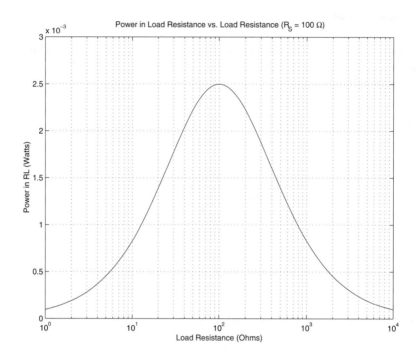

Figure 1-36 *Power in 100 ohm load resistor vs. the load resistor value. Note that the power dissipated in the load resistor is maximum when $R_S = R_L$.*

Complex Loads

If the source impedance is complex and equal to Z_S, maximum power transfer is achieved when the load impedance (Z_L) satisfies the relationship

$$Z_L = Z_S^* \qquad 1.154$$

where Z_S^* is the complex conjugate of the source impedance (see Figure 1-37). If $Z_L = Z_S^*$, then $R_L = R_S$ and $X_L = -X_S$. The reactances cancel and two matched resistors are left.

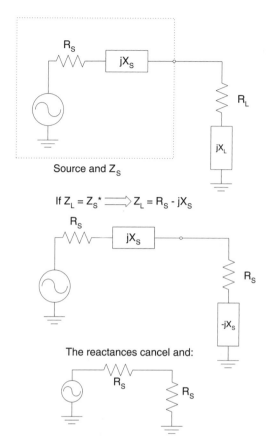

Figure 1-37 *Maximum power transfer for a complex load and source impedances.*

1.14 Matching

Matching usually means setting one impedance equal to another impedance to achieve maximum power transfer. However, there are other valid reasons why we match and there are also situations where matching is not to our advantage.

When to Match

Antennas

We match to achieve maximum power transfer from a source to a load. This is most common in case of an antenna. In a receiving situation, all the signal energy possible should be removed from the antenna. In a transmitting situation, we would like to release as much power as we can into the antenna. Matching allows this effect to occur.

Terminations

We often match to present termination-sensitive devices with a fixed, known impedance. For example, mixers and filters usually require a broadband termination on both their input and output ports. If they do not see a broadband match, their performance will suffer. Some devices are unstable or change their mode of operation radically if an improper termination is presented. For example, amplifiers may break into oscillation or the oscillator's frequency may change.

Transmission Lines

When we discussed transmission lines, we saw that mismatches on either the source or load end of a cable can produce signal reflections. These reflections will distort the signal as well as produce voltage maximums and minimums on the line. Matching helps to avoid all of these problems.

If we use matched loads, we can use arbitrary lengths of transmission lines to connect the different pieces of the systems. If a transmission line is not matched, the performance of the system can change as we change the lengths of the transmission line connecting the components.

Calculations

Matching impedances makes calculations and their associated mathematic operations easier. We do not have to factor the changing impedance levels into the equations if all the source and load impedances are equal.

Time Domain Effects

In digital and other time-domain systems, a poorly terminated transmission line can cause signal reflections as pulses to bounce off of mismatched loads. This causes distortions to the point where they will ruin a system's performance.

When Not to Match

Efficiency

When efficiency is important, matching the source impedance to the load impedance is a poor idea. Figure 1-38 shows a model for an RF amplifier. If the load resistor R_L is dissipating a power of P_{RL} and if $R_{amp,out} = R_L$, then $R_{amp,out}$ must also be dissipating P_{RL}. In other words, 1/2 (or 3 dB) of the amplifier's total output power is lost in the amplifier's output resistor $R_{amp,out}$. The efficiency of the system is defined as

$$\eta = \frac{\text{Power dissipated in } R_L}{\text{Power generated by } V_{amp,out}}$$

$$= (P_{RL}) \frac{R_L + R_{amp,out}}{R_L} \qquad 1.155$$

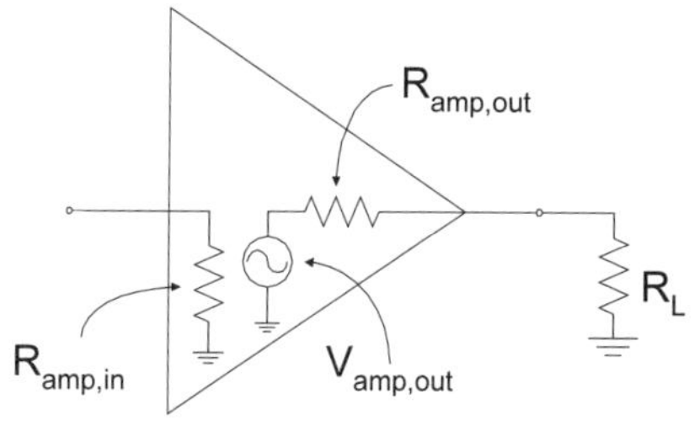

Figure 1-38 *Model of an RF amplifier delivering power to a load resistor.*

Figure 1-39 shows a graph of efficiency vs. $R_{amp,out}$. Maximum efficiency can be achieved when the amplifier's output impedance is 0. Consider 60-hertz wall outlets and the source impedance supplied by the power company. We are not interested in maximum power transfer in this situation but we are interested in a constant load voltage and high efficiency. Making the source impedance as small as possible will allow the voltage to stay almost constant despite the type of load we connect.

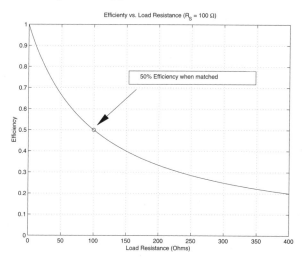

Figure 1-39 *Efficiency vs. load resistance for a 100-ohm load resistance. Note that efficiency is only 50 % when matched. High efficiency occurs when $R_S << R_L$.*

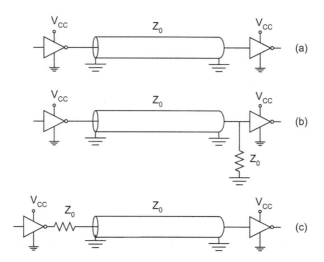

Figure 1-40 Three options for driving a transmission line with digital logic.

Other Considerations

Figure 1-40(a) shows a digital gate driving a transmission line with a characteristic impedance of Z_0. If the line is long in terms of wavelengths, the mismatched impedance of the receiving gate will cause reflections.

One solution is to match the transmission line. Placing a load resistor at the receiving end of the line reduces the reflections [see Figure 1-40(b)]. However, we now have to deal with the drive capability of the digital gate driving the source end of the cable. Is it capable of supplying enough current to handle a 50-ohm load?

Another solution is to source-match, as shown in Figure 1-40(c). Although the mismatched load of the receiving gate will still cause reflections, they will be completely absorbed by the match at the source end of the transmission line. Since the reflections are destroyed at the source end, they have no chance to return down the transmission line and play havoc with the receiving gate.

However, we now are confronted with the threshold levels of the particular logic family. The resistor at the source end of the cable will affect the maximum voltage that the gate at the receiving end of the line will see. We have to make sure that the receiving gate will react reliably to the signals sent by the transmitting gate.

1.15 RF Amplifiers

As part of this introduction to radio frequency systems, some of the characteristics of amplifiers are discussed here. In Chapters 5 and 6, we will investigate characteristics such as noise and linearity in greater detail.

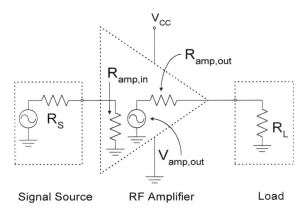

Figure 1-41 Models for a signal source, RF amplifier and load. Under matched conditions, $R_S = R_{amp,in} = R_{amp,out} = R_L$.

Power Gain

An amplifier is placed at a particular location in the system because it provides signal power gain. Figure 1-41 shows the model we will use for a RF amplifier, its source and its load. The signal source can be a signal generator, antenna or another RF amplifier. The signal source is modeled as a voltage source V_S in series with a resistor R_S.

The amplifier in Figure 1-41 accepts power from some external source and that power is dissipated in $R_{amp,in}$. The amplifier measures the power being dissipated in $R_{amp,in}$ and adjusts its internal voltage source V_{amp} so that a fixed multiple of the input power is delivered to R_L. In other words,

$$P_{RL} = G_P P_{in} \qquad 1.156$$

where
   P_{in} = the power delivered to the amplifier's input resistor $R_{amp,in}$ by some external source (in linear units such as watts or mW).
   P_{RL} = the power delivered to the load resistor R_L by the RF amplifier (in linear units).
   G_P = the power gain of the amplifier; usually > 1 (in linear units).

We can write Equation 1.156 in a logarithmic format

$$R_{RL,dBm} = G_{P,dB} + P_{in,dBm} \qquad 1.157$$

where

$P_{in,dBm}$ = the power delivered to the amplifier's input resistor by some external source (expressed in dBm).

$P_{RL,dBm}$ = the power delivered to the load resistor R_L by the RF amplifier (expressed in dBm).

$G_{P,dB}$ = the power gain of the amplifier (usually > 0 dB).

We can also write

$$R_{RL,dBW} = P_{in,dBW} + G_{P,dB} \qquad 1.158$$

where

$P_{in,dBW}$ = the power delivered to the amplifier's input resistor by some external source (expressed in dB).

$P_{RL,dBW}$ = the power delivered to the load resistor R_L by the RF amplifier (expressed in dB).

The power gain of an amplifier (or attenuator) remains the same when we describe the power in dBW, dBm, dBf, and so on. It will change with frequency, temperature and power supply voltage. A typical specification for the power gain of an amplifier is

> Power gain = 15 dB ± 1 dB over a 20 to 500 MHz frequency range and a –20 to +55°C temperature range.

The nominal power gain (15 dB) is indicated. The gain may vary from 14 to 16 dB over a –20 to +55°C temperature range. This specification does not indicate how the power gain will vary with temperature (the gain vs. temperature curve probably will not be linear) and how the device's power gain will vary with frequency. Equations 1.156 through 1.158 also work with lossy devices such as attenuators and passive filters. In the lossy case, the linear gain is less than unity and the decibel gain is less than 0 dB.

Mental Model

> A convenient mental model of an amplifier can be described as follows:
>
> The amplifier measures the input power delivered to $R_{amp,in}$ by some external source. It multiplies the signal power by some power gain G_P and delivers the multiplied power to the load resistor R_L.

This model works for both signals and noise although the amplifier will always add some noise power of its own to the input.

Reverse Isolation

Amplifiers are not perfectly unilateral devices. If we present an amplifier with a signal on its output port, some of the signal will leak through to the amplifier's input port. This is the amplifier's *reverse isolation* (see Figure 1-42).

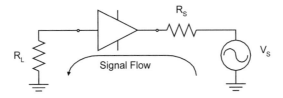

Figure 1-42 *Reverse isolation in an RF amplifier. The signal flows "backwards" through a component.*

Referring to Figure 1-42, the amplifier's reverse isolation is given by

$$\text{Reverse Isolation} = \frac{\text{Power available from the input port}}{\text{Power delivered to the output port}} \qquad 1.159$$

This effect is not intentional and is a general characteristic of most RF and microwave devices. Signals presented to one port of a device will propagate to the other ports despite the the designer's best efforts. These isolation specifications affects the decisions made by receiver designers when they connect devices together in cascade.

Frequency Coverage

Amplifiers will provide their guaranteed specifications only over some given frequency range. Outside the specified frequency range the amplifier may still operate, but its performance specifications cannot be guaranteed over time, temperature, and from unit to unit.

Often it is convenient to distinguish between wideband and narrowband devices. Although the definitions vary greatly, in this book, we will specify that a *wideband system* has a frequency coverage that is greater than an octave (2:1 range). A *narrowband system* has a frequency coverage that is less than an octave.

Signals, Noise and Power Gain

Actual amplifiers add noise to any signal they process. This noise may or may not be a problem depending upon the relative power levels of the noise and signal. An amplifier does not register the difference between signal power and noise power. The external source, for example, an antenna, will deliver both signal power and noise power to the amplifier. The amplifier will add noise of its own to the input, then amplify the total package by its power gain.

Linearity

Note that we have shown the amplifier of Figure 1-41 with connections to both a power supply (V_{cc}) and ground. From a linearity point of view these connections are very important. No matter how small the signal, an amplifier will always distort any signal it processes. Depending upon the final system architecture, distortion may be a problem. There are many linearity specifications such as third-order intercept, second-order intercept and compression point. Their definitions and effects are discussed in Chapter 6.

Input and Output Impedances

When designing an amplifier, we aim for a carefully controlled input and output impedance. If we intend to operate the amplifier in a system whose characteristic impedance is Z_0, we almost universally "match" the amplifier so that $R_S = R_{amp,in} = R_{amp,out} = R_L = Z_0$. Yet the input and output impedance of an amplifier will change with frequency and temperature. A typical specification is

> Input VSWR is < 2.0:1 over a –55 to 85°C temperature range over a 5 to 500 MHz frequency range.

This specification indicates that the input impedance of the amplifier will be such that the input VSWR of the amplifier will not be greater than 2.0:1 over the given temperature and frequency range.

There are similar specifications for the output impedance in terms of VSWR. In practice, there are two types of devices as far as terminal impedances are concerned: some devices present a wideband match to the outside world; other devices demand that the outside world present a wideband match to them. For example, the source and load terminations influence a filter's performance significantly. The filter designer assumes that we will provide a Z_0 resistor on the input and output terminals of the filter. If we do not meet this requirement, the filter will misbehave. On the other hand, an attenuator will usually provide something close to a Z_0 impedance to the outside world. We often use attenuators to quiet a widely varying impedance.

INTRODUCTION | 83

> *War Story — Frequency Synthesizer Design*
> When the impedance an amplifier sees on its output is modified, the amplifier's input impedance changes. Likewise, when we change the load presented to the input of an amplifier, the output impedance of the amplifier changes. This is different than the reverse isolation of the amplifier.
>
> When designing a frequency synthesizer with a *voltage-controlled oscillator* (VCO), we find that the VCO's output frequency changes when we alter the load impedance. The VCO often drives a frequency divider (a modulo-2 prescaler). The prescaler input impedance changes slightly depending upon its output state. This puts an unwanted frequency modulation (FM) on the VCO. One solution is to isolate devices as much as possible. It is common practice to place amplifiers and attenuators between the VCO and the prescaler, which will decrease the impedance changes presented to the VCO.

Output Power and Efficiency

Since an amplifier obtains its operating power from the outside world, it stands to reason that it can only produce so much output voltage before it falls out of specification. The output power specification is important in transmitters and some mixer applications.

The amount of signal power we can draw from an amplifier depends upon the amount of distortion we are willing to tolerate. Generally, higher power means the waveform contains higher levels of distortion. Very linear amplifiers produce much output power but will require a considerable amount of DC power. As a rule of thumb, we have found that a Class A wideband amplifier will deliver a maximum of about 10% of its DC power as RF energy, i.e.,

$$P_{\text{out,max}} \approx \frac{P_{DC}}{10} \qquad 1.160$$

Stability

Any device with power gain has the ability to become unstable and break into self-sustaining oscillation. Amplifier designers have to guarantee their amplifiers are unconditionally stable. This means that the amplifier will not oscillate no matter what impedance is placed on the input and output ports.

For example, amplifiers and other RF devices are normally tested using laboratory-grade signal generators and spectrum analyzers. The manufacturers of the signal generators and spectrum analyzers have ensured that test ports on their equipment appear as a nonreactive Z_0 impedance.

In practice, however, we find that filters, antennas, oscillators and mixers do not present a wideband Z_0 match to the outside world. The terminal impedances of a filter, for example, vary wildly with frequency. In the passband, the filter presents a Z_0 impedance to the outside world. In the stopband, a filter usually presents either a short or open circuit. Similarly, an antenna may look like a voltage source with a Z_0 series impedance in its operating band that changes once we are outside of the operating bandwidth of the antenna.

Figure 1-43 shows how we might build an actual system. The amplifier is store-bought and provides gain over its specified frequency range of 20 to 500 MHz. The antenna covers a range of 250 to 350 MHz, and the filter following the amplifier has a passband of only 270 to 330 MHz. The impedance of each device will be approximately 50 ohms in the 270 to 330 MHz frequency range.

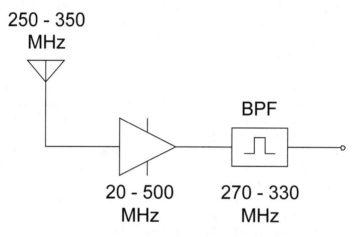

Figure 1-43 A common system configuration utilizing an antenna, amplifier and filter. Each component will present a non-Z_0 impedance to the outside world at frequencies which are outside of their specified operating ranges.

The impedance of the antenna changes rapidly outside the 250 to 350 MHz range. Accordingly, the amplifier input sees a non-Z_0 termination on its input port when the frequency is below 250 MHz and when it is greater than 350 MHz. The amplifier's output port sees a non-Z_0 termination when the frequency is outside the 270 to 350 MHz range. If the amplifier is incorrectly designed, these non-Z_0 terminations can cause the amplifier to oscillate.

Presenting the input and output terminals with the wrong impedance will cause an amplifier to break into oscillation. If the amplifier of Figure 1-43 oscillates, it will usually oscillate at frequencies corresponding to the band edges of the filters or antenna. The input impedance of these devices are changing rapidly as they transverse from their passband into their stopbands. Finally, as the following examples illustrate, an amplifier can be oscillating and still be able to perform all of the functions it was designed to perform.

INTRODUCTION | 85

War Story — Amplifier Stability (I)

A colleague was performing some work on an intermediate frequency (IF) amplifier with *automatic gain control* (AGC), an old design which had been successfully deployed in the field for years. The company had just purchased a new wideband oscilloscope which he was using to characterize the IF amplifier.

When our colleague placed the probe on one of the points in the circuit, he noticed that the trace got just a little fatter. Further investigation revealed that the amplifier was oscillating. In fact, every amplifier he could find in the factory was oscillating. For one final test, our colleague brought in some of the first units ever built (and in use for some 10 years) and found that they were also oscillating. Despite the oscillation, the amplifiers performed every function they were designed to perform. Their gain was flat, the AGC functioned beautifully, and they did not introduce any additional noise into the system.

War Story — Specialized Amplifiers

Several times in the past we have purchased amplifiers with special design characteristics, i.e., very low noise figure or very wide bandwidth. We used these custom amplifiers for several years before we realized they were oscillating after we connected them to antennas.

War Story — Amplifier Stability (II)

The authors were developing a receiving system with a steerable antenna and noticed that some of the received signals would change frequency depending upon the position of the antenna. We were baffled until we discovered that one of the amplifiers connected directly to the antenna feed was oscillating.

The authors reasoned that the impedance seen by the amplifier was a strong function of the antenna position. For example, the amplifier might see a non-Z_0 impedance when the antenna was pointed toward a large metal plate. It might see a Z_0 impedance when the antenna was pointed directly at the sky. Since the impedance seen by the amplifier would change with the environment viewed by antenna, the frequency of oscillation would change with antenna position. Outside signals received by the antenna would mix with the unintended oscillation and arrive in the system's passband. Thus, the received signals would move as the antenna position changed. Redesigning the offending amplifiers fixed the problem.

Any amplifiers we buy should be specified as being unconditionally stable for all values and phases of source and load impedances. Also, the frequency of oscillation does not have to be within the frequency of operation of the amplifier. For example, an amplifier designed to operate in the 10 to 1000 MHz range can oscillate at 2 MHz or at 4560 MHz. As long as the device exhibits power gain, oscillation is possible.

1.16 Signals, Noise and Modulation

The following section serves as an introduction to unmodulated carriers, carriers with just a small amount of modulation, and fully modulated waveforms and the various expressions used to describe them.

Ideal Sine Wave

The simplest waveform is an ideal sine wave, i.e., a mathematically pure sine (or cosine) wave that is free from noise. The signal-to-noise power ratio is infinite. The waveform can be described in the time, frequency and phasor domains.

Time Domain

Figure 1-44 shows the time-domain view of an ideal sine wave. The equation for this waveform is

$$f(t) = V_{pk} \cos(\omega_0 t + \theta) \quad \omega_0 = 2\pi f_0 \quad \quad 1.161$$

We define the period of the sine wave as

$$\text{Period} = T_0 = \frac{1}{f_0} \quad \quad 1.162$$

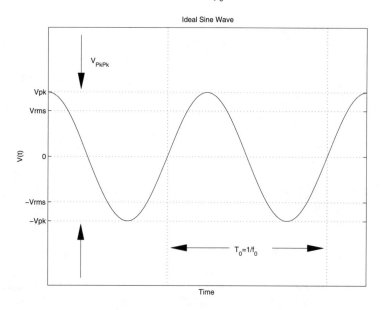

Figure 1-44 *Time-domain representation of a sine wave with common definitions.*

Zero Crossings

When discussing mixers, we can use a sine wave to control a switch. When the instantaneous value of the waveform is greater than zero, the switch will be in one position (for example, open). When the waveform is less than zero, the switch will be in the other position (for example, closed). We are interested in the zero crossings of the waveform and want to describe exactly when and how often the waveform passes through zero volts. This is the function of a *limiter*. Figure 1-45 shows the limiter's output.

$$V_{\text{limit}}(t) = \begin{cases} +1 \text{ when } \cos(\omega_0 t + \theta) \geq 0 \\ -1 \text{ when } \cos(\omega_0 t + \theta) < 0 \end{cases} \qquad 1.163$$

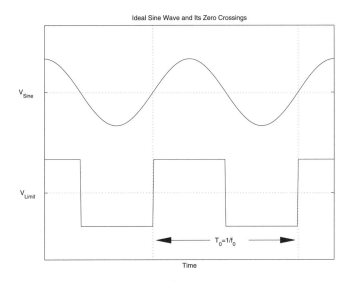

Figure 1-45 *Time-domain representation of a sine wave emphasizing the zero crossings.*

In the case of the ideal sine wave, no information regarding the phase or frequency of the waveform is lost; it now resides in the zero crossings. However, information concerning the amplitude of the original waveform is lost. For example, if V_{pk} is a function of time, the information about that function has disappeared. Note that the zero crossings happen at exact, mathematically deterministic times, i.e., when

$$\cos(\omega_0 t + \theta) = 0 \Rightarrow (\omega_0 t + \theta) = \frac{\pi}{2}, \frac{3\pi}{2}, \frac{5\pi}{2}, \cdots \qquad 1.164$$

Frequency Domain

Figure 1-46 shows an ideal, noiseless sine wave in the frequency domain. The frequency-domain representation is a single impulse function at f_0.

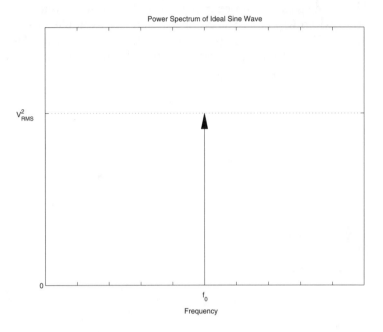

Figure 1-46 *The frequency-domain representation of a sine wave is a single impulse at f_0.*

Phasor

Figure 1-47 shows the phasor representation of a pure, noiseless sine wave. The length of the phasor is V_{pk} rotating about the origin with an angular velocity of exactly ω_0. By definition, the rotation is counterclockwise. A common mental model consists of a mental "strobe light" that freezes the phasor in some particular position. We force the strobe light fires once every T_0 seconds, starting at 0 seconds. The phasor appears to freeze at one position on the complex plane. Since we started the strobe light at 0 seconds, the phasor forms an angle of θ with the real axis. The projection of the phasor on the real axis is the instantaneous amplitude of the time-domain waveform. This project, termed *the real part* of the phasor, represents the waveform seen on an oscilloscope.

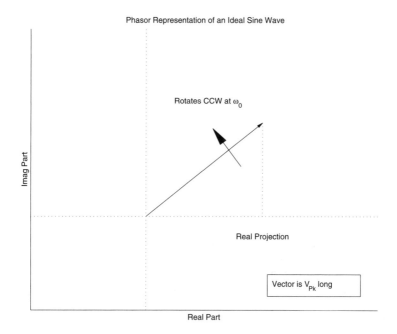

Figure 1-47 *The phasor-domain representation of a sine wave is a single vector rotating counterclockwise at a radian frequency of w_0.*

Two Noiseless Sine Waves

We will now look at expressions describing the arithmetic sum of two sine waves of slightly different frequencies. This insight will be valuable in later sections when we examine modulated and noisy waveforms.

Time Domain

Figure 1-48 shows the time-domain waveform. The equation is

$$f(t) = V_{pk,1} \cos(\omega_1 t + \theta_1) + V_{pk,2} \cos(\omega_2 t + \theta_2) \qquad 1.165$$

where
 ω_1 = radian frequency of the first sine wave,
 ω_2 = radian frequency of the second sine wave and $\omega_1 \neq \omega_2$.

90 | RADIO RECEIVER DESIGN

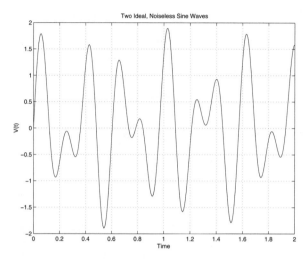

Figure 1-48 *Time-domain representation of two sine waves at slightly different frequencies added together. The frequencies of the two sine waves are f_1 and f_2.*

Zero Crossings

Figure 1-49 shows the zero crossings of the waveform described by Equation 1.165. Due to the noiseless nature of this waveform, the zero crossings occur at mathematically deterministic times, i.e., when

$$V_{pk,1}\cos(\omega_1 t + \theta_1) = -V_{pk,2}\cos(\omega_2 t + \theta_2) \qquad 1.166$$

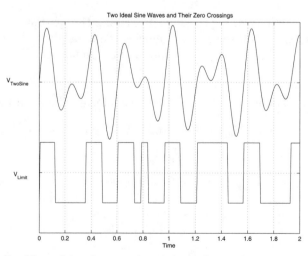

Figure 1-49 *Time-domain representation of two sine waves emphasizing the zero crossings of the composite waveform.*

Frequency Domain

Figure 1-50 shows the frequency-domain representation of Equation 1.168. We see two impulse functions at frequencies f_1 and f_2. Note that the magnitude of each impulse function is a function of $V_{pk,1}$ and $V_{pk,2}$.

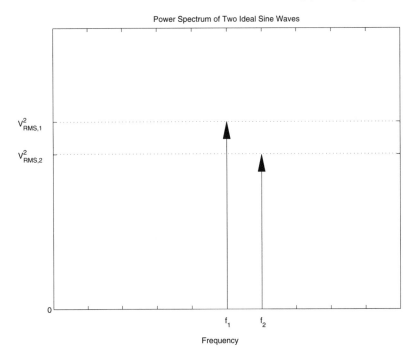

Figure 1-50 *The frequency-domain representation of two sine waves is two impulses. One impulse is at f_1, and the second impulse is at f_2.*

Phasor

Figure 1-51 shows the phasor representation of this waveform. There are two vectors: V_1 and V_2. The length of phasor V_1 is $V_{pk,1}$ and of V_2 is $V_{pk,2}$. Each phasor rotates about the origin. Phasor V_1 rotates with an angular velocity of exactly ω_1; phasor V_2 rotates with an angular velocity of exactly ω_2. Since the two phasors are rotating at different rates, we can freeze only one of the phasors with the mental "strobe light" (see Figure 1-52). Usually, it is best to freeze the phasor with the largest magnitude (V_1 in this case); therefore the strobe light is flashed once every $1/f_1$ seconds. Another mental model of this process is that the viewer is rotating at the same speed and in the same direction as the phasor, which will also cause the phasor to "freeze."

92 | RADIO RECEIVER DESIGN

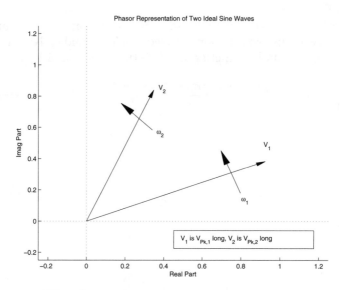

Figure 1-51 *The phasor-domain representation of two sine waves. Each vector rotates counterclockwise. One vector rotates at a radian frequency of ω_1 and the second vector rotates at ω_2.*

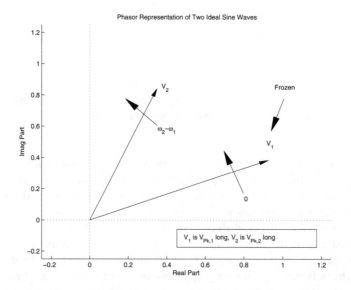

Figure 1-52 *It is often convenient to "freeze" one phasor to understand the relationship between the phasors. In this figure, we have frozen V_1 so this phasor no longer appears to rotate. V_2 rotates at $\omega_2 - \omega_1$ in this reference frame.*

Since we have frozen phasor V_1, phasor V_2 will appear to rotate around the origin counterclockwise with an angular velocity of $\omega_2 - \omega_1$. (V_2 rotates counterclockwise because $\omega_2 > \omega_1$. If $\omega_2 < \omega_1$, the smaller phasor would rotate clockwise.) Figure 1-53 contains exactly the same information as Figure 1-52, but by stacking the phasors we can gain more insight into the situation. For example, we can see the resultant or vector sum of the two phasors more clearly. The vector sum of V_1 and V_2 is

$$\overline{R} = \overline{V_1} + \overline{V_2} \qquad 1.167$$

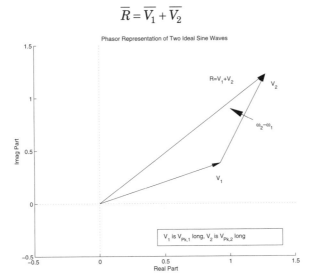

Figure 1-53 *Vectorial sum of two vectors. V_1 is "frozen" and does not rotate. V_2 rotates at $\omega_2 - \omega_1$ about the end of vector V_1.*

Figure 1-54 shows the locus of the resultant vector R over time. The vector addition of two sine waves produces amplitude and phase modulation of the resultant. The resultant is phase-modulated because it moves back and forth between R_{T1} and R_{T2} at a rate of $\omega_2 - \omega_1$. The peak change in phase is $\Delta\phi_{PkPk}$. The geometry of Figure 1-54 reveals

$$\sin\left(\frac{\Delta\phi_{PkPk}}{2}\right) = \frac{V_{pk,2}}{V_{pk,1}} \qquad 1.168$$

For small $\Delta\phi_{PkPk}$, $\sin(\Delta\phi_{PkPk}) = \Delta\phi_{PkPk}$ and

$$\sin\left(\frac{\Delta\phi_{PkPk}}{2}\right) = \frac{V_{pk,2}}{V_{pk,1}} \qquad 1.169$$

Small $\Delta\phi_{pk}$ occurs when $V_{pk,1} \gg V_{pk,2}$.

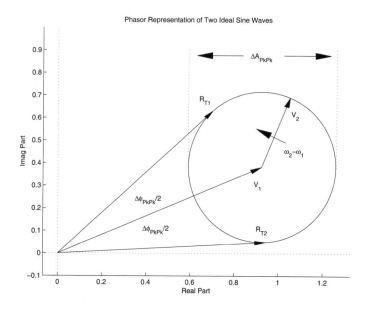

Figure 1-54 *The locus of the resultant vector R over time. The summation of the two vectors (V_1 and V_2) causes the resultant to exhibit a phase deviation $\Delta\phi_{PkPk}$ and an amplitude deviation ΔA_{PkPk}.*

Later we will be dealing with nonsinusoidal signal powers and noise powers. In this situation, RMS values are more common. We can adapt Equation 1.169 for RMS values.

$$\Delta\phi_{RMS} = 2\frac{V_{RMS,2}}{V_{RMS,1}} \qquad 1.170$$

Any angle-modulated waveform can be validly interpreted as either phase or frequency modulation. We can also derive amplitude variation from Figure 1-54, which reads

$$\Delta A_{PkPk} = 2V_{pk,2} \qquad 1.171$$

When we must work with signal power and RMS values, we can write

$$\Delta A_{RMS} = 2V_{RMS,2} \qquad 1.172$$

Finally, the projection of the resultant vector upon the real axis produces the instantaneous amplitude of the time-domain waveform. This projection is the waveform that can be observed on an oscilloscope.

Gaussian Noise

Noise is a statistical phenomenon. Knowing the past values of the noise waveform does not reveal what the instantaneous value of the waveform will be in the future. We can use statistics to describe rather accurately how noise will behave when averaged over a long period of time. When we say that noise is a random, zero-mean Gaussian process, we mean that the values of the time-domain waveform create a Gaussian distribution with no DC component. The standard deviation of the Gaussian process is related to the RMS amplitude of the noise.

Time Domain

In Figure 1-155, we apply broadband, Gaussian noise [$n(t)$] to a bandpass filter. The band-pass filter has a center frequency of f_c and a bandwidth of B. The filtered noise [$n_f(t)$] appears to be a sinusoid of frequency f_c (the center frequency of the filter). However, the amplitude and frequency (phase) both vary randomly over time. The narrower the filter's bandwidth, the slower the frequency and amplitude change. If the filter bandwidth is considerably smaller than the filter's center frequency, or

$$B \ll f_c \qquad 1.173$$

the envelope and frequency of the filtered noise will change very slowly. Both change noticeably over very many cycles of the carrier (which is at f_c). We specify magnitude of the noise voltage in terms of its RMS value ($V_{n,\mathrm{RMS}}$). This RMS value is the standard deviation of the noise and fits directly into the equations describing a Gaussian distribution.

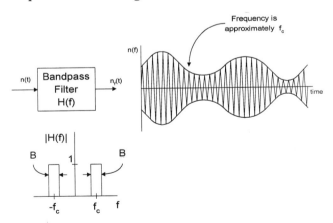

Figure 1-55 Noise passing through a bandpass filter produces a signal centered about the filter's center frequency. The amplitude and phase changes in the signal depend strongly on the filter's bandwidth.

Frequency Domain

The noise generator V_{noise} of Figure 1-56 generates wideband noise with an RMS value of $V_{n,oc,\text{RMS}}$. We will assume that the spectrum of this noise is infinite. The two-sided *power spectral density* (PSD) of the noise is $\eta/2$ (in watts/Hz). The one-sided power spectral density of the noise is η in watts/Hz (see Figure 1-57). In this book, we will switch between the one-sided and two-sided PSD depending upon the problem.

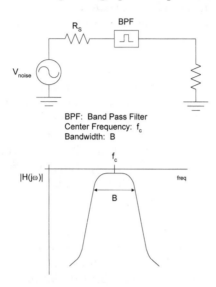

Figure 1-56 *White noise passing through a bandpass filter produces a signal whose spectral shape is identical to the bandpass filter.*

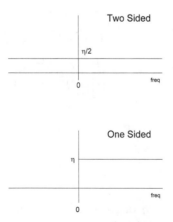

Figure 1-57 *Definitions of one-sided and two-sided noise power spectral density.*

From a practical point of view, we will invariably be dealing with band-limited noise (see model in Figure 1-56). Noise will always be viewed through a band-pass filter. In Figure 1-59, the band-pass filter passes only the components of the noise present within the bandwidth of the filter. We can interpret the frequency spectrum of noise as an infinite sum of distinct sine waves (see Figure 1-58) and use the concepts for noiseless sine waves.

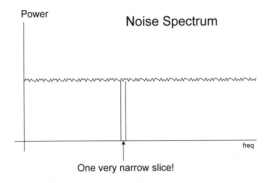

Figure 1-58 *Dividing white noise into small frequency slices allows us to analyze the effects of the noise using the methods developed in the "Two Noiseless Sine Waves" section.*

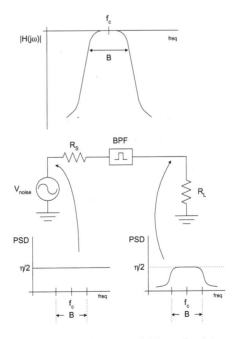

Figure 1-59 *The power spectral density of filtered white noise.*

Zero Crossings

Since noise is a statistical process, statistics are used to describe its zero crossings. For white random noise with frequency limits f_U and f_L, the expected number of zero crossings per second of either the positive- or negative-going sense is

$$\frac{\text{Zero Crossings}}{\text{Second}} = \sqrt{\frac{f_U^3 - f_L^3}{3(f_U - f_L)}}$$

$$= \sqrt{\frac{1}{3}\left(f_U^2 + f_U f_L + f_L^2\right)} \qquad 1.174$$

For a low-pass filter, $f_L = 0$ and

$$\frac{\text{Zero Crossings}}{\text{Second}} = 0.577 f_U \qquad 1.175$$

For narrow band-pass filters (bandwidths of less than 10% of the center frequency), $f_U \cong f_L$, we can write

$$\frac{\text{Zero Crossings}}{\text{Second}} = \frac{f_U + f_L}{2} \qquad 1.176$$

which is the arithmetic center frequency of the band-pass filter. The values from Equations 1.174 through 1.176 are the expected values averaged over a long period of time.

Phasor

Figure 1-60 shows the phasor representation of bandlimited noise. The noise is passed through a band-pass filter with a center frequency of f_c and a bandwidth of B. The "strobe light" is set to flash every $1/f_c$ seconds.

The tip of the resultant vector takes a random path around the origin. Since we have frozen the vector at $1/f_c$ seconds (we flash the strobe light every $1/f_c$ seconds), there is no bias to either clockwise or counterclockwise rotation. In other words, the mean angular velocity is ω_c: the center frequency of the band-pass filter.

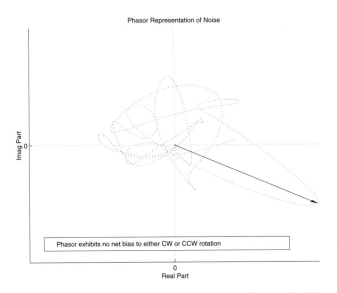

Figure 1-60 *The phasor representation of bandpassed white noise. The amplitude of the resultant vector is a Rayleigh-distributed random variable. The phase of the resultant is a uniformly distributed random variable.*

Ideal Sine Wave and Gaussian Noise

Figure 1-61 shows the model we will use to describe the behavior of an ideal sine wave in Gaussian noise. Note that we can observe noise only inside of some bandwidth. In practice, the noise that adds directly to the signal of interest is called noise *additive white Gaussian noise* (AWGN).

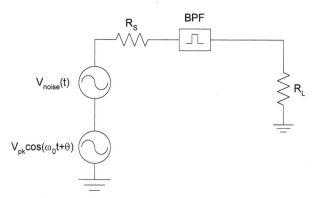

Figure 1-61 *Generating an ideal sine wave corrupted with additive white Gaussian noise (AWGN).*

Time Domain

Figure 1-62 shows the time-domain waveform of the noisy sine wave.

$$f(t) = V_{pk} \cos(\omega_0 t + \theta) + V_{noise}(t) \qquad 1.177$$

where
  $V_{noise}(t)$ represents the time-domain noise waveform,
  $V_{noise}(t)$ is a random, zero-mean process.

The standard deviation of $V_{noise}(t)$ is $V_{noise,RMS}$. The larger $V_{noise,RMS}$, the larger the noise power.

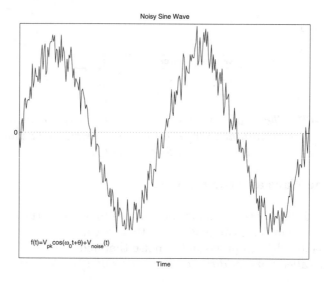

Figure 1-62 *Ideal sine wave corrupted with (AWGN).*

Zero Crossings

Figure 1-63 shows a close-up view of a zero crossing point. For high signal-to-noise ratios (where $V_{pk} \gg V_{noise}$), we can see that the zero crossings of the noisy sine wave occur at approximately the same times as the ideal sine wave. The additive noise causes some uncertainty when the waveform actually crosses zero. The zero crossings have to be described statistically.

The random variable Δt in Figure 1-63 represents the uncertainty in the zero crossings. Several terms are used to describe the uncertainty in the zero crossings, for example, *clock jitter, phase noise, incidental phase modulation* and *frequency drift*. It can be expected that a lower SNR corresponds to a larger Δt. The average number of either positive- or negative-going zero crossings per second at the output of a narrowband band-pass filter of rectangular shape when the input is a sine wave in Gaussian noise is

$$\frac{\text{Positive Zero Crossings}}{\text{Second}} = \frac{\text{Negative Zero Crossings}}{\text{Second}}$$

$$= f_0 \sqrt{\frac{S/N + 1 + \left(\frac{B^2}{12 f_0^2}\right)}{S/N + 1}} \qquad 1.178$$

where
- f_0 = the center frequency of the filter,
- B = the filter bandwidth,
- S/N = the signal-to-noise power ratio in linear format.

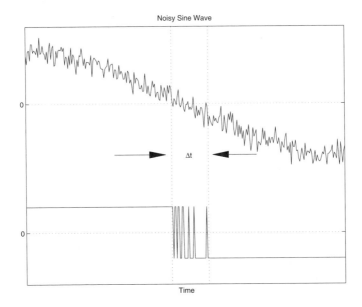

Figure 1-63 *Zero crossings of a noisy sine wave.*

Frequency Domain

Figure 1-64 shows the frequency domain representation of a sine wave in filtered Gaussian noise. Again, the noise has been passed through a band-pass filter. In the case of the ideal sine wave, there is an infinite signal power-to-noise power ratio. When we add noise to the ideal signal, a finite signal power-to-noise power ratio is produced.

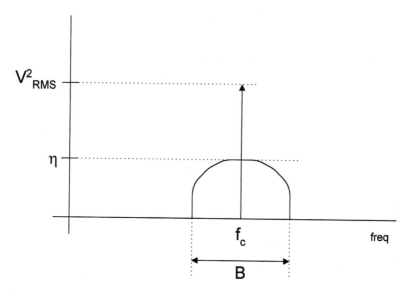

Figure 1-64 *Frequency domain representation of a sine wave corrupted with bandlimited noise.*

Phasor

Let the signal vector be S and let the noise vector be N. The vector sum of S and N will be the resultant vector R. We can write

$$\overline{R} = \overline{S} + \overline{N} \qquad 1.179$$

The magnitude of the signal vector S will be the peak value of the signal from Equation 1.177 or V_{pk}, which is a peak value. The magnitude of the noise vector N will be the RMS value of $V_{noise}(t)$ from Equation 1.177, which is a RMS value. If we hope to obtain meaningful results from the analysis, we have to bear in mind this distinction.

Figure 1-65 shows the vector sum of S plus N graphically. Using the techniques we developed earlier, we place the random noise vector at the end of the deterministic sine-wave vector. We can draw the position of the noise vector for various values of standard deviations. The magnitude of the noise sector is Rayleigh distributed. Using the cumulative Rayleigh PDF of Figure 1-66, we can see that the resultant noise vector will reside inside the $z = 1$ circle about 38% of the time. It will reside inside the $z = 2$ locus about 86% of the time and inside the $z = 3$ locus for 98% of the time.

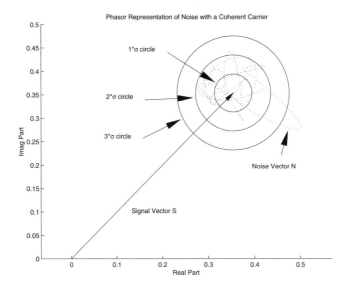

Figure 1-65 *Phasor representation of a sine wave corrupted with AWGN.*

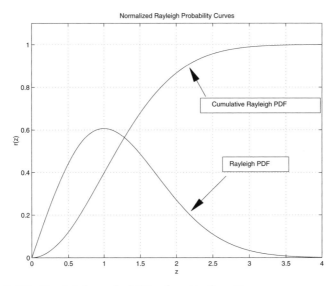

Figure 1-66 *Rayleigh probability density functions.*

The amplitude and phase of the resultant vector can be found using Figure 1-65, which allows us to determine the amplitude and phase variations the noise vector impresses upon the signal of interest. Again, we have to keep in mind which quantities are peak quantities and which quantities are RMS values.

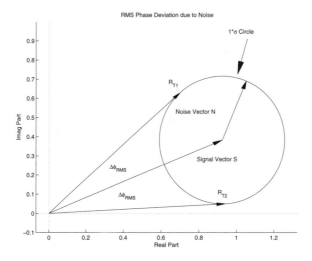

Figure 1-67 *Amplitude and phase uncertainties present in a sine wave corrupted by AWGN.*

Figure 1-67 shows the phase modulation present on the signal due to the AWGN. The AWGN is a zero-mean, random process. As an example, we will characterize the phase deviation present on the carrier using the $z = 1$ standard deviation circle. We will also use the RMS value of the sine wave

$$V_{RMS} = \frac{V_{pk}}{\sqrt{2}} \qquad 1.180$$

This also applies to nondeterministic data signals. Referring to Figure 1-67, we can write

$$\sin(\Delta\phi_{RMS}) = \frac{V_{N,RMS}}{V_{S,RMS}} \qquad 1.181$$

Using the small angle approximation produces

$$\Delta\phi_{RMS} = \frac{V_{N,RMS}}{V_{S,RMS}} \qquad 1.182$$

This analysis produces the RMS phase deviation. Since, over time, the value of the noise vector N can take on any value, the angle $\Delta\phi$ can take on a variety of values. Note again that phase modulation implies frequency modulation. We can use the phase and frequency modulation concepts developed elsewhere to analyze this behavior. Figure 1-67 also gives us some insight into the amplitude modulation present on the signal due to the AWGN. The AM present on the resultant vector is

$$\Delta A_{RMS} = 2V_{N,RMS} \qquad 1.183$$

Double Sideband Amplitude Modulation (DSBAM)

In DSBAM, we vary the amplitude of a carrier sine wave with the information waveform. The carrier frequency (f_c) is usually much greater than the modulation frequency (f_m). The distinguishing characteristics of DSBAM are

- Both the upper and lower sidebands are present.
- The carrier is present.
- The RF bandwidth is twice the bandwidth of the modulating signal.

Time Domain

The equation describing a DSBAM-modulated signal is

$$V_{DSBAM}(t) = V_{pk}\left[1 + m_a \cos(\omega_m t)\right]\cos(\omega_c t) \qquad 1.184$$

where
$\omega_m = 2\pi f_m$ = modulation frequency (contains the information),
$\omega_c = 2\pi f_c$ = carrier frequency,
m_a = AM modulation index and $0 <= m_a <= 1.0$ for DSBAM.

f_m is often in the audio range and may equal 5 kHz; f_c is often in the RF range and may equal 1 MHz. Rewriting Equation 1.187 reveals

$$\begin{aligned}V_{DSBAM}(t) &= V_{pk}\cos(\omega_c t) + V_{pk}m_a\cos(\omega_m t)\cos(\omega_c t) \\ &= V_{pk}\cos(\omega_c t) \\ &\quad + \frac{V_{pk}m_a}{2}\cos\left[(\omega_c + \omega_m)t\right] \\ &\quad + \frac{V_{pk}m_a}{2}\cos\left[(\omega_c - \omega_m)t\right]\end{aligned} \qquad 1.185$$

where
$\cos(\omega_c t)$ is the carrier.
$\cos[(\omega_c - \omega_m)t]$ is the lower sideband (LSB). The quantity $\omega_c - \omega_m$ is often referred to as the *difference frequency* or *beat frequency*.

$\cos[(\omega_c + \omega_m)t]$ is the upper sideband (USB). The quantity $\omega_c + \omega_m$ is also called *sum frequency*.

Equation 1.185 indicates that a DSBAM waveform consists of three separate sinusoidal waveforms. Figure 1-68 shows a time-domain plot of a DSBAM waveform for m_a = 0.1, 0.5 and 1.0.

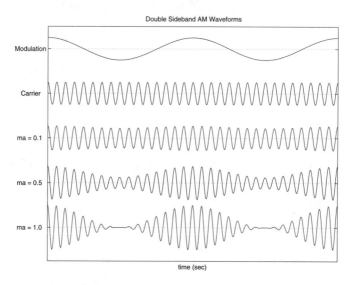

Figure 1-68 *Time-domain representation of a double sideband AM signal for m_a = 0.1, 0.5 and 1.0.*

Frequency Domain

Figure 1-69 shows the spectrum of DSBAM. The carrier (i.e., the tone at f_c) is always present and its power does not change with modulation index (m_a). The tone below the carrier (at $f_c - f_m$) is the *lower sideband* (LSB). The tone above the carrier at $f_c + f_m$ is the *upper sideband* (USB). As Equation 1.185 reveals, the carrier power is always constant regardless of the value of m_a. The power in the USB and LSB changes with the modulation index m_a. As the modulation index is increased, the power in the two sidebands increases. The power level of each sideband is

$$P_{\text{Sideband}} = P_{\text{Carrier}} - 20\log\left(\frac{m_a}{2}\right) \qquad 1.186$$

where $0 <= m_a <= 1.0$. For 100% modulation, the sideband levels are

$$20\log\left(\frac{1}{2}\right) = -6\ dB \qquad 1.187$$

or 6 dB below the carrier.

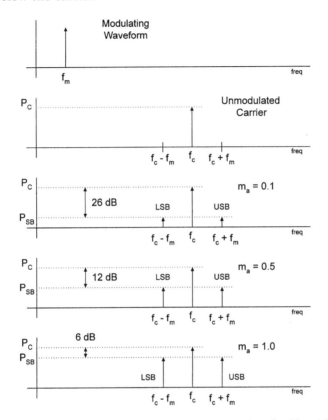

Figure 1-69 *Frequency-domain representation of a double sideband AM signal for $m_a = 0.1, 0.5$ and 1.0.*

Zero Crossings

Figure 1-70 shows the zero crossing performance of DSBAM. Applying amplitude modulation to a waveform does not change the zero crossings as long as $0 <= m_a < 1.0$. When $m_a = 1.0$, the waveform goes to zero for a brief period one every $1/f_m$ seconds. This affects the zero crossings and also causes some practical implementation problems.

Note that the zero crossings are at the same points as the unmodulated carrier. In practice, the amplitude variations are often removed from a signal to recover the carrier. This process is called *limiting*.

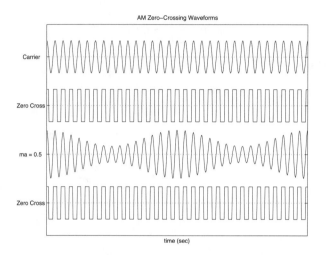

Figure 1-70 Zero crossings of a double sideband AM signal. The zero crossings do not contain information about the modulation applied to the carrier.

Phasor

Figure 1-71 shows a phasor diagram of a DSBAM wave modulated with a single sinusoidal carrier. Again, the carrier is frozen in position. The USB appears to rotate counterclockwise at a rate of ω_m. The LSB rotates in a clockwise direction at a rate of ω_m. Analysis of Equation 1.185 shows that the angle θ_{LSB} always equals the angle θ_{URB} in Figure 1-71.

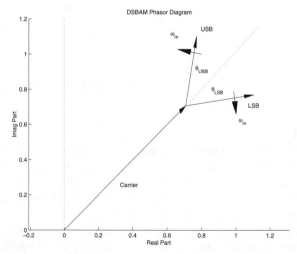

Figure 1-71 Phasor representation of a double sideband AM signal. Note the symmetry of the USB and the LSB with respect to the carrier.

Figure 1-72 shows the composition of the envelope of an AM signal by phasors. The vector sum of the LSB and USB phasors always forms a resultant that is directly in line with the carrier.

Figure 1-72 *A different view of the phasor representation of a double sideband AM signal. The symmetry of the USB and LSB produces only amplitude modulation of the carrier.*

The phasor diagram of DSBAM provides insight into the effects of distortion on the AM waveform. For example, Figure 1-73 shows an AM waveform passing through a medium that exhibits a time delay that is a function of frequency, i.e., the lower-frequency components of the signal experience a different time delay than the higher-frequency components (*unequal group delay*).

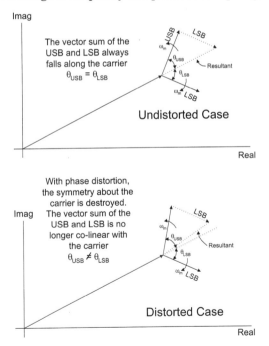

Figure 1-73 *Phasor representation of a double sideband AM signal. Group delay distortion destroys the symmetry of the sidebands and can introduce phase modulation into the waveform.*

The group delay response of the medium causes the LSB to be delayed in time more than the USB. Without phase distortion, the vector sum of the USB and LSB vectors always falls along the line of the carrier vector because

$$\theta_{USB} = \theta_{LSB} \qquad 1.188$$

However, when one of the sidebands is delayed more than the other, the symmetry about the carrier no longer exists and the vector sum of the USB and LSB vectors is no longer colinear with the carrier vector. This distorts the envelope of the AM waveform. It also causes the entire waveform to exhibit phase modulation. The carrier phase will move unsteadily at a rate of f_m. For DSBAM systems, this conversion of AM to PM usually is not a problem because most DSBAM demodulators are insensitive to phase modulation on the carrier.

RF Bandwidth

Figure 1-74 shows the spectrum of a general baseband signal. When we use the signal as the modulating waveform in a DSBAM system, both the USB and the LSB are as wide as the original modulating signal. Consequently, the RF spectrum appears twice as wide as the baseband signal. For example, if the highest frequency component in the modulating waveform is $f_{m,max}$, then the RF waveform will extend from $f_c - f_{m,max}$ to $f_c + f_{m,max}$ for a total RF bandwidth of $2f_{m,max}$. Also, note that the LSB has experienced a frequency inversion.

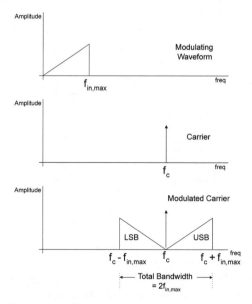

Figure 1-74 *Frequency-domain representation of a double sideband AM signal for nonsinusoidal modulation.*

Angle Modulation

Angle modulation covers both frequency and phase modulation, which are, as will be shown, very similar.

Frequency Modulation (FM)

When we frequency-modulate a carrier with a modulating waveform, we change the instantaneous frequency of the carrier in step with the instantaneous amplitude of the modulating wave. Figure 1-75 illustrates some of the variables we will use.

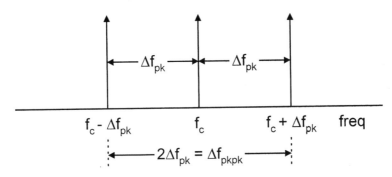

Figure 1-75 *Definition of the variables used in the frequency modulation equations.*

where
 f_c = the frequency of the unmodulated carrier,
 Δf_{pk} = the peak frequency deviation,
 Δf_{pkpk} = the peak-to-peak frequency deviation.

From Figure 1-75 we can see

$$\Delta f_{PkPk} = 2(\Delta f_{Pk}) \qquad 1.189$$

Time Domain

First we will examine sinusoidal modulation. For frequency modulation, the instantaneous frequency of the carrier should be determined by the modulating waveform. The instantaneous frequency of the FM signal is

$$Freq_{FM}(t) = f_c + \Delta f_{Pk}\cos(\omega_c t) \qquad 1.190$$

Note that the minimum and maximum frequencies are

$$Freq_{FM,min} = f_c - \Delta f_{Pk}$$
$$Freq_{FM,max} = f_c + \Delta f_{Pk} \qquad 1.191$$

Phase is the integral of frequency, or

$$\begin{aligned}
\phi_{FM}(t) &= 2\pi \int Freq_{FM}(t)\, dt \\
&= 2\pi \int \left[f_c + \Delta f_{Pk} \cos(\omega_m t) \right] dt \\
&= 2\pi f_c t + 2\pi \Delta f_{Pk} \frac{1}{2\pi f_m} \sin(\omega_m t) \qquad 1.192 \\
&= 2\pi f_c t + \frac{\Delta f_{Pk}}{f_m} \sin(\omega_m t)
\end{aligned}$$

The time-domain equation for a frequency-modulated waveform is

$$V_{FM}(t) = V_{Pk} \cos\left[\omega_c t + \frac{\Delta f_{Pk}}{f_m} \sin(\omega_m t) \right] \qquad 1.193$$

We substitute

$$\beta = \frac{\Delta f_{Pk}}{f_m} \qquad 1.194$$

and Equation 1.193 becomes

$$V_{FM}(t) = V_{Pk} \cos\left[\omega_c t + \beta \sin(\omega_m t) \right] \qquad 1.195$$

The quantity β is the modulation index. The value of β coupled with the value of f_m leads into an expression for the RF bandwidth and other FM characteristics. Figure 1-76 shows the time-domain plots for several values of β. Note that the amplitudes of all the waveforms are constant.

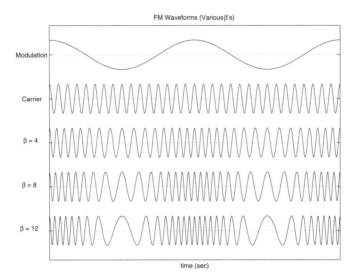

Figure 1-76 Time-domain representation of a frequency-modulated waveform for β = 4, 8 and 12.

Zero Crossings

A FM waveform passes all of its information in the instantaneous frequency of its carrier. It can be processed with a limiter without losing any information (see Figure 1-77).

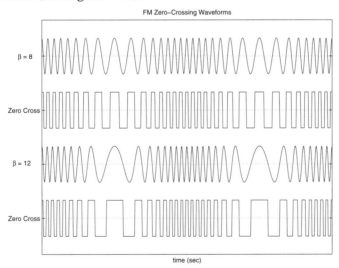

Figure 1-77 Zero crossings of a frequency-modulated waveform for β = 4, 8 and 12. The modulation is preserved in the zero crossings.

For example, if an angle-modulated waveform accidentally acquires unwanted AM (due to multipath, fading, poor filtering, and so on), it can easily be removed. Also, many of the FM demodulation processes depend upon the FM waveform having a constant envelope. However, limiters can adversely affect the signal-to-noise ratio of a signal. designers often filter after the limiter to remove the harmonics of the carrier frequency and restore the time-domain waveform to a sine wave.

Frequency Domain

Applying Fourier analysis techniques to Equation 1.195 produces

$$\begin{aligned}V_{FM}(t) = &V_{Pk}J_0(\beta)\cos(\omega_c t)\\ &+V_{Pk}J_1(\beta)\{\cos[(\omega_c+\omega_m)t]-\cos[(\omega_c-\omega_m)t]\}\\ &+V_{Pk}J_2(\beta)\{\cos[(\omega_c+2\omega_m)t]+\cos[(\omega_c-2\omega_m)t]\}\\ &+V_{Pk}J_3(\beta)\{\cos[(\omega_c+3\omega_m)t]-\cos[(\omega_c-3\omega_m)t]\}\\ &+V_{Pk}J_4(\beta)\{\cos[(\omega_c+4\omega_m)t]+\cos[(\omega_c-4\omega_m)t]\}\\ &+\cdots\end{aligned} \quad 1.196$$

where J_n is the Bessel function of the first kind and order n.

Figure 1-78 shows the typical spectrum of a FM waveform. Several observations are of interest.

- All of the spectral components are f_m apart.

- There are an infinite number of spectral components even for the simple case of sinusoidal modulation.

- The spectral components are clustered around the carrier frequency (at f_c).

- The spectral components are symmetrical about the carrier frequency. In other words, the power level of the spectral component at $f_c + nf_m$ equals the power level of the spectral component at $f_c + nf_m$ where n is any integer.

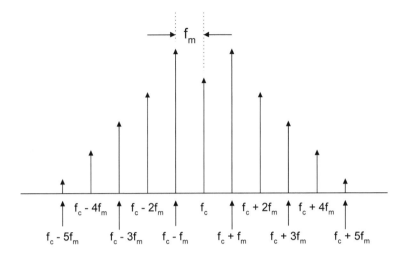

Figure 1-78 *Frequency domain representation a frequency-modulated waveform under sinusoidal modulation. Note the symmetry of the sidebands.*

Bessel Functions

Equation 1.196 provides the levels of the spectral components of a FM waveform in terms of Bessel functions. Figures 1-79 and 1-80 show the first 5 Bessel functions. Remember that the 0th Bessel function describes the amplitude of the carrier at f_c. The 1st Bessel function describes the amplitude of the two sidebands at $f_c \pm f_m$; the 2nd Bessel function describes the amplitude of the sidebands at $f_c \pm 2 \cdot f_m$; the 3rd Bessel function describes the sidebands at $f_c \pm 3 \cdot f_m$, and so on. In general, the nth Bessel function describes the amplitude of the sideband at $f_c \pm n \cdot f_m$. In a 1-ohm system, the power level of the unmodulated carrier is

$$P_{FM,\text{Unmodulated Carrier}} = \left(\frac{V_{Pk}}{\sqrt{2}}\right)^2 \quad 1.197$$

The power level of the spectral component at $f_c \pm n \cdot f_m$ is

$$P_{FM}(f_c + nf_m) = P_{FM}(f_c - nf_m)$$
$$= \left[\frac{V_{Pk}}{\sqrt{2}} \cdot J_n(\beta)\right]^2 \quad 1.198$$

The ratio of the power level any single spectral component to the power in the unmodulated carrier is

$$\frac{P_{FM}(f_c + nf_m)}{P_{FM,\text{Unmodulated Carrier}}} = \frac{P_{FM}(f_c - nf_m)}{P_{FM,\text{Unmodulated Carrier}}}$$

$$= \left[\frac{V_{Pk}}{\sqrt{2}} J_n(\beta)\right]^2 \bigg/ \left[\frac{V_{Pk}}{\sqrt{2}}\right]^2 \qquad 1.199$$

$$= J_n^2(\beta)$$

$$= 20\log\left|J_n(\beta)\right| \quad \text{in dB}$$

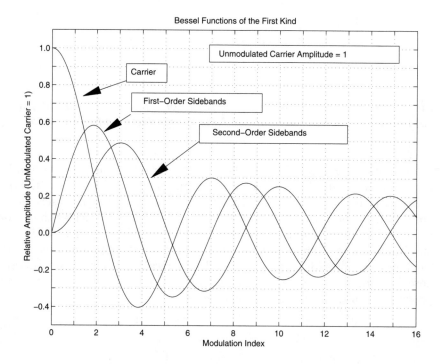

Figure 1-79 *The first three Bessel functions, J_0, J_1 and J_2. The magnitude of the sidebands at f_c, $f_c \pm f_m$ and $f_c \pm 2f_m$ of a sinusoidally modulated FM signal are given by these functions.*

INTRODUCTION | 117

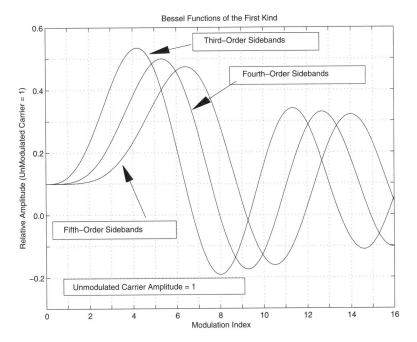

Figure 1-80 *The Bessel functions, J_3, J_4 and J_5. The magnitude of the sidebands at $f_c \pm 3f_m$, $f_c \pm 4f_m$ and $f_c \pm 5f_m$ of a sinusoidally modulated FM signal are given by these functions.*

Example 1.23 — Bessel Functions and FM

The FCC restricts commercial FM radio stations to $\Delta f_{pk} = 75$ kHz and $f_m = 15$ kHz. Find

a. the ratio of the power level of the signal at f_c to the power of the unmodulated carrier,

b. the ratio of the power level of the signal at $f_c + 3f_m$ to the power of the unmodulated carrier.

Solution —

We know

$$\beta = \frac{\Delta f_{Pk}}{f_m} = \frac{75}{15}$$
$$= 5$$

1.200

a. Using Figure 1-79, the value of $J_0(5) = -0.18$ and

$$\frac{P_{FM}(f_c)}{P_{FM,\text{Unmodulated Carrier}}} = 20\log[J_n(\beta)]$$
$$= 20\log|J_0(5)|$$
$$= 20\log|-0.18|$$
$$= -14.9 \ dB$$

1.201

The level of the tone at f_c is 14.9 dB below the level of the unmodulated carrier.

b. Using Figure 1-79, the value of $J_3(5) = 0.36$ and

$$\frac{P_{FM}(f_c + 3f_m)}{P_{FM,\text{Unmodulated Carrier}}} = 20\log|J_n(\beta)|$$
$$= 20\log|J_3(5)|$$
$$= 20\log|0.36|$$
$$= -8.9 \ dB$$

1.202

The level of the tone at $f_c + 3f_m$ is 8.9 dB below the level of the unmodulated carrier.

Phasor

Looking at Equation 1.195, which describes the time-domain waveform of a FM wave, and at the time-domain plots of Figure 1-76, we notice that the magnitude of the FM waveform is always constant. In other words, the amplitude of the wave never changes as it goes through its modulation cycle. In the phasor domain, the resultant of the FM wave must always have a constant magnitude although its angle will change. Figure 1-81 shows the phasor diagrams of a FM waveform. The plot shows the addition of the various Bessel components over several points in the modulating cycle. The vectorial sum of the vector components always add up to a resultant with a constant magnitude.

INTRODUCTION | 119

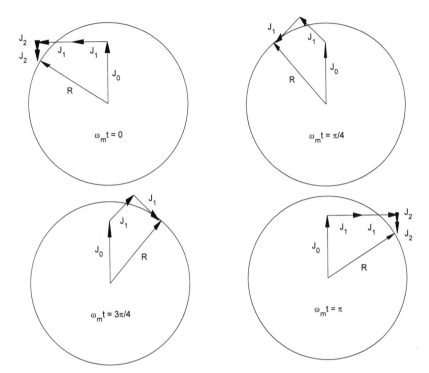

Figure 1-81 *Phasor representation of a sinusoidally modulated FM signal. The amplitude and phase of the sidebands always forces the end of the resultant vector to lie on a circle. This effect produces a waveform with constant amplitude.*

FM RF Bandwidth (Carson's Rule)

The final RF bandwidth of a FM-modulated wave depends upon the maximum frequency the carrier is modulated with and upon Δf_{pk}, the amount of carrier frequency deviation. Figure 1-82 shows the effects of varying the Δf_{pk} while keeping the modulation frequency f_m constant. Notice that the spectral components are always spaced f_m apart and that, as Δf_{pk} increases, more of the spectral components acquire significant energy. Notice also that as Δf_{pk} increases, the power in each individual spectral component becomes smaller. Because there are more components with significant power, the total signal power remains constant.

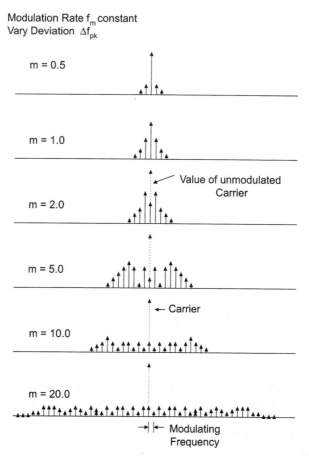

Figure 1-82 *Spectrum of an FM signal. Modulation rate FM is held constant. Various values of deviation Δf_{Pk}.*

If we keep the modulation frequency constant while varying Δf_{pk}, the RF bandwidth will increase. Figure 1-83 shows the effects of keeping the peak frequency deviation constant, but varying the modulation frequency. The spectral components are always f_m apart. As f_m decreases, the spacing between the tones becomes smaller. The generally accepted rule of thumb for the RF bandwidth of a FM signal is given by Carson's rule.

$$B_{FM} = 2(\Delta f_{Pk} + f_m) \qquad 1.203$$
$$= 2f_m(\beta + 1)$$

This bandwidth includes approximately 95% of the energy available in the FM waveform. The theoretical 100% bandwidth is infinite.

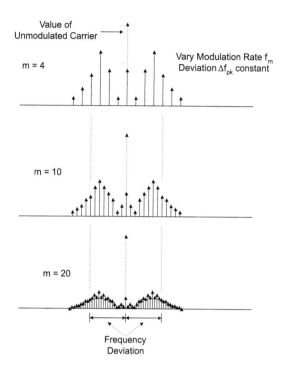

Figure 1-83 *Spectrum of an FM signal. Modulation rate FM is variable. Deviation Δf_{PK} is held constant.*

Example 1.24 — FM Bandwidth
What is the RF bandwidth of a commercial FM radio station? Remember that $\Delta f_{pk} = 75$ kHz and $f_m = 15$ kHz.

Solution —
Using Equation 1.203, we can write

$$B_{FM} = 2(\Delta f_{Pk} + f_m)$$
$$= 2(75 + 15) \qquad \text{1.204}$$
$$= 180 \text{ kHz}$$

Phase Modulation (PM)

In phase modulation, the instantaneous phase of the carrier depends upon the instantaneous amplitude of the modulating wave. Phase modulation is very similar to frequency modulation and results presented for frequency modulation are useful here.

Time Domain

In a phase-modulated waveform, the instantaneous phase of the carrier is determined by the modulating waveform. The instantaneous phase of the PM signal is

$$\phi_{PM}(t) = 2\pi f_c t + \Delta\phi_{Pk} \cos(2\pi f_m t) \qquad 1.205$$

where
Δf_{pk} = the peak phase shift,
f_m = the modulating frequency.

The time-domain equation for a phase-modulated wave is

$$\begin{aligned} V_{PM}(t) &= V_{Pk} \cos\left[2\pi f_c t + \Delta\phi_{Pk} \cos(2\pi f_m t)\right] \\ &= V_{Pk} \cos\left[\omega_c t + \Delta\phi_{Pk} \cos(\omega_m t)\right] \end{aligned} \qquad 1.206$$

Equations 1.206 and 1.195 are identical except for the substitution of Δf_{pk} for β and the substitution of a cosine for a sine. Figure 1-84 shows the time-domain plots for several values of Δf_{pk}.

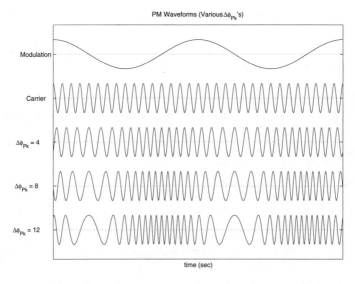

Figure 1-84 *Time-domain representation of a phase-modulated waveform, for $\Delta\phi_{PK}$ = 4, 8 and 12.*

Comparison of FM and PM Waveforms

Figure 1-85 shows a FM waveform and an equivalent PM waveform on the same graph. Both waveforms are modulated by the same cosine wave and both have a deviation of eight.

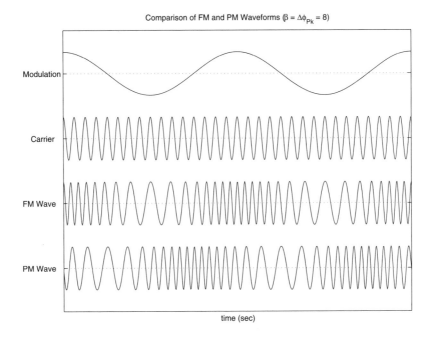

Figure 1-85 *Comparison of frequency- and phase-modulated waveforms. ($\beta = \Delta\phi_{PK} = 8$). Maximum frequency of FM signal occurs when modulating waveform is maximum. Maximum frequency of PM signal occurs when the derivative of the modulating waveform is maximum.*

$$f_{m,FM} = f_{m,PM}$$
$$\text{and}$$
$$\Delta f_{Pk} = \Delta\phi_{Pk} = 8$$

1.207

The frequency of the FM waveform is the highest when the modulating waveform has a maximum value. The frequency of the PM waveform is the highest when the derivative of the modulating waveform has a maximum value.

Figure 1-86 shows a modulating waveform that emphasizes this point. The frequency of the FM wave changes directly with the value of the modulating waveform. The frequency of the PM wave stays constant except where the modulating waveform changes.

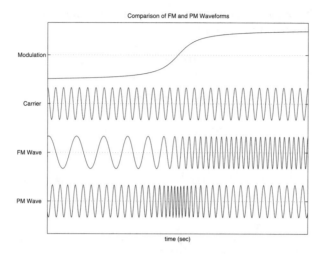

Figure 1-86 *Comparison of frequency- and phase-modulated waveforms. Maximum frequency of FM signal occurs when modulating waveform is maximum. Maximum frequency of PM signal occurs when the derivative of the modulating waveform is maximum.*

Zero Crossings

Similar to FM, we can pass a PM waveform through a limiter without losing information (see Figure 1-87). If an angle-modulated waveform accidentally acquires unwanted AM due to multipath, fading, poor filtering and such, it can be easily removed.

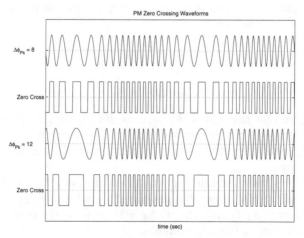

Figure 1-87 *Zero crossings of a phase-modulated waveform, for $\Delta\phi_{PK} = 8$ and 12. The modulation is preserved in the zero crossings.*

Also, many of the PM demodulation processes depend upon the PM waveform having a constant envelope. We can use limiters to remove the unwanted AM and filter after the limiter to remove the harmonics of the carrier frequency and restore the time-domain waveform to a sine wave.

Frequency Domain

Applying Fourier analysis techniques to Equation 1.206 produces a similar result as the FM case. Equation 1.206 can be rewritten as

$$V_{PM}(t) = V_{Pk}J_0(\Delta\phi_{Pk})\cos(\omega_c t)$$
$$+V_{Pk}J_1(\Delta\phi_{Pk})\{\sin[(\omega_c+\omega_m)t]+\sin[(\omega_c-\omega_m)t]\}$$
$$-V_{Pk}J_2(\Delta\phi_{Pk})\{\cos[(\omega_c+2\omega_m)t]+\cos[(\omega_c-2\omega_m)t]\}$$
$$-V_{Pk}J_3(\Delta\phi_{Pk})\{\sin[(\omega_c+3\omega_m)t]+\sin[(\omega_c-3\omega_m)t]\}$$
$$+V_{Pk}J_4(\Delta\phi_{Pk})\{\cos[(\omega_c+4\omega_m)t]+\cos[(\omega_c-4\omega_m)t]\}$$
$$+\cdots$$

1.208

Equation 1.208 is almost exactly identical to Equation 1.199, which describes the FM spectrum. Thus, for most purposes, we can consider the spectrum of a PM wave to be identical to the spectrum for the FM wave. All the spectrum characteristics described above for FM are valid for PM.

Example 1.25 — Bessel Functions and PM

Given a digital data transmission system with $\Delta\phi_{pk} = \pi/2$ radians and $f_m = 30$ kHz. Find
a. the ratio of the power level of the signal at $f_c - f_m$ to the power of the unmodulated carrier,
b. the ratio of the power level of the signal at $f_c + 2f_m$ to the power of the unmodulated carrier.

Solution —

Since $\Delta\phi_{pk}$ in Equation 1.208 is equivalent to β of Equation 1.196, we can substitute $J_n(\Delta\phi_{pk})$ for $J_n(\beta)$ in Equation 1.199.
a. Using Figure 1-79, the value of $J_1(\pi/2) = J_1(1.57) = 0.58$ and

$$\frac{P_{FM}(f_c)}{P_{FM,Un\,mod ulated\ Carrier}} = 20\log[J_n(\Delta\phi_{Pk})]$$

$$= 20\log\left|J_1\left(\frac{\pi}{2}\right)\right| \qquad 1.209$$

$$= 20\log|0.58|$$

$$= -4.7\ dB$$

The level of the tone at $f_c - f_m$ is 4.7 dB below the level of the unmodulated carrier.

b. Using Figure 1-79, the value of $J_2(1.57) = 0.23$ and

$$\frac{P_{FM}(f_c + 2f_m)}{P_{FM,Un\,mod ulated\ Carrier}} = 20\log|J_n(\Delta\phi_{Pk})|$$

$$= 20\log\left|J_2\left(\frac{\pi}{2}\right)\right| \qquad 1.210$$

$$= 20\log|0.23|$$

$$= -12.8\ dB$$

Accordingly, the level of the tone at $f_c + 2f_m$ is 12.8 dB below the level of the unmodulated carrier.

Phasor

Equation 1.206 and Figure 1-84 show that the magnitude of the FM waveform is always constant. The amplitude of the wave does not change as it goes through its modulation cycle. In the phasor domain, the vector sum of the individual components must always exhibit a constant magnitude. These concepts are also valid for PM waveforms.

RF Bandwidth

The final RF bandwidth of a PM-modulated wave depends upon the maximum frequency the carrier is modulated with and upon Δf_{pk}, the maximum phase deviation. The generally accepted rule of thumb for the RF bandwidth of a PM signal is

$$B_{PM} = 2f_m(\Delta\phi_{Pk} + 1) \qquad 1.211$$

This bandwidth includes approximately 95% of the energy available in the FM waveform. The theoretical 100% bandwidth is infinite.

Small β Approximations

We will examine waveforms possessing frequency or phase modulation with very low values of beta. Analysis of small beta conditions are required to understand the approximations made when analyzing phase noise and signals with high SNR's. Figure 1-88 shows a close-up view of the Bessel functions of Figure 1-79. For small values of β (i.e., $\beta < 0.2$ radians), we can make several important approximations.

1. The value of $J_0(\beta)$ is very close to unity, or

$$J_0(\beta) \approx 1 \ \ for \ \beta \leq 0.2 \quad\quad 1.212$$

2. The value of $J_1(\beta)$ is

$$J_1(\beta) \approx \frac{\beta}{2} \ \ for \ \beta \leq 0.2 \quad\quad 1.213$$

3. The values of the rest of the Bessel functions $J_2(\beta)$ through $J_n(\beta)$ are zero.

$$J_n(\beta) \approx 0 \ \ for \begin{cases} \beta \leq 0.2 \\ n = 2,3,4,\ldots \end{cases} \quad\quad 1.214$$

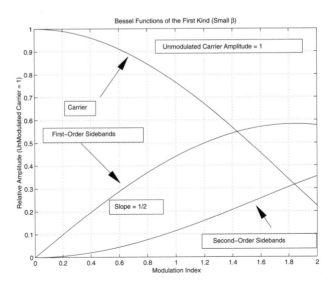

Figure 1-88 *The first three Bessel functions for small values of modulation indices. This range of modulation indices represents the small β conditions.*

Time Domain

Figure 1-89 shows a time-domain plot of a FM waveform with a low value of β. Note that the waveform does not change considerably with the modulating waveform and appears similar to the unmodulated carrier.

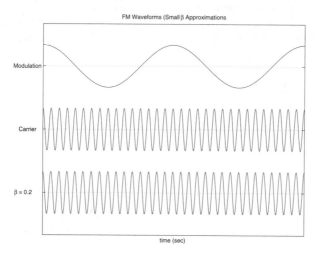

Figure 1-89 *Time domain representation of a frequency-modulated waveform under small β conditions ($\beta = 0.2$ in this plot). The modulated waveform (below) is not significantly different from the unmodulated carrier (above).*

Frequency Domain

Figure 1-90 shows the RF spectrum of a FM or PM wave when the small β approximations are valid. Since $J_0(\beta) \cong 1$, the power of the component at f_c is approximately equal to the power of the unmodulated carrier. Equation 1.208, coupled with the small β approximations, produces an expression for the sideband power.

$$\frac{P_{FM}(f_c + f_m)}{P_{FM,\text{Unmodulated Carrier}}} = \frac{P_{FM}(f_c - f_m)}{P_{FM,\text{Unmodulated Carrier}}}$$

$$= J_1^2(\beta) \qquad \qquad 1.215$$

$$= \frac{\beta^2}{4} \quad \text{for } \beta \leq 0.2$$

$$= 20\log(\beta) - 6 \ dB \quad \text{for } \beta \leq 0.2$$

Note that the narrowband FM spectrum is similar to the AM spectrum except that the LSB of the FM spectrum is 180° out of phase with the LSB of the AM spectrum. Since this power spectrum does not show phase, the AM and FM spectra appear identical.

INTRODUCTION | 129

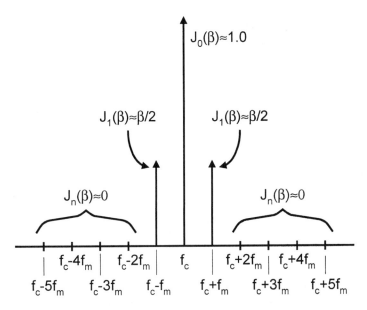

Figure 1-90 The spectrum of a frequency-modulated signal under small β conditions. Only the carrier and first sidebands (at f_c and $f_c \pm f_m$) contain significant power.

Example 1.26 — Small β and FM
For a modulation index of 0.2,
a. what is the level of the tones at f_c and
b. what is the level of each of the two sidebands at $f_c \cong f_m$?

Solution—
a. For small β, the level of the tone at f_c is approximately equal to the level of the unmodulated carrier.
b. Using Equation 1.215, we know

$$\frac{P_{FM}(f_c + f_m)}{P_{FM,\text{Unmodulated Carrier}}} = \frac{P_{FM}(f_c - f_m)}{P_{FM,\text{Unmodulated Carrier}}} \quad 1.216$$

$$= 20\log(0.2) - 6\ dB \quad \text{for } \beta \le 0.2$$

$$= -20\ dB$$

Each tone at $f_c - f_m$ is 20 dB below the level of the unmodulated carrier. Since $\beta = 0.2$ is the upper limit of the small β approximation, this is the highest the sideband levels will ever get under small β conditions.

RF Bandwidth for Small Beta

As Figure 1-90 shows, the RF bandwidth of an FM- or PM- modulated wave under small β conditions is

$$B_{FM} = B_{PM} = 2f_m \quad for \; \beta \leq 0.2 \qquad 1.217$$

Example 1.27 — FM Bandwidth Under Small β Conditions

What is the RF bandwidth of a system with a modulation index of 0.1 and $f_m = 30$ Hz?

Solution —
Using Equation 1.203, we can write

$$\begin{aligned} B_{FM} &= (2)(30) \\ &= 60 \text{ Hz} \end{aligned} \qquad 1.218$$

Phasor

Figure 1-91 shows a phasor diagram of a FM or PM wave under small β conditions. Note that this phasor diagram is slightly exaggerated to help explain the concepts. Figure 1-92 shows an enlargement of the area around the end of the resultant vector of Figure 1-91. If we accounted for all of the Bessel functions of the FM wave, the final resultant vector would always end on the arc, implying that the resultant has a constant magnitude.

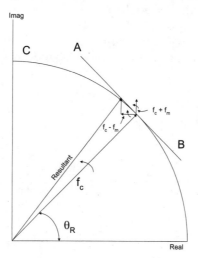

Figure 1-91 *The phasor representation of an angle-modulated signal under small β conditions. Only the carrier and first sidebands (at f_c and $f_c \pm f_m$) are significant.*

INTRODUCTION | 131

If we include all the Bessel components in the analysis, the resultant vector will always lie on circle C of Figures 1-91 and 1-92. If we ignore the Bessel components but those at f_c and $f_c \pm f_m$, the resultant vector will always terminate on the line perpendicular to the carrier (line AB in Figures 1-91 and 1-92). Because of the small β approximations, the amplitude will change slightly and the final angle θ_R of Figure 1-92 will not be exactly right. However, as β decreases, the magnitude of the two sideband vectors also decrease and the resultant will more exactly follow the true arc.

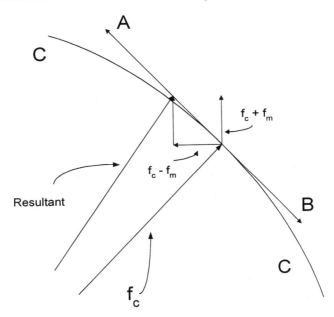

Figure 1-92 *A closer view of Figure 1-91. The vector sum of the carrier and the two sidebands (the resultant) always lies on line AB. If we include all the sidebands, the resultant will lie on circle C.*

Looking at a Spectrum

When we analyze sinusoidal modulation, the boundary between the small and large β regions is mathematically deterministic. When working with complex, random signals, we must integrate the signal power to find the boundaries. We can produce a graph that describes this boundary given signal characteristics.

Figure 1-93 shows a graph of the approximate boundary between the large and small β regions. The line has a slope of –10 dB per decade and passes through the 1-hertz offset at –30 dBc/Hz. It represents a peak phase deviation of approximately 0.2 radians integrated over any one decade of offset frequency.

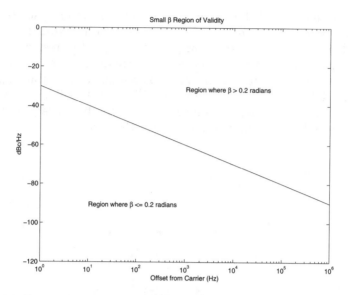

Figure 1-93 *Regions of validity for the small β conditions [11].*

Example 1.28 — Small β and Spectrum Plots

Figure 1-94 shows an oscillator spectrum. Is the modulation index < 0.2 at a 10 kHz offset?

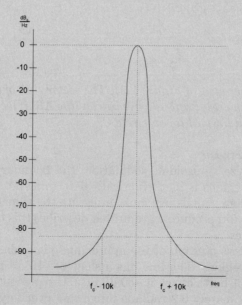

Figure 1-94 *Oscillator spectrum for Example 1.28.*

> **Solution —**
> In Figure 1-93, the small β threshold is –70 dBc/Hz at 10 kHz. The oscillator of Figure 1-94 measures –82dBc/Hz at an offset of 10 kHz.

1.17 Introduction Data Summary

dB's, dBm and dBW

dB
One dB change represents 26% change in a quantity. A change of 0.1 dB represents a 2.6% change in a quantity. A hundredth of a dB (0.01 dB) is a 0.26% change.

dBm
dBm is power referenced to one milliwatt or

$$P_{dBm} = 10\log\left(\frac{P_{Watts}}{0.001}\right)$$
$$= 10\log(P_{mW})$$

In a 50-ohm system, a 1 mW or 0 dBm sine wave has a RMS voltage of 0.224 V_{RMS}.

dBW
dBW is power referenced to one watt or

$$P_{dBW} = 10\log(P_{Watts})$$

In a 50-ohm system, a 1 ohm or 0 dBW sine wave has a RMS voltage of 7.07 V_{RMS}.

Relationship
dBm is related to dBW by

$$P_{dBw} = P_{dBm} - 30$$

Frequency and Wavelength

Frequency, wavelength and propagation velocity are related by

$$\lambda f = v$$

In free space,

$$\lambda_0 f = c$$

where
λ = the wavelength of the signal,
v = the velocity of propagation,
f = the frequency of the signal,
c = the velocity of light in free space.

The Speed of Light

The speed of light is

$$c = 2.9979 \cdot 10^8 \; \frac{\text{meters}}{\text{second}}$$

$$\approx 3 \cdot 10^8 \; \frac{\text{meters}}{\text{second}}$$

This implies

$$c = \left(2.998 \cdot 10^8 \; \frac{\text{meters}}{\text{second}}\right)\left(39.37 \; \frac{\text{inches}}{\text{meter}}\right)$$

$$= 11.80 \; \frac{\text{inches}}{\text{nanosecond}}$$

or

$$c \approx 1 \; \frac{\text{ft}}{\text{nanosecond}}$$

With less than a 2% error, we can say that light travels about one foot in one nanosecond (or conversely, it takes about a nanosecond for light to travel one foot).

Transmission Line Relationships

Impedances are defined with respect to the system's characteristic impedance (often 50 ohms). The impedance relationships are the following:

Reflection Coefficient
For a real load R_L,

$$\rho = \frac{R_L - Z_0}{R_L + Z_0} \quad -1 \leq \rho \leq +1$$

For a complex load, $Z_L = R_L + jX_L$, the reflection coefficient is

$$\rho = |\rho|\angle\theta_\rho$$
$$= \frac{Z_L - Z_0}{Z_L + Z_0}$$

and

$$Z_L = Z_0 \frac{1+\rho}{1-\rho}$$

The return loss is given by

$$\text{Return Loss}_{\text{dB}} = 10\log\left(\frac{P_{\text{Reflected}}}{P_{\text{Incident}}}\right)$$
$$= 20\log\left|\frac{V_{\text{Reflected}}}{V_{\text{Incident}}}\right|$$
$$= 20\log|\rho|$$
$$= 20\log\left|\frac{Z_L - Z_0}{Z_L + Z_0}\right|$$
$$\leq 0 \text{ dB}$$

VSWR

VSWR is related impedance by

$$\text{VSWR} = \frac{V_{\text{max}}}{V_{\text{min}}}$$
$$= \frac{1+|\rho|}{1-|\rho|}$$
$$\geq 1$$

VSWR is related to the reflection coefficient by

$$\text{VSWR} = \frac{1+|\rho_L|}{1-|\rho_L|} \quad 1 \leq \text{VSWR} \leq \infty$$

and

$$|\rho| = \frac{VSWR - 1}{VSWR + 1}$$

VSWR is related to impedance by

$$R_L = Z_0(VSWR) \quad \text{or} \quad R_L = \frac{Z_0}{VSWR}$$

Mismatch Loss

Mismatch loss is defined as

$$\text{Mismatch Loss} = \frac{P_{\text{Available}} - P_{\text{Reflected}}}{P_{\text{Available}}}$$

$$= \frac{P_{\text{Delivered}}}{P_{\text{Available}}}$$

$$= 1 - |\rho|^2$$

$$\text{Mismatch Loss}_{\text{dB}} = 10\log\left(1 - |\rho|^2\right)$$

where
  $P_{\text{Available}}$ = the power that is delivered to the matched load,
  $P_{\text{Delivered}}$ = the power that is delivered to the unmatched load,
  $P_{\text{Reflected}}$ = the power that is reflected off of the unmatched load.

Mismatch loss is yet another way of describing the relationship between the characteristic impedance of a transmission line and its load. Mismatch loss is usually specified in decibels.

$$\text{Mismatch Loss}_{\text{Load,dB}} = 10\log\left(1 - |\rho_L|^2\right)$$

Velocity Factor

The wavelength inside of a transmission line is related to the wavelength in free space through the velocity factor.

$$\text{Velocity Factor} = v_f = \frac{\lambda_g}{\lambda_0} \leq 1$$

S-Parameters

Scattering parameters are

$$|s_{11}|^2 = \frac{\text{Power reflected from the network input}}{\text{Power incident on the network input}}$$

$$|s_{22}|^2 = \frac{\text{Power reflected from the network output}}{\text{Power incident on the network output}}$$

$$|s_{21}|^2 = \frac{\text{Power delivered to a } Z_0 \text{ load}}{\text{Power available from a } Z_0 \text{ load}}$$
$$= \text{Transducer power gain with } Z_0 \text{ load and source}$$

$$|s_{12}|^2 = \text{Reverse transducer power gain with } Z_0 \text{ load and source}$$
$$= \text{Reverse isolation}$$

The terminal impedances s_{11} and s_{22} are the input and output reflection coefficients when the system is embedded in a Z_0 system

$$\left. \begin{array}{l} s_{11} = \rho_{in} \\ s_{22} = \rho_{out} \end{array} \right\} \text{when the 2-port is placed in a } Z_0 \text{ environment}$$

Modulated Signals

DSBAM

The Equation describing the time-domain behavior of a DSBAM signal is

$$V_{DSBAM}(t) = V_{pk}\cos(\omega_c t) + V_{pk} m_a \cos(\omega_m t)\cos(\omega_c t)$$
$$= V_{pk}\cos(\omega_c t)$$
$$+ \frac{V_{pk} m_a}{2}\cos[(\omega_c + \omega_m)t]$$
$$+ \frac{V_{pk} m_a}{2}\cos[(\omega_c - \omega_m)t]$$

where
$\omega_c = 2\pi f_c$ = the carrier frequency.
m_a = the AM modulation index and $0 < = m_a < = 1.0$ for DSBAM.
$\cos[(\omega_c - \omega_m)t]$ is the lower sideband (LSB). The quantity $\omega_c - \omega_m$ is often referred to as the *difference frequency* or *beat frequency*.
$\cos[(\omega_c + \omega_m)t]$ is the upper sideband (USB). The quantity $\omega_c + \omega_m$ is also called *sum frequency*.

The power in each sideband is

$$P_{\text{Sideband}} = P_{\text{Carrier}} - 20\log\left(\frac{m_a}{2}\right)$$

FM

The time-domain equation for an FM waveform is

$$V_{FM}(t) = V_{Pk}\cos\left[\omega_c t + \frac{\Delta f_{Pk}}{f_m}\sin(\omega_m t)\right]$$
$$= V_{Pk}\cos\left[\omega_c t + \beta\sin(\omega_m t)\right]$$

where β is the *modulation index* and

$$\beta = \frac{\Delta f_{Pk}}{f_m}$$

Applying Fourier analysis techniques produces

$$\begin{aligned}V_{FM}(t) = &\ V_{Pk}J_0(\beta)\cos(\omega_c t)\\ &+ V_{Pk}J_1(\beta)\{\cos[(\omega_c+\omega_m)t] - \cos[(\omega_c-\omega_m)t]\}\\ &+ V_{Pk}J_2(\beta)\{\cos[(\omega_c+2\omega_m)t] + \cos[(\omega_c-2\omega_m)t]\}\\ &+ V_{Pk}J_3(\beta)\{\cos[(\omega_c+3\omega_m)t] - \cos[(\omega_c-3\omega_m)t]\}\\ &+ V_{Pk}J_4(\beta)\{\cos[(\omega_c+4\omega_m)t] + \cos[(\omega_c-4\omega_m)t]\}\\ &+ \cdots\end{aligned}$$

where J_n is the Bessel function of the first kind and order n. The power in any sideband with respect to the power in the unmodulated carrier is

$$\frac{P_{FM}(f_c + nf_m)}{P_{FM,\text{Unmodulated Carrier}}} = \frac{P_{FM}(f_c - nf_m)}{P_{FM,\text{Unmodulated Carrier}}}$$
$$= \left[\frac{V_{Pk}}{\sqrt{2}}J_n(\beta)\right]^2 \Big/ \left[\frac{V_{Pk}}{\sqrt{2}}\right]^2$$
$$= J_n^2(\beta)$$
$$= 20\log|J_n(\beta)| \text{ in dB}$$

The approximate bandwidth of a FM signal is given by Carson's Rule.

$$B_{FM} = 2(\Delta f_{Pk} + f_m)$$
$$= 2f_m(\beta + 1)$$

PM

The time-domain equation for a phase-modulated wave is

$$V_{PM}(t) = V_{Pk} \cos\left[2\pi f_c t + \Delta\phi_{Pk} \cos(2\pi f_m t)\right]$$
$$= V_{Pk} \cos\left[\omega_c t + \Delta\phi_{Pk} \cos(\omega_m t)\right]$$

where
Δf_{pk} = the peak phase shift,
f_m = the modulating frequency.

In terms of Bessel functions, the phase-modulated signal is

$$V_{PM}(t) = V_{Pk} J_0(\Delta\phi_{Pk}) \cos(\omega_c t)$$
$$+ V_{Pk} J_1(\Delta\phi_{Pk})\{\sin[(\omega_c + \omega_m)t] + \sin[(\omega_c - \omega_m)t]\}$$
$$- V_{Pk} J_2(\Delta\phi_{Pk})\{\cos[(\omega_c + 2\omega_m)t] + \cos[(\omega_c - 2\omega_m)t]\}$$
$$- V_{Pk} J_3(\Delta\phi_{Pk})\{\sin[(\omega_c + 3\omega_m)t] + \sin[(\omega_c - 3\omega_m)t]\}$$
$$+ V_{Pk} J_4(\Delta\phi_{Pk})\{\cos[(\omega_c + 4\omega_m)t] + \cos[(\omega_c - 4\omega_m)t]\}$$
$$+ \cdots$$

Small β Conditions

Small β conditions occur when β or Δf_{pk} is less than 0.2 radians. The small β conditions are

$$\left. \begin{array}{l} J_0(\beta) \approx 1 \\ J_1(\beta) \approx \dfrac{\beta}{2} \\ J_{2-n}(\beta) \approx 0 \end{array} \right\} \text{ for } \beta \leq 0.2$$

Under small β conditions, frequency- and phase-modulated waveforms are of the form

$$V_{FM}(t) = V_{Pk} \cos(\omega_c t)$$
$$+ V_{Pk} \frac{\beta}{2}\{\cos[(\omega_c + \omega_m)t] - \cos[(\omega_c - \omega_m)t]\}$$

and

$$V_{PM}(t) = V_{Pk}\cos(\omega_c t)$$
$$+V_{Pk}\frac{\Delta\phi_{Pk}}{2}\left\{\sin[(\omega_c+\omega_m)t]+\sin[(\omega_c-\omega_m)t]\right\}$$

1.18   References

1. Anderson, Richard W. "S-Parameter Techniques for Faster, More Accurate Network Design." *Hewlett-Packard Application Note* 95-1, 2 (1967).

2. Bowick, Chris. *RF Circuit Design*. Indianapolis: Howard W. Sams, 1982.

3. Feynman, Richard P. *What Do You Care What Other People Think?* New York: Bantam Doubleday Bell, 1992.

4. *Guide to RF Connectors*. Amp Incorporated, Catalog 80-570 (Streamlined 5/90).

5. *HP-41 User's Library Solutions — Antennas*. Hewlett-Packard, 1983.

6. Johnson, Walter C. *Transmission Lines and Networks*. New York: McGraw-Hill, 1950.

7. Kraus, Herbert L., Charles W. Bostian, and Frederick H. Raab. *Solid-State Radio Engineering*. New York: John Wiley and Sons, 1980.

8. Montgomery, David. "Borrowing RF Techniques for Digital Design." *Computer Design*. no. 5 (1982): 207-217.

9. Noether, Gottfried E. *Introduction to Statistics: A Nonparametric Approach*. 2nd ed. Boston: Houghton Mifflin, 1976.

10. *Reference Data for Radio Engineers*. 5th ed. Edited by H.P. Westman, et al. Indianapolis: Howard W. Sams, 1968.

11. "RF and Microwave Phase Noise Seminar Notes." *Hewlett-Packard*, 1991.

12. Royle, David. "Designer's Guide to Transmission Line and Interconnections." Parts 1, 2 and 3. *EDN Magazine*. 23 June 1988: 131-136, 143-148, 155-160.

13. Schroenbeck, Robert J. *Electronic Communications: Modulation and Transmission*. New York: Macmillan, 1992.

14. Sklar, Bernard. *Digital Communications: Fundamentals and Applications*. Englewood Cliffs, NJ: Prentice-Hall, 1988.

15. "S-Parameter Design." *Hewlett-Packard Application Note* 154 (April 1972).

16. "Spectrum Analysis: Amplitude and Frequency Modulation." *Hewlett-Packard Application Note* 150-1 (November 1971).

17. Taub, Herbert and Donald L. Schilling. *Principles of Communications Systems*. New York: McGraw-Hill, 1971.

18. Terman, Frederick E. *Electronic and Radio Engineering*. New York: McGraw-Hill, 1955.

19. Wolff, Edward A. and Roger Kaul. *Microwave Engineering and Systems Applications*. New York: John Wiley and Sons, 1988.

2

Filters

Still a man hears what he wants to hear and disregards the rest.
—Simon and Garfunkel, *The Boxer*

2.1 Introduction

In the frequency domain, we differentiate between wanted and unwanted signals. A filter removes the unwanted signals while retaining (and not distorting) the signals we want. Thus, relevant questions that will be addressed in this chapter include:

- How many unwanted signals can we tolerate? A filter will not remove the unwanted spectra entirely; it will attenuate them by some finite amount. How much attenuation do we actually need?

- How much insertion loss can we tolerate in the filtering process? As a desired filter removes the unwanted signals, its insertion loss will also attenuate the signals we want to preserve.

- How does the filter alter the wanted signal? The filter may change the magnitude of the wanted signal unevenly over its bandwidth or it may produce more subtle changes by altering the signal's phase.

2.2 Linear Systems Review

Calling a system *linear* has several specific implications.

- If we excite a linear system at a particular frequency, we will observe only that frequency in the system. We may observe phase and amplitude changes, but we will not find any frequencies present in the system that were not intentionally included.

- The transfer function of the system does not depend upon the magnitude of the input signal. If the system performance varies with signal level, the system is not linear.

- Superposition works. We can excite a system with a signal S_1 and observe an output O_1, then excite the system with a signal S_2 and observe an output O_2. If we excite the system with a linear combination of S_1 and S_2, the output will be a linear combination of O_1 and O_2.

- The equations that describe linear systems are well understood and friendly to use. Nonlinear mathematics can be unwieldy.

Note that we can approximate any device or system as linear when the input signal level is small enough. This concept is the basis for *s*-parameters and small signal models.

Pole/Zero Review — Series RLC

Figure 2-1 shows a series RLC circuit along with the equation that describes the circuit.

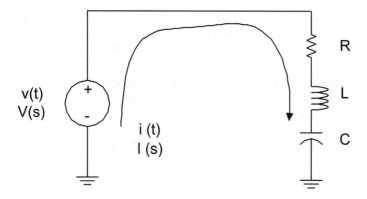

Figure 2-1 *Simple series RLC circuit*

The voltage $v(t)$ or $V(s)$ is the input and $i(t)$ or $I(s)$ is the the output. The transfer function is

$$\frac{I(s)}{V(s)} = \frac{s/L}{s^2 + s\frac{R}{L} + \frac{1}{LC}} \qquad 2.1$$

The behavior of the circuit in Figure 2-1 is completely described by Equation 2.1. Equation 2.1 is exactly specified by the roots of the polynomials in the numerator and the roots of the polynomial in the denominator. Concerning the behavior of the circuit in Figure 2-1 (or any transfer function), we only need to specify the roots of the numerator and the roots of the denominator. Equation 2.1 then becomes

$$\frac{I(s)}{V(s)} = \frac{s - s_{Z1}}{(s - s_{P1})(s - s_{P2})} \qquad 2.2$$

Combining Equations 2.1 and 2.2, we obtain the single root of the numerator.

$$s = s_{Z1} \qquad 2.3$$

We will derive the roots of the denominator from

$$s^2 + s\frac{R}{L} + \frac{1}{LC} = 0 \qquad 2.4$$

which produces

$$s_{P1}, s_{P2} = -\frac{R}{2L} \pm \frac{1}{2}\sqrt{\left(\frac{R}{L}\right)^2 - \frac{4}{LC}} \qquad 2.5$$

Let us look at the numerator and the denominator separately. As long as the denominator is nonzero, the transfer function (Equation 2.2) will be zero whenever $s = s_{Z1}$. The roots of the numerator are called the *zeroes* of the transfer function. Similarly, as long as the numerator is nonzero, the transfer function will be infinite whenever $s = s_{P1}$ or $s = s_{P2}$. Hence, the roots of the denominator are called the *poles* of the transfer function.

In summary, for most of the situations described in this book, we can write the transfer function of device as the ratio of two polynomials

$$H(s) = \frac{(s - s_{Z1})(s - s_{Z2})(s - s_{Z3})K}{(s - s_{P1})(s - s_{P2})(s - s_{P3})K} \qquad 2.6$$

where

s_{ZN} = the Nth zero of the transfer function,
s_{PN} = the Nth pole of the transfer function.

Let us assume that the denominator has complex roots, i.e.,

$$\left(\frac{R}{L}\right)^2 < \frac{4}{LC} \qquad 2.7$$

which forces the quantity under the square root to be negative. When we plot the complex roots on the *s*-plane, we obtain a picture that exactly describes the behavior of the series-RLC circuit (see Figure 2-2). We can use the roots plotted here to gain insight into the behavior of the transfer function.

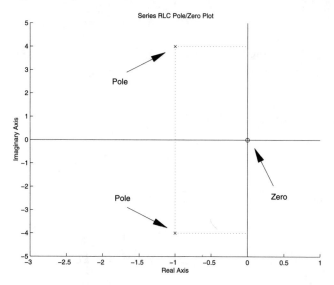

Figure 2-2 *S-plane description of simple series RLC circuit design.*

Magnitude of $H(j\omega)$

How can we describe the magnitude of a transfer function from its pole/zero plot? If we are only interested in the magnitude, we can rewrite Equation 2.6 as

$$|H(s)| = \frac{|s - s_{Z1}||s - s_{Z2}||s - s_{Z3}|\mathrm{K}}{|s - s_{P1}||s - s_{P2}||s - s_{P3}|\mathrm{K}} \qquad 2.8$$

For a sinusoidal input signal at frequency (f), $s = j\omega$ where $\omega = 2\pi f$. The term $|s - s_{ZN}|$ represents the distance between the evaluation frequency s and the Nth zero. Likewise, the term $|s - s_{PN}|$ is the distance between the evaluation frequency s and the Nth pole. Thus, Equation 2.8 can be written as

$$|H(s)| = \frac{[dist(s - s_{Z1})][dist(s - s_{Z2})][dist(s - s_{Z3})]K}{[dist(s - s_{P1})][dist(s - s_{P2})][dist(s - s_{P3})]K} \qquad 2.9$$

where
$dist(s - s_{ZN})$ = the distance between the evaluation frequency (s) and the Nth zero.
$dist(s - s_{PN})$ = the distance between the evaluation frequency (s) and the Nth pole.

Figure 2-3 shows the redrawn RLC pole/zero plot of Figure 2-2, emphasizing the magnitude characteristics. For the RLC circuit of Figure 2-1, Equation 2.9 becomes

$$|H(j\omega)| = \frac{dist(Z_1)}{[dist(P_1)][dist(P_2)]} \qquad 2.10$$

where
$dist(Z_1)$ = the distance from zero #1 to the evaluation frequency,
$dist(P_1)$ = the distance from pole #1 to the evaluation frequency,
$dist(P_2)$ = the distance from pole #2 to the evaluation frequency,
$s = j\omega$ = the evaluation frequency.

As the evaluation frequency $(s = j\omega)$ changes, the distances between s and the various poles and zeroes of the transfer function change. As $j\omega$ approaches a transfer function zero, the distance between $j\omega$ and the zero decreases. This distance will decrease if the transfer function zero rests directly on the $j\omega$-axis. Since this distance is expressed in the numerator of the transfer function, the magnitude of the transfer function approaches zero whenever $j\omega$ approaches a transfer function zero.

Alternately, as the evaluation frequency $(s = j\omega)$ approaches either of the two poles, the distance between $j\omega$ and the pole decreases. This distance will also approach zero when the pole is very close to the $j\omega$-axis. Since this distance is in the denominator of the transfer function, the magnitude of the transfer function will increase as $j\omega$ approaches a pole.

Thus,

As $j\omega$ approaches any zero, $|H(j\omega)|$ approaches 0

2.11

As $j\omega$ approaches any pole, $|H(j\omega)|$ approaches ∞

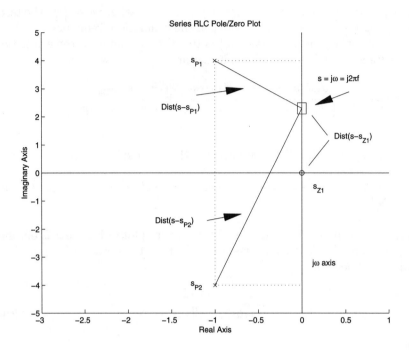

Figure 2-3 *S-plane description of simple series RLC circuit transfer function (emphasizing magnitude characteristics).*

Large $j\omega$

Figure 2-4 shows the RLC pole/zero plot when $j\omega$ is very large with respect to the pole/zero constellation. The distance from the evaluation frequency to each pole and zero is about the same, or

$$dist(s - s_{Z1}) \approx dist(s - s_{P1}) \approx dist(s - s_{P2})$$

$$\Rightarrow |H(j\omega)| = \frac{dist(Z_1)}{[dist(P_1)][dist(P_2)]} \approx \frac{1}{[dist(P_1)]} \approx \frac{1}{[dist(P_2)]} \quad 2.12$$

If $|H(j\omega_1)|$ is the magnitude of the transfer function at ω_1 and $|H(j2\omega_1)|$ is the magnitude at $2\omega_1$, we can write

$$\frac{|H(j\omega_1)|}{|H(j2\omega_1)|} \approx \frac{1/j\omega_1}{1/j2\omega_1} = 2 \quad 2.13$$

If we double the evaluation frequency, we halve the magnitude of the transfer function for the simple RLC circuit. The magnitude response of any transfer function will drop off at a constant rate under large $j\omega$ conditions. The *ultimate roll-off* of the transfer function is given by

$$\text{Ultimate Rolloff} = \left(6\frac{\text{dB}}{\text{Octave}}\right)(\#\text{Poles} - \#\text{Zeroes})$$

$$= \left(20\frac{\text{dB}}{\text{Decade}}\right)(\#\text{Poles} - \#\text{Zeroes})$$

2.14

where
    an *octave* refers to a doubling in frequency (i.e., going from f_1 to $2f_1$)
    a *decade* refers to increasing the frequency by an order of magnitude (i.e., going from f_1 to $10f_1$).

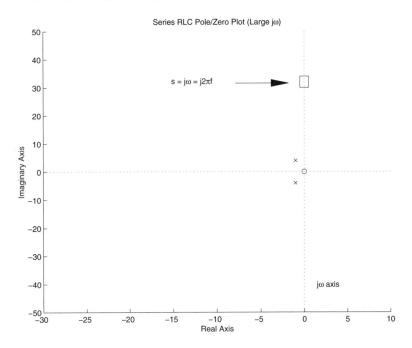

Figure 2-4 *Simple series RLC circuit transfer function, magnitude characteristics, with large $j\omega$.*

Example 2.1 — *Transfer Function Ultimate Roll-Off Rate*
Given the number of poles and zeroes for the low-pass filters described in a. through d., predict the ultimate roll-off of each filter.
a. 6 poles, 5 zeroes,
b. 7 poles, 6 zeroes,

c. 7 poles, 2 zeroes,
d. 3 poles, 1 zero.

Solution —
Using Equation 2.14, we find
a. Ultimate roll-off = 6dB/octave (6 poles, 5 zeroes) = 6 dB/octave,
b. Ultimate roll-off = 6dB/octave (7 poles, 6 zeroes) = 6 dB/octave,
c. Ultimate roll-off = 6dB/octave (7 poles, 2 zeroes) = 30 dB/octave,
d. Ultimate roll-off = 6dB/octave (3 poles, 1 zero) = 12 dB/octave.

Angle of H(jω)

We can also use the pole/zero plot to glean information about the angle of the transfer function. Figure 2-5 shows the pole/zero plot of the RLC circuit. The figure to emphasizes the angle relationships between the evaluation frequency $(s = j\omega)$ and the various poles and zeroes of the transfer function. Equation 2.6 becomes

$$\angle H(j\omega) = \left[\angle(s-s_{Z1}) + \angle(s-s_{Z2}) + \angle(s-s_{Z3})K \right] \\ - \left[\angle(s-s_{P1}) + \angle(s-s_{P2}) + \angle(s-s_{P3})K \right] \quad 2.15$$

where
$\angle(s-s_{ZN})$ = the angle between the evaluation frequency s and the Nth zero,
$\angle(s-s_{PN})$ = the angle between the evaluation frequency s and the Nth pole.

In other words, the angle of $H(s), \angle H(s)$, is the sum of the angles from the zeroes to the evaluation frequency $(s = j\omega)$ minus the sum of the angles from the poles to the evaluation frequency. Note the definition of the angles in Figure 2-5. We measure the angle between a line parallel to the real axis and the line connecting the pole or zero to the evaluation frequency. Counterclockwise represents a positive angle.

From Figure 2-5, the angle of the series RLC transfer function is

$$\angle H(j\omega) = \angle(s-s_{Z1}) - \left[\angle(s-s_{P1}) + \angle(s-s_{P1}) \right] \quad 2.16$$

where
$\angle(s - s_{Z1})$ = the angle between the zero and the evaluation frequency,
$\angle(s - s_{P1})$ = the angle between the first pole and the evaluation frequency,
$\angle(s - s_{P2})$ = the angle between the second pole and the evaluation frequency.

Each angle changes with changing frequency. When the evaluation frequency is close to a pole or a zero, the pole or zero will have a significant impact upon the phase response of the transfer function. (See **Dominant Poles and Zeroes** for more information).

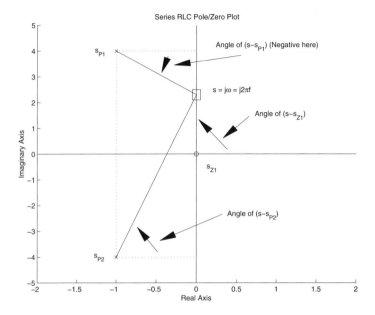

Figure 2-5 *Simple series RLC circuit transfer function (emphasizing phase characteristics).*

Large jω

If $j\omega$ is very large with respect to the pole/zero constellation, we can make approximations regarding the ultimate phase of the transfer function. Figure 2-6 shows Figure 2-5 redrawn under large $j\omega$ conditions. Figure 2-6 suggests that, for large $j\omega$, the angles between s (the frequency of interest) and the poles of the transfer function all approach 90 degrees. The same is true for the angles between s and the zeroes of the transfer function. Equation 2.2, which describes a simple RLC circuit, can be rewritten as

$$\angle H(j\omega) = \angle(s - s_{Z1}) - \left[\angle(s - s_{P1}) + \angle(s - s_{P1})\right]$$
$$\approx 90° - [90° + 90°]$$
$$= -90°$$

2.17

Ultimate phase describes the value the transfer function phase ($\angle H(j\omega)$) attains for large $j\omega$. The ultimate phase of a general transfer function is

$$\text{Ultimate Phase} = -90° \, [\#Poles - \#Zeroes]$$

2.18

152 | RADIO RECEIVER DESIGN

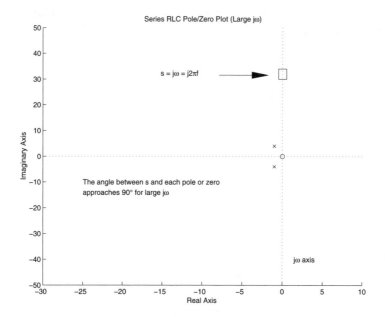

Figure 2-6 *Simple series RLC circuit transfer function, phase characteristics, with large $j\omega$.*

Example 2.2 — Transfer Function Ultimate Phase
Given the number of poles and zeroes for the low-pass filters described below, predict the ultimate phase of each filter.
a. 6 poles, 5 zeroes,
b. 7 poles, 6 zeroes,
c. 7 poles, 2 zeroes,
d. 3 poles, 1 zero.

Solution —
Using Equation 2.18, we find
a. Ultimate phase = $-90°$ (6 poles, 5 zeroes) = $-90°$,
b. Ultimate phase = $-90°$ (7 poles, 6 zeroes) = $-90°$,
c. Ultimate phase = $-90°$ (7 poles, 2 zeroes) = $-450° = -90°$,
d. Ultimate phase = $-90°$ (3 poles, 1 zero) = $-180°$.

Dominant Poles and Zeroes

Figure 2-7 shows the pole/zero plot of a transfer function containing five poles. These poles are very close to the $j\omega$-axis. The magnitude response of any transfer function is

$$|H(j\omega)| = \frac{\sum \text{Distance from } j\omega \text{ to the Zeroes}}{\sum \text{Distance from } j\omega \text{ to the Poles}} \qquad 2.19$$

For Figure 2-7, Equation 2.19 becomes

$$|H(j\omega)| = \frac{1}{[dist(P_1)][dist(P_2)][dist(P_3)][dist(P_4)][dist(P_5)]} \qquad 2.20$$

where
$dist(P_N)$ = the distance from pole #N to the evaluation frequency $s = j\omega$.

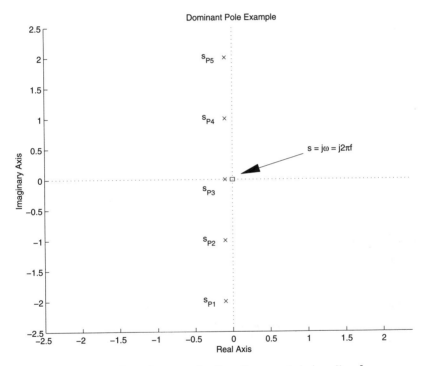

Figure 2-7 *Pole/zero plot of a transfer function containing 5 poles.*

Consider Figure 2-7 where $j\omega$ is near zero. The transfer function contains a pole s_{P3} which is very close to $\omega = 0$, so the distance between s_{P3} and $s = j\omega$ will be very small when $\omega = 0$. Equations 2.19 and 2.20 suggest that, when the distance between one of the poles and the evaluation frequency is small, $H(j\omega)$ will be large regardless of the distances to the other poles.

In Figure 2-7, the magnitude of the transfer function is almost entirely controlled by s_{P3} when $j\omega$ is in the neighborhood of s_{P3}. Figure 2-8 shows the case when $j\omega$ is in the neighborhood of s_{P5}. The distance from $s = j\omega$ to s_{P5} is very small and Equation 2.20 indicates that the magnitude of the transfer function will be very large.

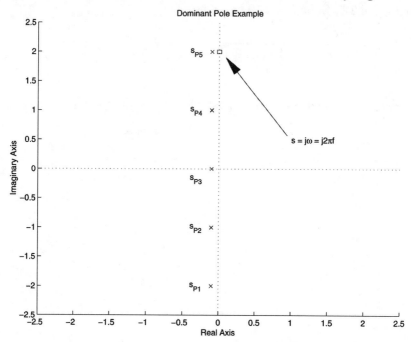

Figure 2-8 Pole/zero plot containing 5 poles. Frequency $j\omega$ is in the neighborhood of pole s_{p5}.

The concept of a *dominant pole* applies whenever the distance between $j\omega$ and any pole is small in relation to the other distances on the pole/zero plot. When this is true, then the behavior of the transfer function is almost entirely dominated by the closest, or dominant, pole. s_{P1} is a dominant pole whenever

$$\begin{array}{ll} d_{P1} \ll d_{P2} & d_{P1} \ll d_{Z1} \\ d_{P1} \ll d_{P3} & d_{P1} \ll d_{Z2} \\ d_{P1} \ll d_{P4} \quad \text{and} & d_{P1} \ll d_{Z3} \\ \quad \vdots & \quad \vdots \\ d_{P1} \ll d_{Pn} & d_{P1} \ll d_{Zk} \end{array} \qquad 2.21$$

where

d_{PN} = the distance from pole #N to the evaluation frequency,
d_{ZN} = the distance from zero #N to the evaluation frequency.

Similarly, the concept of a *dominant zero* applies whenever the distance between the evaluation frequency (s) and a zero is much smaller than the distances on the pole/zero plot. In other words, Z_1 is a dominant zero if

$$\begin{array}{ll} d_{Z1} << d_{P1} & d_{Z1} << d_{Z2} \\ d_{Z1} << d_{P2} & d_{Z1} << d_{Z3} \\ d_{Z1} << d_{P3} \quad \text{and} & d_{Z1} << d_{Z4} \\ \quad M & \quad M \\ d_{Z1} << d_{Pn} & d_{Z1} << d_{Zk} \end{array} \qquad 2.22$$

where

d_{PN} = the distance from pole #N to the evaluation frequency,
d_{ZN} = the distance from zero #N to the evaluation frequency.

Magnitude Characteristics

Figure 2-9 shows the pole/zero diagram of a transfer function with several dominant poles and no zeroes. The figure also includes the magnitude of the transfer function. Note the more pronounced peak of the transfer function as ω increases from s_{P1} (at $\omega = -2$) to s_{P5} (at $\omega = +2$). The transfer function reaches its peak when ω is near a dominant pole because the distance between the evaluation frequency and dominant pole changes rapidly over frequency. The distances between the evaluation frequency and the other poles/zeroes are almost constant.

Figure 2-10 shows an all-zero transfer function that exhibits dominant-zero behavior. Note the areas of high attenuation in the transfer function magnitude response when ω is near a dominant zero.

Careful study of Figures 2-9 and 2-10 reveals that we can use the pole/zero concept to machine any arbitrary transfer function by judiciously placing poles and zeros, which is, in fact, how many filters are realized. Figures 2-11 and 2-12 show the pole-zero and magnitude plots of an elliptic filter. The positions of the poles and zeroes have been carefully manipulated to produce the desired amplitude response. Sections of the transfer function of Figure 2-12 are labeled with the pole or zero most responsible for the characteristic.

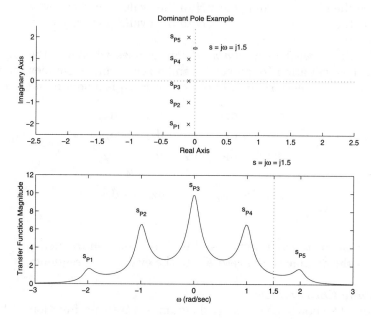

Figure 2-9 *Transfer function with dominant poles (no zeros).*

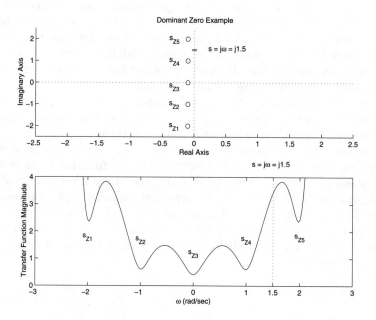

Figure 2-10 *Transfer function with dominant zeros (no poles).*

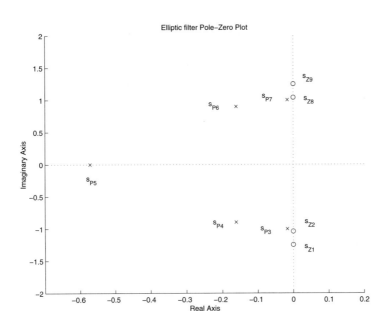

Figure 2-11 *Pole/zero plot of an elliptical filter.*

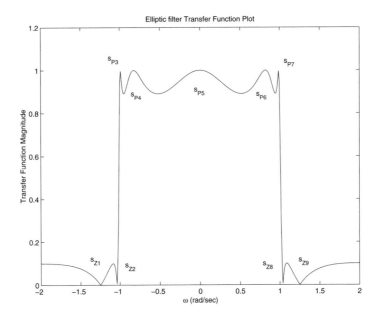

Figure 2-12 *Magnitude response of an elliptical filter.*

Phase Characteristics

Let us examine the phase response of a transfer function as the input frequency passes by a dominant pole or zero. Figure 2-13 shows the $j\omega$-axis with a dominant zero at s_{Z1}. An expanded version of the $j\omega$-axis around s_{Z1} is also shown. The evaluation frequency is set at point [1] in Figure 2-13, which is well away from s_{Z1} and the zero is not dominant at this frequency. The angle θ_z represents the amount of phase shift contributed by the zero to the overall transfer function. At point [1], the zero at s_{Z1} contributes almost -90° of phase shift to the transfer function.

As the frequency increases and we move from position [1] to position [2], the phase of the transfer function changes from –90° to –45°. The frequency has to be changed considerably to change by 45°. Looking at the enlarged portion of Figure 2-13, as we change the frequency from [2] to [3] to [4], we change the phase shift from -45 (at [2]) to 0° (at [3]) to +45° (at [4]). Thus, the total phase shift has been changed by 90° in a very small frequency change. We moved the entire distance from [1] to [2] in order to produce the same 45° phase shift. In the neighborhood of a dominant pole or zero, the phase of the transfer function changes very rapidly with frequency. Finally, we move from position [4] to position [5] and change another 45°. When we are at position [1] or position [5], we are in the large $j\omega$–region of the graph.

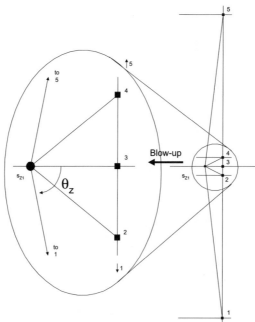

Figure 2-13 *Transfer function behavior (magnitude) in the neighborhood of a dominant pole.*

Figure 2-14 shows the phase response of the transfer function as we move from position [1] to position [5]. Note how the phase changes rapidly as the frequency changes from [2] to [3] to [4]. The phase change is fairly small from [1] to [2] and from [4] to [5]. The phase changes over a 180° range as the frequency moves from [1] to [5].

We have assumed that the zero at f_{z1} is a dominant zero, i.e., that the distances from the evaluation frequency to all of the other poles and zeroes have not changed significantly. In conclusion we find that very rapid phase changes occur when we are dealing with poles or zeroes which are close to the $j\omega$-axis, i.e., high Q poles or zeroes.

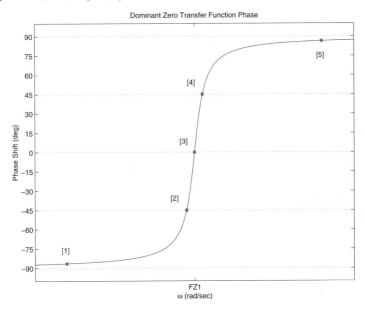

Figure 2-14 Transfer function behavior (phase) in the neighborhood of a dominant pole.

Example 2.3 — Dominant Poles and Zeroes in Filter Responses

Figures 2-11 and 2-12 show the magnitude response of a low-pass filter and its associated pole/zero plot.
a. Reconcile the pole/zero plot with the magnitude response.
b. Are the poles or zeroes responsible for the sharp drops in the magnitude response?

Solution —

This is an elliptic filter, which uses transfer function zeroes to produce a quick transition from the filter's passband to its stopband.

a. The filter exhibits severe attenuation when the evaluation frequency is near any of the zeroes (at low frequencies and high frequencies). The filter shows only a small amount of attenuation when the frequency is in the neighborhood of the transfer function poles. The pole/zero plot corresponds to the magnitude response.

b. Comparing the pole/zero plot with the filter's magnitude response, we can see that the zeroes are responsible for the drop in the magnitude response.

2.3 Evaluating Pole/Zero Plots

A basic understanding about how poles and zeroes work can be applied to gain insight into a filter's behavior.

Thought Experiment

Imagine you are outside on a dark and foggy night, driving along a long, straight road (the $j\omega$-axis) with a large field (the left-hand s-plane) off to your left. The poles and zeroes of a filter lay somewhere in the field. Mentally attach a bright light to each of the filter poles.

Figure 2-15 shows one possible pole/zero layout. All poles are very close to the road. As you approach the field from a distance, you will first notice a diffuse glow ahead and to the left. Due to the thick fog and the relatively tight grouping of the poles from far away, no one pole will stand out. As you approach the first pole from the bottom of Figure 2-15, most of the brightness is emanating from pole closest to you (s_{p1}). You may see some light from the other poles but, because s_{p1} is the closest and the light recedes quickly in the dense fog, the nearest pole will determine how much light you see. In filter terms, you are now at a peak in the transfer function. If you continue to drive upwards (Figure 2-15), past pole s_{p1} and toward the second pole (s_{p2}), the light will gradually decrease until you are about halfway between the poles. The total amount of light you see is now due to about equal contributions of s_{p1} and s_{p2}. You are now at a valley of the transfer function. As you continue to drive up the road, the amount of light around you will peak and valley as you arrive at and pass the separate poles. This rising and falling light is exactly analogous to the behavior of a "peaky" transfer function with lots of dominant poles. Figure 2-16 shows what this transfer function might look like.

FILTERS | 161

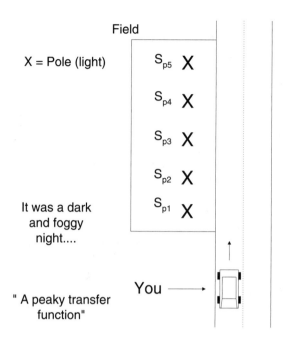

Figure 2-15 *Pole/zero thought experiment, "peaky" magnitude response.*

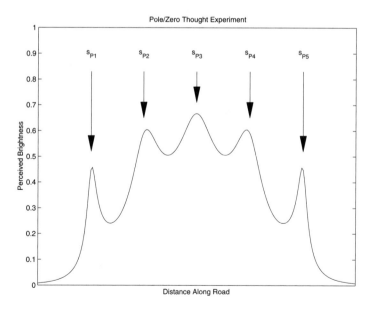

Figure 2-16 *Magnitude response derived from Figure 2-15.*

162 | RADIO RECEIVER DESIGN

Let us rearrange the poles and try another experiment. Figure 2-10 shows the same road and field on the same dark and foggy night. We have moved the filter poles to the left-most portion of the field as far away from the road as we can. Again, we will begin the drive at the bottom of Figure 2-10. As you approach the field, you will still see a diffuse glow off in the distance and to the left. At this distance, no one pole stands out because of the tight grouping of the poles (remember that you are in the "large $j\omega$" region of the pole/zero plot).

As you approach the field, you can see that the poles are farther off to the left than before. Yet, the distance and thick fog do not allow you to distinguish any specific pattern. As you pass the field, the characteristic of the light does not change noticeably but continues to emanate from one diffuse blob to the left. No one pole stands out. Figure 2-17 shows the light intensity you might experience.

This behavior is analogous to a low Q or gentle transfer function with no dominant poles. The first "peaky" example is similar to a Chebychev filter; the low Q example is analogous to a Bessel filter. We can also perform these experiments with the zeroes. We can attach lights of different colors to the poles and the zeroes. Close to a zero, the magnitude of the transfer function will decrease rather than increase.

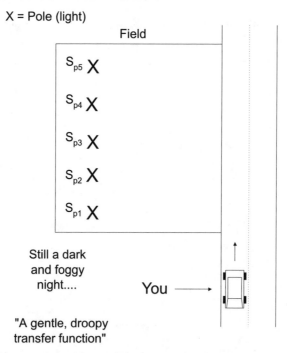

Figure 2-17 *Pole/zero thought experiment, gentle magnitude response.*

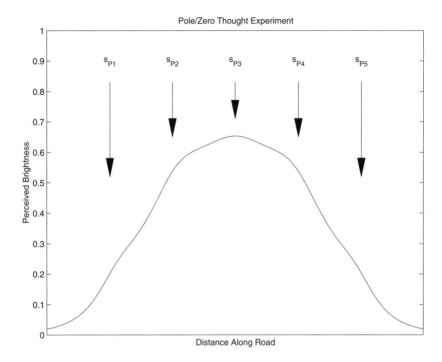

Figure 2-17 *Magnitude response derived from Figure 2-10.*

2.4   Filters and Systems

A filter's transfer function is characterized by three key aspects: the *magnitude response*, the *phase response* and the *group delay response*. All three are meaningful only in connection with the steady-state performance of a filter (although we can often infer how the filter will behave in a transient situation from its steady-state filter response).

Figure 2-18 shows the circuit diagram under discussion. The signal source consists of a voltage source V_s and an internal source resistor R_s. This model is reasonably accurate for signal generators, receiving antennas and most other signal sources. We connect the input port of the filter to the signal source and the output port of the filter to a load resistor R_L. The voltage present at the input of the filter is V_{in} and the voltage across R_L is V_{out}.

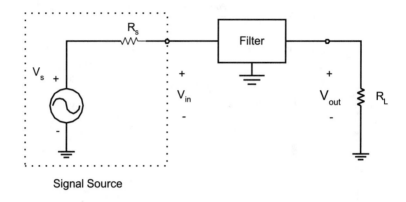

P_{avail} = Maximum power available from signal source

P_{RL} = Power dissipated in R_L
   = V^2_{out}/R_L

Figure 2-18 *Filter model.*

In a radio-frequency environment, we are usually interested in *power transfer* — not voltage or current. The relevant magnitude response is

$$\frac{P_{RL}}{P_{Avail}} = \frac{\text{Power Dissipated in } R_L}{\text{Maximum Power Available From the Signal Source}} \qquad 2.23$$

where
  Maximum Power Available From the Signal Source = the signal power measured in a matched load resistor connected directly to the signal source.

The exact wording of this definition is important. In Chapter 1 (*Matching and Maximum Power Transfer*), we saw that we will receive the most power from the signal source when the input impedance of the filter is matched to the source resistor (R_s). If the input of the filter is not matched, all the available signal power from the source cannot be accepted. This definition considers the filter's input impedance. Similarly, if the output impedance of the filter is not matched to R_L, the power delivered to R_L will be less than the maximum power available from the filter.

This definition also considers the filter's internal losses. If both the input and output are properly matched yet the filter dissipates some of the signal power internally, the loss will appear in Equation 2.23.

2.5 Filter Types and Terminology

Filter Terminology

Listed are three common filters. These terms refer to a filter's magnitude response.

Passband. The passband refers to a band of frequencies that a filter will pass with minimum attenuation. The frequencies within the passband usually contain the signals of interest.

Stopband. The stopband refers to a band of frequencies that the filter attenuates severely. Frequencies in the stopband are not signals of interest and can cause problems if they propagate further into a system.

Transition band. The transition band marks the band of frequencies between the passband and the stopband and lies between a filter's passband and its stopband. Ideally, we would like a filter to transition immediately from its passband (where the filter passes signals without attenuation) to its stopband (where the filter attenuates signals significantly).

Figure 2-19 illustrates the above terms. The boundaries between the passband, transition band and the stopband are not always clearly defined. In Figure 2-19, we arbitrarily drew the boundaries at the filter's 3 dB and 20 dB attenuation levels.

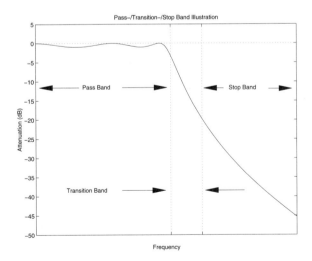

Figure 2-19 *Definition of filter terms.*

Filter Types

Four types of filters, which are labeled according to their magnitude responses, will be discussed. As you read the following paragraphs, refer to Figures 2-20 through 2-23.

Low-Pass Filters

Low-pass filters (LPF) pass only the lower frequency components of a signal and attenuate the higher frequency components. The frequency marking the boundary between the low-frequency passband and high-frequency stopband is the *cutoff frequency* (f_c). Figure 2-20 shows the magnitude response of a low-pass filter.

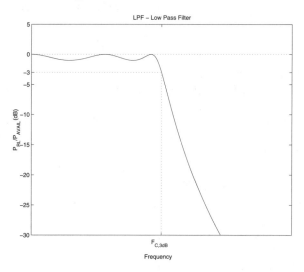

Figure 2-20 *Low-pass filter ideal magnitude response.*

Due to realization effects, the magnitude response of a LPF usually misbehaves at frequencies well above its cutoff frequency. Often we have to specify just how high up in frequency we want the filter to behave.

High-Pass Filters

A *high-pass filter* (HPF) allows the higher-frequency components of a signal to pass while it severely attenuates its lower-frequency components. The *cutoff frequency* (f_c) marks the boundary between the high- and low-frequency bands of the filter. Figure 2-21 shows the magnitude response of a high-pass filter.

As the LPF, the magnitude response of a HPF does not extend to infinite frequency. When building or buying a high-pass filter, we often have to specify how high up in frequency we need the filter to function.

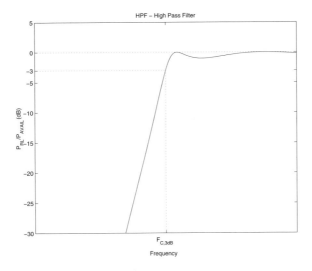

Figure 2-21 *Highpass filter ideal magnitude response.*

Band-Pass Filters

Band-pass filters (BPF) pass only a band of frequencies while attenuating signals both below and above the filter's passband. The center frequency of the filter is f_c, and the lower and upper cutoff frequencies are f_{low} and f_{high}, respectively. A band-pass filter has often an intrinsic power loss at the center frequency, which is referred to as the *insertion loss* of the filter. Figure 2-22 depicts the magnitude response of a band-pass filter.

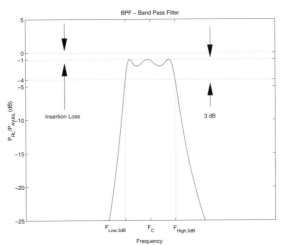

Figure 2-22 *Band-pass filter ideal magnitude response.*

Band-Stop Filters

Band-stop filters pass all frequencies except for a particular band where the filter should provide attenuation. The center frequency is f_c and the lower and upper frequencies are f_{low} and f_{high}. Figure 2-23 shows the magnitude response of a band-stop filter.

A band-stop filter's magnitude response will not extend arbitrarily high in frequency. Again, we usually must specify some upper operating frequency when we build or buy a band-stop filter. Both band-pass and band-stop filters have two transition bands.

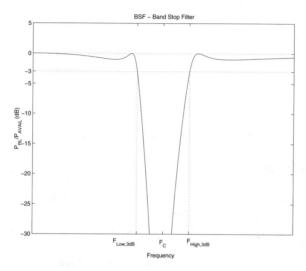

Figure 2-23 *Band-stop filter ideal magnitude response.*

2.6 Generic Filter Responses

We will look at all three filter responses and how they relate to the filter's pole/zero plot.

Magnitude Response

Figure 2-24 shows a generalized magnitude response of a low-pass filter. The terms used in Figure 2-24 also apply to high-pass, band-pass and band-stop filters with minor modifications. Note that areas of the amplitude response [(1) through (4)] refer to particular sections of the filter's response. We have included the same numbers on the plots for the filter's pole/zero plot (Figure 2-25), its phase response plot (Figure 2-26) and the group delay response (Figure 2-27).

FILTERS | 169

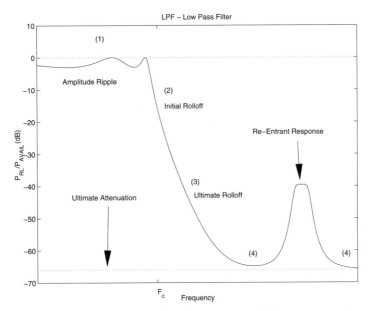

Figure 2-24 *Low-pass filter magnitude response with important regions labeled.*

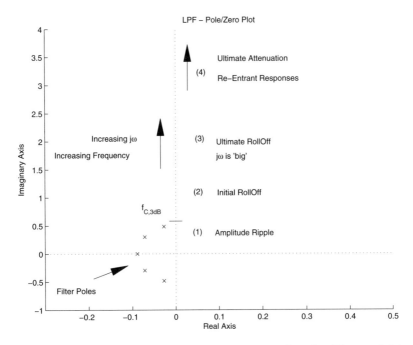

Figure 2-25 *Low-pass filter pole/zero response related to Figure 2-24.*

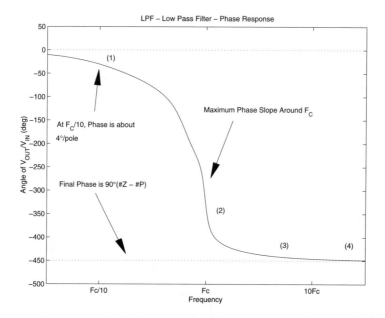

Figure 2-26 *Low-pass filter phase response with important regions labeled (related to Figures 2-24 and 2-25).*

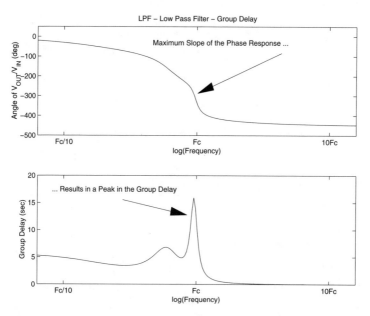

Figure 2-27 *Phase response of a filter and its corresponding group delay.*

FILTERS | 171

Amplitude ripple [area (1)]. Amplitude ripple refers to variations in the magnitude response of the filter's transfer function, particularly in the passband. Chebychev filters contain passband ripple by design.

Initial roll-off [area (2)]. The initial roll-off rate measures how fast the magnitude response of the filter initially drops off just above the filter's cutoff frequency (f_c) in the transition band. The more amplitude ripple we allow in the filter's passband, the steeper the initial roll-off of the filter.

Ultimate roll-off [area (3)]. Eventually, the initial roll-off rate gives way to the filter's ultimate roll-off rate. Equation 2.14 gives an expression for the ultimate roll-off of an ideal filter. Due to realization effects, we may never reach the ultimate roll-off rate given in Equation 2.14.

Ultimate attenuation [area (4)]. Mathematically, most filter design equations predict that a filter will roll off indefinitely. In reality, the filter eventually reaches a point where it will provide only so much attenuation and no more. This *ultimate attenuation* can range from 30 dB in commercial, physically small band-pass filters to 110 dB in large, carefully packaged units.

Re-entrant response [area (5)]. At one or more frequencies in the stopband, the filter's attenuation may suddenly change from 90 dB, for example, to 20 dB. Similar to ultimate attenuation, re-entrant responses are purely a realization problem (i.e., how the filter was built, what kind of components were used, how they placed, and so on). We cannot predict a re-entrant response or its frequency mathematically.

Pole/Zero Plot

Figure 2-25 shows the pole/zero plot of the filter in Figure 2-24. The areas numbered (1) through (4) correspond to the same areas of the magnitude plot shown in Figure 2-24.

Amplitude ripple [area (1)]. Area (1) in Figure 2-25 corresponds to area (1) in Figure 2-24, i.e., the filter's passband. The poles of this particular filter are positioned on an ellipse whose long axis runs along the $j\omega$-axis. Equivalently, both foci of the ellipse lie on the $j\omega$-axis. As $j\omega$ increases, it passes by each of the poles in turn. The closest pole at a particular $j\omega$ becomes slightly dominant and, as $j\omega$ passes each pole, the magnitude response of the filter increases sharply. This is exactly the condition we see in the magnitude response of Figure 2-24.

Initial roll-off [area (2)]. Area (2) corresponds to the initial roll-off portion of Figure 2-24. As $j\omega$ begins to pull away from the constellation of poles, the magnitude response drops very quickly because the distances to all of the poles are increasing rapidly. However, $j\omega$ has not increased to the point where we are experiencing the ultimate rolloff because we cannot yet make the approximation that the distances to all of the poles and zeroes are equal.

Ultimate roll-off [area (3)]. This section of the pole/zero plot corresponds to the ultimate roll-off area of Figure 2-24. We can assume that the distances from $j\omega$ to each pole or zero are equal.

Ultimate attenuation, re-entrant response [area (4)]). The simple pole/zero plot does not predict these effects. According to Figure 2-25, the magnitude response should continue to fall off at the ultimate roll-off rate.

Component losses, stray capacitance and inductance, and component coupling all play their part in these effects. The component vagaries combine to produce transfer function poles that were not in the original design. However, these poles are at relatively high frequencies and do not affect the transfer function until $j\omega$ is large in relation to the filter's cutoff frequency. Figure 2-28 shows the pole/zero plot of a transfer function with these parasitic poles added. Note the scale change between Figures 2-25 and 2-28.

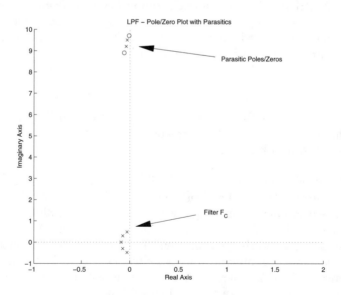

Figure 2-28 *Pole/zero plot of a low-pass filter containing parasitic poles and zeros.*

Phase Plot

Figure 2-26 shows the phase plot of the transfer function we have been assuming. This plot deals with the voltage transfer function of the filter (V_{out}/V_{in}) as opposed to the magnitude response which deals with the power transfer function (P_{RL}/P_{AVAIL}). The areas of interest are labeled (1) through (4).

Area (1). This is the filter's passband. For a typical filter, we will normally see some phase change beginning at one-tenth of f_c. This is long before any significant amplitude changes occur. For example, a Butterworth filter exhibits about 4° of phase shift per pole at one-tenth of f_c.

Area (2). As the frequency increases, the slope of the phase transfer function approaches a maximum in the general neighborhood of the cutoff frequency (f_c). Also, the phase is approximately one-half of the final phase at f_c.

Area (3). As the frequency of interest increases beyond f_c, the phase response slope decreases and the phase value approaches the ultimate phase value given by Equation 2.18.

Group Delay

Let us examine the circuit diagram of Figure 2-18 again. At a frequency ω_0, we will assume the filter produces a phase shift of $-\beta_0$ (the minus sign signifies that V_{out} exits the filter delayed in time from V_{in}). Figure 2-29 shows the steady-state time-domain plots of V_{out} and V_{in}. We can interpret the phase shift β_0 as a time delay (t_{pd}). We know that one cycle of a sine wave takes $1/f_0$ seconds to complete and that there are 2π radians in one cycle. Thus we can write

$$\frac{-\beta_0}{t_{pd}} = \frac{2\pi}{1/f_0} = 2\pi f_0 = \omega_0 \qquad 2.24$$

Rearranging Equation 2.24 produces

$$t_{pd} = -\frac{\beta_0}{\omega_0} \qquad 2.25$$

The quantity t_{pd} is the *phase delay time*, or *carrier delay time* and refers to the time required for a sine wave to pass through a filter under steady-state conditions. Figure 2-29 shows that β_0 and t_0 represent the same amount of time in different formats. Note that β_0 can be larger than 360°.

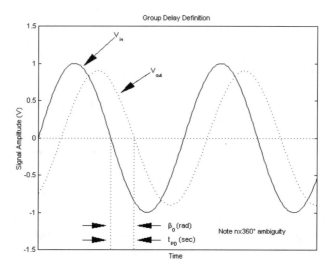

Figure 2-29 *Group delay definition in the time domain.*

The absolute amount of time it takes for a signal to pass through a filter is usually irrelevant. The *differential time delay* through a filter is of importance, however. If some components of a signal take longer to pass through a filter than others, the phase relationship of the Fourier components of the signal is destroyed. This can cause severe distortion especially in complex signals.

Group delay measures the differential time delay caused by a filter, i.e., it indicates if certain frequency components will be delayed more than other components and by how much. It is defined as

$$\text{Group Delay} = t_{gd} = -\frac{\partial \beta}{\partial \omega} \qquad 2.26$$

Equation 2.26 represents the slope of the V_{out}/V_{in} phase vs. frequency curve. In other words, the greater the slope of the V_{out}/V_{in} phase curve, the higher the value of group delay.

A flat group delay curve means that all the frequency components of a signal pass through the filter in the same amount of time without differential delay. An uneven group delay curve indicates that some components will take longer to pass through the filter than other components. At the output of the filter the phase relationships among the signal's components will be upset and the signal will be distorted.

Figure 2-27 shows the phase response of a filter and its corresponding group delay plot. Again, note that the group delay plot is the derivative of the phase plot with respect to frequency. If there are ripples in a filter's phase plot, there will be lumps in the group delay plot. Group delay usually peaks at or near a low-pass filter's cutoff frequency. In a band-pass filter, the group delay usually peaks near the edges of the passband. Group delay does not directly suggest the transient response of a filter. Group delay is a steady-condition; transient measurements speak to non-steady-state conditions. Figure 2-30 illustrates some of the fundamental differences between group delay and transient response measurements.

Filter delay and attenuation characteristics are interdependent. The narrower a filter's transition band, the larger the delay peaks. Generally, filters with numerous poles and filters with close-in stopband zeroes have large delay peaks. On the other hand, low selectivity filters with large transition bands tend to have smaller delay peaks.

Rise Time Measurement

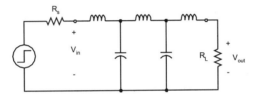

o Input is a step or impulse function.

o Before applying the input, all inductor currents and capacitor voltages = 0

o Apply input and plot output voltage in the time domain.

Group Delay Measurement

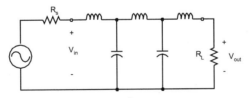

o Input is a sine wave source that has been on long enough for all transients to have died down.

o All I_c and V_c are in their steady state conditions

o Measure $\angle \frac{V_{out}}{V_{in}}$ @ different frequencies and plot the phase and group delay responses

Figure 2-30 Group delay versus transient response measurements.

Effects of Group Delay

Inappropriate group delay can create unexpected distortion in modulated signals. *Amplitude modulation* (AM) is fairly immune from the effects of group delay because of the way the signal is demodulated. Employing an AM signal, we can show how group delay can affect modulation. Using Figure 2-18 as a circuit model, let us assume the signal source consisting of V_s in series with R_s is an AM source and V_s takes the form of

$$V_s = V_{PK}\left[1 + m_a \cos(2\pi f_m t)\right]\cos(2\pi f_c t) \qquad 2.27$$

where
m_a = the modulation index ($0 <= m_a <= 1$),
f_c = the RF carrier frequency,
f_m = the modulation frequency.

Further, let us assume the modulation index (m_a) equals 0.5, the carrier frequency f_c equals 1 MHz and the modulating frequency f_m is 1 kHz. Figure 2-31 shows V_s in the time-domain, the frequency domain and as a phasor.

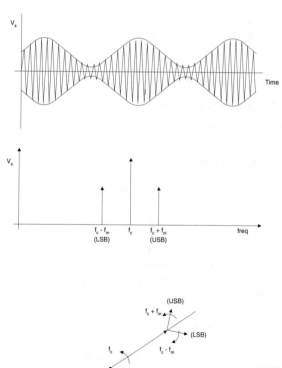

Figure 2-31 *Amplitude modulation in the time-, frequency- and phasor domains.*

A Fourier analysis of the input waveform V_s shows the signal consists of three separate sine waves:

- the carrier is at f_c,
- the lower side band (LSB) is at a frequency of $(f_c - f_m)$,
- the upper side band (USB) is at a frequency of $(f_c + f_m)$.

The phasor interpretation of V_s will be the most useful to the present discussion (see Figure 2-32). The carrier is stationary and functions as a reference. The USB is placed at the end of the arrow end and rotates counter-clockwise about the end of the carrier vector. The rotation rate is f_m. When the LSB is positioned at the end of the carrier, it rotates clockwise around the end of the carrier at the same rate.

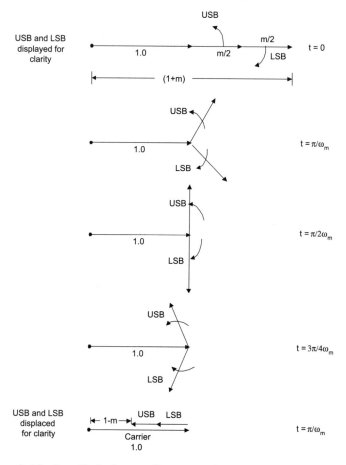

Figure 2-32 *Detailed phasor diagram of amplitude modulation.*

Before the AM signal was distorted with poor group delay, the phases of the LSB and USB appeared in a way that the two vectors were always symmetrical to the carrier. The upper and lower sidebands rotate about the end of the carrier at the same rate but in opposite directions. Jointly, they lengthen and shorten the carrier.

The symmetry of the LSB and USB is critical. If one sideband experiences a time delay (or equivalently, a phase shift) that the other sideband does not experience, the AM waveform will be distorted. Figure 2-33 shows an AM waveform passing through a filter with a nonsymmetrical group delay. Since the filter's group delay has destroyed the symmetry of the USB and LSB with respect to the carrier, the instantaneous amplitude of the waveform is distorted. Since the angles of the USB and LSB no longer cancel out, the carrier is now shifting in phase at a rate of f_m.

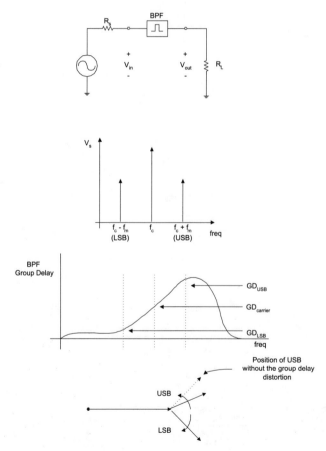

Figure 2-33 *Effects of filter group delay on amplitude modulation.*

Although group delay distortion is easy to visualize in the AM system, it usually is not a problem with this modulation. It can be a considerable problem in frequency- or phase-modulated systems. Like AM, the sidebands of FM and PM systems must add vectorially to produce the proper composite waveform. If they do not, the distortion irrevocably damages the signal.

2.7 Classes of Low-Pass Filters

In this section, four common types of filters will be discussed. All filters are 5-pole low-pass filters. The low-pass characteristics of each filter apply to the equivalent high-pass, band-pass and band-stop filters.

The filter performance will be graphed on a logarithmic frequency x-axis to linearize the plot and make visible aspects of the plots that would not be obvious if performances with linear frequencies were plotted on the x-axis. Filter cutoff frequencies are specified as either $f_{c,3dB}$ (the 3 dB or half-power cutoff frequence which is the most common specification), or, $f_{c,ER}$ (the equal-ripple cutoff frequency normally associated with Chebychev filters).

Butterworth Low-Pass Filters

The Butterworth filter is the most common filter. Since its characteristics are simple to describe mathematically, we will use the Butterworth as the calculation standard. Whenever we are interested in calculating the attenuation of a filter in a system, we will assume it is a Butterworth filter. The cutoff frequency of the Butterworth filter is the frequency at which the output power has dropped by 1/2 or 3 dB from its maximum value. This is the 3 dB or half-power frequency ($f_{c,3dB}$).

Figure 2-34 shows the magnitude response of the Butterworth filter. At low frequencies, the filter does not attenuate the signal at all. As the frequency increases, the attenuation of the filter increases slowly at first, then more rapidly as the frequency increases beyond $f_{c,3dB}$. This type of response is called a *monotonically decreasing*, or *maximally flat response*, which means that it does not have any ripples in it and exhibits a smooth transition from the passband to the stopband and beyond. In the stopband, the filter reaches its ultimate roll-off rate almost immediately beyond $f_{c,3dB}$.

180 | RADIO RECEIVER DESIGN

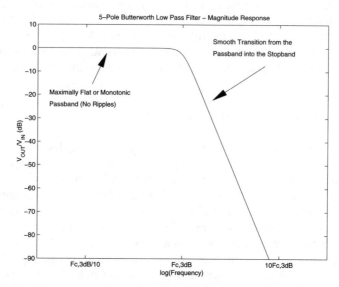

Figure 2-34 *The magnitude response of a 5-pole Butterworth low-pass filter.*

Figure 2-35 shows the pole/zero plot for a 5-pole, Butterworth low-pass filter. Simple Butterworth filters consist only of poles (no zeroes) and the poles all lie on a circle centered about the origin. No one pole is ever dominant; a smooth magnitude response can be observed.

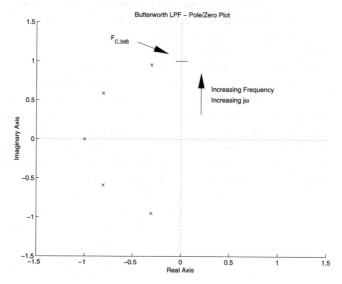

Figure 2-35 *Pole/zero plot of a 5-pole Butterworth low-pass filter.*

Figure 2-36 shows the phase and group delay responses for a 5-pole Butterworth low-pass filter. The plots exhibit several noteworthy features.

- The phase response begins to change long before the magnitude response shows any significant attenuation. For example, at $f_{c,3dB}/10$, a Butterworth low-pass filter will typically exhibit a phase shift of about 4° per pole. The 5-pole Butterworth LPF will show about 20° of phase shift at $f_{c,3dB}/10$ (The measurement on an actual Butterworth filter showed less than 0.1 dB of attenuation and 18° of phase shift at $f_{c,3dB}/10$).

- The phase slope is steepest in the neighborhood of the 3 dB cutoff frequency $f_{c,3dB}$. Since the group delay is the derivative of the phase plot, it reaches a peak around the same spot.

- When the frequency increases to well beyond $f_{c,3dB}$, the large $j\omega$ approximations we discussed earlier are valid and the filter approaches the ultimate phase given by Equation 2.18. A 5-pole LPF exhibits a final phase of $-90 \cdot (5 - 0) = -450$.

- At very high frequencies (i.e., large $j\omega$), the filter reaches its ultimate phase. However, at $10f_{c,3dB}$, we still have not reached the final phase given by Equation 2.18. At $10f_{c,3dB}$, we are about 4° per pole away from the final phase.

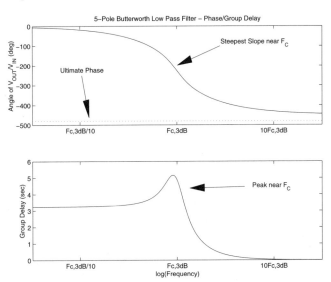

Figure 2-36 *Phase and group delay responses of a 5-pole Butterworth low-pass filter.*

Chebychev Low-Pass Filters

The Chebychev (pronounced "Cheby-shev") filter is the second most common filter in use. As Figure 2-37 indicates, the Chebychev's magnitude response displays ripples in the passband and a steeper initial roll-off than the Butterworth. It is acceptable to allow ripples in the passband (which is undesirable) if this results in a steeper initial roll-off (which is desirable).

Nomenclature

Some authors distinguish two types of Chebychev filters: Chebychev I and Chebychev II. Figure 2-37 shows a Chebychev I filter. The passband exhibits amplitude ripple, and the theoretical stopband attenuation increases monotonically. Figure 2-38 shows a Chebychev II filter. The passband is flat but the stopband exhibits amplitude ripple and is not monotonic. In this book, we will refer to the Chebychev I filter simply as Chebychev filter. Any filter with a nonmonotonic stopband response will be classified as an *elliptic* filter.

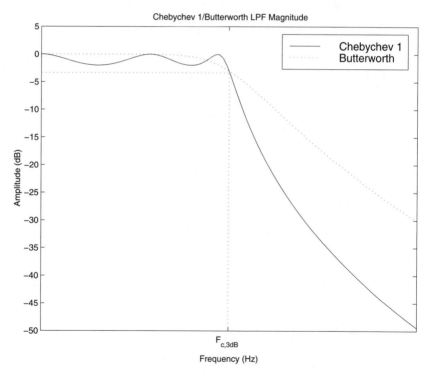

Figure 2-37 *Magnitude response of a 5-pole Chebychev I low-pass Filter.*

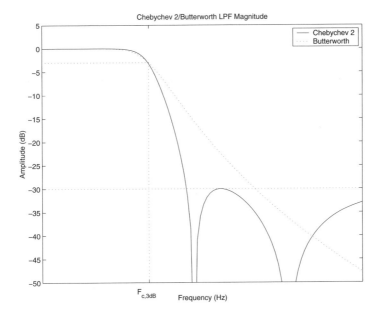

Figure 2-38 Magnitude response of a 5-pole Chebychev II low-pass filter.

Chebychev Characteristics

A Chebychev filter lets us trade passband ripple for steepness of the initial roll-off. The more passband ripple we allow, the faster the initial roll-off. The two cutoff frequencies associated with a Chebychev filter are $f_{c,ER}$ and $f_{c,3dB}$. The $f_{c,3dB}$ is the same half-power cutoff frequency we associated with the Butterworth filter.

We might specify a Chebychev filter with the following statement: a 5-pole 0.1 dB ripple Chebychev low-pass filter with a 25 MHz equal-ripple cutoff frequency. The "0.1 dB" indicates that the passband will exhibit 0.1 dB of ripple. The passband ripple has to be tolerated in order to obtain the steeper initial roll-off the Chebychev will provide. The "25 MHz equal-ripple cutoff frequency" indicates that when the filter passes from the passband into the stopband, it will exhibit 0.1 dB of attenuation at 25 MHz. The attenuation will increase as the frequency increases from that point.

Sometimes the terms *cutoff frequency* (of a Chebychev low-pass filter) and *bandwidth* (of a Chebychev band-pass filter), are used synonymously with *equal-ripple cutoff frequency* and *equal-ripple bandwidth*. In this book, we will specify in each instance which we mean.

Similar to the Butterworth filter, the Chebychev also has a 3 dB cutoff frequency (see Figure 2-39). Since Chebychev filters with more than 3 dB of passband ripple are uncommon, it holds that

$$f_{c,3dB} \geq f_{c,ER} \qquad 2.28$$

The difference between $f_{c,ER}$ and $f_{c,3dB}$ is important when we want to compare a Chebychev filter to a Butterworth. Figure 2-39 shows a 5-pole Chebychev low-pass filter and a 5-pole Butterworth low-pass filter. Both have the same 3 dB cutoff frequency.

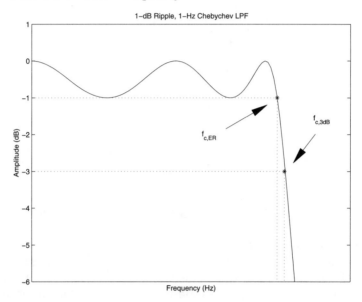

Figure 2-39 *Passband magnitude response of a 5-pole Chebychev low-pass filter.*

As Figure 2-40 shows, the initial roll-off of the Chebychev filter is greater than the Butterworth's. However, the ultimate roll-off of both filters is the same because both are 5-pole filters. At any given frequency in the stopband, the Chebychev filter will offer more attenuation than the Butterworth filter. The difference in attenuation increases if we allow more ripple in the passband of the Chebychev.

Figure 2-41 shows the pole/zero plot of a 5-pole Chebychev low-pass filter. As the Butterworth filter, the Chebychev is an all-pole filter. The poles of a Chebychev low-pass filter lie on an ellipse whose long axis coincides with the $j\omega$-axis. This pole configuration gives the Chebychev its "ripply" magnitude response. As the frequency rises and passes by each pole, the closest pole appears slightly dominant. This in turn causes the magnitude ripple. The closer the poles are to the $j\omega$-axis, the more passband ripple the filter will exhibit. Remember that the poles of the Butterworth low-pass filter lie on a circle centered about the origin, which gives the filter its smooth magnitude plot.

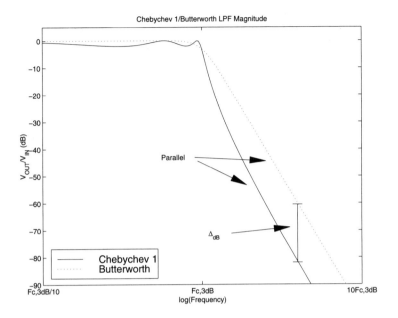

Figure 2-40 *Comparison of the magnitude responses of a 5-pole Butterworth low-pass filter and a 5-pole Chebychev I low-pass filter.*

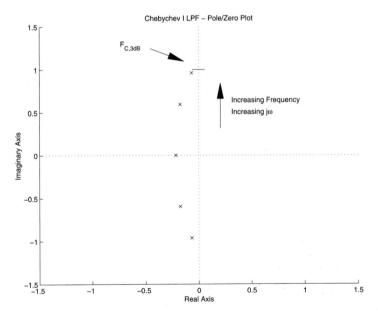

Figure 2-41 *Pole/zero plot of a 5-pole Chebychev I low-pass filter.*

Figure 2-42 shows the pole placement for several Chebychev filters with differing amounts of passband ripple. The poles of the filters with the larger passband ripple are closer to the $j\omega$-axis.

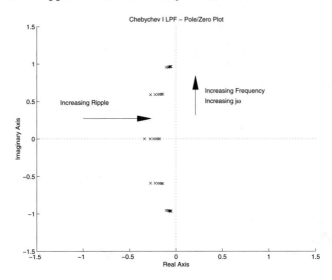

Figure 2-42 *Relationship between Chebychev pole placement and passband ripple.*

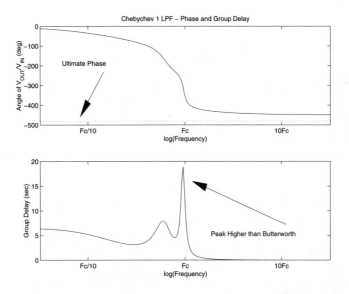

Figure 2-43 *Phase and group delay responses of a typical 5-pole Chebychev I low-pass filter.*

Figure 2-43 shows the phase and group delay responses of a typical Chebychev low-pass filter. Overall, these graphs are similar to the phase and group delay plots of the Butterworth filter.

- There is some measurable phase shift at $f_{c,3dB/1}$, long before any significant attenuation occurs.

- The phase slope is at its steepest in the neighborhood of the 3 dB cut-off frequency $f_{c,3dB}$. Since the group delay is the derivative of the phase plot, it peaks at $f_{c,3dB}$.

- Similar to the Butterworth, when the frequency increases to well beyond $f_{c,3dB}$, the Chebychev filter will eventually reach the ultimate phase given by Equation 2.18.

- Unlike the Butterworth, the phase and group delay response of the Chebychev exhibits ripples due to the pole spacing. The phase ripples cause the group delay of a Chebychev filter to exhibit peaks and valleys. Also, the main group delay peak (around $f_{c,3dB}$) is higher than the one of the Butterworth filter.

- Similar to the Butterworth, the Chebychev filter will eventually reach its ultimate phase at very high frequencies. At $10 f_{c,3dB}$, we have not reached the final phase given by Equation 2.18, and we will still be about 4° per pole away from the ultimate phase at $10 f_{c,3dB}$.

Filters for the Time Domain

The following exemplary discussion of Bessel filters applies also to other time-domain filter types such as Gaussian, Bessel and equi-ripple group delay filters.

Bessel Filters

The magnitude responses of Bessel Filters are poor, but their phase and group delay characteristics are excellent. This means a Bessel filter handles transient phenomena very well. Bessel filters exhibit little or no ringing or overshoot, and their rise time is nearly optimal.

The Bessel filter has a maximally flat group delay response, similar to the magnitude response of a Butterworth filter. It stays flat throughout the filter's passband, then drops off slowly beyond the filter's cutoff frequency. Figure 2-44 shows the phase and group delay response of a Bessel filter.

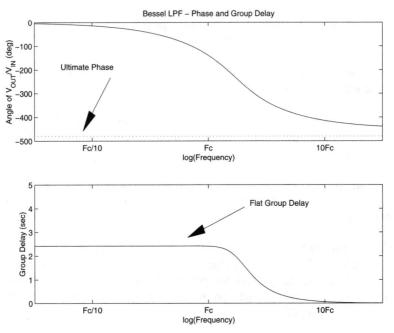

Figure 2-44 *Phase and group delay responses of a typical 5-pole Bessel low-pass filter.*

The magnitude response of a Bessel filter is gentle and rounded. Figure 2-45 shows the magnitude responses of both the Bessel and Butterworth filters for comparison. The Butterworth filter exhibits less attenuation throughout the passband and more attenuation in the stopband than the Bessel filter. We use the 3 dB cutoff frequency ($f_{c,3dB}$) to specify the filter.

Figures 2-45 and 2-44 show the magnitude, phase and group delay responses of a Bessel filter, which is another all-pole filter. Figure 2-46 shows the pole constellation for a 5-pole Bessel filter. Similar to the Chebychev filter, the poles of the Bessel filter lie on an ellipse whose axis is situated along the $j\omega$-axis. Figure 2-47 shows the pole locations for a 5-pole Butterworth; a 5-pole Bessel filter;, a 5 pole 1.0 dB ripple Chebychev; and a 5-pole, 1 dB ripple elliptic filter with four zeroes. As the Butterworth, the pole configuration of the Bessel filter insures no pole is ever dominant and the magnitude response remains gentle and sloping.

Although the magnitude response is poor, the Bessel filter excels in terms of phase and group delay performance (see Figure 2-44). The phase response is very linear throughout the passband. The group delay stays flat until beyond f_c, then drops off gently to zero. Because they exhibit quick rise and fall times, Bessel filters are ideal for handling pulses.

FILTERS | 189

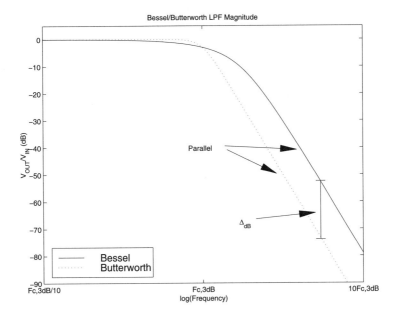

Figure 2-45 *Comparison of the magnitude responses of a typical Butterworth and Bessel low-pass filter.*

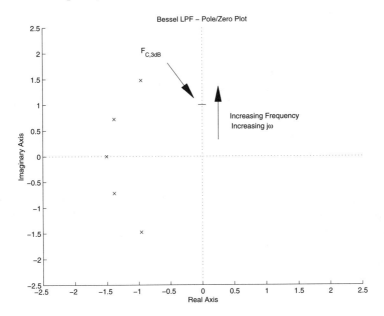

Figure 2-46 *Pole/zero plot of a 5-pole Bessel low-pass filter.*

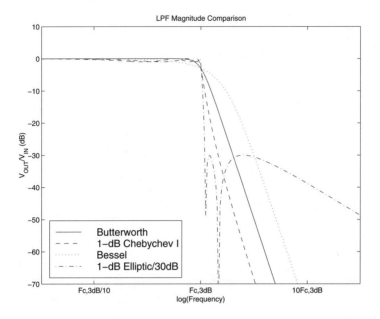

Figure 2-47 Comparison of various low-pass filter magnitude responses.

Other Filters for the Time Domain

The Bessel filter is only one of many filters designed for its time-domain response. Others include

- *Gaussian magnitude response filters.* Both the magnitude response and impulse response of this filter resemble the familiar Gaussian bell curve of statistics, which means there is no ringing or overshoot for transient inputs. However, the magnitude response does not roll off as quickly as a simple Butterworth filter.

- *Equal-ripple group delay filters.* These filters are designed to show a flat group delay response. In other words, the group delay is nearly constant except for a tightly controlled amount of ripple. The group delay response of this filter vaguely resembles the magnitude response of a Chebychev filter. An equal-ripple group delay filter exhibits some of the time-domain properties of a Gaussian or Bessel filter (i.e., good pulse handling capabilities), but the magnitude response is similar to the Butterworth filter. In short, this filter is a compromise between time domain and frequency domain characteristics.

- *Transitional filters.* Transitional filters are arrangements of flat group delay and a rectangular magnitude response. For example, we might design a low-pass filter with a Gaussian passband response. When the attenuation of the filter reaches, for example, 12 dB, we design the filter to exhibit a Chebychev-type roll-off response. The result is a Gaussian-Chebychev filter. Another type of transitional filter is the *Thompson-Butterworth filter.* This filter has a Thompson- or Bessel-type passband response coupled with a Butterworth stopband roll-off. The performance of these filters lies somewhere between the two filters used to generate the transitional filter.

- *Finite-impulse response filters (FIR).* FIR filters are usually built using the tapped delay line topology of Figure 2-48. The signal is fed into a delay line input and sampled at various points along the line. The signal from each delay tap is multiplied by a weight (W_I), then the delayed, weighted signals are added.

We tailor the filter's response by adjusting the complex weights present at each tap. By appropriately adjusting the weights, we can build filters with absolutely flat group delay and very good magnitude characteristics at the same time. In the analog world, we build FIR filters using crystals for the delay lines. Surface acoustic wave or SAW filters are FIR filters. FIR filters are also very common in the digital world.

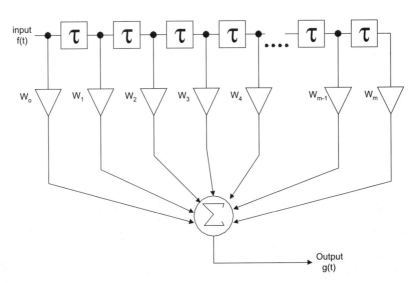

Figure 2-48 Topology of a finite-impulse response (FIR) filter.

Elliptic Filters

The Butterworth, Chebychev I and Gaussian filters we have studied are all-pole filters because their transfer functions contain only poles and no zeroes. All-pole filters have a very high theoretical attenuation at high frequencies, yet their transition from passband to stopband is not always as steep as it is desirable. At times, we can forgo attenuation at very high frequencies to obtain a quicker transition from the passband to the stopband. Note that we have gathered all filters with transfer function zeroes in the elliptic category. Some authors refer to filters exhibiting passband ripples as *Chebychev I*, filters with stopband ripples as *Chebychev II* and filters with both passband and stopband ripples as *elliptic* filters. Elliptic filters are also known as *Cauer* filters.

Elliptic filters contain both poles and zeroes in their transfer functions. The zeroes help the filter achieve a sharp transition into the stopband at the expense of reduced stopband attenuation. An elliptic filter usually requires more components than an equivalent all-pole design.

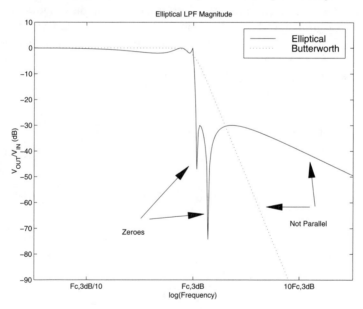

Figure 2-49 *Comparison of the magnitude response of a typical Butterworth low-pass filter and a typical elliptic low-pass filter.*

Figure 2-49 shows the magnitude response of a typical elliptic filter. The passband can contain ripples like the Chebychev or it can be flat like the Butterworth filter. The transfer function zeroes are responsible for the two points of high attenuation in the stopband. By properly placing the poles and zeroes, we can assign almost any amount of initial roll-off.

Figure 2-49 also contains a Butterworth filter response. Both filters shown are 5-pole filters with the same $f_{c,3dB}$, but the elliptical filter contains 4 zeroes in the stopband and has a much smaller transition band than the Butterworth. However, the Butterworth's stopband rejection is superior at high frequencies. The elliptic filter is more suitable when attenuation is needed very close to the passband.

The ultimate roll-off of an elliptic filter is given by Equation 2.14. We used the same equation for the Butterworth, Chebychev and Bessel filters but now have a nonzero number of transfer function zeroes. Since the exemplary elliptic filter has 5 poles and 4 zeroes, its ultimate roll-off will be only 6 dB/octave. The other 5-pole filters produced a 30 dB/octave ultimate roll-off.

Figure 2-50 depicts the pole/zero plot of a 5-pole, 4-zero low-pass filter. For the most part, the pole positions determine the passband characteristics, and the zero positions determine where the points of maximum attenuation occur. The zeroes are positioned on or very near the $j\omega$-axis. This causes the zero to become dominant whenever $j\omega$ is nearby. The dominant zero is responsible for the high attenuation.

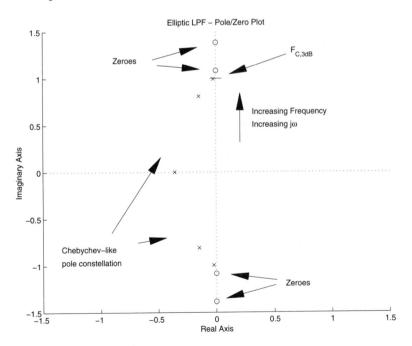

Figure 2-50 *Pole/zero plot of a 5-pole, 4-zero elliptic low-pass filter.*

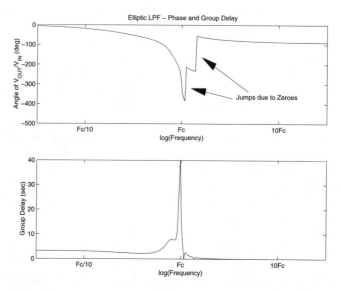

Figure 2-51 *Phase and group delay responses of a 5-pole, 4-zero elliptic low-pass filter.*

The phase response shows sudden 180° jumps at the frequencies of the zeroes, f_{z1} and f_{z2}. Since the zeroes are dominant when $j\omega$ is in the neighborhood of f_{z1} and f_{z2}, they cause a rapid phase change as well as a significant magnitude change. The sudden leaps in the phase response at f_{z1} and f_{z2} produce jumps in the group delay response at the same frequencies. However, since these group delay spikes occur at the frequencies of maximum attenuation, they are not likely to present a problem. Depending on the computer program used to design the filters, the magnitude response of an elliptic filter can be specified in any one of several ways.

- *Specify minimum stopband attenuation*. We can specify that the filter always provides at least A_{min} of rejection throughout the stopband. In Figure 2- 49, we chose 30 dB. The filter algorithm positions the zeroes in a way that by the time the frequency is past the first zero and the attenuation is decreasing, the next zero takes effect, ensuring that the filter's stopband attenuation is always greater than A_{min}.

- *Specify initial roll-off*. First, we specify the filter's cutoff frequency, then we specify a frequency f_{attn} above the filter's cutoff frequency where we want to achieve at least A_{attn} dB of rejection. Then the filter algorithm will put the stopband zeroes wherever it is necessary to achieve the desired attenuation at the desired frequency.

- *Specify the frequencies of the zeroes.* Sometimes it is convenient to be able to specify the frequencies of the zeroes directly. For example, we may want to place the filter zeroes at the image frequency of a receiver. Other possibilities include using one of the zeroes to improve the IF rejection or LO radiation performance of the same receiver. Note that the stopband ripples may not be of uniform depth and the A_{min} will vary. In addition, we no longer have control over the initial roll-off.

- *Complicated.* In the worst cases, we may have to obtain the filter response in a complicated way (perhaps to meet some FCC specification). The problem may place different rejection requirements on certain frequency bands. Advanced computer programs allow us to place the poles and zeroes of the filter directly to achieve almost any arbitrary response.

2.8 Low-Pass Filter Comparison

To complete this discussion, Figures 2-47, 2-52, 2-53 and 2-54 show the magnitude, pole/zero, phase and group delay plots of the following filters:

- Butterworth LPF
- Chebychev I LPF with 1 dB of passband ripple
- Bessel LPF
- Elliptic LPF with 1 dB of passband ripple and 30 dB of stopband attenuation.

In Figure 2-47, we can see that the filters all have the same $f_{c,3dB}$. On this scale, it is difficult to discern any significant differences in their passband responses. However, in the stopband, we can see that the Butterworth, Chebychev and Bessel filters all have the same ultimate roll-off (30 dB/octave). We can also see that although the elliptic filter has a smaller ultimate roll-off (6 dB/octave), it provides the most attenuation immediately above the filter's passband. As advertised, the elliptic filter trades attenuation at very high frequencies for attenuation immediately after $f_{c,3dB}$.

The Butterworth performs in a middle-of-the-road fashion. It is not quite as good as the Chebychev, but better than the Bessel. Figure 2-52 shows the pole/zero plots of the four filters on the same scale. In the Butterworth and Bessel filters, no single pole becomes dominant and the magnitude responses of these filters are smooth and quiet. The poles of both the Chebychev and elliptic filters are closer to the $j\omega$-axis than in the Butterworth or Bessel case. Single poles become slightly dominant and produce passband ripple. The zeroes in the elliptic filter cause the magnitude response of the filter to change to its stopband very quickly. Note the increased group delay and a decreased amount of attenuation at high frequencies.

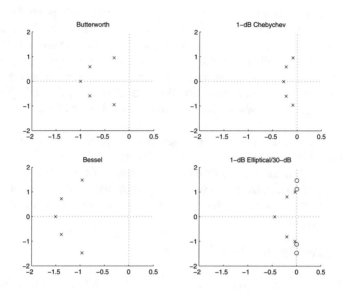

Figure 2-52 Comparison of pole/zero constellations of four low-pass filters.

Figure 2-53 presents the phase response of all four filters. At $f_{c,3dB}/10$, the filters exhibit between 15° and 27° of phase shift. The Bessel filter is the smoothest, followed by the Butterworth and the 1 dB Chebychev. The most abrupt phase plot is the 1.0 dB ripple elliptic filter.

The elliptic filter's phase response shows a sudden 180° jump as the frequency passes through the frequencies of the two zeroes. The zeroes are very close to the $j\omega$-axis and behave as dominant zeroes when the input frequency is close to either of the zero frequencies.

Figure 2-54 shows a plot of the passband group delay of the four filters under discussion. The Bessel filter has the best group delay response; it remains flat well into the filter's stopband. The Butterworth filter's group delay is the next flattest, followed by the 1.0 dB Chebychev filter then the 1.0 dB ripple elliptic filter.

Except for the Bessel filter, the group delay tends to become uneven as we approach the filter's cutoff frequency. In addition, the group delay usually peaks at a cutoff frequency of approximately 3 dB. The 1.0 dB ripple Chebychev filter exhibits group delay ripple. The low-ripple elliptic filter, the Butterworth and the Bessel filter rise only once to hit their single peak at approximately $f_{c,3dB}$.

The sudden phase jumps we observed at the zeroes of Figure 2-53 should produce group delay peaks at the same frequencies. These group delay peaks are not displayed due to an anomaly in the circuit analysis program but do exist. However, since these group delay leaps occur at frequencies of maximum attenuation, they are almost never a problem.

Figure 2-55 shows a close up view of the passband responses of the four filters. The plot shows clearly that the filters all have the same 3 dB cutoff frequency ($f_{c,3dB}$). The Butterworth filter stays reasonably flat throughout most of the passband. The Bessel filter exhibits measurable attenuation even at very low frequencies (for example, below $f_{c,3dB}/5$).

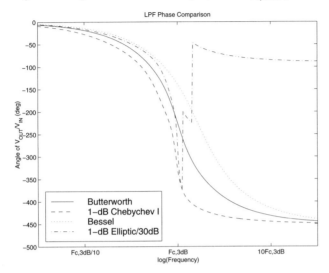

Figure 2-53 *Comparison of phase responses of four low-pass filters.*

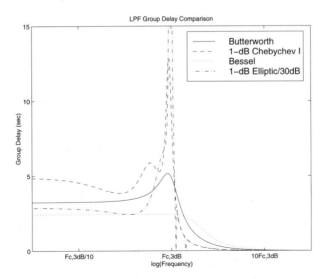

Figure 2-54 *Comparison of passband group delay of four low-pass filters.*

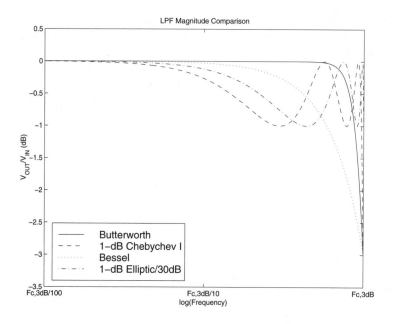

Figure 2-55 *Comparison of passband magnitude responses of four low-pass filters.*

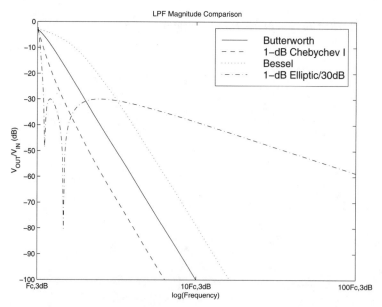

Figure 2-56 *Comparison of stopband magnitude responses of four lowpass filters.*

Figure 2-56 shows the stopband performance of the four filters from $f_{c,3dB}$ to 100 $f_{c,3dB}$. The horizontal scale represents the logarithmic frequency and the vertical scale attenuation. 100 dB is almost certainly beyond the ultimate attenuation of any common filter; 70 to 80 dB is a good rule of thumb. We can easily compare the elliptic filter's stopband performance with the all-pole filters.

The Butterworth, Chebychev and Bessel filters are 5-pole, no-zero filters and exhibit an ultimate roll-off of 30 dB/octave. The 5-pole, 4-zero elliptic filter provides 6 dB/octave of ultimate roll-off. These results are consistent with Equation 2.14.

2.9 Filter Input and Output Impedances

Circuit Realizations

Electronic filter design programs offer several circuit topologies. For example, Figure 2-57 shows two possible realizations of a 0.5 dB ripple Chebychev filter with a 3 dB cutoff frequency of 100 MHz. Both filters are designed to work in a 50 ohm system (i.e., $R_s = R_L = 50$ ohms). Let us assume the components in the filter are lossless and the filter cannot dissipate any energy. Figures 2-58 through 2-60 show the magnitude, phase and transient responses of the two Chebychev filters. The two topologies perform identical functions.

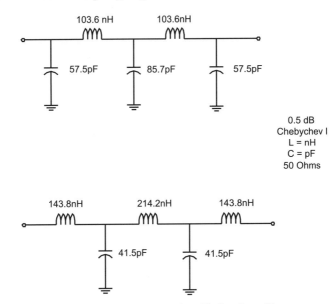

Figure 2-57 *Two practical realizations of a Chebychev filter.*

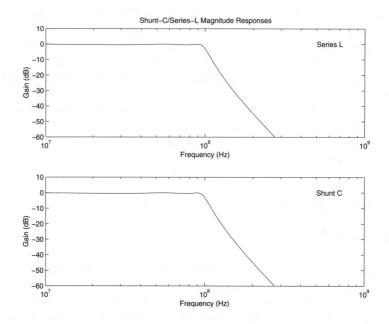

Figure 2-58 *Magnitude response of the two filters of Figure 2-57.*

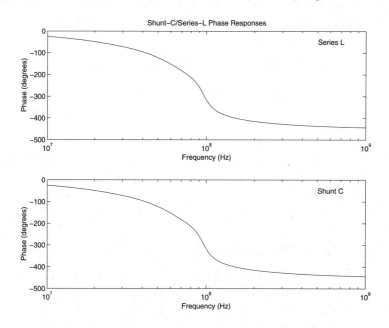

Figure 2-59 *Phase responses of the two filters of Figure 2-57.*

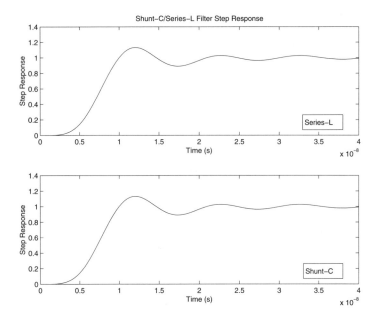

Figure 2-60 *Transient responses of the two filters of Figure 2-57.*

Filter Element Impedances

Consider the impedances of the filter components above and below cutoff. Figure 2-61 shows the series L filter and the reactance of each element when the frequency is well into the passband ($f_{c,3dB}/100 = 1$ MHz). Figure 2-62 shows the same filter when the frequency is well into the stopband ($100 f_{c,3d} = 10$ GHz)

At 1 MHz (Figure 2-61), the series elements exhibit low impedances while the parallel elements show high impedances. As the frequency decreases, the series elements will short-circuit while the shunt elements will open-circuit. At low frequencies, the source and load resistor are connected directly. There is a perfect impedance match between the source and load and maximum power transfer can be achieved.

As Figure 2-62 shows, the series elements become high impedance at 10 GHz. The shunt elements are low impedances and short-circuit. At this frequency, the series elements in the filter stop the signal from propagating. Any signal that passes through the high series impedance enters a node connected to ground through a low impedance. This process of high series impedance, low shunt impedance repeats as the signal continues to move from the source to the load.

This fundamental mechanism of filtering occurs in all lossless filters, including low-pass, high-pass, band-pass and band-stop filters. In the passband, the series elements short-circuit while the shunt element open-circuit. The source and load resistors both see 50 ohms when they look into the filter. The filter is matched to 50 ohms on both the input port and output port of the filter. Since the filter is lossless, maximum power transfer from the source to the load resistor can be experienced.

In a filter's stopband, the series element open-circuit while the shunt element short-circuit. The source and load both see a high impedance when they look into the filter because the first element is a series element. Since the filter is severely mismatched on both the input and output ports, maximum power transfer cannot be achieved. The degree of mismatch determines the amount of power that passes through the filter and also regulates the stopband rejection the filter provides.

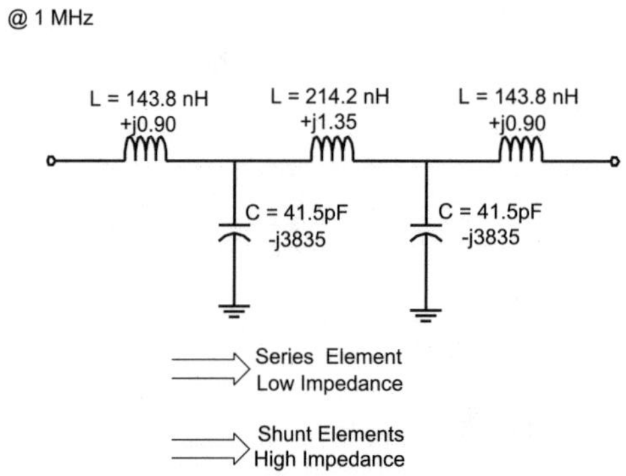

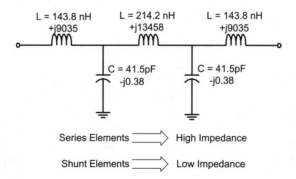

Figure 2-61 *Element impedances of the series-L filter below the cutoff frequency.*

Figure 2-62 *Element impedances of the series-L filter above the cutoff frequency.*

Series Element First, Shunt Element First

When features for a 100 MHz 0.5 dB ripple Chebychev low-pass filter for a 50-ohm system are electronically selected, the program displays the two candidates shown in Figure 2-57. The first filter has a shunt element closest to the source and will present a low impedance in the stopband. The second filter has a series element closest to the source and will present a high impedance to the source in the stopband.

Most filters perform their filtering action through frequency-selective matching. Figures 2-58, 2-59 and 2-60 show that the two filters in Figure 2-57 behave identically. Figure 2-63 shows the input impedance of the two filters plotted over frequency. One filter exhibits a high impedance in its stopband while the other filter exhibits a low impedance. These mismatches reject power from the source and keep power from the load. In summary, we find that

- In the passband, where maximum power transfer through the filter is desirable, the filter "opens up" and connects the load resistor to the source resistor. The filter is well matched on both its input and output ports.

- In the stopband, a filter presents a severe mismatch to both the source and the load resistors. Since the filter ports are no longer matched to the source and load, the filter will accept neither the power from the source nor will it efficiently deliver any power it accepts to the load.

- This is a fundamental filtering mechanism. It is present in low-pass, high-pass, band-pass and band-stop filters.

- The input and output impedances of a filter normally have both resistive and reactive components and change with frequency. The impedances may be expressed in terms of VSWR, return loss, resistance and reactance, or magnitude and phase angle as well as return loss or VSWR.

- Most commonly used filter designs are based on reflective rather than absorptive theory. For example, a lossless filter can have no resistance to absorb power but must attenuate by reflecting power. At the 3 dB passband edge, half of the incident power is reflected; the return loss is already reduced to 3 dB and the VSWR is 5.8:1.

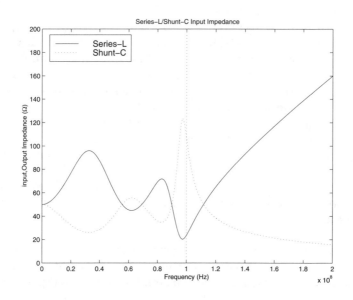

Figure 2-63 *Input impedances of the two filters of Figure 2-57.*

Example 2.4 — Diplexers

Figure 2-64 shows a diplexer that consists of a low-pass filter in parallel with a high-pass filter. It directs signals higher than the cutoff frequency to one port while directing signals lower than the cutoff frequency to the other port.

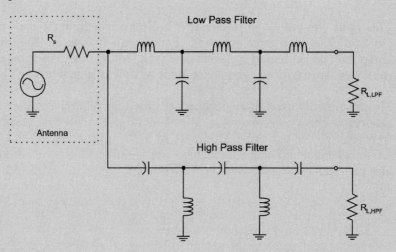

Figure 2-64 *Diplexer consisting of a parallel low-pass and high-pass filter.*

Both the high-pass and low-pass filters have the same f_c. Below f_c, the high-pass filter presents a high impedance to the source, and the low-pass filter is effectively the only device in the circuit. All the energy below f_c is directed to $R_{L,LPF}$. Similarly, above f_c, the low-pass filter becomes high impedance and removes itself from the circuit. All signals pass through the high-pass filter and go to $R_{L,HPF}$. Due to the high-pass/low-pass arrangement, the antenna always sees a matched load when it looks into the diplexer.

Example 2.5 — Elliptic Filter I

Figure 2-65 shows the circuit diagram for an elliptic filter.

a. Is this a high-pass filter or a low-pass filter?
b. In the filter's stop-band, does this filter present a high- or low-impedance load to the source resistor R_s?
c. In the stopband, does the load resistor see a high- or low-impedance when it looks into the filter?

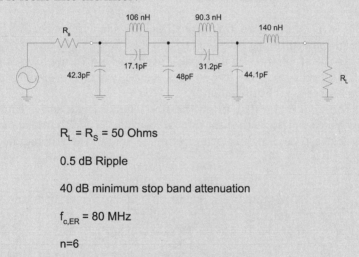

$R_L = R_S = 50$ Ohms

0.5 dB Ripple

40 dB minimum stop band attenuation

$f_{c,ER} = 80$ MHz

n=6

Figure 2-65 *Elliptic filter circuit diagram (Example 2.5).*

Solution —

Bearing in mind that the impedance of a capacitor approaches zero at high frequencies and the impedance of an inductor approaches infinity at high frequencies, we know that

a. The three shunt capacitors and the series inductor next to R_L indicate that this is a low-pass filter.
b. The shunt capacitor on the filter's source end indicates the filter will present a low impedance to the source in the stopband.

c. The series inductor on the filter's load end means the filter will present a high impedance to the load in the stopband.

Example 2.6 — Elliptic Filter II
Figure 2-66 shows the circuit diagram for an elliptic filter.

a. Is this a low-pass or a high-pass filter?
b. In the filter's stopband, does this filter present a high- or low-impedance to the source resistor R_s?
c. In the filter's stopband, does this filter present a high- or low-impedance to the load resistor R_L?

Solution —
Bearing in mind that the impedance of a capacitor approaches zero at high frequencies and the impedance of an inductor approaches infinity at high frequencies, we know that

a. The three shunt inductors and the series capacitor next to R_L indicate a high-pass filter.
b. Since the stopband of a high-pass filter is at low frequencies, the behavior of the filter for low frequencies is significant. The element closest to the source is a shunt inductor.
c. At the load end of the filter, the first element is a series capacitor. At low frequencies, a capacitor appears similar to a high impedance, so the filter presents a high impedance to the load in the stopband.

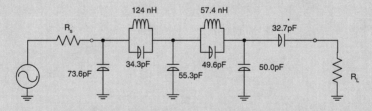

R_L = 31 Ohms

R_S = 50 Ohms

0.25 dB ripple

35 dB min stopband attenuation

$f_{c,ER}$ = 108 MHz

Figure 2-66 *Elliptic filter circuit diagram (Example 2.6).*

Example 2.7 — Band-Pass Filter
Figure 2-67 shows a Butterworth band-pass filter. A band-pass filter has two stopbands: one below and one above the center frequency of the filter.

a. Does the filter present a high or low impedance to the outside world in the filter's lower stopband?
b. Does the filter present a high or low impedance to the outside world into the filter's upper stopband?

Solution —
a. The shunt-C/series-C configuration on the filter's input and output ports present a high impedance in the filter's lower stopband".
b. In the upper stopband, the shunt-C/series-C configuration present a low impedance to the world.

No matter how complex the filter or how it is realized (transmission line inductors and cavity resonators, for example), we can usually determine its out-of-band characteristics by examining the elements closest to the source and load.

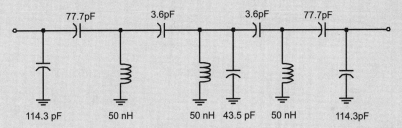

Figure 2-67 Butterworth band-pass filter circuit diagram (Example 2.7).

2.10 Transient Response of Filters

Figure 2-68 shows a typical transient event measurement. The signal source V_s has been turned off for a long time and the transients in the filter have receded. When the signal source is turned on quickly we observe the voltage across R_L.

Figure 2-69 shows another typical occurrence. We have been feeding a signal into the filter for a long time. We then abruptly change the input signal. The frequency of the new input signal is f_2, the new amplitude is A_2 and the new phase is θ_2. The two situations are equivalent. In order to to determine how the filter reacts to signal changes, several questions are relevant.

- What is the shape of the output waveform? Does it rise slowly, without ringing, or does it ring for a long time?

- How fast can we expect the filter to react to input changes? How long before the output is a reasonable representation of the filter's input?

- What transient responses can we expect out of the different filter types? How do they differ? What is the best filter to use in a particular situation?

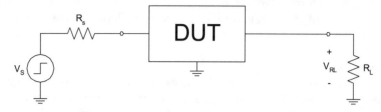

DUT = Device Under Test
     = The low pass filter we are investigating

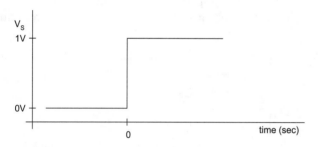

Examine V_{RL} in the time domain

$R_L = R_s = 50$ Ohms

Figure 2-68 One type of transient event.

Transient Response of Low-Pass Filters

Figure 2-70 shows the output characteristics of the four low-pass filters: a Butterworth, a 1.0 dB ripple Chebychev I, a 0.1 dB ripple elliptic filter with 55 dB of stopband attenuation and a Bessel filter. All filters have a 100 MHz 3 dB cutoff frequency. The Bessel filter is the quickest to rise. It approaches its final value quickly and with very little ringing. It has a relatively poor magnitude response and does not filter very well. The Butterworth reaches its final value after the Bessel, followed closely by the 0.1 dB passband ripple elliptic filter and the 1.0 dB ripple Chebychev. Each of these three filters exhibits significant ringing.

FILTERS | 209

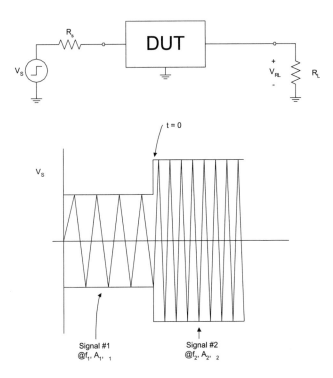

Figure 2-69 *A second type of transient event.*

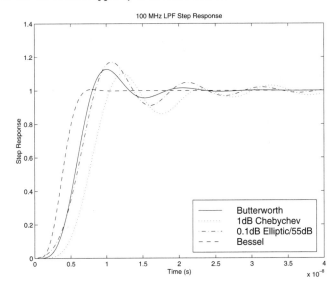

Figure 2-70 *Comparison of the transient responses of four low-pass filters.*

210 | RADIO RECEIVER DESIGN

Rules of Thumb

Generally, the more ripple the filter exhibits in its passband, the longer the filter will take to respond to a change in its input. In Figure 2-70, the Bessel settles out first with a minimum of ringing, followed by the Butterworth (with no passband ripple). Then the 0.5 dB ripple elliptic filter settles followed by the 1.0 dB ripple Chebychev. Also, the sharper the transition band, the longer a filter will take to settle out after its input changes. Note that the stopband response can play a role in the settling time. The elliptic filter has the highest overshoot but still settles quicker than the Chebychev. A filter built with high Q-poles and zeroes will ring longer than one built from low Q poles and zeroes. A high Q-pole is a pole placed very close to the $j\omega$-axis.

Williams [9], Blinchikoff and Zverev [1], and Zverev [10] provide many graphs depicting transient responses of various filters. Their books are an excellent source of information on filtering of all types.

Transient Response of Band-Pass Filters

In various situations, the transient response of a band-pass filter becomes an important issue. In a spectrum analyzer, for example, the settling time of the narrowest filter usually determines how fast we can sweep through a piece of spectrum. A radar receiver must also process pulsed signals to produce its output.

Figure 2-71 shows the schematic diagrams of four band-pass filters. As in the low-pass filter case, we have a Butterworth, 1.0 dB ripple Chebychev, 0.1 dB ripple elliptical with 40 dB of minimum attenuation and a Bessel filter. Each filter contains three poles; all have a 10 MHz 3 dB bandwidth with a 100 MHz center frequency. We designed all of the filters to work in a 50- ohm system. Figure 2-72 shows the magnitude responses of these four filters.

Figure 2-73 is a diagram of the transient response test. We will apply an impulse to the filter at time T = 0 and observe the output of the filter. Figures 2-74 through 2-77 show the voltage across the load resistor for the four filter cases. The time-domain responses of the four band-pass filters are strikingly similar to the time-domain responses of the four low-pass filters. References [9], [10] and [1] expand on filter transient responses and filters in general.

FILTERS | 211

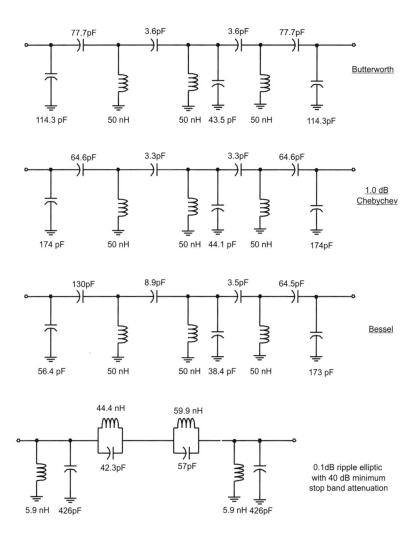

Figure 2-71 *Circuit diagrams of four band-pass filters.*

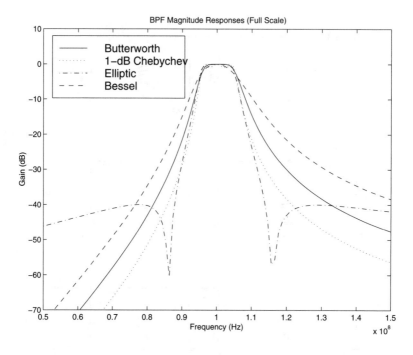

Figure 2-72 *Comparison of the magnitude responses of four low-pass filters*

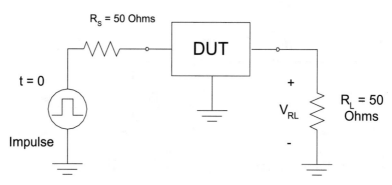

DUT = Device Under Test
= Filter whose transient response we want to measure

Figure 2-73 *Transient response test schematic.*

FILTERS | 213

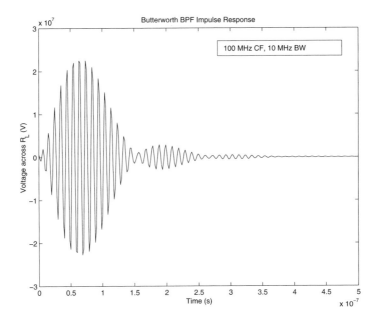

Figure 2-74 *Impulse response of Butterworth band-pass filter.*

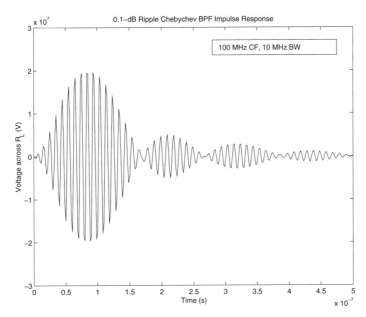

Figure 2-75 *Impulse response of Chebychev band-pass filter.*

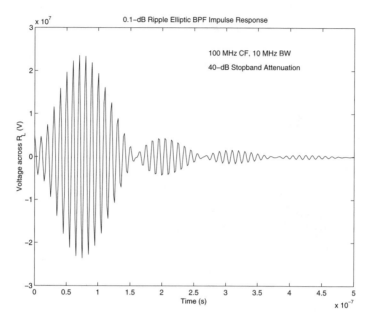

Figure 2-76 *Impulse response of elliptic band-pass filter.*

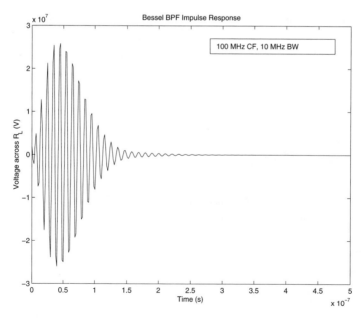

Figure 2-77 *Impulse response of Bessel band-pass filter.*

2.11 Band-Pass Filters

Band-pass filters are by far the most common filters used in receiver design.

Band-Pass Filter Terminology

Figures 2-78 and 2-79 show the magnitude response of a typical band-pass filter. We will define the following terms:

- IL_{dB}. Insertion loss in dB. This is the minimum amount of attenuation present in the passband. The minimum value for IL_{dB} usually lies in the neighborhood of the filter's center frequency.

- $f_{U,3dB}$, $f_{L,3dB}$. These are the upper and lower 3 dB passband frequencies of the filter. They are the frequencies where the filter provides 3 dB of attenuation below the insertion loss of the filter (see Figure 2-78).

- $f_{U,ER}$, $f_{L,ER}$. These are the upper and lower equal-ripple passband frequencies of a Chebychev band-pass filter. They are equivalent to the equal-ripple cutoff frequency of a Chebychev low-pass filter. Note that the upper and lower equal-ripple frequencies are defined with the filter's insertion loss in mind (see Figure 2-79).

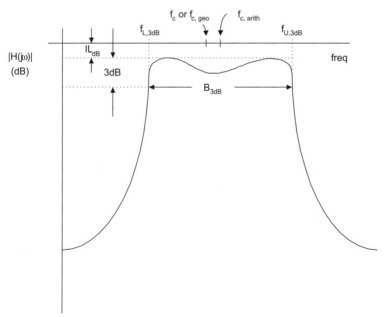

Figure 2-78 Band-pass filter 3 dB bandwidth definitions.

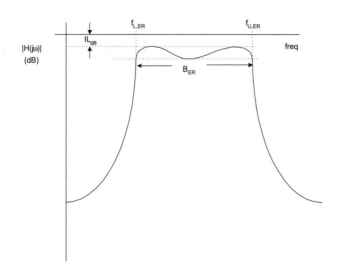

Figure 2-79 *Band-pass filter equal-ripple bandwidth definitions.*

Band-Pass Filter Center Frequencies

- f_c, $f_{c,geo}$. The filter's geometric center frequency. The geometric center frequency of a band-pass filter is not equal to the filter's arithmetic center frequency. The geometric center frequency of a band-pass filter follows the geometric relationship

$$\frac{f_U}{f_{c,geo}} = \frac{f_{c,geo}}{f_L} \qquad 2.29$$

which can be simplified to

$$f_c = f_{c,geo} = \sqrt{f_U f_L} \qquad 2.30$$

The upper and lower frequencies f_U and f_L specified in Equations 2.29 and 2.30 can be either the 3 dB frequencies or the equal-ripple frequencies.

- $f_{c,arith}$. The filter's arithmetic center frequency (often confused with the geometric center frequency).

$$f_{c,arith} = \frac{f_U + f_L}{2} \qquad 2.31$$

Again, the f_U and f_L values in Equation 2.31 can be either the 3 dB values or the equal ripple values. Note that

$$f_{c,arith} > f_{c,geo} \qquad 2.32$$

Band-Pass Filter Bandwidths

Figure 2-80 illustrates several of the most common bandwidth definitions.

- B_{3dB}. This is the 3 dB bandwidth of the filter. The frequencies where the filter's attenuation is 3 dB greater than its insertion loss are $f_{L,3dB}$ and $f_{U,3dB}$. The 3 dB bandwidth is

$$B_{3dB} = |f_{U,3dB} - f_{L,3dB}| \qquad 2.33$$

- B_{6dB}. This is the 6 dB bandwidth of the filter. We find the frequencies that are 6 dB down from the filter's insertion loss point to determine $f_{L,6dB}$ and $f_{U,6dB}$. The 6 dB bandwidth is

$$B_{6dB} = |f_{U,6dB} - f_{L,6dB}| \qquad 2.34$$

- B_{ER}. This is the equal-ripple bandwidth of a Chebychev filter. The equal-ripple bandwidth of the filter is

$$B_{ER} = |f_{U,ER} - f_{L,ER}| \qquad 2.35$$

where $f_{L,ER}$ and $f_{U,ER}$ are the equal-ripple cutoff frequencies.

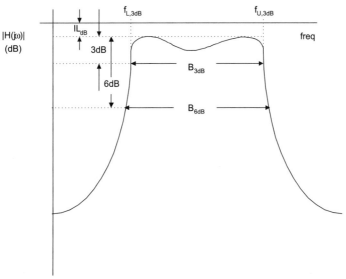

Figure 2-80 *Definition of a band-pass filter 3 dB and 6 dB bandwidths.*

218 | RADIO RECEIVER DESIGN

Figure 2-81 shows the most general case of bandwidth definition. At any particular bandwidth defined by points of equal attenuation (x dB in the case of Figure 2-81), we can write

$$\frac{f_{U,X-dB}}{f_{c,geo}} = \frac{f_{c,geo}}{f_{L,X-dB}} \qquad 2.36$$

or

$$f_c = f_{c,geo} = \sqrt{f_{U,X-dB} f_{L,X-dB}} \qquad 2.37$$

We can also write

$$B_{X-dB} = |f_{U,X-dB} - f_{L,X-dB}| \qquad 2.38$$

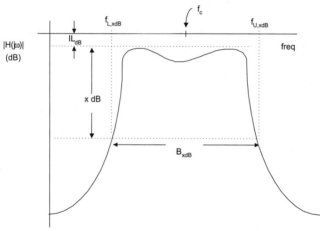

Figure 2-81 *Definition of band-pass filter x dB bandwidth.*

Example 2.8 — Band-Pass Filters

Given a filter with a geometric center frequency (f_c) of 510 MHz, a 75 MHz 3 dB bandwidth and a 2 dB insertion loss, find
a. $f_{L,3dB}$ and $f_{U,3dB}$
b. $f_{c,arith}$.

Solution —
See Figure 2-82.
a. Substituting f_c = 510 MHz and B_{3dB} = 75 MHz into Equations 2.30 and 2.31 produces

$$f_{L,3dB}^2 + 75f_{L,3dB} - 510 = 0$$
$$\Rightarrow f_{L,3dB} = 473.9 \, MHz \quad \quad 2.39$$

Substituting $f_{L,3dB}$ = 473.9 MHz into Equation 2.33 produces $f_{U,3dB}$ = 548.9 MHz. Using Equation 2.30 reveals

$$\sqrt{(473.9)(548.9)} = 510 \, MHz \quad \quad 2.40$$

b. Using equation 2.31, we find

$$f_{c,arith} = \frac{(f_{L,3dB} + f_{U,3dB})}{2} \quad \quad 2.41$$
$$= \frac{(473.9 + 548.9)}{2}$$
$$= 511 \, MHz$$

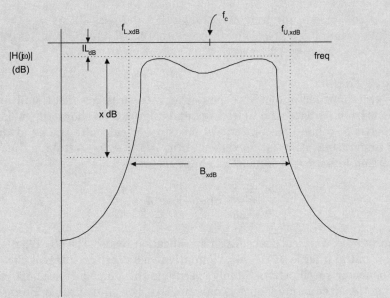

Figure 2-82 *Figure for Example 2.8.*

Example 2.9 — Band-Pass Filters

The following band-pass filters all have a geometric center frequency of 700 MHz but different 3 dB bandwidths. Find the difference between $f_{c,geo}$ and $f_{c,arith}$:

a. $f_{L,3dB}$ = 682.7 MHz, $f_{U,3dB}$ = 717.7 MHz. This filter's 3 dB bandwidth is 5% of its center frequency,

b. $f_{L,3dB} = 665.9$ MHz, $f_{U,3dB} = 735.9$ MHz. This filter's 3 dB bandwidth is 10% of its center frequency,
c. $f_{L,3dB} = 633.5$ MHz, $f_{U,3dB} = 773.5$ MHz. This filter's 3 dB bandwidth is 20% of its center frequency,
d. $f_{L,3dB} = 432.6$ MHz, $f_{U,3dB} = 1133$ MHz. This filter's 3 dB bandwidth is 100% of its center frequency.

Solution —
Using Equation 2.31, we can show
a. $f_{c,arith} = 700.2$ MHz. The difference between f_c and $f_{c,arith}$ is 0.2 MHz = 0.029%.
b. $f_{c,arith} = 700.9$ MHz. The difference between f_c and $f_{c,arith}$ is 0.9 MHz 0.13%.
c. $f_{c,arith} = 703.5$ MHz. The difference between f_c and $f_{c,arith}$ is 3.5 MHz = 0.50%.
d. $f_{c,arith} = 782.8$ MHz. The difference between f_c and $f_{c,arith}$ is 82.8 MHz = 11.8%.

In conclusion, for narrow bandwidth band-pass filters, $f_{c,geo} = f_{c,arith}$ with less than 0.5% error. For filters with 100% bandwidth or less, $f_{c,geo} = f_{c,arith}$ with less than 12% error.

Shape Factor

The shape factor of a band-pass filter measures how fast the filter transitions from its passband to its stopband. Figure 2-83 shows the magnitude response of a band-pass filter. We measure or calculate the bandwidth of the filter at two attenuation values: $Attn_{1,dB}$ and $Attn_{2,dB}$ (taking the filter's insertion loss into account). The shape factor is

$$\frac{Attn_{1,dB}}{Attn_{2,dB}} \text{ Shape Factor } = \frac{B_2}{B_1} \qquad 2.42$$

For example, a typical shape factor specification reads "The 60 dB/3 dB shape factor of the filter is 3.3." The attenuation values are an integral part of the shape factor specification. Simply specifying the shape factor as 4.4 without giving the attenuation values is meaningless. For all realizable filters,

$$\text{Shape Factor} > 1 \qquad 2.43$$

The 60 dB/3 dB shape factor and the 60 dB/6 dB shape factor are two common specifications.

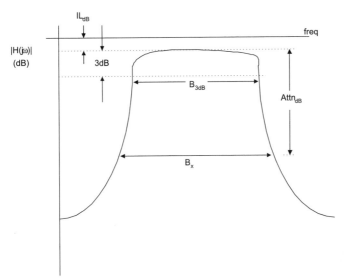

Figure 2-83 Band-pass filter shape factor.

Example 2.10 — Shape Factor of a Band-Pass Filter

Figure 2-84 shows the magnitude plot of a Chebychev band-pass filter. We designed this filter to pass the 88 to 108 MHz commercial FM broadcast band. It is a 7-pole 0.5 dB ripple filter with an equal-ripple bandwidth of 20 MHz. This model includes the effects of component Q (or internal filter losses). Find
a. the 60 dB/3 dB shape factor,
b. the 60 dB/6 dB shape factor.

Solution —

Measuring from Figure 2-84, we find that the 3 dB bandwidth is about 41.25 MHz, the 6 dB bandwidth is about 43.1 MHz and the 60 dB bandwidth is 86.25 MHz.
a. 60 dB/3 dB shape factor is

$$\frac{60\ dB}{3\ dB}\text{ Shape Factor} = \frac{86.25}{41.25} = 2.1 \qquad 2.44$$

b. 60 dB/6 dB shape factor is

$$\frac{60\ dB}{6\ dB}\text{ Shape Factor} = \frac{86.25}{43.1} = 2.0 \qquad 2.45$$

The shape factor is used to describe filters that have actually been built.

As such, shape factor considers actual effects such as insertion loss and component tolerance.

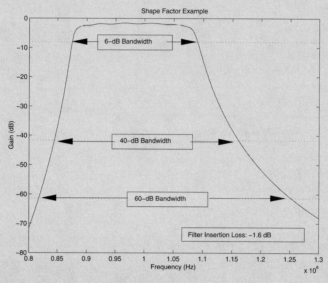

Figure 2-84 *Figure for Example 2.10.*

Comparison of Band-Pass Filter Types

We will examine a Butterworth, a 1.0 dB ripple Chebychev, a Bessel and an elliptic filter. The filters are all 5 pole filters centered at 100 MHz with a 3 dB bandwidth of 10 MHz. We designed the filters for a 50-ohm system. The elliptic filter contains 5 poles, 4 zeroes and has 40 dB of minimum stopband attenuation (A_{min}). These design criteria produce transmission zeroes at 87.5 MHz and 118 MHz.

Figure 2-85 shows the magnitude responses of the four filters swept from 50 to 150 MHz. We can see that the band-pass filters mirror the behavior of their low-pass equivalents. All of the comparisons we made when we discussed the corresponding low-pass filters apply to band-pass filters as well. Note the steep roll-off of the elliptical filter and its narrow transition bands.

Figure 2-86 shows the passband characteristics of the four band-pass filters. Figure 2-87 shows the stopband characteristics. Figure 2-88 shows the group delay responses of the filters. As in the low-pass case, the Bessel is the flattest in the passband and the 1.0 dB Chebychev is the lumpiest.

Note that the characteristics of a particular low-pass filter carry over into the band-pass realizations. The characteristics also carry into the high-pass and band-stop cases.

FILTERS | 223

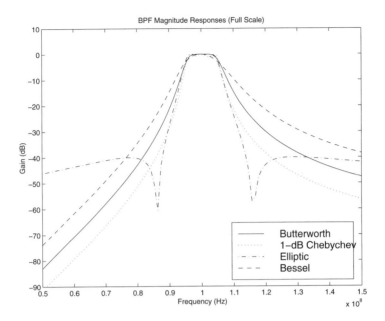

Figure 2-85 *Comparison of the magnitude responses of four band-pass filters.*

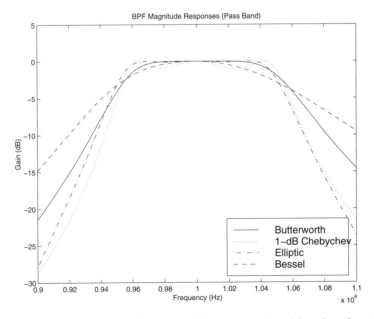

Figure 2-86 *Comparison of passband characteristics of four band-pass filters.*

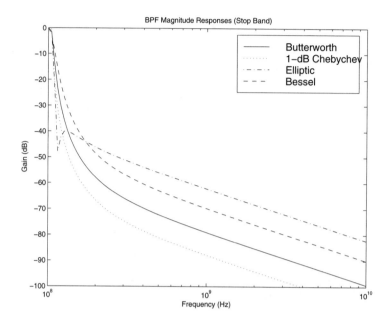

Figure 2-87 Comparison of stopband characteristics of four band-pass filters.

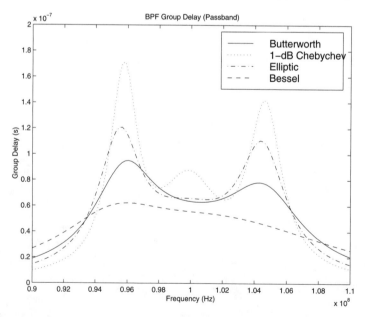

Figure 2-88 Comparison of group delay characteristics of four band-pass filters.

2.12 Other Filters

Some filtering problems may not fit neatly into one of the Butterworth, Chebychev, Gaussian or elliptic filter solutions we have presented here and other filter realizations have to be found.

Band-Pass and Band-Stop Filter

Consider a system that contains a fixed-tuned receiver and a transmitter. The receiver tunes to 650 MHz while the transmitter tunes to 750 MHz (see Figure 2-89). Due to system constraints, the antennae of the receiver and transmitter are in close proximity; accordingly, the receiving antenna will collect a considerable amount of energy at 750 MHz.

We would like to place a filter between the receiver and antenna that has a minimum insertion loss at 650 MHz, yet provides high isolation at 750 MHz. Clearly, this filter does not fit conveniently into any of the standard filters we have discussed and should be a combination band-pass/band-stop filter. We have not placed any restrictions on the type of passband response we require.

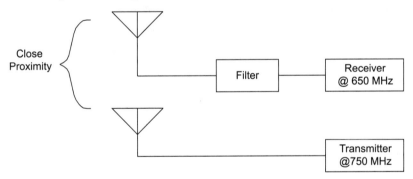

Figure 2-89 Receiver/transmitter block diagram.

Figure 2-90 shows the schematic diagram of the starting point, a Butterworth band-pass filter centered at approximately 650 MHz. The 3 dB bandwidth was approximately 30 MHz, or 5%. Figure 2-91 shows the magnitude response. The filter exhibits only 10 dB of rejection at 750 MHz, which is insufficient.

Each 2.5 pF capacitor in Figure 2-90 was first series-resonated with an inductor to produce the circuit of Figure 2-92. We set the resonant frequency to 750 MHz because, at resonance, a series LC configuration becomes a short circuit. The two circuit nodes were grounded at 750 MHz while passing 650 MHz. This procedure placed transmission zeroes at 750

MHz (see Figure 2-93). Although high attenuation was achieved at 750 MHz, the passband has shifted down in frequency. This new filter also exhibits an undesired passband at 1700 MHz. Figure 2-94 shows the next iteration. We have added a low-pass filter and used the computer to optimize the components for the best overall response. The final response of the filter is shown in Figure 2-95. Although this filter is far from being ideal, it nevertheless illustrates that we do not have to be restricted to one of the filter types we have described here. If necessary, an individualized filter can be created to achieve exactly the desired response.

650 MHz Band Pass Filter

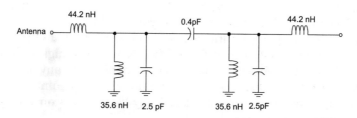

Figure 2-90 Circuit diagram of a 650 MHz band-pass filter.

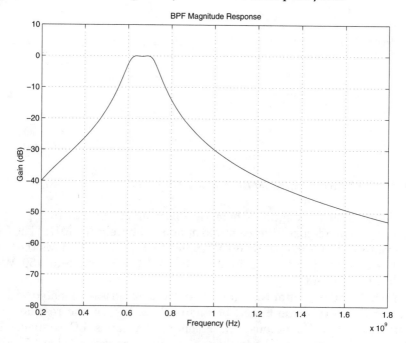

Figure 2-91 Magnitude response of the filter in Figure 2-90.

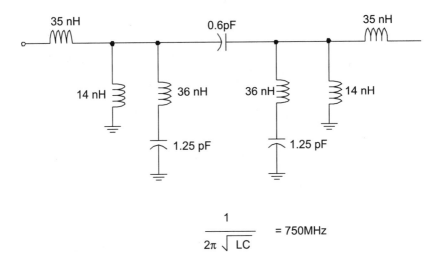

$$\frac{1}{2\pi \sqrt{LC}} = 750 \text{MHz}$$

Figure 2-92 *650 MHz band-pass filter with transmission zeros at 750 MHz.*

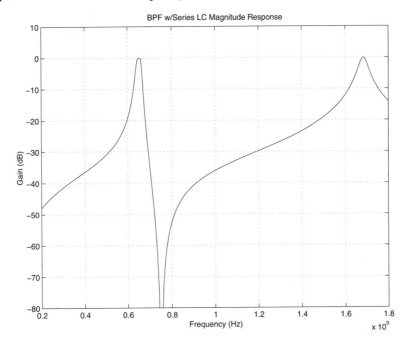

Figure 2-93 *Magnitude response of the filter in Figure 2-92.*

228 | RADIO RECEIVER DESIGN

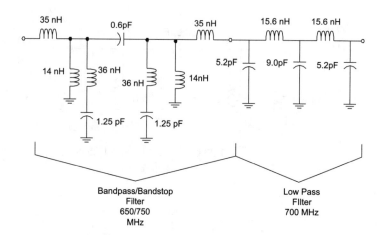

Figure 2-94 *Filter in Figure 2-92 with follow-on low-pass filter.*

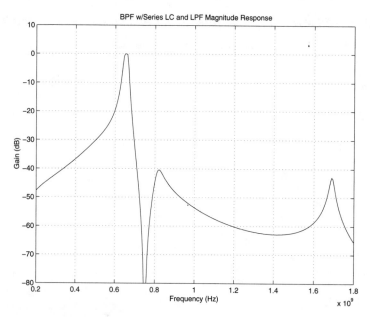

Figure 2-95 *Magnitude response of the filter in Figure 2-94.*

Amplitude Equalizer

A filter can be used to equalize the amplitude response of a system with a nonflat response. For example, Figure 2-96(a) shows the amplitude response of a channel. The attenuation increases with increasing frequency. Figure 2-96(b) shows the appropriate amplitude equalizer. The slope of the equalizer is the inverse of the slope of the channel.

Figure 2-96(c) shows the equalized channel. Note that the total insertion loss has increased over the entire band, but the equalized channel has a flat passband.

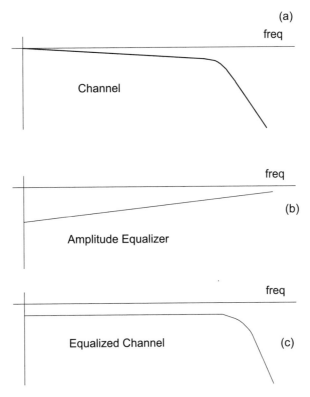

Figure 2-96 Amplitude Equalizer (a) unequalized channel, (b) equalizer response, (c) equalized channel.

2.13 Noise Bandwidth

A radio receiver's ultimate sensitivity or noise floor is a strong function of the noise bandwidth of the receiver's final IF filter. *Noise bandwidth* is a filter parameter. It allows us to determine how much noise power a filter will pass based upon its passband shape. Although low-pass, high-pass and band-stop filters also possess noise bandwidths, the concept is most useful when we discuss band-pass filters. Many of the concepts introduced here are covered in more depth in Chapters 5 and 7.

In this section, we assume we are dealing with spectrally flat, Gaussian noise. Figure 2-97 shows the problem in a simplified block diagram. The signal source and its associated source resistor generate a signal and noise, respectively. The spectrum present at the input of the filter is shown in Figure 2-98. The filter allows the signal and a certain amount of noise to pass to the load resistor. The spectrum present at the load resistor R_L is shown in Figure 2-99. The band-pass filter is wide enough to pass the entire signal. We want to know the amount of noise that passes through the filter. This is the noise that will compete with the signal for control of the demodulator. Clearly, the amount of noise present at the load resistor is a function of the filter's passband shape. Figure 2-100 shows the spectral shape of the noise dissipated in R_L for various types of band-pass filters.

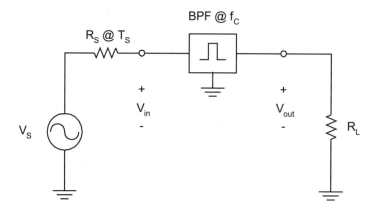

Figure 2-97 *Noise analysis of a band-pass filter.*

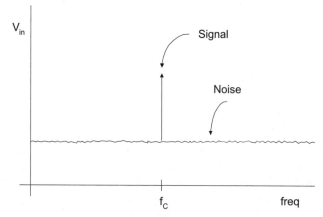

Figure 2-98 *Signal and noise present at the input of the band-pass filter of Figure 2-97.*

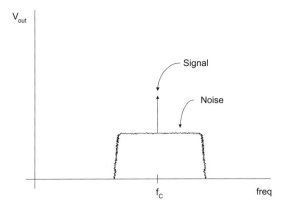

Figure 2-99 Signal and noise present across the load resistor (at the output of the band-pass filter) of Figure 2-97.

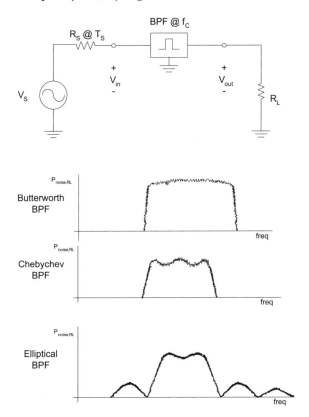

Figure 2-100 Noise present across the load resistor for various types of band-pass filters.

For the sake of analysis, the arbitrary band-pass filter in Figure 2-100 is replaced with a filter that has the same insertion loss yet a perfectly rectangular passband (see Figure 2-101). If this rectangular filter has the same noise bandwidth (B_n) as the first filter, the load resistor will dissipate the same amount of noise energy. In other words, the noise bandwidth of any arbitrary filter equals the bandwidth of an ideal rectangular band-pass filter with the same insertion loss. The ideal rectangular filter will allow the same amount of noise power to pass as the original filter allowed. In Figure 2-101, the load resistor will dissipate the same amount of noise power in cases (a) and (b). Given an arbitrary filter shape, we are interested in finding the noise bandwidth B_n.

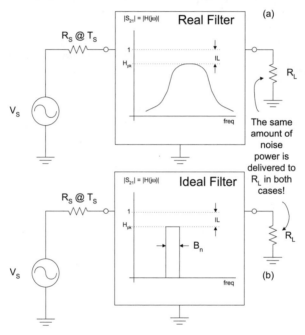

Figure 2-101 *Definition of noise bandwidth.*

Noise Bandwidth Calculation

Figure 2-101 (a) shows a filter of arbitrary shape presented with noise that has a one-sided power spectral density η. The noise presented to the load resistor is

$$N_{arbitrary} = \int_0^\infty \eta |H(j\omega)| \partial f$$
$$= \eta \int_0^\infty |H(j\omega)| \partial f \qquad 2.46$$

When $|H(j\omega)|$ is a perfect rectangular filter [see Figure 2-101(b)], the noise power in the load resistor is

$$N_{\text{rectangular}} = \eta \int_{f_c - B_n/2}^{f_c + B_n/2} H_{pk} \partial f$$
$$= \eta H_{pk} B_n$$

(2.47)

Equating 2.46 with Equation 2.47 produces

$$B_n \left(\frac{\text{rad}}{\text{sec}}\right) = \frac{1}{H_{pk}} \int_0^\infty |H(j\omega)| \partial \omega$$

$$B_n \text{ (Hz)} = \frac{1}{H_{pk}} \int_0^\infty |H(j\omega)| \partial f$$

(2.48)

Equation 2.48 is valid if the filter's transfer function is presented to us as a power ratio (i.e., $|S_{21}|^2$ or $20\log|S_{21}|$). If the data represents a voltage ratio or a voltage transfer function, i.e., $|S_{21}|$, the noise bandwidth is

$$B_n \left(\frac{\text{rad}}{\text{sec}}\right) = \frac{1}{H_{pk}} \int_0^\infty |H(j\omega)|^2 \partial \omega$$

$$B_n \text{ (Hz)} = \frac{1}{H_{pk}} \int_0^\infty |H(j\omega)|^2 \partial f$$

(2.49)

The most difficult part of evaluating the integral is deciding which type of data we have, then verifying we have the correct units.

Example 2.11 — Noise Bandwidth Calculation
Figure 2-102 shows the linear voltage transfer function of a 10.7 MHz IF band-pass filter with a 3 dB bandwidth of around 350 kHz. Find the noise bandwidth by graphical integration.

Solution —
Since we have a voltage transfer function plot, we use Equation 2.49. We will perform the integration by finding the height of each 50 kHz column and squaring the height. Then we will multiply the height by the width (in hertz) of each column. Finally, we will add these numbers to obtain the noise bandwidth. Figure 2-103 shows some of the definitions that are used.

$$H_{avg} = \frac{a+b}{2}$$
and
$$W_{Hz} = 50 \text{ kHz}$$

(2.50)

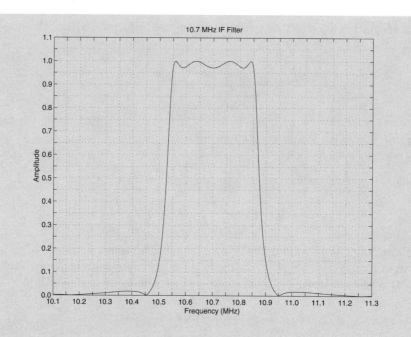

Figure 2-102 *10.7 MHz IF filter magnitude response (linear units).*

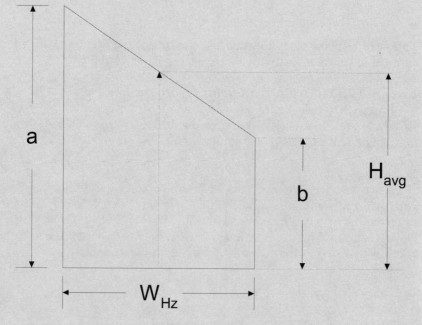

Figure 2-103 *Area of trapezoidal for trapezoid integration.*

Table 2-1 Numerical integration of filter passband.

	a	b	H_{avg}	$H_{avg}^2 \cdot W_{Hz}$
1	0.0033	0.0008	0.0021	0.2205
2	0.0008	0.0023	0.0016	0.128
3	0.0023	0.0062	0.0043	0.9245
4	0.0062	0.0107	0.0085	3.6125
5	0.0107	0.0155	0.0131	8.5805
6	0.0155	0.0176	0.0166	13.778
7	0.0176	0.0022	0.0099	4.9005
8	0.0022	0.1357	0.069	238.05
9	0.1357	0.9716	0.5537	15329.1845
10	0.9716	0.9779	0.9748	47511.752
11	0.9779	0.9958	0.9869	48698.5805
12	0.9958	0.9716	0.9837	48383.2845
13	0.9716	0.9967	0.9842	48432.482
14	0.9967	0.9770	0.9869	48698.5805
15	0.9770	0.9716	0.9743	47463.0245
16	0.9716	0.1472	0.5594	15646.418
17	0.1472	0.0009	0.0741	274.5405
18	0.0009	0.0173	0.0091	4.1405
19	0.0173	0.0162	0.0168	14.112
20	0.0162	0.0119	0.0141	9.9405
21	0.0119	0.0076	0.0098	4.802
22	0.0076	0.0038	0.0057	1.6245
23	0.0038	0.0007	0.0023	0.2645
24	0.0007	0.0018	0.0013	0.0845
Sum				320743.01

The noise bandwidth of this filter is about 320 kHz.

Noise Bandwidth of Various Band-Pass Filters

Figure 2-104 shows the noise bandwidth of various band-pass filters. The following rule of thumb can be derived.

Often the noise bandwidth of any arbitrary band-pass filter is approximated by the filter's 3 dB bandwidth. In truth, the noise bandwidth is usually somewhere between the filter's 3 dB and 6 dB points.

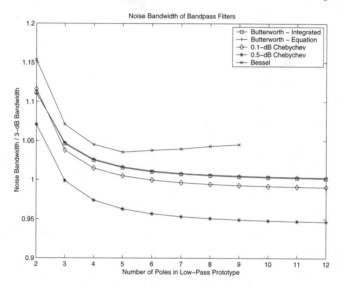

Figure 2-104 *Noise bandwidth of various band-pass filters.*

2.14 Butterworth Filters in Detail

Butterworth filters are simple and common filters. Their equations are straightforward and generally intuitive. This is a great benefit when things are not working quite right and a design has to be improved. We can easily answer questions like, How many more poles do I need? or, What is the attenuation at this frequency?

- Although the numbers we receive will not be exact, they will usually be accurate enough. If we want to be more accurate, we can always resort to computer simulation.

- Without actually going to the trouble of building your filter or simulating it on a computer, it is nearly impossible to judge what effect realization factors, such as component losses or realization approximations, will have on the filter. Even if exact equations were used, they probably would not accurately reflect reality.

- Filter synthesis algorithms may indicate that unrealizable components will be required and the filter topology may have to be changed with various circuit transforms to produce a more realizable design. As a result, however, the mathematical purity of the filter is lost. Although the passband has not changed much in appearance, the stopband can change dramatically. For example, the transformed filter might have better rejection at low frequencies than the original design but poorer rejection at higher frequencies. Since the filter's performance has changed, and since there is not a convenient or precise way to describe the new filter mathematically, it is necessary to use the Butterworth equations.

Butterworth Low-Pass Filters

Pole Positions

The poles of a Butterworth filter lie on a circle centered about the origin. This pole configuration will not allow any pole to be dominant, so the magnitude response of a Butterworth filter is maximally flat or monotonic. The positions of the poles for a Butterworth low-pass filter with a 3 dB cutoff frequency of f_c are

$$p_k = 2\pi f_{c,3dB} \left[-\sin\left(\frac{(2k-1)\pi}{2N}\right) + j\cos\left(\frac{(2k-1)\pi}{2N}\right) \right] \qquad 2.51$$

where
N = the number of poles in the filter
$k = 1, 2, ..., N$.

The circle containing the Butterworth poles intersects the $j\omega$-axis at the radian 3 dB cutoff frequency $\omega_{c,3dB} = 2\pi f_{c,3dB}$.

Example 2.12 — Butterworth Pole Positions

Find the pole positions for a 5-pole Butterworth low-pass filter with a 50 MHz 3 dB cutoff frequency.

Solution —

Using $f_{c,3dB}$ = 50 MHz, $N = 5$ and $k = 1...5$, Equation 2.51 produces
$p_1 = -97.08 \cdot 10^6 + j298.8 \cdot 10^6$
$p_2 = -254.2 \cdot 10^6 + j184.7 \cdot 10^6$
$p_3 = -314.2 \cdot 10^6 + j0.000$

$p_4 = -254.2 \cdot 10^6 - j184.7 \cdot 10^6$
$p_5 = -97.08 \cdot 10^6 - j298.8 \cdot 10^6$

Figure 2-105 shows a plot of these poles. Note that the circle containing the filter poles intersects the $j\omega$-axis at $\omega_{c,3dB} = 314.2 \cdot 10^6$ radian/second $= 2\pi(50 \cdot 10^6)$ hertz.

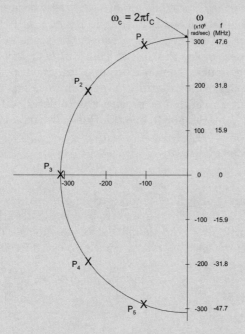

Figure 2-105 *Pole positions for a 5-pole, 50 MHz Butterworth low-pass filter.*

Magnitude Response

Figure 2-106 shows the magnitude response of a Butterworth filter. The response proceeds without ripples and the attenuation increases steadily as the operating frequency increases beyond the 3 dB cutoff frequency $f_{c,3dB}$. The attenuation for a lossless low-pass Butterworth filter is

$$Attn_{dB} = 10\log\left[1 + \left(\frac{f_x}{f_{c,3dB}}\right)^{2N}\right] \qquad 2.52$$

where
 N = the number of poles in the filter,
 $f_{c,3dB}$ = the filter's 3 dB cutoff frequency,
 f_x = the frequency of interest.

Equation 2.52 describes the attenuation characteristic for a lossless Butterworth low-pass filter. The lossless approximation is usually accurate enough for practical, realizable filters, even those with insertion loss. However, insertion loss usually has to be considered when we are dealing with band-pass filters.

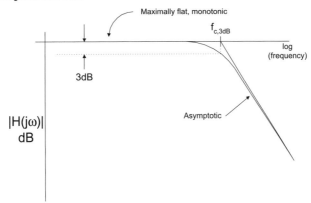

Figure 2-106 Magnitude response of a Butterworth low-pass filter.

Example 2.13 — Butterworth LPF Attenuation
We find a 6-pole Butterworth low-pass filter with 3 dB of attenuation at 120 MHz. Find the attenuation at 650 MHz.

Solution —
We know
N = the number of filter poles = 6,
$f_{c,3dB}$ = the 3 dB cutoff frequency = 120 MHz,
f_x = the frequency of interest = 650 MHz.
Substituting these values into Equation 2.52, we find

$$Attn_{dB} = 10\log\left[1+\left(\frac{650}{120}\right)^{2(6)}\right] \qquad 2.53$$

$$= 88 \text{ dB}$$

Example 2.14 — Butterworth LPF Attenuation
We have measured a Butterworth LPF with $f_{c,3dB}$ = 6.6 MHz. We have also measured an attenuation of 30 dB at about 20.9 MHz. How many poles are in the filter?

Solution —
We know that the filter's attenuation at 20.9 MHz is 30 dB, so

$f_{c,3dB}$ = the filter's 3 dB cutoff frequency = 6.6 MHz,
f_x = the evaluation frequency = 20.9 MHz,
$Attn_{dB}$ = 30 dB at 20.9 MHz.

Substituting these values into Equation 2.52 and solving for N, we find

$$30 \text{ dB} = 10\log\left[1 + \left(\frac{20.9}{6.6}\right)^{2N}\right] \quad \text{2.54}$$

$$\Rightarrow 999 = (3.167)^{2N}$$

$$\Rightarrow N = 3 \text{ Poles}$$

Butterworth Band-Pass Filters

Magnitude Response

Using Figure 2-107 as a guide, the absolute attenuation for a lossy Butterworth band-pass filter is

$$Attn_{Abs,dB} = 10\log\left[1 + \left(\frac{B_x}{B_{3dB}}\right)^{2N}\right] + IL_{dB} \quad \text{2.55}$$

where

$Attn_{Abs,dB}$ = the absolute amount of attenuation the filter provides at B_x,
N = the number of poles in the filter,
B_{3dB} = the 3 dB bandwidth of the filter,
B_x = the bandwidth of interest,
IL_{dB} = the insertion loss of the filter in dB.

The relative attenuation is the difference between the absolute attenuation and the insertion loss or

$$Attn_{Rel,dB} = Attn_{Abs,dB} - IL_{dB}$$

$$\Rightarrow Attn_{Rel,dB} = 10\log\left[1 + \left(\frac{B_x}{B_{3dB}}\right)^{2N}\right] \quad \text{2.56}$$

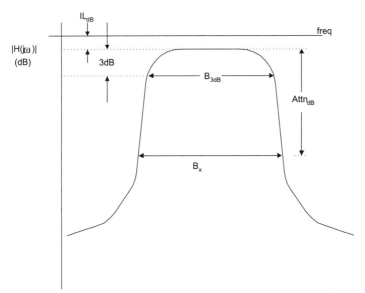

Figure 2-107 Magnitude response of a lossy Butterworth band-pass filter.

Example 2.15 — Butterworth BPF Attenuation

Given a 7 pole band-pass filter with a center frequency of 1.2 GHz, a 3 dB bandwidth of 340 MHz and an insertion loss of 3.4 dB, find the absolute and relative attenuation at a bandwidth of
a. 500 MHz,
b. 780 MHz.

Solution —

Using Equations 2.55 and 2.56 with $N = 7$, $B_{3dB} = 340$ MHz, $IL_{dB} = 3.4$ dB and
a. $B_x = 500$ MHz. The absolute attenuation is

$$Attn_{Rel,dB} = 10 \log \left[1 + \left(\frac{500}{340} \right)^{2(7)} \right] + 3.4 \qquad 2.57$$
$$= 26.9 \text{ dB}$$

We use Equation 2.56 to determine the relative attenuation.

$$Attn_{Rel,dB} = 26.9 - 3.4$$
$$= 23.5 \text{ dB} \qquad 2.58$$

b. $B_x = 780$ MHz. The absolute attenuation is

$$Attn_{\text{Abs,dB}} = 10\log\left[1 + \left(\frac{780}{340}\right)^{2(7)}\right] + 3.4 \qquad 2.59$$
$$= 53.9\,\text{dB}$$

The Equation 2.56 can also be used to to determine the relative attenuation.

$$Attn_{\text{Rel,dB}} = 53.9 - 3.4 \qquad 2.60$$
$$= 50.5\,\text{dB}$$

Example 2.16 — Shape Factor of a Butterworth Band-Pass Filter

Given a 6 pole Butterworth band-pass filter with a 275 MHz 3 dB bandwidth,
a. find the 60 dB/6 dB shape factor,
b. find the 40 dB/6 dB shape factor.

Solution —

Since this is a relative measurement, we will use the equation for relative attenuation. Substituting $N = 6$, $B_{3dB} = 275$ MHz and $Attn_{dB} = 6$ dB, 40 dB and 60 dB into equation 2.56 and solving for B_{6dB}, B_{40dB} and B_{60dB} produces

$$6\,\text{dB} = 10\log\left[1 + \left(\frac{B_{6dB}}{275}\right)^{2(6)}\right]$$
$$\Rightarrow B_{6dB} = 301\,\text{MHz}$$
$$40\,\text{dB} = 10\log\left[1 + \left(\frac{B_{40dB}}{275}\right)^{2(6)}\right] \qquad 2.61$$
$$\Rightarrow B_{6dB} = 592\,\text{MHz}$$
$$60\,\text{dB} = 10\log\left[1 + \left(\frac{B_{60dB}}{275}\right)^{2(6)}\right]$$
$$\Rightarrow B_{6dB} = 870\,\text{MHz}$$

a. The 60 dB/6 dB shape factor is

$$\frac{B_{60dB}}{B_{6dB}} = \frac{870}{301} = 2.89 \qquad 2.62$$

b. The 40 dB/6 dB shape factor is

$$\frac{B_{40\text{dB}}}{B_{6\text{dB}}} = \frac{592}{301} = 1.97 \qquad 2.63$$

Equations 2.55 and 2.56 are accurate, but only marginally useful. Normally, the filter's geometric center frequency (f_c or $f_{c,geo}$) and the filter's 3 dB bandwidth (B_{3dB}) are known. We have to calculate the filter's attenuation at some frequency (f_x). Combining Equations 2.38, 2.55 and 2.56, we can show that the absolute attenuation of a Butterworth band-pass filter at any frequency f_x is

$$Attn_{Abs,dB} = 10\log\left[1 + \left(\frac{\left|f_x - \frac{f_{c;geo}^2}{f_x}\right|}{B_{3dB}}\right)^{2N}\right] + IL_{dB} \qquad 2.64$$

The relative attenuation is

$$Attn_{Rel,dB} = 10\log\left[1 + \left(\frac{\left|f_x - \frac{f_{c;geo}^2}{f_x}\right|}{B_{3dB}}\right)^{2N}\right] \qquad 2.65$$

where
$Attn_{Abs,dB}$ = the absolute amount of attenuation the filter provides at f_x,
$Attn_{Rel,dB}$ = the relative amount of attenuation the filter provides at f_x,
N = the number of poles in the filter,
B_{3dB} = the 3 dB bandwidth of the filter,
$f_{c,geo}$ = the band-pass filter's center frequency,
f_x = the frequency of interest,
IL_{dB} = the insertion loss of the filter in dB. The insertion loss is the minimum amount of signal loss the filter will provide.

Example 2.17 — *Butterworth Filter Equations*
Derive Equation 2.64.

Solution —
Looking at Equations 2.55 and 2.64, we need to express B_x in terms of $f_{c,geo}$ and f_x (see Figure 2-108). Note that there are two frequencies, $f_{U,X\text{-}dB}$

and $f_{L,X\text{-}dB}$, which apply to the bandwidth B_x and

$$B_x = f_{U,X-dB} - f_{L,X-dB} \qquad 2.66$$

Let us examine the upper frequency ($f_{U,X\text{-}dB}$). Using Equation 2.30 and Equation 2.66, we find

$$f_{L,X-dB} = \frac{f_{c,geo}^2}{f_{U,X-dB}} \qquad 2.67$$

and

$$B_x = f_{U,X-dB} - \frac{f_{c,geo}^2}{f_{U,X-dB}} \qquad 2.68$$

Combining Equation 2.68 with Equation 2.55 produces

$$Attn_{\text{Abs,dB}} = 10\log\left[1 + \left(\frac{\left|f_{U,X-dB} - \dfrac{f_{c,geo}^2}{f_{U,X-dB}}\right|}{B_{3dB}}\right)^{2N}\right] + IL_{dB} \qquad 2.69$$

When we follow a similar derivation and express Equation 2.68 in terms of $f_{L,X\text{-}dB}$ we can write

$$Attn_{\text{Abs,dB}} = 10\log\left[1 + \left(\frac{\left|\dfrac{f_{c,geo}^2}{f_{L,X-dB}} - f_{L,X-dB}\right|}{B_{3dB}}\right)^{2N}\right] + IL_{dB} \qquad 2.70$$

Replacing $f_{L,X\text{-}dB}$ and $f_{U,X\text{-}dB}$ by f_X and taking the absolute value of the appropriate expression produces

$$Attn_{\text{Abs,dB}} = 10\log\left[1 + \left(\frac{\left|f_x - \dfrac{f_{c,geo}^2}{f_x}\right|}{B_{3dB}}\right)^{2N}\right] + IL_{dB} \qquad 2.71$$

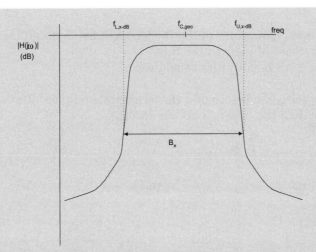

Figure 2-108 Magnitude response of a band-pass filter.

Example 2.18 — Lossy Butterworth Band-Pass Filter

Find the total attenuation a signal at 1.75 GHz will experience after passing through a 3-pole Butterworth band-pass filter. The insertion loss of the filter is 4 dB at the center frequency, the center frequency is 2.7 GHz and the filter's 3 dB bandwidth is 1.1 GHz.

Solution —

The problem is shown in Figure 2-109. Using Equation 2.64 with $N = 3$, $f_{c,geo} = 2.7$ GHz, $IL_{dB} = 4$ dB, we can write

$$Attn_{Abs,dB} = 10\log\left[1 + \left(\frac{\left|1.75 - \frac{2.7^2}{1.75}\right|}{1.1}\right)^{2N}\right] + 4 \qquad 2.72$$

$$= 20.5 + 4$$
$$= 24.5 \text{ dB}$$

Noise Bandwidth

We will assume we are using Butterworth filters when we are evaluating a system. The noise bandwidth of a Butterworth band-pass filter is

$$B_n = B_{3dB} \frac{\pi/(2N)}{\sin(\pi/(2N))} > B_{3dB} \qquad 2.73$$

where
B_n = the noise bandwidth of the Butterworth band-pass filter,
N = the number of poles in the filter,
B_{3dB} = the 3 dB bandwidth of the filter.

Note that the noise bandwidth of a Butterworth band-pass filter is always slightly larger than the filter's 3 dB bandwidth.

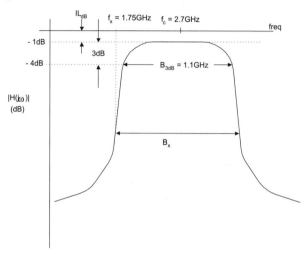

Figure 2-109 Magnitude response of a band-pass filter.

Example 2.19 — Butterworth Noise Bandwidth
Using Equation 2.73, we built the following table.

Table 2-2 Noise bandwidth of Butterworth filters.

# of Poles	B_n/B_{3dB}
1 (1-Pole RC Filter)	1.57
2	1.11
3	1.05
4	1.03
5	1.02
6	1.01
7	1.008
8	1.006
9	1.005
10	1.004

Butterworth High-Pass Filter

Magnitude Response

Figure 2-110 shows the magnitude response for a Butterworth high-pass filter. We find that

$$Attn_{dB} = 10\log\left[1 + \left(\frac{f_{c,3dB}}{f_x}\right)^{2N}\right] \qquad 2.74$$

where
 N = the number of poles in the filter,
 $f_{c,3dB}$ = the filter's 3 dB cutoff frequency,
 f_x = the frequency of interest.

Equation 2.74 is very similar to Equation 2.52 which describes the attenuation characteristics of a Butterworth low-pass filter.

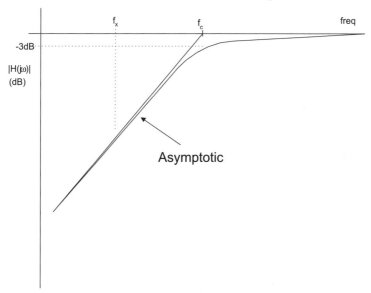

Figure 2-110 Magnitude response of a Butterworth highpass filter.

Example 2.20 — Butterworth High-Pass Filter

Given a 3 pole Butterworth high-pass filter with a 3 dB cutoff frequency of 110 MHz, find the attenuation at
a. the CB band (27 MHz),
b. the FM radio station at 88.3 MHz.

Solution —

Applying Equation 2.74 with $f_c = 110$ MHz, $N = 3$ and

a. $f_x = 27$ MHz produces

$$Attn_{dB} = 10\log\left[1 + \left(\frac{110}{27}\right)^{2(3)}\right] \qquad 2.75$$

$$= 36.6 \text{ dB}$$

b. $f_x = 88.3$ MHz produces

$$Attn_{dB} = 10\log\left[1 + \left(\frac{110}{88.3}\right)^{2(3)}\right] \qquad 2.76$$

$$= 6.8 \text{ dB}$$

Butterworth Band-Stop Filters

Magnitude Response

The attenuation of a Butterworth band-stop filter is

$$Attn_{dB} = 10\log\left[1 + \left(\frac{B_{3dB}}{B_x}\right)^{2N}\right] \qquad 2.77$$

or,

$$Attn_{dB} = 10\log\left[1 + \left(\frac{B_{3dB}}{\left|f_x - \frac{f_{c,geo}^2}{f_x}\right|}\right)^{2N}\right] \qquad 2.78$$

where
  $Attn_{dB}$ = the absolute amount of attenuation the filter provides at f_x,
  N = the number of poles in the filter,
  B_{3dB} = the 3 dB bandwidth of the filter,
  $f_{c,geo}$ = the band-pass filter's center frequency,
  f_x = the frequency of interest.

Note the similarity to Equations 2.56 and 2.65 for the Butterworth band-pass filter (see Figures 2-111 and 2-112 for details on the variables).

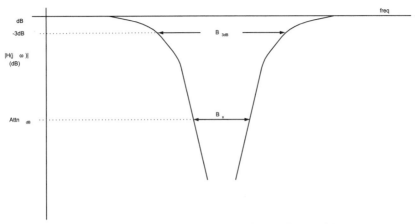

Figure 2-111 FM broadcast band Butterworth band-stop filter parameters.

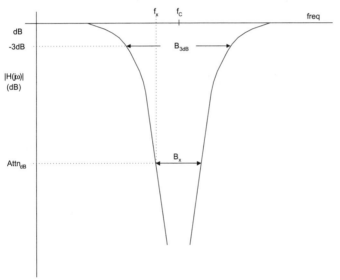

Figure 2-112 Bandwidth and frequency definitions for Equation 2.77.

Example 2.21 — *Butterworth Notch Filter*
We need a notch filter to provide at least 10 dB of rejection throughout the FM broadcast band. If we use an 8-pole Butterworth filter, what is the 3 dB bandwidth of the filter?

Solution —
The FM broadcast band covers 88 to 108 MHz. We need at least 10 dB of rejection at both 88 and 108 MHz so the 10 dB bandwidth is

$$B_{10dB} = 108 - 88$$
$$= 20 \text{ MHz} \qquad (2.79)$$

We use Equation 2.77 to determine B_{3dB}.

$$10 \text{ dB} = 10 \log\left[1 + \left(\frac{B_{3dB}}{20}\right)^{2(8)}\right] \qquad (2.80)$$

$$\Rightarrow B_{3dB} = 22.9 \text{ MHz}$$

Insertion loss is usually not a problem in low-pass, high-pass or notch filters. If necessary, the insertion loss can be added to any of the attenuation equations.

2.15 Filter Technologies and Realizations

As we write this (in the late 1990s), filter technologies are changing at a rapid pace. New materials and technologies have made filters smaller, exhibit less loss and perform better than ever. Any advice given regarding filter realization would be dated before this book got to press. Thus, we ask you to turn to the filter vendors and their catalogs for realization advice. Some classic technologies, however, will remain workhorses over time. We have included an introduction of those techniques and their characteristics. Be aware that this data may be dated by the time you read this.

Characteristics of Classic Realizations

We have chosen the following six design technologies to spotlight:

- LC filters,
- Monolithic crystal filters,
- Discrete crystal filters,
- Cavity filters,
- Ceramic Resonator filters,
- Surface Acoustic Wave (SAW) filters
- Ceramic filters.

Although there are other techniques available such as BAW, mechanical and stripline, these technologies cover a center frequency range from 10 kHz to 18 GHz with bandwidths from 100 Hz to 10 GHz.

LC Filters

LC filters are the workhorses of the filter industry because they are versatile, easy to build on the bench and well-understood. Moreover, LC filters are also very cost-effective. LC filters do not exhibit the close-in spurious responses that plague crystal filters and will normally be smaller (but lossier) than distributed-element filters. However, LC filters do not possess the stability or the high Q of crystal filters or distributed element filters, which limits their minimum bandwidth capability.

Crystal Filters

The resonators of crystal filters consist of quartz crystals. They offer very low loss (i.e., high Q or quality factor) and high stability. These resonator characteristics allow crystal filters to exhibit very narrow bandwidth band-pass filters with reasonably low loss. Crystal filters can be either monolithic or discrete. The resonant elements in monolithic filters are built on the same piece of quartz ; the elements in a discrete filter are packaged separately and then connected to form the filter.

Monolithic Crystal Filters
Monolithic crystal filters are usually smaller and less expensive than discrete crystal filters because they use fewer crystal elements for the same number of resonators. They also tend to be more reliable because they use fewer components. Additionally, monolithic crystal filters do not use hybrid coils and tend to be more stable and have less flat loss than most discrete crystal filters. In the VHF range, the monolithic approach allows bandwidths on overtones that can only be realized with costly fundamental mode resonators if a discrete resonator design is used.

Discrete Crystal Filters
Discrete-resonator crystal filters have better power handling capabilities than monolithic crystal filters. They are usually a better design choice for very narrow or very wide bandwidths. In addition, discrete crystal filters allow more flexibility concerning the choice of network topology. This allows the design of networks that have sharply asymmetrical performance (required for single sideband work), which is difficult to achieve with monolithic designs.

Spurious Responses in Crystal Filters
Figure 2-113 shows the passband response of a typical crystal filter. The 30 MHz passband is well-defined. This particular filter exhibits re-entrant responses at 340 and 400 MHz. These spurious responses are common in crystal band-pass filters and are caused by unwanted resonances in

the crystals that were used to build the filter. The AT-crystal cut, which is most commonly used for filters, has a family of unwanted responses at frequencies slightly above the desired resonance and at approximately odd harmonic frequencies of the fundamental resonance. We can often suppress the overtone responses (which cause the response at three times the filter center frequency) with additional LC filtering but this increases the loss, complexity and size of the filter. However, we cannot filter the close-in spurious responses.

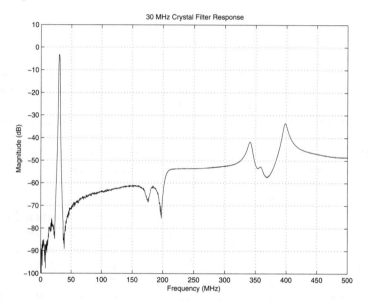

Figure 2-113 Bandwidth and frequency definitions for Equation 2.78.

Ceramic Filters

Ceramic filters are very similar to crystal filters. The resonant elements are realized with piezoelectric materials although the quality factors of ceramics are much smaller than those of quartz. Ceramic filters excel in the low-cost, physically small arena. They are designed for consumer goods such as FM radios, televisions and cordless telephones. A ceramic filter should be the choice if a filter at a standard center frequency and bandwidth is needed.

Cavity Filters

A cavity filter replaces the resonant LC structure of a discrete filter with a resonant cavity. We couple the energy from cavity to cavity by placing apertures between the cavities. The center frequency of the filter is inversely proportional to the size of the cavity; the bandwidth is proportional to the aperture size between cavities. Thus the ultimate limit on center frequency, bandwidth and stability is directly related to manufacturing tolerances and mechanical stability over time and temperature. Cavity filters exhibit higher Qs and therefore lower loss and more selectivity than LC filters. They also tend to be more stable than LC filters and are easier to manufacture than combline filters.

Ceramic Resonator Filters

Ceramic resonator filters are similar to cavity filters except that the cavity is filled with dielectric material to decrease the size of the resonator. We can couple the resonators using the same technique utilized in cavity filters, or we can externally couple the elements, which affords a wider choice of filter topologies. The dielectric increases the loss of the filter over the cavity filter. We can tune some ceramic resonator filters by moving a dielectric "slug" into and out of the resonant cavity. Ceramic resonator filters are very popular for front-end preselection filters in consumer electronics such as cellular telephones and wireless LANs.

Surface Acoustic Wave (SAW) Filters

Surface acoustic wave filters convert the signal to be filtered into an acoustic waveform and then launch the acoustic energy into a crystal medium. The response of the filter results from the structure used to launch and collect the acoustic energy. Since these structures are deposited directly on the crystal medium, the designer determines the properties of the filter.

SAW filters can operate in any of several modes; consequently, the bandwidth, insertion loss and center frequency relationships are quite complicated. However, we can realize rather sophisticated designs using SAW filters. Building a custom filter can be expensive and time-consuming but many old designs can help to accommodate most needs.

2.16 Miscellaneous Items

Isolators

High-performance filters are often equipped with isolators on both the input and output ports. The isolators assure that the filter is always terminated in the proper impedance. This guarantees that the critical performance of the filter will not be spoiled by outside forces.

Nonlinearities

Like any other component, a filter acts as a linear device if the input signal power is small enough. However, the filter may exhibit nonlinear behavior when the input signals are strong. Intermodulation occurs when a filter acts in a nonlinear manner, which creates new signals whose frequencies are based upon the frequencies of the input signals. These new signals are called *intermodulation products*. Intermodulation products are the most troublesome in receiver applications when a very large input signal is present simultaneously with a very small signal.

Crystal filters produce the most intermodulation distortion of the filter types we have discussed here. The crystals themselves can exhibit strong nonlinearities for large signal input levels. At low signal levels, the linearity of the filter is determined by surface defects associated with the resonator manufacturing process.

Crystal filters in particular can exhibit gain compression, i.e., the insertion loss of the filter may change with drive level. At high power levels, quartz resonators become nonlinear causing an increase in the filter loss. At low drive levels, resonator processing becomes critical in order to maintain constant insertion loss.

Power-Handling

Power-handling in a filter is closely related to the factors determining the filter's nonlinearities. The following occurrences can limit the amount of power passed by a filter.

- Insertion loss may cause the filter to heat up and dissipate signal power. For example, a filter with a 1 dB insertion loss absorbing 100 watts from a source will dissipate about 20 watts internally.

- At high current levels, the magnetic materials used in HF and low VHF filters can saturate. Consequently, the inductors behave as resistors and the insertion loss increases.

- Many filters, particularly narrow band-pass filters, have very high impedance internal nodes. Even at moderate power levels, peak voltages high enough to cause capacitor breakdown or corona may occur, which leads to higher insertion loss and degraded filter performance.

- Crystal filters can exhibit unexpected nonlinearities. At low power levels, the filter may perform beautifully, but even moderate power changes may cause the filter to behave differently.

In general, if a filter must pass more than 1 watt of power, the filter's power handling capability has to be considered.

Vibration Sensitivity

If the components that comprise a filter change when subjected to vibration, the filter's performance will change as well. Thus, a filter performance depends upon the filter's mechanical stability.

Vibration-induced sidebands may appear on a signal passing through a filter when the filter is subject to acceleration forces due to vibration. Quartz crystal resonators, being piezoelectric devices, convert mechanical to electrical energy. Therefore, the resonant frequency of a crystal is modulated at the frequency of vibration. The source of vibration can be an internal equipment fan, a transformer, or a speaker. Many manufacturers of precision signal generators shock-mount their spectrum clean-up crystal filters to avoid this problem.

War Story — Filter Vibration

One satellite modem manufacturer noticed bursts of bit errors occurring only at certain times of the week. After much research, the company correlated the appearance of the bit error bursts with the routine maintenance performed on some other unrelated equipment. It turned out that the bit error bursts occurred when the maintenance people slammed the doors to the racks containing the satellite modems. Again, shock-mounting the critical filters solved the problem.

2.17 Filter Design Summary

This section contains quick reference data from this chapter. Similar data is placed at the end of each chapter.

Ultimate Filter Roll-Off

$$\text{Ultimate Rolloff} = \left(6\frac{\text{dB}}{\text{Octave}}\right)(\#Poles - \#Zeroes)$$
$$= \left(20\frac{\text{dB}}{\text{Decade}}\right)(\#Poles - \#Zeroes)$$

Ultimate Phase of a Filter

$$\textit{Ultimate Phase} = -90°\,[\#Poles - \#Zeroes]$$

Noise Bandwidth of a Filter

If $H(j\omega)$ is a power ratio (i.e., $|S_{21}|^2$).

$$B_n\left(\frac{\text{rad}}{\text{sec}}\right) = \frac{1}{H_{pk}}\int_0^\infty |H(j\omega)|\,\partial\omega$$

$$B_n(\text{Hz}) = \frac{1}{H_{pk}}\int_0^\infty |H(j\omega)|\,\partial f$$

If $H(j\omega)$ is a voltage ratio, i.e., $|S_{21}|$

$$B_n\left(\frac{\text{rad}}{\text{sec}}\right) = \frac{1}{H_{pk}}\int_0^\infty |H(j\omega)|^2\,\partial\omega$$

$$B_n(\text{Hz}) = \frac{1}{H_{pk}}\int_0^\infty |H(j\omega)|^2\,\partial f$$

where
H_{pk} is the maximum filter response.

Magnitude Response of Butterworth Filters

Low-Pass

$$Attn_{dB} = 10\log\left[1+\left(\frac{f_x}{f_{c,3dB}}\right)^{2N}\right]$$

where
N = the number of poles in the filter,
$f_{c,3dB}$ = the filter's 3 dB cutoff frequency,
f_x = the frequency of interest.

Band-Pass

The absolute attenuation for lossy Butterworth band-pass filters is

$$Attn_{Abs,dB} = 10\log\left[1+\left(\frac{B_x}{B_{3dB}}\right)^{2N}\right] + IL_{dB}$$

The absolute attenuation of a Butterworth band-pass filter at any frequency (f_x) is

$$Attn_{Abs,dB} = 10\log\left[1+\left(\frac{\left|f_x - \frac{f_{c;geo}^2}{f_x}\right|}{B_{3dB}}\right)^{2N}\right] + IL_{dB}$$

where
$Attn_{Abs,dB}$ = the absolute amount of attenuation the filter provides at f_x,
N = the number of poles in the filter,
B_{3dB} = the 3 dB bandwidth of the filter,
$f_{c,geo}$ = the band-pass filter's center frequency,
f_x = the frequency of interest,
B_x = the bandwidth of interest,
IL_{dB} = the insertion loss of the filter in dB.

High-Pass

$$Attn_{dB} = 10\log\left[1+\left(\frac{f_{c,3dB}}{f_x}\right)^{2N}\right]$$

where
N = the number of poles in the filter,
$f_{c,3dB}$ = the filter's 3 dB cutoff frequency,
f_x = the frequency of interest.

Band-Stop
The attenuation of band-stop filters is

$$Attn_{dB} = 10\log\left[1+\left(\frac{B_{3dB}}{B_x}\right)^{2N}\right]$$

or

$$Attn_{dB} = 10\log\left[1+\left(\frac{B_{3dB}}{\left|f_x - \frac{f_{c,geo}^2}{f_x}\right|}\right)^{2N}\right]$$

where
  $Attn_{dB}$ = the absolute amount of attenuation the filter provides at f_x,
  N = the number of poles in the filter,
  B_{3dB} = the 3 dB bandwidth of the filter,
  $f_{c,geo}$ = the band-pass filter's center frequency,
  f_x = the frequency of interest.

Noise Bandwidth of a Butterworth Band-Pass Filter

The noise bandwidth of a Butterworth band-pass filter is

$$B_n = B_{3dB} \frac{\pi/(2N)}{\sin(\pi/(2N))} > B_{3dB}$$

where
  B_n = the noise bandwidth of the Butterworth band-pass filter,
  N = the number of poles in the filter,
  B_{3dB} = the 3 dB bandwidth of the filter.

2.18 References

1. Blinchikoff, Herman J. and Anatoli I. Zverev. *Filtering in the Time and Frequency Domains*. New York: John Wiley and Sons, 1976.

2. Ferrand, Michael. "Practical Microwave Filter Design." *Applied Microwave Magazine*, vol. 1, no. 5 (1989): 120.

3. Freeman, Roger L. *Reference Manual for Telecommunications Engineering*. New York: John Wiley and Sons, 1985.

4. *Filter Catalog*. Piezo Technology, Inc., 1990.

5. Gardner, Floyd M. *Phaselock Techniques*. 2nd ed. New York: John Wiley and Sons, 1979.

6. Orchard, H. J. "The Phase and Envelope Delay of Butterworth and Tchebycheff Filters." *IRE Transaction Circuit Theory* CT-7, no. 6 (1960): 180-181.

7. Porter, Jack. "Noise Bandwidth of Chebyshev Filters." *RF Design Magazine*. (summer 1980): 19.

8. White, Donald R. J. *A Handbook on Electrical Filters*. Don White Consultants, Inc., 1980.

9. Williams, Arthur B. *Electronic Filter Design Handbook*. New York: McGraw-Hill, 1981.

10. Zverev, Anatoli. I. *Handbook of Filter Synthesis*. New York: John Wiley and Sons, 1967.

3

Mixers

> A translation is no translation, he said, unless it will give you the music of a poem along with the words of it.
> — John Millington Stone
>
> Translation increases the faults of a work and spoils its beauties.
> —Voltaire

3.1 Introduction

A mixer translates signals from one center frequency to another while keeping the modulation intact. The translation is performed for the following reasons:

Filtering

When building a receiver, we need to filter the signal of interest from other signals before we apply the signal to a demodulator. Filtering reduces noise and attenuates unwanted signals. For example, if we are building a commercial FM broadcast receiver, we want to separate the desired station from all of the other stations. It is much easier to move the signal of interest from its original center frequency to some *intermediate frequency* (IF) and perform the filtering at the IF. This method also has the advantage that the IF can be chosen based upon filtering needs. It is often less problematic to realize narrowband filters at low center frequencies and better to handle large bandwidth signals at high center frequencies. The realizable insertion loss, percent bandwidth, and physical size of the filter are all strong functions of the filter center frequency.

Frequency Assignments

On-the-air frequencies are almost always assigned by government agencies such as the FCC. These frequencies are often chosen based on external requirements that have nothing to do with filtering or demodulator requirements. To build the best system possible, we perform all the complex receiver functions at a frequency we choose, then translate the signal of interest to that center frequency.

For example, moving data through a satellite system is expensive. The transmitter's power, center frequency and bandwidth are tightly controlled. In order to pass the most data through this channel, satellite transmitters use complex modulation schemes to limit the transmitted bandwidth while keeping the data rate as high as possible. The transmitter performs the modulation at one frequency (usually 70 MHz), then a separate piece of equipment translates the modulated waveform to the desired center frequency. This method produces a system with carefully controlled characteristics because the modulator has to perform at only one frequency.

Antenna Size

The physical size of an antenna in terms of wavelengths determines the electrical properties of the structure. For example, antennas at very low frequencies are very large. On the other hand, the mechanical tolerances required for very high frequency devices may be difficult or expensive to realize. If we have to build a device with a small antenna, such as a missile or an electronic pager, it is preferable to use a high frequency in order to minimize the antenna size.

Propagation

Propagation characteristics change with frequency. The attenuation, multipath and reflection characteristics of the propagation environments are all strong functions of frequency. The atmosphere exhibits points of high attenuation due to water, oxygen and other atmospheric components. For example, there is a point of high attenuation at 60 GHz due to energy absorption by oxygen (the "oxygen line"). Normally, we want to avoid transmitting at absorption frequencies because of the high attenuation. However, if we do not want anyone to listen to personal communications and only have to transmit a short distance, these frequencies are ideal. Some military communications links operate at 60 GHz for just this purpose. Some satellite-to-satellite links operate at 60 GHz to avoid interference from earth-bound communications. The signal passes easily through the vacuum of space but will not pass through the atmosphere.

Realization Effects

It is often easier to achieve performance goals over a single, narrow frequency range than over a large range of frequencies. For example, radio receivers almost universally move a signal at some tuned frequency to some intermediate frequency before processing. The demodulator has to function at only one frequency, rather than over a range of frequencies. Translating the signal to a common IF also helps to ensure that the receiver's characteristics do not change as the receiver is turned.

Component Availability

Many people desire cheaper and better components for mass-marketed electronic systems. Once a standard develops, economic factors allow the design of components that work within the standard. For example, most commercial FM radios use a 10.7 MHz IF, a frequency for which many cheap electronic components are designed. Similarly, the television industry has developed amplifiers, filters and other useful components that perform at 45 MHz. The satellite and military industries favor 70 MHz IFs.

3.2 Frequency Translation Mechanisms

When moving signals from one frequency to another, we want to avoid distortion and keep the modulation of the signal intact.

Amplifier Distortion

A nonlinear device will produce output signals that are not present at the input. Figure 3-1 shows the input and output spectrum of a typical nonlinear device.

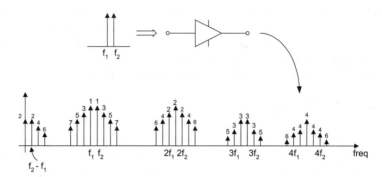

Figure 3-1 *The input and output spectra of a nonlinear device.*

The spectrum of Figure 3-1 has the following characteristics:

- The output frequencies are given by

$$f_{out} = |nf_1 \pm mf_2| \qquad 3.1$$

where
$n = 0, 1, 2, \ldots$
$m = 0, 1, 2, \ldots$
$n + m =$ the order of the response.

- If we space the two input signals so that $f_1 \approx f_2$, the output signals tend to "bunch" together about the harmonics of f_1 and f_2. The "bunched" signals are Δf apart where $\Delta f = |f_1 - f_2|$.

- The higher the order, the smaller the output power of the signal. In other words, the *conversion loss* increases with increasing order.

Second-Order Response

The second-order nonlinear response is preferable because it produces the strongest output signals, i.e., it exhibits the least conversion loss. The second-order operation is

$$V_{out,2}(t) = k_2 V_{in}^2(t) \qquad 3.2$$

We will assume V_{in} is the sum of two signals: a modulated signal V_{mod} and an unmodulated cosine wave V_{unmod}. We can write

$$V_{mod}(t) = A(t)\cos[\omega_1 t + \phi(t)] \qquad 3.3$$

and

$$V_{\text{unmod}}(t) = \cos(\omega_2 t) \qquad 3.4$$

where
  $A(t)$ = AM information waveform,
  $\phi(t)$ = PM or FM information waveform,
  $\omega_1 = 2\pi f_1$ = carrier frequency of the modulated wave,
  $\omega_2 = 2\pi f_2$ = carrier frequency of the unmodulated wave.

The second-order output is

$$\begin{aligned}V_{out,2}(t) &= k_2\left[V_{\text{mod}}(t) + V_{\text{unmod}}(t)\right]^2 \\ &= k_2\left\{A(t)\cos[\omega_1 t + \phi(t)] + \cos(\omega_2 t)\right\}^2 \\ &= \frac{k_2\left[1 + A^2(t)\right]}{2} + \frac{k_2 A^2(t)}{2}\cos[2\omega_1 t + 2\phi(t)] \\ &\quad + \frac{k_2}{2}\cos(2\omega_2 t) \\ &\quad + k_2 A(t)\cos[(\omega_1 + \omega_2)t + \phi(t)] \\ &\quad + k_2 A(t)\cos[(\omega_1 - \omega_2)t + \phi(t)]\end{aligned} \qquad 3.5$$

The last two terms of Equation 3.5 at $\omega_1 \pm \omega_2$ are the desired signals. They contain the undistorted AM and FM modulation and are located at a different center frequency. The signal at $\omega_1 + \omega_2$ is the "sum" product; the signal at $\omega_1 - \omega_2$ is the "difference" product.

We use filters to remove the unwanted second-order products and pass the desired signals. A conversion scheme is designed so that none of the unwanted products are too close to the wanted signals and too difficult to filter.

Amplifier Difficulties

Figure 3-2 shows how we might use a nonlinear amplifier as a mixer in a receiver. We connect the antenna to a band-pass filter (BPF_1) to limit the number of signals into the amplifier. The output of BPF_1 is connected to the input port of the amplifier.

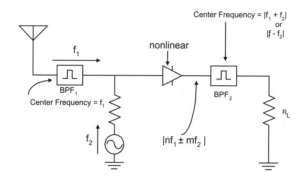

Figure 3-2 *Using a nonlinear amplifier as a mixer.*

We also connect the amplifier's input port to a locally generated sine wave, i.e., a *local oscillator* (LO). The frequency of the LO signal is f_2. The local oscillator signal must be strong enough to force the amplifier to generate nonlinear output products. The signals present on the output of the amplifier will be f_1 and its harmonics, f_2 and its harmonics, and the nonlinear intermodulation products described by Equation 3.1.

The two signals that are commonly useful are at $f_1 + f_2$ and $f_1 - f_2$ (the sum and difference frequencies). In Figure 3-2, we use BPF_2 to select either the sum or the difference signal. This architecture is widely employed in commercial equipment although it has several flaws.

The output of the amplifier contains many signals that are not desirable, including f_1, f_2 and all their sum and difference products. Since we want to ensure we are operating the amplifier in its nonlinear region, we have to make the locally generated signal at f_2 very large. This means we will see high-power signals at the amplifier's output at the LO frequency and its harmonics. A typical LO can be 90 dB higher than the desired signals.

The signals at f_1 and f_2 experience the amplifier's first-order power gain and will likely be larger than the desired signals at $f_1 + f_2$ and $f_1 - f_2$. A large portion of the local oscillator will travel out of the antenna. This raises interference questions and wastes power.

The first three items place high demands on BPF_2. The filter must adequately reject the unwanted signals while passing the desired signal with minor attenuation. Much of the finesse of receiver design lies in arranging the conversion scheme so the unwanted signals are easy to remove by filtering.

In summary, we can use the second-order distortion characteristics of an amplifier to generate the signals we want but it is not a very high-performance solution. Many of these shortcomings can be overcome with a different nonlinear operation.

Time-Domain Multiply

A table of trigonometric identities provides us with the following equation:

$$\cos(A)\cos(B) = \frac{1}{2}\cos(A+B) + \frac{1}{2}\cos(A-B) \qquad 3.6$$

If we multiply two signals in the time domain, we produce only two output signals: one at the sum frequency and one at the difference frequency. None of the other frequency components are present. With one modulated signal and one unmodulated signal, the input signals will be

$$V_{mod}(t) = A(t)\cos[\omega_1 t + \phi(t)] \qquad 3.7$$

and

$$V_{unmod}(t) = \cos(\omega_2 t) \qquad 3.8$$

Combining these two equations with Equation 3.6 produces

$$\begin{aligned} V_{mult}(t) &= V_{mod}(t) V_{unmod}(t) \\ &= A(t)\cos[\omega_1 t + \phi(t)]\cos(\omega_2 t) \\ &= \frac{A(t)}{2}\cos[(\omega_1 + \omega_2)t + \phi(t)] \\ &\quad + \frac{A(t)}{2}\cos[(\omega_1 - \omega_2)t + \phi(t)] \end{aligned} \qquad 3.9$$

Equation 3.9 illustrates a fundamental property of mixing. The $A(t)/2$ terms indicate that we will lose at least 6 dB in the conversion process.

Except for a scaling factor, both the AM and PM modulation pass through the multiply unscathed. The preferred method of frequency translation is the time-domain multiply. At high frequencies, it is difficult to perform the multiplication in a mathematically precise way but we can approximate the operation. These implementation details result in higher levels of spurious output signals than Equation 3.9 indicates.

3.3 Nomenclature

Figure 3-3 shows the block diagram of a frequency converter, or mixer.

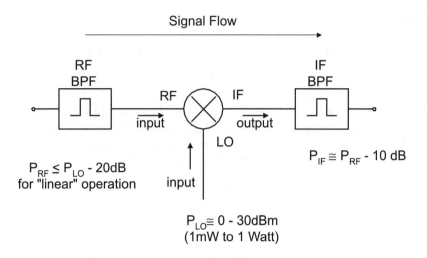

Figure 3-3 A typical mixer used to convert signals at the radio frequency (RF) to some intermediate frequency (IF) using a local oscillator (LO).

Ports

A mixer has three ports.

Radio Frequency (RF) Port

The *radio frequency (RF)* port is often the input port of the mixer. The signal carrying the information enters through this port. A mixer changes the center frequency of the signal present at the RF port to the sum and difference frequencies and places those signals on the IF port. The IF filter then selects either the sum or difference signal and passes it further upstream. As a rule of thumb, the signal power entering the mixer's RF port should be at least 20 dB below the LO power. If the RF power is any larger than this number, the mixer will still operate but it will generate more unwanted signals than necessary.

Intermediate Frequency (IF) Port

The mixer's intermediate frequency port functions usually as the output port of the mixer. The sum and difference frequencies appear at this port along with the unwanted spurious signals. The IF port is almost universally connected to a band-pass filter in order to remove the unwanted signals and pass the desired signals. The sum and difference products present at the IF port are generally about 10 dB below the level of the RF signal applied to the RF port. In other words, common mixers have about 10 dB *conversion loss*.

Local Oscillator (LO) Port

The local oscillator, which is generated internally in the receiver, is another mixer input port. The frequency of the local oscillator, along with the RF and IF band-pass filters, determine which signals present at the RF port are converted to the IF frequency. The local oscillator power usually ranges from about 0 dBm (1 mW) to 30 dBm (1 watt). As a rule, the LO power should be the strongest signal present in the mixer by at least 20 dB for reasons we will be discussing later. In a high-performance receiver, the local oscillator should be free of amplitude and phase noise.

Port Interchangeability

Most passive mixers have ports labeled "RF," "IF" and "LO." Despite the labels, these ports are usually interchangeable, i.e., the RF signal can be applied to the IF port, the LO can be applied to the IF port and the signal can be removed from the RF port. This is convenient to do when the frequency ranges of the three ports are different. If we operate a mixer using its ports for different purposes, the performance will likely be different from the data sheet's description.

The ports are not always interchangeable on specialized mixers. Some mixers contain active devices or special circuitry to cancel undesired signals, so their ports cannot be interchanged. When we use the term "RF," for example, we will be referring to the signal present at the RF frequency, not to a specific mixer port. The same holds true for the terms "IF" and "LO."

Frequency Translation Equations

Two theoretical processes, *second-order nonlinearities* and *time-domain multiplication*, can be used to translate signals in frequency. In both cases, the equation that quantifies the relationship between the signals present at a mixer's RF, IF and LO ports is

$$f_{IF} = f_{LO} \pm f_{RF} \qquad 3.10$$

Figure 3-4 illustrates the conversion process graphically. We will apply a single tone at f_{RF} into the RF port of the mixer. A second tone will be placed at f_{LO} into the mixer's LO port. We will now examine the spectrum present on the output port.

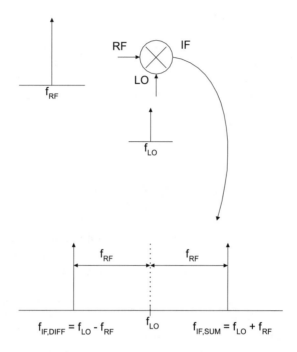

Figure 3-4 *A frequency conversion and the resulting spectrum.*

For now, we will ignore everything leaving the mixer's IF port except the two signals at the sum and difference frequencies. One signal will be at the difference frequency of $f_{IF,DIFF} = f_{LO} - f_{RF}$. The second signal will be at the sum frequency of $f_{IF,SUM} = f_{LO} + f_{RF}$. Often, the signal at the sum frequency is referred to as the *upper sideband* and the signal at the difference frequency as the *lower sideband*.

Figure 3-4 shows the graphical symmetry between the two IF frequencies and the LO frequency. If we start at the LO frequency and move down by f_{RF} hertz, we will find the signal at the difference frequency. If we start at the LO frequency and move up by f_{RF} hertz, we will find the signal at the sum frequency.

Zero Hertz

Assuming that f_{LO} and f_{RF} are both positive (which we can force by definition), zero hertz will lie below the lower sideband when $f_{RF} < f_{LO}$ (see Figure 3-5). Zero hertz will lie between f_{LO} and the lower sideband (at $f_{IF,DIFF}$) when $f_{RF} > f_{LO}$, as shown in Figure 3-6(a). What is the most useful way to interpret a negative frequency?

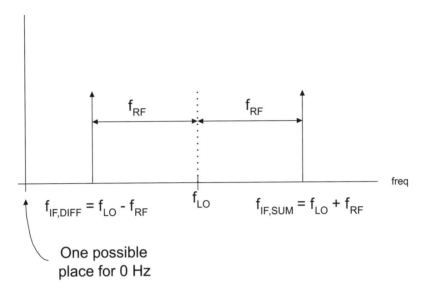

Figure 3-5 *Conversion spectra. One possible position for zero hertz (or DC).*

Absolute Value

Any table of trigonometric functions shows

$$\cos(-\omega t) = \cos(\omega t) \qquad 3.11$$

We can simply form the absolute value of any frequency produced by the conversion equations. With this in mind, we can write Equation 3.10 as

$$f_{IF} = |f_{LO} \pm f_{RF}| \qquad 3.12$$

If we accept only positive frequencies and consider Equation 3.11, the spectrum in Figure 3-6(a) is equivalent to the spectrum in Figure 3-6(b). The lower sideband, which was formerly below zero, has been mirrored off the y-axis and is now in the positive frequency domain. Note that the distance from f_{LO} to the y-axis plus the distance from the y-axis to the lower sideband is still f_{RF}. We will consider the sign of the frequency whenever it is useful to solve the problem. Likewise, we will ignore the sign when it is not necessary.

272 | RADIO RECEIVER DESIGN

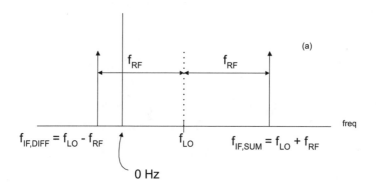

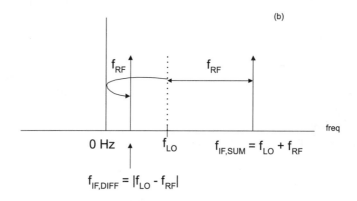

Figure 3-6 *When one sideband of the mixing process falls below zero hertz, we simply use the absolute value of the frequency for all calculations.*

Three Forms

Equation 3.12 can be expressed as three separate equations when we fully expand the "±" signs. If we know the LO and RF frequencies, then the RF signal will be simultaneously converted to two different IFs. These two IFs are given by

$$f_{IF} = f_{LO} \pm f_{RF} \qquad 3.13$$

If we know the LO and IF frequencies, we can find the two possible RFs, which will be converted to the IF by the mixer.

$$f_{RF} = f_{LO} \pm f_{IF} \qquad 3.14$$

The third equation yields two possible LO frequencies we can use to convert some RF signal to a given IF.

$$f_{LO} = f_{RF} \pm f_{IF} \qquad 3.15$$

Figure 3-7 shows these equations graphically. A useful way to interpret the graphs is by comparing starting points and distances traveled. For example, the top of Figure 3-7 illustrates Equation 3.13. If we begin at f_{LO} and move a distance f_{RF} to either side, we find the two possible solutions of the equation.

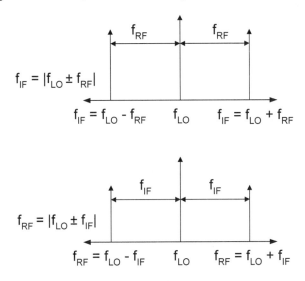

Figure 3-7 *The frequency conversion Equations 3.13 through 3.15 expressed graphically.*

The "±" sign in Equations 3.13 through 3.15 means that there are always two possible solutions anytime we solve for one of the variables. This can be both a convenience and an annoyance, depending upon the problem to be solved.

Example 3.1 — Conversion Equations

a. What two possible LO frequencies can be used to convert a commercial FM radio station broadcasting at 106.9 MHz to a 10.7 MHz IF?
b. What are the center frequencies of the two possible signals which can be converted to an IF of 70 MHz using a LO frequency of 935 MHz?
c. If a signal centered at 215.75 MHz is mixed with a 256.25 MHz local oscillator, what are the center frequencies of the two resulting signals?

Solution —

a. Equation 3.15 produces

$$f_{LO} = f_{RF} \pm f_{IF}$$
$$= 106.9 \pm 10.7$$
$$= 117.6 \text{ or } 96.2 \text{ MHz}$$

(3.16)

b. This calls for Equation 3.14.

$$f_{RF} = f_{LO} \pm f_{IF}$$
$$= 935 \pm 70$$
$$= 1005 \text{ or } 865 \text{ MHz}$$

(3.17)

c. Equation yields

$$f_{IF} = f_{LO} \pm f_{RF}$$
$$= 256.25 \pm 215.75$$
$$= 472.0 \text{ or } 40.5 \text{ MHz}$$

(3.18)

Frequency Translation and Filters

Figure 3-8 illustrates the conversion process in practice. The antenna receives many different signals at different frequencies, but we are interested in only one at a time. Ideally, we would like to remove all the unwanted signals from the antenna using the RF BPF but, because of practical concerns, we usually cannot do so. Instead, the RF filter will remove only those signals that are grossly different in frequency from the desired signal. The filtered RF spectrum then travels to the mixer's RF port and all the signals undergo the frequency conversion. Two copies of the RF spectrum are generated by the mixing process: one centered about $f_{IF,DIFF}$ and one centered about $f_{IF,SUM}$. Ultimately, the IF filter passes the one signal we want to process.

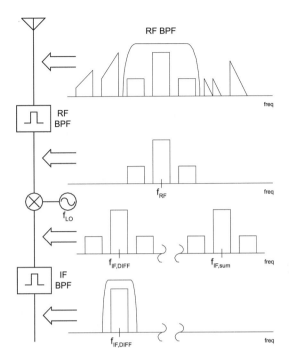

Figure 3-8 *The conversion and filtering processes in a single-conversion receiver.*

Practical Considerations

Figure 3-8 lends itself to a discussion of some of the practical considerations. Why, you might ask, do we go through all this trouble of converting the signal of interest to the IF before filtering it? The answer is that we either cannot or do not want to filter the signal adequately at its RF frequency. Let us look at the FM broadcast band as an example. A commercial FM broadcast receiver should be able to process all of the 101 possible FM stations. We can

- Build 101 band-pass filters just wide enough to pass one particular FM station, then switch to a new filter whenever the user changes stations. It would be difficult and also expensive to build filters of such small percentage bandwidths, especially on a production basis.

- Build a narrow, tunable RF band-pass filter and change its center frequency as the user changes stations. This is a difficult-to-realize and expensive solution because of the small percentage bandwidths and the extra circuitry involved.

- Let the RF band-pass filter pass all 101 stations, then filter out the particular station we want at a later stage. This is exactly the solution used in practice.

Receiver design involves solving filtering problems. If you cannot filter a signal at a particular frequency, move the signal to a frequency where you can filter it. In a commercial FM broadcast receiver, we convert the signal of interest to 10.7 MHz, no matter what the RF frequency was originally. When we want to change the station, we only have to change the LO frequency and the mixer will convert a different station down to the 10.7 MHz IF. This technique has the advantage that everything following the IF filter has to work only at the IF frequency.

Conversion Loss

When a mixer converts a signal from its RF to some IF, the signal at the IF has usually lost some of its power. This is due to the mixer's conversion loss. Figure 3-9 illustrates the concept.

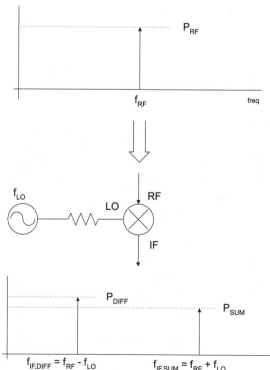

Figure 3-9 Definition of mixer conversion loss.

If, after the conversion, we will be using the lower sideband IF signal (at $f_{IF,DIFF}$), the mixer's conversion loss is

$$CL_{LSB} = \frac{\text{Signal Power at } f_{RF} - f_{LO}}{\text{Signal Power at } f_{RF}} \quad \quad 3.19$$

$$= \frac{P_{DIFF}}{P_{RF}}$$

If we will be using the IF signal at the upper sideband frequency ($f_{IF,SUM}$), then the mixer's conversion loss is

$$CL_{USB} = \frac{\text{Signal Power at } f_{RF} + f_{LO}}{\text{Signal Power at } f_{RF}} \quad \quad 3.20$$

$$= \frac{P_{SUM}}{P_{RF}}$$

Note that P_{SUM} does not necessarily equal P_{DIFF} although they are often very close. Mixer data sheets usually give only one conversion loss specification, implying that $CL_{SUM} = CL_{DIFF}$ and that $P_{SUM} = P_{DIFF}$.

Example 3.2 — Mixer Conversion Loss

The power of the RF signal applied to the mixer's input port is –45 dBm. The power measured at the mixer's IF port at the sum frequency is –60 dBm, and the power measured at the difference frequency is –63 dBm. What is the mixer's conversion loss?

Solution—

The upper sideband conversion loss is

$$CL_{USB} = \frac{P_{SUM}}{P_{RF}}$$

$$= P_{SUM,dB} - P_{RF,dB} \quad \quad 3.21$$

$$-45 - (-60)$$

$$= 15 \text{ dB}$$

The lower sideband conversion loss is

$$CL_{LSB} = \frac{P_{DIFF}}{P_{RF}}$$

$$= P_{DIFF,dB} - P_{RF,dB} \quad \quad 3.22$$

$$-45 - (-63)$$

$$= 18 \text{ dB}$$

We can divide conversion loss into two categories: single-sideband conversion loss and double-sideband conversion loss. A mixer will convert a signal RF signal into two separate IF signals and, in most cases, the receiver designer will use only one of the IF signals. *Single-sideband conversion loss* refers to the signal power of only one IF signal. *Double-sideband conversion loss* means that the designer is using both IF signals in some way. Most receiver designers use the single-sideband conversion loss. Whenever you see a reference simply to *conversion loss*, the writer usually means the signal sideband conversion loss.

Port-to-Port Isolation

Due to realization effects, signals applied to one port of a mixer will leak through the mixer to the other two ports. This occurs without any frequency translation. In other words, if a signal is applied at f_{LO} to the LO port of a mixer, we will see a signal at f_{LO} on both the IF and RF ports of the mixer. Identical effects occur when we apply a signal to both the RF and IF ports. Leakage through the mixer (as specified by the interport isolation specifications) is often the major driver in the design of IF and RF BPFs. The three mixer isolation specifications are as follows (see Figure 3-10).

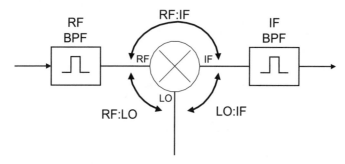

Figure 3-10 Mixer port-to-port isolation.

LO:RF Isolation

This is the amount of signal attenuation between a mixer's LO and RF ports. For passive mixers, the isolation is usually bilateral; it applies for signals travelling from the LO port to the RF port and for signals travelling from the RF port from the LO port. We are more interested in how much LO power leaks onto the mixer's RF port than how much of the RF signal leaks to the LO port. With that in mind, we will define the LO:RF isolation as

$$Isol_{LO:RF} = \frac{LO \text{ power present at the RF port } (at\ f_{LO})}{LO \text{ power entering the LO port } (also\ at\ f_{LO})} \quad 3.23$$

MIXERS | 279

Since the LO is a high-power signal (usually > 7 dBm), even a mixer with a large LO:RF isolation can have a large LO signal present at the RF port. The RF filter in Figure 3-10 helps to attenuate the LO leaving the RF port.

LO:IF Isolation

This is the amount of attenuation between the LO and IF ports of the mixer. Similar to the LO:RF isolation, this attenuation is usually bilateral. We are mostly interested in how much LO power appears on the IF port of the mixer. Thus, the LO:IF isolation is defined as

$$Isol_{LO:IF} = \frac{LO \text{ power present at the IF port } (at\ f_{LO})}{LO \text{ power entering the LO port } (also\ at\ f_{LO})} \qquad 3.24$$

Since the LO is a relatively high-power signal, even slight LO:IF leakage places a large LO signal at the mixer's IF port. The IF band-pass filter in Figure 3-10 must be designed to attenuate this LO leakage.

RF:IF Isolation

To measure the RF:IF isolation, we apply a signal at f_{RF} to the RF port of the mixer, then measure how much RF signal is present on the IF port of the mixer (still at f_{RF}). The leakage has not undergone the frequency conversion process.

The RF:IF isolation applies to signals present at the same frequency on the RF and IF ports; conversion loss applies to signals at different frequencies (f_{RF} on the RF port and $f_{RF} \pm f_{LO}$ on the IF port). In equation form, the RF:IF isolation is

$$Isol_{RF:IF} = \frac{RF \text{ power present at the IF port } (at\ f_{RF})}{RF \text{ power entering the RF port } (still\ at\ f_{RF})} \qquad 3.25$$

Example 3.3—Mixer Conversion Loss and Isolation

Given the system in Figure 3-11 and the following mixer specifications:
$CL_{dB} = 8$ dB,
RF:IF Isolation = 30 dB (about average),
LO:IF Isolation = 45 dB (unusually high),
LO:RF Isolation = 25 dB (about average) find

a. the strength of the signals at 407 MHz, 243 MHz, 82 MHz and 325 MHz on the IF port,
b. the strength of the signal at 325 MHz on the RF port.

280 | RADIO RECEIVER DESIGN

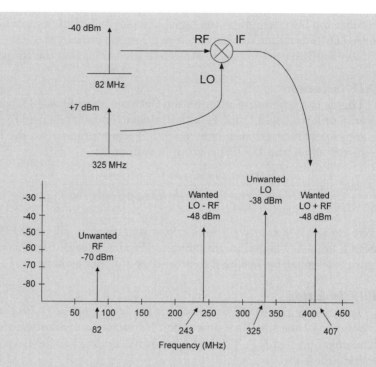

Figure 3-11 *Example 3.3. Wanted and unwanted signal present at a mixer's IF port.*

Solution —
Using Equation 3.13, the sum and difference frequencies present on the IF port are

$$f_{IF} = f_{LO} \pm f_{RF}$$
$$= 325 \pm 82$$
$$= 243 \text{ MHz and } 407 \text{ MHz}$$

3.26

The problem did not specify whether the conversion loss was the single-sideband conversion loss or the double-sideband conversion loss. (This ambiguity is typical.) We will assume it is a single-sideband conversion loss.
a. The conversion loss applies to the signals at 243 and 407 MHz. The level of these signals is

$$P_{IF,243} = P_{IF,407}$$
$$= P_{RF,82,dBm} - CL_{dB}$$
$$= -40 - 8$$
$$= -48 \text{ dBm}$$

3.27

MIXERS | 281

The RF:IF isolation applies to signals at the RF port leaking through the mixer to the IF port (without changing frequency), so the level of the signal at 82 MHz is

$$P_{IF,82} = P_{RF,82,dBm} - Isol_{RF:IF,dB}$$
$$= -40 - 30$$
$$= -70 \text{ dBm}$$
3.28

The LO:IF isolation applies to leakage from the LO port to the IF port. The level of the signal at the IF port at 325 MHz is

$$P_{IF,325,dBm} = P_{LO,325,dBm} - Isol_{LO:IF,dB}$$
$$= 7 - 45$$
$$= -38 \text{ dBm}$$
3.29

Figure 3-11 shows the spectrum of the signals present at the IF port. Note that the signal due to LO leakage (at 325 MHz and –38 dBm) is stronger than the desired signals at 243 and 407 MHz and –48 dBm.

b. The LO:RF isolation applies to LO leakage to the RF port. At the RF port, the strength of the signal at 325 MHz is

$$P_{RF,325,dBm} = P_{LO,325,dBm} - Isol_{LO:RF,dB}$$
$$= 7 - 25$$
$$= -18 \text{ dBm}$$
3.30

Practical Effects of Mixer Inter-port Isolation

The following discussion shows how to apply mixer isolation specifications.

Antenna Radiation

Figure 3-12 shows that the LO:RF isolation of the mixer can allow the LO to radiate out of the antenna port. By receiver standards, the LO is a very strong signal and can interfere with other receivers. For example, the LO of a commercial US FM radio receiver runs from about 99 MHz to about 119 MHz. It happens that the commercial US aircraft band is about 108 MHz to 132 MHz. Note that the two bands overlap.

There is a distinct possibility that the LO of the FM receiver could interfere with aircraft communications equipment if everything is just right. This is the reason that commercial airlines do not allow their passengers to use FM radio or other electronic equipment during take-off and landings. If a passenger's equipment radiates a signal that is near one of the important aircraft frequencies, it could interfere with the pilot's communications.

Let us assume that a mixer in a particular receiver requires an LO power of +7 dBm and its mixer will provide a LO:RF isolation of 40 dB. (This is an unusually large number.) A –33 dBm signal, which is a strong signal, will leave the antenna.

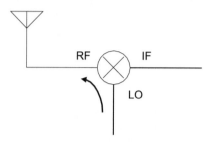

Figure 3-12 *LO:RF isolation causes a strong LO signal to radiate from a receiver's antenna.*

Intermixer Isolation

Receivers with two separate mixers and two separate local oscillators are called *dual conversion* receivers. Figure 3-13 shows a simple block diagram of a dual conversion receiver.

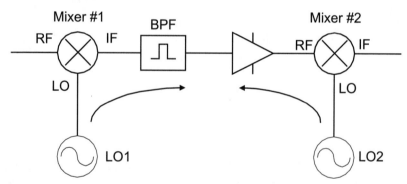

Figure 3-13 *Isolation problem in a dual conversion receiver. Good design keeps LO_1 out of mixer #2 and keeps LO_2 out of mixer #1.*

In a poorly designed receiver, significant amounts of LO_1 will leak out of mixer #1 (via its LO:IF isolation) and enter the RF port of mixer #2. Likewise, LO_2 can leave mixer #2 (via its LO:IF isolation) and make its way into mixer #1. Any signal that enters a mixer will participate in the conversion process. In mixer #1, for example, LO_2 will mix with LO_1 as well as any RF signals present in the mixer. In mixer #2, LO_1 will mix with LO_2 as well as any other signal present in the mixer.

As a result, many different kinds of signals will be generated at many different frequencies. These signals are one cause of *internally generated spurious responses* or "birdies." "Birdies," to the casual observer, appear to be coming from the receiver's antenna port. Internally generated spurious signals can be minimized by designing a proper conversion scheme and carefully selecting filters and other components.

The receiver LOs are not the only source of unwanted signals that cause internally generated spurious signals. Other oscillator or signal source present in the receiver can contribute signals. The receiver's microprocessor and its power switching supply are also major contributors.

Mixers in Cascade

A mixer used in a cascade is treated in the same way any other component is treated. It has a loss or a gain, a noise figure, a TOI and a SOI. For the purposes of the cascade calculations, we ignore the frequency translation aspects of the device.

Example 3.4 — Mixers in Cascade
Find the gain, noise figure, and ITOI of the cascade shown in Figure 3-14.

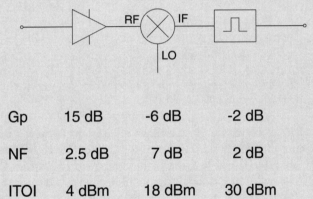

Gp	15 dB	-6 dB	-2 dB
NF	2.5 dB	7 dB	2 dB
ITOI	4 dBm	18 dBm	30 dBm

Figure 3-14 Example 3.4. Mixers in cascade calculations.

Solution —
Using the cascade equations developed earlier, we use the mixer's conversion loss as its gain. The equations resolve to

$G_{p,cas} = 7$ dB,
$NF_{dB} = 3.0$ dB,
$ITOI_{dBm} = 0.4$ dBm.

3.4 Block vs. Channelized Systems

Most of the frequency translation tasks can be divided into two different types: block conversions and channelized conversions. We will often perform them both in the same system.

Block Conversions

In a block conversion, we move an entire band of signals from one center frequency to another using a fixed, single-frequency local oscillator. Figure 3-15 shows the conversion scheme of a commercial *satellite television receive only* (TVRO) system. The TVRO receiver collects television signals from geostationary satellites. The satellite's C-band transponders transmit 24 television channels at 3700 MHz to 4200 MHz (a 500 MHz bandwidth centered at 3950 MHz). After the outdoor antenna captures the TVRO signal, we must move the signal to the house, which can be several hundred feet away. This is a commercial application and, to keep costs down, we can use cheap lossy cable between the antenna and the house.

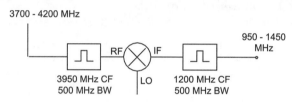

Figure 3-15 Television Receive-Only (TVRO) frequency conversion scheme. This is a block conversion.

To solve the problem, we place a block converter at the receive antenna's feed point to convert the 3950 MHz signal to a 1200 MHz IF. We then move the 1200 MHz IF signal to the house, taking advantage of the lower cable losses at 1200 MHz. No one channel from the 24 possible channels has been selected yet — only a swatch of spectrum was moved from one center frequency to another using a single-frequency LO. Since we are moving the entire spectrum into the house, several users can access the same antenna at once.

Channelized Conversions

A channelized system uses a tuning local oscillator to pick just one signal out of a crowded spectrum. For example, Figure 3-16 shows the architecture of a commercial FM receiver. The FM broadcast band in the United States runs from about 88 MHz to 108 MHz. In an FM receiver, we apply the entire 20 MHz FM spectrum into the mixer and tune the LO. We select the frequency of the LO so that only the desired signal will be centered in the IF filter. All of the other stations are severely attenuated by this filter.

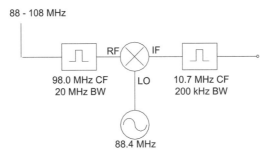

Figure 3-16 Commercial FM receiver frequency conversion scheme. This is a channelized conversion.

Suppose we wanted to listen to a station broadcasting at 99.1 MHz. The 88.4 MHz LO frequency of Figure 3-16 translates the desired signal from 99.1 MHz to 10.7 MHz (the center frequency of the IF filter). The desired signal, now at 10.7 MHz, passes through the IF filter without attenuation. The stations below 98.9 MHz and above 99.3 MHz are considerably reduced in power. After the IF filter, the strongest signal comes from the station we want to receive. This FM demodulator then processes the signal further downstream and we hear music, music, music!

3.5 Conversion Scheme Design

Designing a conversion scheme is an iterative process. We start with an initial guess, examine the consequences and problems, adjust the premises, reexamine the problems, and so on. This process is continued until a workable system is established. We usually start with several questions or specifications.

- What is the frequency of the RF spectrum we want to convert? This specification will contribute to the choice of the first IF and the selection of the first LO frequencies.

- Do we want to perform a block conversion or a channelized conversion? For a block conversion, the LOs are usually fixed whereas a channelized system demands tunable LOs.

- If we are performing a channelized conversion, what is the bandwidth of the RF signals? This affects the final IF selection as well as the accuracy of the LOs.

- What kind of modulation is present on the RF signals? This places demand on the phase noise of the LO which, in turn, can limit the selection of the IF center frequency. (High IFs tend to require high frequency LOs which can have poor phase noise.)

- Are there any cost, manufacturability, physical size, power or other special features? This category tends to override many of the other specifications of the radio. Often, these miscellaneous considerations force the designer to use a technique or component that he or she would rather avoid.

For example, if the receiver is battery-powered, a power supply to convert the changing battery voltage into a stable, well-regulated supply for the receiver must be designed. The switching power supplies used will generate large, low-frequency current spikes around the system, which will take some effort to filter properly.

Small physical size forces many packaging issues such as heat removal and the physical size of the components. The component size issue compels us to consider common IF frequencies because of the variety of physically small components available at these frequencies.

Television Receive-Only (TVRO) Example

Figure 3-17 shows a block diagram of a satellite TVRO receiver. The signal from the satellite covers 3700 to 4200 MHz (a 3950 MHz center frequency). We would like to convert this band of frequencies to a 1200 MHz center frequency. These details determine the bandwidths and center frequencies of both the RF and IF band-pass filters. We now calculate the LO frequency needed for this conversions scheme.

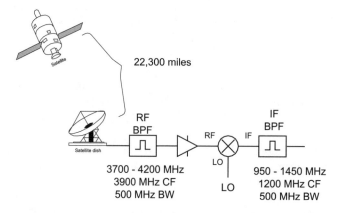

Figure 3-17 Television receive-only (TVRO) receiving system.

Equation 3-15 yields the two possible LO frequencies that can be used to convert a signal from some RF to an IF. For now, we will only look at the center frequencies of both the RF and the IF. Applying Equation 3.15 produces

$$f_{LO} = f_{RF} \pm f_{IF}$$
$$= 3950 \pm 1200 \text{ MHz} \qquad 3.31$$
$$= 2750 \text{ MHz or } 5150 \text{ MHz}$$

We can select either 2750 MHz or 5150 MHz as the LO frequency. Both will convert a signal from 3950 MHz to 1200 MHz with equal efficiency. Is there a difference?

Band Edges

We now examine the effect of the LO on the RF signal at 3700 MHz and 4200 MHz. Using $f_{LO} = 2750$ MHz and Figure 3-18 (a) as a guide, Equation 3.13 produces

$$\text{Lower Band Edge} \Rightarrow 3700 - 2750 = 950 \text{ MHz} \qquad 3.32$$
$$\text{Upper Band Edge} \Rightarrow 4200 - 2750 = 1450 \text{ MHz}$$

A signal at 3700 MHz converts to 950 MHz; a signal at 4200 MHz converts to 1450 MHz. The IF spectrum is identical to the RF spectrum when $f_{LO} = 2750$ MHz.

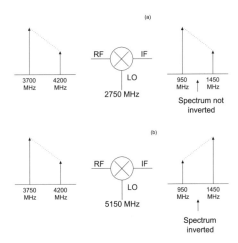

Figure 3-18 *The TVRO RF and IF spectra for the two possible LO frequencies.*

Let us look at the effect of the other LO frequency, $f_{LO} = 5150$ MHz. Equation 3.13 and Figure 3-18(b) produce

$$\text{Lower Band Edge} \Rightarrow 5150 - 3700 = 1450 \text{ MHz}$$
$$\text{Upper Band Edge} \Rightarrow 5150 - 4200 = 950 \text{ MHz}$$
3.33

This local oscillator selection produces a frequency inversion. Signals that were above the center frequency at the RF port of the mixer are now below the center frequency at the IF port. This is usually not a problem as long as we are aware of the effect. We have not altered the information content of the signal; we have simply moved it.

High-Side and Low-Side LO

We have the choice of two LO frequencies to convert a signal at the RF to a signal at the IF. The two LO possibilities are often defined as

Low-Side LO (LSLO). The LO frequency is less than the RF frequency.
High-Side LO (HSLO). The LO frequency is greater than the RF frequency.

In equation form, these definitions correspond to

$$\text{Low-Side LO} \Rightarrow f_{LO} < f_{RF}$$
$$\text{High-Side LO} \Rightarrow f_{LO} > f_{RF}$$
3.34

Occasionally, the terms *low side injection* for LSLO and *high side injection* for HSLO are used.

LO Frequency Calculation

In a channelized system, several similar signals (or channels) are spread over some frequency range. Each individual signal should be convertible to some common IF, where the signal will be processed or demodulated. The desired signal can be selected by changing the frequency of the LO.

Given the channelized conversion scheme of Figure 3-19, we can determine the HSLO and LSLO frequency ranges. In the following discussion, we define

f_{RFL} = the lowest frequency at the RF port to be converted to the IF center frequency.
f_{RFH} = the highest frequency at the RF port to be converted to the IF center frequency.
f_{IFCF} = the center frequency of the IF BPF.

For the low-side LO,
$f_{LSLO,L}$ = the lowest LO frequency required to perform the conversion,
$f_{LSLO,H}$ = the highest LO frequency required to perform the conversion.

For the high-side LO,
$f_{HSLO,L}$ = the lowest LO frequency required to perform the conversion,
$f_{HSLO,H}$ = the highest LO frequency required to perform the conversion.

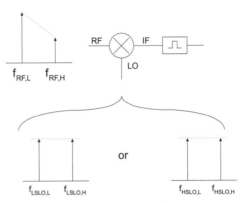

Figure 3-19 *To convert a given band of RF to some IF center frequency, there are two possible choices of LO frequencies.*

Figure 3-20 clarifies the definitions of $f_{RF,L}$ and $f_{RF,H}$. From here on, we will show only the lower and upper RF frequencies to keep the drawings from becoming too cluttered.

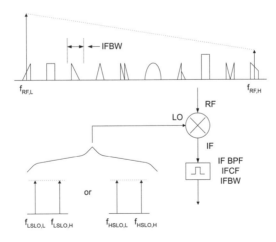

Figure 3-20 *RF spectrum and the two possible LO choices.*

290 | RADIO RECEIVER DESIGN

Equation 3.15 and Figure 3-21 provide some insight into the LO selection process. Figure 3-21 shows that the low-side LO frequency is a distance of f_{IF} below the RF. The high-side LO frequency is a distance f_{IF} above the RF.

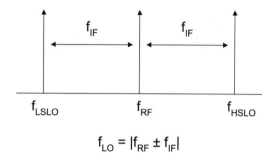

Figure 3-21 *Graphical interpretation of Equation 3.15.*

Low-Side LO

Combining Equation 3.15 with Figures 3-20 and 3-21 results in expressions for the low-side LO frequencies [see Figure 3-22(a)].

$$f_{LSLO,L} = f_{RF,L} - f_{IFCF}$$
$$f_{LSLO,H} = f_{RF,H} - f_{IFCF}$$
3.35

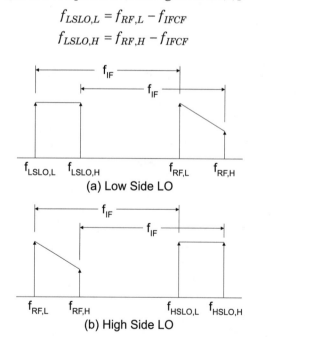

Figure 3-22 *Graphical expression of Equation 3.15. Given the required RF passband, find the two possible ranges of LO frequencies.*

High-Side LO

Similarly, the expressions for the high side LO frequencies are [see Figure 3-22(b)]

$$f_{HSLO,L} = f_{RF,L} + f_{IFCF}$$
$$f_{HSLO,H} = f_{RF,H} + f_{IFCF}$$
3.36

Example 3.5 — High-Side and Low-Side LO

In the United States, the cellular telephone system operates at 825 to 890 MHz. Each channel is 30 kHz apart and about 20 kHz wide. Use an IF of 45 MHz and find the *high-side LO* (HSLO) and *low-side LO* (LSLO) frequency ranges for this channelized system. Figure 3-23(a) shows the problem and Figure 3-23(b) shows the two LO ranges.

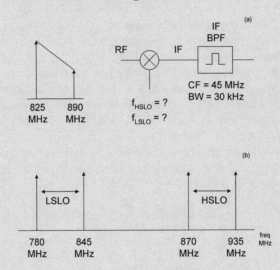

Figure 3-23 *Example 3.5. Given the RF range, find the two possible LO frequency ranges.*

Solution —

Looking only at the band edges of the RF, we find the LSLO using Equation 3.35.

$$f_{LSLO,L} = f_{RF,L} - f_{IFCF}$$
$$= 825 - 45 = 780 \text{ MHz}$$
$$f_{LSLO,H} = f_{RF,H} - f_{IFCF}$$
$$= 890 - 45 = 845 \text{ MHz}$$
3.37

The HSLO comes from Equation 3.36.

$$f_{HSLO,L} = f_{RF,L} + f_{IFCF}$$
$$= 825 + 45 = 870 \text{ MHz}$$
$$f_{HSLO,H} = f_{RF,H} + f_{IFCF}$$
$$= 890 + 45 = 935 \text{ MHz}$$

3.38

3.6 Frequency Inversion

Some conversion schemes produce a spectral or frequency inversion, which nevertheless does not distort the signal. For example, in a *digital frequency shift keying* (FSK) system, transmitting a tone below center frequency may correspond to a binary zero, and transmitting a tone above center frequency may represent a binary one. A system with frequency inversion will invert all the bits. The data is still present. We just have to know enough to place an inverter somewhere in the data path to make things right again. Which conversion schemes produce a frequency inversion and why?

Suppose we are given a LO frequency and a RF and want to examine the IF spectrum. We can apply Equation 3.13. Figure 3-24 shows Equation 3.13 in a graphical format. If we start at f_{LO} and move up in frequency a distance f_{RF}, we find one output sideband of the mixer. If we move f_{RF} down in frequency from f_{LO}, we find the other sideband. If we ignore the position of zero hertz, Figure 3-24 indicates that a frequency inversion will occur if the lower sideband is selected. However, if the lower sideband falls below zero hertz, the spectrum will be re-inverted.

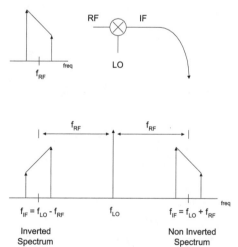

Figure 3-24 *A single conversion. If we ignore the position of zero hertz, the lower sideband is always spectrally inverted.*

LSLO

We will look at the low-side LO first (see Figure 3-25). In a LSLO conversion, the local oscillator frequency is always less than the RF and thus zero hertz lies between f_{LO} and $f_{IF} = f_{LO} - f_{RF}$. The position of zero hertz indicates that the lower sideband at $f_{IF} = f_{LO} - f_{RF}$ is made up entirely of negative frequencies. When we take the absolute values of the lower and upper band edges, we receive the noninverted spectrum shown at the bottom of Figure 3-25. The implied absolute value operation in Equation 3.13 re-inverts the spectrum of the lower sideband if it falls below zero hertz. Since the lower sideband always lies below zero hertz when a LSLO is used, a frequency inversion is not possible

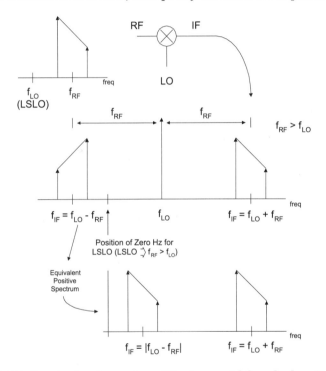

Figure 3-25 *Spectral re-inversion of the lower sideband when it falls below zero hertz. This conversion scheme uses a LSLO.*

HSLO

Figure 3-26 shows the situation with a high-side LO. With a HSLO, f_{LO} is always greater than f_{RF}, so the position of zero hertz is always below the lower sideband (i.e., it is below $f_{IF} = f_{LO} - f_{RF}$). Since the lower sideband spectrum is always above zero hertz, the spectrum remains inverted.

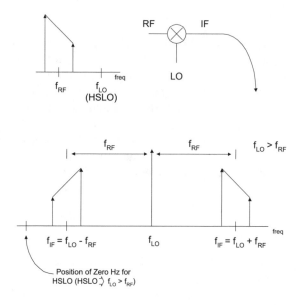

Figure 3-26 *The lower sideband of a HSLO never falls below zero hertz. The lower sideband does not experience a frequency re-inversion.*

Conclusion

A frequency inversion occurs only when we use a high-side LO and select the lower sideband (at $f_{IF} = f_{LO} - f_{RF}$). In equation form,

$$\text{Spectrum Inversion when} \begin{cases} \text{Using HSLO} \\ \text{and} \\ \text{Selecting the Lower Sideband} \end{cases} \quad 3.39$$

Example 3.6 — Cellular Telephone and Spectrum Inversion
In Example 3.5, we found that the United States cellular telephone system works with an RF frequency of 825 to 890 MHz. The IF is 45 MHz. We found two possible LO ranges that could be used to perform this conversion: the LSLO of 780 to 845 MHz and the HSLO of 870 to 935 MHz. Do either of these conversions produce a spectrum inversion?

Solution —
Equation 3.39 indicates that a frequency inversion occurs only with the HSLO and only when the lower sideband is used. The LSLO does not produce a frequency inversion. We can answer this problem looking only at the

midband HSLO and RF frequencies. The midband LO frequency is 902.5 MHz, and the midband RF is 857.5 MHz. Figure 3-26 shows that the sum and difference products of this conversion are 45 MHz (the desired IF) and 1760 MHz. Since the desired IF is the lower sideband, we conclude that we will see a frequency inversion.

Figure 3-27 shows one channel centered at 829.230 MHz.

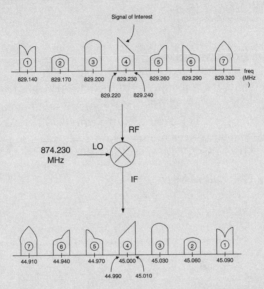

Figure 3-27 Example 3.6. Cellular telephone conversion scheme showing frequency inversion. The scheme uses a HSLO and the IF BPF selects the lower sideband.

This channel is about 20 kHz wide and the band edges are 829.220 and 829.240 MHz. We will mix this channel with a HSLO at

$$f_{HSLO} = f_{RF} + f_{IF}$$
$$= 829.230 + 45$$
$$= 874.230 \text{ MHz}$$

3.40

The lower RF band edge (at f_{RF} = 823.220 MHz) of this one channel emerges from the mixer at

$$f_{IF} = f_{HSLO} - f_{RF}$$
$$= 874.230 - 829.220$$
$$= 45.010 \text{ MHz}$$

3.41

Using a similar procedure, we find that the signal at the upper band edge of the channel at f_{RF} = 823.220 MHz emerges from the mixer at f_{IF} = 44.990 MHz. We will see a frequency inversion.

Every signal present at the RF port of the mixer combines with the LO and is frequency-converted into approximately 45 MHz. In this IF, or another lower-frequency IF, we will use a band-pass filter to select the one signal we are interested in receiving.

3.7 Image Frequencies

Looking at the TVRO example (Figure 3-28), we have to take another problem into consideration. A 3950 MHz signal is emitted by the satellite. We have selected a LSLO of 2750 MHz to convert the incoming TVRO signal to a 1200 MHz center frequency. Note that, for the purposes of example, we have omitted the RF filter in Figure 3-28.

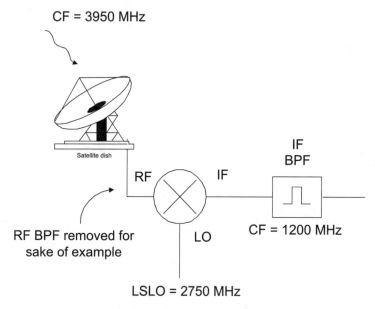

Figure 3-28 TVRO conversion scheme. The RF BPF has been removed to illustrate the image response of the system.

We know the LO frequency and the IF frequency (The IF frequency is set by the IF band-pass filter). Since we have omitted the RF band-pass filter, we will treat the RF as an unknown and use Equation 3.14 to analyze the system. Equation 3.14 produces

$$f_{RF} = f_{LO} \pm f_{IF}$$
$$= 2750 \pm 1200$$
$$= 1550 \text{ MHz} \quad \text{or} \quad 3950 \text{ MHz}$$
3.42

This analysis shows that there are two frequencies which, when placed into the RF port of this mixer, will convert to 1200 MHz. One is the desired frequency of 3950 MHz, the other is the image frequency at 1550 MHz. The design goal for this system was to convert 3950 MHz down to 1200 MHz. However, because of the way the conversion equations work out, we have also designed a system that will convert a signal at 1550 MHz down to 1200 MHz with the same efficiency. Any energy at 1550 MHz, including noise or an interfering signal, will be converted to 1200 MHz. Once an unwanted signal is converted to an IF, it cannot be removed. This *image response* is the reason we place a filter on the RF port of a mixer. One of the functions of the RF filter is to attenuate the image frequency so that signals or noise present at the image frequency are not converted to the IF along with the desired signals.

Locating Image Frequencies

To calculate the image frequencies we use Equation 3.14.

$$f_{RF} = f_{LO} \pm f_{IF}$$
3.43

Given a f_{IF} and a f_{LO}, any f_{RF} that satisfies Equation 3.14 will mix down to f_{IF} at the mixer's output.

LSLO Image Frequencies
Combining Equation 3.14 with Equation 3.35 produces an expression for the LSLO image frequencies.

$$f_{LSIM,L} = f_{LSLO,L} - f_{IF}$$
$$f_{LSIM,H} = f_{LSLO,H} - f_{IF}$$
3.44

Figure 3-29(a) shows this relationship graphically and makes another relationship apparent.

$$f_{LSIM} = f_{RF} - 2f_{IF}$$
3.45

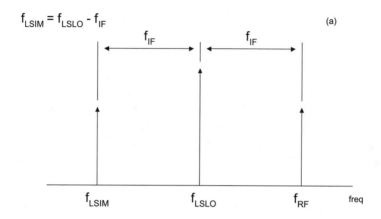

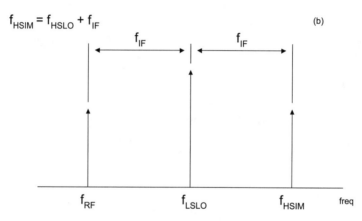

Figure 3-29 *Graphical relationship between the RF, LO and image frequencies.*

HSLO Image Frequencies

The expression for the HSLO image frequency is derived from Equations 3.14 and 3.36.

$$f_{HSIM,L} = f_{HSLO,L} + f_{IF}$$
$$f_{HSIM,H} = f_{HSLO,H} + f_{IF} \qquad 3.46$$

Figure 3-29(b) shows this relationship graphically. Again, we can see a second relationship among the variables.

$$f_{HSIM} = f_{RF} + 2f_{IF} \qquad 3.47$$

Example 3.7 — Commercial FM Radio

The commercial US FM radio band covers a RF frequency range of about 88 to 108 MHz. Normally, each station is converted to a 10.7 MHz IF for demodulation. Find
a. the HSLO range,
b. the range of HSLO image frequencies,
c. the LSLO range,
d. the range of LSLO image frequencies.

Solution —

a. Using Equation 3.36 for f_{HSLO}, we find

$$f_{HSLO,L} = f_{RF,L} + f_{IFCF}$$
$$= 88 + 10.7$$
$$= 98.7 \text{ MHz}$$

3.48

$$f_{HSLO,H} = f_{RF,H} + f_{IFCF}$$
$$= 108 + 10.7$$
$$= 118.7 \text{ MHz}$$

b. Equation 3.46 yields an expression for the high-side LO image frequencies.

$$f_{HSIM,L} = F_{HSLO,L} + f_{IF}$$
$$= 98.7 + 10.7$$
$$= 109.4 \text{ MHz}$$

3.49

$$f_{HSIM,H} = f_{HSLO,H} + f_{IF}$$
$$= 118.7 + 10.7$$
$$= 129.4 \text{ MHz}$$

Figure 3-30(a) shows the relationships among the high-side quantities.

c. Equation 3.35 yields f_{LSLO}.

$$f_{LSLO,L} = f_{RF,L} - f_{IFCF}$$
$$= 88 - 10.7$$
$$= 77.3 \text{ MHz}$$

3.50

$$f_{LSLO,H} = f_{RF,H} - f_{IFCF}$$
$$= 108 - 10.7$$
$$= 97.3 \text{ MHz}$$

d. Equation 3.44 yields the LSLO image frequencies.

$$f_{LSIM,L} = f_{LSLO,L} - f_{IF}$$
$$= 77.3 - 10.7$$
$$= 66.6 \text{ MHz}$$

3.51

$$f_{LSIM,H} = f_{LSLO,H} - f_{IF}$$
$$= 97.3 - 10.7$$
$$= 86.6 \text{ MHz}$$

Figure 3-30(b) shows the low-side relationships.

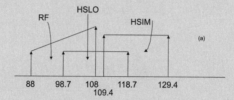

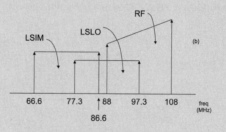

Figure 3-30 *Graphical relationship between the RF:LO and image frequencies in a commercial FM radio using (a) a HSLO (b) a LSLO.*

Image Noise

Mixers can add extra noise to a system in a way that is not obvious. This effect, called *image noise*, can also cause incorrect results when the mixer's noise figure is measured. Figure 3-31 illustrates the problem. The local oscillator frequency is 200 MHz and the center frequency of the IF filter is 45 MHz. Let us assume that the bandwidth of the RF amplifier is 20 to 1000 MHz and that the signal of interest from the antenna resides at 245 MHz.

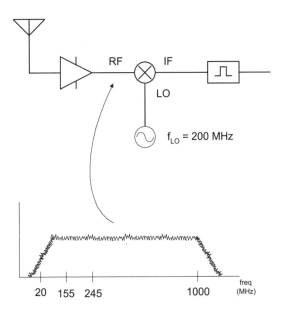

Figure 3-31 *The image response of the conversion process can translate out-of-band noise into the IF filter, increasing the noise contribution of the mixer.*

Figure 3-31 also shows the noise spectrum present at the RF port of the mixer. The architecture of the system assures that any energy present on the mixer's RF port at 245 MHz will be converted to 45 MHz. However, because of the "±" present in the conversion equation, we will also convert energy centered at 155 MHz on the mixer's RF port to the 45 MHz IF. The result is that there will be more noise present at the mixer's IF port than anticipated and, if the noise power at 245 MHz is equal to the noise power present at 155 MHz, the noise on the mixer's IF port will be 3 dB greater than anticipated.

3.8 Other Mixer Products

In practice, we build a receiver by first deciding on a conversion scheme, then placing the appropriate RF and IF band-pass filters in position. Normally, the RF band-pass filter is wider than desirable and passes more than just the signal of interest. For example, in the case of commercial FM broadcast reception, we allow the entire 20 MHz worth of RF spectrum to come through although we only want to listen to one station at a time. Finally, we apply the LO to the mixer and examine the signals which emerge from the mixer's IF port.

TVRO Example

Figure 3-32 shows the LSLO conversion scheme for the TVRO downconverter.

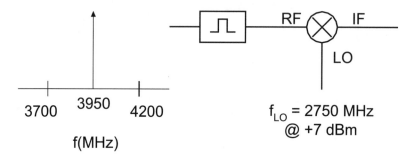

Figure 3-32 *TVRO downconverter with one RF signal and one LO frequency.*

We apply the LO and RF to the mixer and examine the signals present on the IF port. Equation 3.13 indicates the frequency of the desired output signals.

$$f_{IF} = f_{LO} \pm f_{RF} \\ = 2750 \pm 3950 \\ = 6700 \text{ MHz and } 1200 \text{ MHz}$$ 3.52

Two signals are emerging from the mixer due to the mathematically derived mixing function. However, because of practical considerations, signals at other frequencies will emerge from the mixer including the RF signal (still at f_{RF}) and its harmonics, the LO signal (at f_{LO}) and its harmonics and every combination of

$$|mf_{LO} \pm nf_{RF}| \quad for \begin{cases} m = 0,1,2,... \\ n = 0,1,2,... \end{cases}$$ 3.53

including the desired signal and its image.

Table 3-1 lists some of the frequencies that will be present on the output of the mixer.

Table 3-1 Mixer spurious components.

m	n	mf_{LO} (MHz)	nf_{RF} (MHz)	•$mf_{LO}+nf_{RF}$• (MHz)	•$mf_{LO}-nf_{RF}$• (MHz)
0	0	0	0	0	0
0	1	0	3950	3950	3950
0	2	0	7900	7900	7900
0	3	0	11850	11850	11850
1	0	2750	0	2750	2750
1	1	2750	3950	6700	1200
1	2	2750	7900	10650	5150
1	3	2750	11850	14600	9100
2	0	5500	0	5500	5500
2	1	5500	3950	1550	9450
2	2	5500	7900	2400	13400
2	3	5500	11850	6350	17350
3	0	8250	0	8250	8250
3	1	8250	3950	12200	4300
3	2	8250	7900	16150	350
3	3	8250	11850	20100	3600

The output spectrum of the TVRO mixer will resemble Figure 3-33. In reality, there will probably be more signals present at the mixer's output than those shown.

Many of these output products can be easily removed by the IF filter. Others will fall either inside the IF filter or close enough to the filter's passband that they will not experience significant attenuation. We are interested in the frequencies of each mixer product and its expected power level.

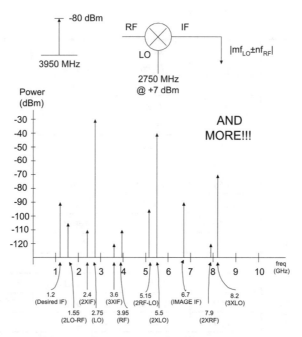

Figure 3-33 *Possible signals present at the IF port of the mixer in Figure 3-32.*

Mixer Spur Tables

Mixer data sheets often include just such information. Table 3-2 shows the data measured for one particular mixer used in one particular conversion scheme at specific levels of RF and LO powers. The RF and LO frequencies are fixed. The power of the spurious products present on the mixer's output is a strong function of the LO and RF signal power. Spurious performance also changes with RF and LO frequencies.

Table 3-2 shows relative power levels of the various $mf_{LO} \pm nf_{RF}$ mixing products for one particular mixer and conversion scheme. The numbers in the table represent the number of dB the undesired product (at $mf_{LO} \pm nf_{RF}$) is below the desired product (at $f_{LO} \pm f_{RF}$). This data was gathered under the following conditions:

- f_{RF} = 500 MHz at –4 dBm
- f_{LO} = 470 MHz at 7 dBm
- f_{IF} = 30 MHz, measured to be –11 dBm.

The numbers in Table 3-2 are fairly typical although the RF power is quite high. Mixer catalogs often contain similar spurious power tables.

Table 3-2 Power in mixer spurious components.

		RF Harmonic (n)									
		0	1	2	3	4	5	6	7	8	9	10
LO Harmonic (m)	0		17	43	44	76	66	69	72	>85	>85	84
	1	32	0	60	34	68	75	76	71	>85	83	84
	2	23	39	45	49	56	67	83	73	81	84	83
	3	36	17	56	35	72	53	84	74	>85	83	85
	4	43	33	51	49	56	57	70	73	85	84	84
	5	35	37	60	36	61	48	71	67	>85	>85	83
	6	44	38	65	45	57	58	64	64	83	83	84
	7	36	39	52	50	65	47	75	63	76	77	84
	8	61	39	58	52	76	71	64	63	73	71	83
	9	46	47	41	57	75	70	78	61	76	71	81
	10	63	51	69	50	66	67	79	71	71	71	81

Table 3-2 shows the amount of power in every n by m spurious output product (up to 10 by 10). If we change the input conditions (their frequency and especially the power of the LO and RF signals), this table will not be accurate.

The desired output signal of the mixer (at $f_{LO} \pm f_{RF}$) is at the intersection of RF harmonic (n) = LO harmonic (m) = 1. The table entry at that spot is zero.

Example 3.8 — Mixer Spurious Products
The desired signal present at the IF port of a mixer (at $f_{RF} - f_{LO}$) has a frequency of 30 MHz and an amplitude of –11 dBm. The RF is f_{RF} = 500 MHz and LO frequency f_{LO} is 470 MHz. Using Table 3-2 above, find the levels of the following signals.
a. $3f_{LO} - 4f_{RF}$ = 590 MHz,
b. $2f_{LO} - 2f_{RF}$ = 60 MHz,
c. $6f_{LO} + 7f_{RF}$ = 680 MHz,
d. $-4f_{LO} + 2f_{RF}$ = 880 MHz.

Solution —
Using Table 3-2, we find
a. Looking up LO harmonic = 3, RF harmonic = 4, we find this output product is 72 dB below the desired output product at –11 dBm. The level of the signal at 590 MHz is –11 – 72 = –83 dBm.
b. For LO harmonic = 2, RF harmonic = 2, the output product is 45 dB below the desired output product at –11 dBm. The level of the signal at 60 MHz is –11 – 45 = –56 dBm.

c. The table tells us the suppression of the product with LO harmonic = 6, RF harmonic = 7 is 64 dB. The power of the signal at 680 MHz is –11 – 64 = –75 dBm.
d. Finally, the $m = 4$, $n = 2$ product is suppressed by 51 dB. The output power at 880 MHz is –11 – 51 = –62 dBm.

Double-Balanced Mixers

Table 3-2 characterizes a *double-balanced mixer* (DBM), the most common mixer used in high-end receiving equipment. From Table 3-2 and other mixer tables describing double-balanced mixers, we can draw some general conclusions about the spurious output products of these mixers.

- *Even-by-even products*. Examples of even by even products are 2 by 6, 8 by 4 and so on. Examination of many mixer tables indicates that the even-by-even products tend to be the weakest spurious components generated in the mixer. As such, they are the least worry.

- *Even-by-odd and odd-by-even products*. Examples are 3 by 2, 4 by 5 and 1 by 7. These responses are generally higher power than the even-by-even products.

- *Odd-by-odd products*. Examples are 3 by 7 and 1 by 5. These are generally the highest-level undesired products. Consequently, they are the most trouble.

- *Order of the Undesired Product*. The order of the product is

$$\text{Order} = m + n \qquad 3.54$$

It follows that the higher the order, the lower the power level of the product. Low-order spurious output signals tend to have larger magnitudes than high-order spurious products. With this in mind, we tend to be more concerned about low-order products than about high-order products. In general, the more odd products the undesired component is associated with, the higher the output power. These characteristics are a direct result of the topology of the double-balanced mixer. The power of the spurious products generated inside of a mixer is a strong function of the input power levels and the characteristics of the particular mixer. The spurious output power is a weaker function of the frequency of operation. In short, if any of the input characteristics are changed, the mixer table probably will not be accurate.

3.9 Spurious Calculations

Suppose we have found a candidate conversion scheme for a system we want to build. We know the range of RF signals we want to process, we have decided on an LO range and know the IF center frequency and bandwidths. Figure 3-34 shows the system complete with RF and IF filters. Now we would like to analyze this system to find out how well it performs from a spurious signal perspective.

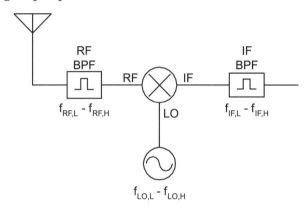

Figure 3-34 *A candidate receiver architecture, showing the definitions of frequencies used in the analysis of the receiver's spurious performance.*

In the simplest case, the signals present on the IF port of the mixer will be

$$f_{IF} = mf_{LO} \pm nf_{RF} \qquad 3.55$$

where
f_{IF} = the frequencies of the signals present at the mixer's output port,
f_{LO} = the LO frequency,
f_{RF} = the RF frequency,
$m = 0, 1, 2, 3 \ldots$,
$n = 0, 1, 2, 3 \ldots$

The RF and LO signals present at the ports of the mixer in Figure 3-34 now represent frequency bands rather than single tones. In Equation 3.55, f_{LO} now represents the ranges of $f_{LO,L}$ to $f_{LO,H}$, and f_{RF} represents $f_{RF,L}$ to $f_{RF,H}$.

We begin by asking ourselves the following questions: Do any of the mixer's spurious products fall within the passband of the IF filter? If some spurious products from the mixer do find their way into the IF filter, what is their power level? Are they low-order signals (implying that they will be relatively high power) or are they high-order signals (and thus, tending to be low power)?

Assumptions

To simplify the analysis of Figure 3-34, we will assume the following:

- The RF band-pass filter is a brick wall filter, i.e., it passes all the frequencies between $f_{RF,L}$ and $f_{RF,H}$ while completely attenuating everything else.

- The IF band-pass filter is also a brick wall filter. It will pass all the signals from $f_{IF,L}$ to $f_{IF,H}$ while completely attenuating signals outside of this range. Alternately, we will sometimes specify the IF band-pass filter with its center frequency f_{IFCF} and its bandwidth (B_{IF}).

- This conversion scheme dictates that the local oscillator will tune from $f_{LO,L}$ to $f_{LO,H}$.

- In an operational system, the RF band-pass filter will be connected to an antenna so we will have no control over the signals entering the RF port of the mixer. We only know that their frequencies must lie between $f_{RF,L}$ to $f_{RF,H}$. The worst-case scenario is to assume that the entire RF band is flooded with signals. In other words, we will assume that every frequency in the range of $f_{RF,L}$ to $f_{RF,H}$ is present at the RF port of the mixer.

- As the receiver tunes from one channel to the next, the frequency of the LO will change. In an operational system, the LO can be constantly changing in frequency, or it could sit on one frequency for months at a time. All we know is that the LO frequency will be somewhere between $f_{LO,L}$ and $f_{LO,H}$. The worst-case assumption is to assume that every frequency from $f_{LO,L}$ to $f_{LO,H}$ is present at the LO port of the mixer. In reality, only one frequency at a time is present on the LO port of the mixer.

Analysis Procedure

Figure 3-35 illustrates the analysis procedure. We have a band of signals extending from $f_{RF,L}$ to $f_{RF,H}$ on the mixer's RF port. We also have a band of signals from $f_{LO,L}$ to $f_{LO,H}$ on the mixer's LO port. We then want to calculate the signals present on the output port of the mixer with special attention to those signals falling between $f_{IF,L}$ and $f_{IF,H}$. We must perform this analysis for every value of m (the LO harmonic number) and n (the RF harmonic number). Any spurious signals that pass through the IF filter without significant attenuation may present a problem.

MIXERS | 309

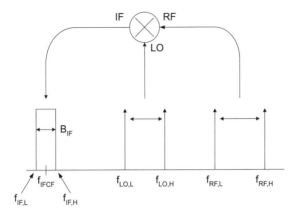

Figure 3-35 *Graphical illustration of the spurious analysis procedure. We repeat this analysis for each value of m and n.*

Derivation

We want to find the range of frequencies present on the mixer's IF port for each value of m and n. Equation 3.55, repeated here, describes the frequencies of the spurious IF output signals given the RF and LO input frequencies.

$$f_{IF} = mf_{LO} \pm nf_{RF} \qquad 3.56$$

where
f_{LO} = any frequency from $f_{LO,L}$ to $f_{LO,H}$,
f_{RF} = any frequency from $f_{RF,L}$ to $f_{RF,H}$,
$m = 0, 1, 2, 3 ...$,
$n = 0, 1, 2, 3 ...$.

Figure 3-36 indicates that the LO spectrum from $f_{LO,L}$ to $f_{LO,H}$ will mix with the RF spectrum from $f_{RF,L}$ to $f_{RF,H}$ to produce two frequency bands on the mixer's IF port. The two bands fall in the ranges of $f_{IF,-,L}$ to $f_{IF,-,H}$ (the lower sideband) and $f_{IF,+,L}$ to $f_{IF,+,H}$ (the upper sideband). Figure 3-36 depicts the operation for only one value of m and one value of n. If we change either m or n, we have to recalculate because the upper and lower IF sidebands will shift in frequency.

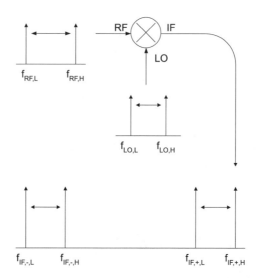

Figure 3-36 *Given the possible RF and LO ranges, find the possible outputs present on the mixer's IF port. Perform this analysis for every value of m and n.*

We can find expressions for $f_{IF,-,L}$, $f_{IF,-,H}$, $f_{IF,+,L}$ and $f_{IF,+,H}$ by applying Equation 3.55 to the band edges of the RF and LO spectrums. For the lower sideband using $f_{IF} = mf_{LO} - nf_{RF}$, we find

$$f_{IF,-,L} = mf_{LO,L} - nf_{RF,H}$$
$$f_{IF,-,H} = mf_{LO,H} - nf_{RF,L}$$
 3.57

For the upper sideband, using $f_{IF} = mf_{LO} + nf_{RF}$, we find

$$f_{IF,+,L} = mf_{LO,L} + nf_{RF,L}$$
$$f_{IF,+,H} = mf_{LO,H} + nf_{RF,H}$$
 3.58

Example 3.9 — Cellular Telephone

Find the range of the two frequency bands emerging from a mixer's IF port for the second harmonic of the LO combined with the third harmonic of the RF. Assume the system is a cellular telephone receiver covering 825 to 890 MHz. The HSLO is 870 to 935 MHz. If the IF is 45 MHz and we have a 1 MHz wide band-pass filter, do either of these output bands represent a possible problem?

Solution —
Equation 3.57 is used to find the range of the lower sideband output.

$$f_{IF,-,L} = mf_{LO,L} - nf_{RF,H}$$
$$= (2)(870) - (3)(890) = -930 \text{ MHz}$$
$$f_{IF,-,H} = mf_{LO,H} - nf_{RF,L}$$
$$= (2)(935) - (3)(825) = -605 \text{ MHz}$$

3.59

The lower sideband will exist between 605 and 930 MHz. Now we use equation 3.58 to find the range of the upper sideband.

$$f_{IF,+,L} = mf_{LO,L} + nf_{RF,L}$$
$$= (2)(870) + (3)(825) = 4215 \text{ MHz}$$
$$f_{IF,+,H} = mf_{LO,H} + nf_{RF,H}$$
$$= (2)(935) + (3)(890) = 4540 \text{ MHz}$$

3.60

The upper sideband lies between 4215 and 4540 MHz. Figure 3-37 is a graphical interpretation of these results. It shows that neither output band overlaps the IF band-pass filter. In fact, they are both well removed from the filter's center frequency. We conclude that this particular response will not pose a problem.

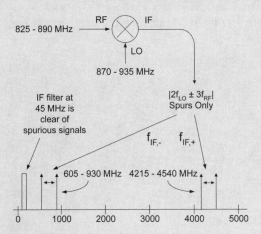

Figure 3-37 Spurious analysis of a cellular telephone conversion scheme for m = 2 and n = 3. This combination of m and n does not produce signals which will pass through the IF filter. The m = 2, n = 3 spurious product is therefore not a concern.

Example 3.10 — Cellular Telephone
Repeat the problem for the third harmonic of the LO and the third harmonic of the RF.

Solution —

Equation 3.57 can be used to find the range of the lower sideband output.

$$f_{IF,-,L} = mf_{LO,L} - nf_{RF,H}$$
$$= (3)(870) - (3)(890) = -60 \text{ MHz} \quad\quad 3.61$$
$$f_{IF,-,H} = mf_{LO,H} - nf_{RF,L}$$
$$= (3)(935) - (3)(825) = 330 \text{ MHz}$$

Note that this band runs from −60 MHz to 330 MHz (not from 60 to 330 MHz). If we consider only positive frequencies, this response covers 0 to 330 MHz. We use Equation 3.58 to find the range of the upper sideband.

$$f_{IF,+,L} = mf_{LO,L} + nf_{RF,L}$$
$$= (3)(870) + (3)(825) = 5085 \text{ MHz} \quad\quad 3.62$$
$$f_{IF,+,H} = mf_{LO,H} + nf_{RF,H}$$
$$= (3)(935) + (3)(890) = 5475 \text{ MHz}$$

Figure 3-38 shows that the IF passband does fall inside the range of the lower sideband. This combination of LO and RF products may produce a spurious response. The next step is to calculate the likely power level of these signals.

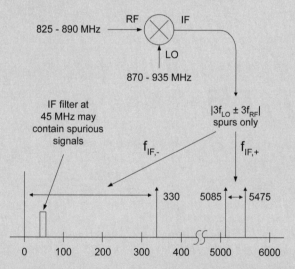

Figure 3-38 *Spurious analysis of a cellular telephone conversion scheme for m = 3 and n = 3. This combination of m and n might produce signals which will pass through the IF filter. The m = 3, n = 3 spurious product is therefore a concern.*

To analyze a conversion scheme properly, this calculation should be performed for every value of m and n that could produce a troublesome product. We usually run the analysis for all orders up to 15. Computer assistance makes it easy to compare many different candidate conversions schemes.

3.10 Mixer Realizations

So far we have looked at mixers as black boxes placed at appropriate points in the receiver cascade. To appreciate the strengths and weaknesses of these devices, it is important to understand how mixers work and how they are built. We will look at several different topologies and explore the advantages and problems with each of them.

3.11 Single-Ended Mixers (SEM)

The term *single-ended mixer* (SEM) generally refers to the technique of using a nonlinear device to perform frequency conversion. One input signal is the RF, the other is the LO. The nonlinear device can be a diode, a transistor, or an amplifier driven into saturation. In this example, we will describe a single-diode mixer. This inexpensive technique is common in commercial equipment.

Operation

Figure 3-39 shows a simplified diagram of a single-ended mixer. The RF, LO and IF band-pass filters increase the interport isolation, and the diode is the nonlinear element that performs the time-domain multiply.

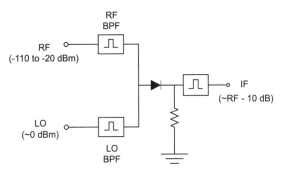

Figure 3-39 Schematic diagram of a simple single-ended mixer (SEM). The LO switches the diode and gates the RF to the IF port. The bandpass filters work to improve interport isolation.

Conceptually, it is helpful to think of the diode as a switch which is controlled by the local oscillator voltage (see Figure 3-40). When the instantaneous LO voltage is greater than the diode's turn-on voltage, current flows through the diode, forward-biasing it. The switch in Figure 3-40 is closed, and the RF signal passes through to the IF filter. When the instantaneous LO voltage is less than the diode's turn-on voltage, the diode is reversed-biased and the RF port is isolated from the IF port.

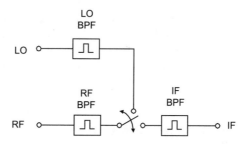

Figure 3-40 A single-ended mixer consists of a switch controlled by the LO voltage. When the LO is greater than the diode's turn-on voltage, the RF is passed to the IF port. When the LO is less than the diode's turn-on voltage, the RF is isolated from the IF port.

Figure 3-41 shows the appropriate waveforms. (V_{IF} is the voltage present at the input to the IF filter.)

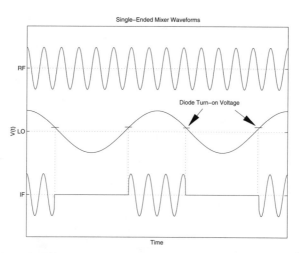

Figure 3-41 Time-domain waveforms present in a single-ended mixer. The switching action of the diode performs an approximation of the time-domain multiply between the LO voltage and the RF voltage.

Instead of performing a time-domain multiply between the LO and RF waveforms, we are multiplying the RF waveform by either zero or unity at a rate determined by the LO. In other words, when the LO voltage is greater than a diode drop, we multiply the RF by unity; when the LO voltage is less than a diode drop, we multiply the RF by zero. In equation form, the multiplying waveform, $W_{LO}(t)$, derived from the LO is

$$W_{LO}(t) = \begin{cases} 0 \text{ when } V_{LO}(t) \leq \text{Diode Turn-on Voltage} \\ 1 \text{ when } V_{LO}(t) > \text{Diode Turn-on Voltage} \end{cases} \quad 3.63$$

Fourier analysis reveals an equivalent expression for $W_{LO}(t)$.

$$W_{LO}(t) = a_0 + \sum_{k=1}^{\infty} a_k \cos(k\omega_{LO} t + \theta_k) \quad 3.64$$

Ideally, we multiply the RF waveform by the LO waveform which produces only the sum and difference products. In the single-ended mixer, we multiply the RF waveform by a constant and all the harmonics of the LO. This still produces the desired sum and difference products but it also produces many other unwanted (spurious) mixing products.

Example 3.11 — DC and RF:IF Rejection
Show that the DC term in Equation 3.64 is responsible for the poor RF:IF isolation of the single-ended and single-balanced mixers.

Solution —
Equation 3.64 is the Fourier transform of the LO-derived multiplying waveform. This waveform equals +1 when the instantaneous LO voltage is greater than zero and the waveform equals 0 when the instantaneous LO voltage is less than zero. In equation form, this waveform and its Fourier transform are

$$W_{LO}(t) = \begin{cases} 0 \text{ when } V_{LO}(t) \leq \text{Diode Turn-on Voltage} \\ 1 \text{ when } V_{LO}(t) > \text{Diode Turn-on Voltage} \end{cases}$$

$$= a_0 + \sum_{k=1}^{\infty} a_k \cos(k\omega_{LO} t + \theta_k) \quad 3.65$$

In the mixing process, we form a time-domain multiplication between this waveform and the RF signal. The result of this multiplication is

$$V_{IF}(t) = \left[\cos(\omega_{RF}t)\right]\left[a_0 + \sum_{k=1}^{\infty} a_k \cos(k\omega_{LO}t + \theta_k)\right]$$

$$= a_0 \cos(\omega_{RF}t) + \left[\cos(\omega_{RF}t)\right]\left[\sum_{k=1}^{\infty} a_k \cos(k\omega_{LO}t + \theta_k)\right] \quad 3.66$$

The first term in this equation represents the RF:IF feedthrough of the mixer. The second term cannot produce a signal at the f_{RF}.

The astute reader may notice that the multiplying waveform $W_{LO}(t)$ is almost a square wave with a 50% duty cycle. This approximation approaches reality when the LO voltage is much greater than the turn-on voltage of the diode. For a large LO voltage, we can approximate $W_{LO}(t)$ as

$$W_{LO}(t) = \frac{1}{2} + \sum_{k=1}^{\infty} \frac{1}{2k-1} \sin\left[(2k-1)\omega_{LO}t\right] \quad 3.67$$

This expression contains only the odd harmonics of the LO frequency. Since the even harmonics are not present in the multiplying waveform, they will not cause spurious products. For large LO powers, the single-ended mixer suppresses the spurious products generated by the even harmonics of the LO.

Due to realization effects, the time-domain waveforms shown in Figure 3-41 are not exactly accurate. Every diode has a forward voltage drop, a forward series resistance, a reverse current, interterminal capacitance and a nonzero switching time. Consequently, the waveform present at the input to the IF filter is different from how we have drawn it. The actual waveform is decidedly nonsymmetrical and contains all of the LO harmonics (not just the odd ones) and all of the harmonics of the RF signal. Instead of multiplying the RF waveform by the LO, all of the harmonics of the LO are multiplied by all of the harmonics of the RF. In other words, the spectrum of the signal present at the input port of the IF filter contains all of the $|mf_{LO} \pm nf_{RF}|$ spectral components in significant amounts.

The SEM can be a spurious product nightmare. However, if the RF, IF and LO signals are widely separated, then the RF, IF and LO band-pass filters can provide excellent interport isolation. We can often tolerate the poor spurious performance of a SEM in narrowband systems such as commercial FM radio receivers.

SEM LO Power

When we discussed single-ended mixers, we assumed that the diode switch was completely controlled by the LO voltage and opened and closed exactly at the zero crossing of the LO signal. This is approximately true when the LO voltage is much greater than the RF voltage. As the magnitude of the RF voltage approaches the magnitude of the LO voltage, the exact switching instant of the diode will change. Equations 3.63 and 3.64 will be less accurate as the RF power increases. As a result, the spurious output levels of the mixer will rise although the desired mixing action still occurs. In practice, we make sure that the RF power level entering the mixer is at least 20 dB below the LO power level or

$$P_{RF,dBm} \leq P_{LO,dBm} - 20 \qquad 3.68$$

A single-ended mixer typically requires at least 0 dBm of LO power to forward-bias the mixing diode. This limits the maximum RF signal power to approximately –20 dBm.

Diode Resistance

In practice, we do not have to supply enough LO voltage to turn the diode from its fully nonconducting state to its fully conducting state. We can get by with much less LO voltage. Figure 3-42 shows a more accurate model of the single-ended mixer.

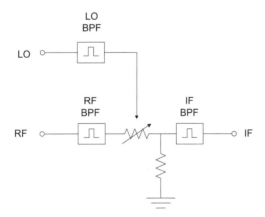

Figure 3-42 A single-ended mixer that accounts for the non-zero diode resistance when the diode is in its conducting state. This model also accounts for the finite resistance of the diode when the diode is in its nonconducting state.

The diode is now modeled as a resistor whose resistance varies with the instantaneous LO voltage. The voltage divider action between the diode and the shunt resistor at the input to the IF filter forms the basis for the multiplying action. Since the change in diode impedance can vary widely with temperature and DC bias current, it is difficult to describe the LO-derived multiplying function mathematically. It varies at a rate determined by the LO. As the LO power decreases, the change in diode resistance also decreases. Consequently, for decreasing LO power, the conversion loss of the mixer increases (see Figure 3-43).

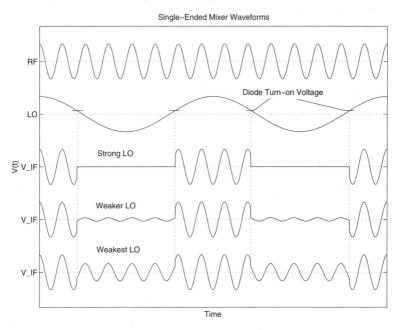

Figure 3-43 *The effect of insufficient LO power on the waveforms in a single-ended mixer. Insufficient LO power causes increased RF:IF feedthrough, higher conversion loss and higher spurious power.*

SEM Advantages and Disadvantages

Single-ended mixers are on the low end of the performance spectrum yet cost-efficient. As such, these mixers are usually operated with little concern for the spurious performance, LO power or anything other than cost. One of the chief advantages of a single-ended mixer is that the nonlinear element can be a simple diode or transistor.

LO and RF Power

There are various trade-offs in selecting the LO power suitable for a single-ended mixer. A large LO power causes the switching waveform to approach a square wave and suppresses the spurious products involving the odd-order LO harmonics. However, a large LO may be difficult to generate or may waste power. The minimum power required by a SEM is relatively low (usually 0 dBm or less) since we only have to drive a single diode or transistor. This simplifies the LO design and is easier on battery-powered systems.

In any mixer, we would like the LO to be the only waveform that switches the mixer diodes on and off. With this in mind, it is a good idea to keep the RF power applied to the mixer at least 20 dB below the LO power level. When a low-power LO is used, the amount of RF power we can apply to the mixer is limited, which reduces the dynamic range of the mixer.

Interport Isolation

The LO:RF, LO:IF and RF:IF isolations depend almost entirely upon the three band-pass filters placed around the nonlinear element. For high isolation, the frequencies of the RF, LO and IF should be well separated, which puts constraints on the possible conversion schemes that can be realized with a single-ended mixer.

Suppression of Spurious Products

The spurious performance of a SEM is poor. For small LO powers, there is no mechanism to suppress any of the

$$|mf_{LO} \pm nf_{RF}| \qquad 3.69$$

products. We can normally accept this in narrowband systems, but we must consider a more sophisticated topology in a wideband system, or in cases where more spurious signal suppression is needed.

Port Impedances

The input and output impedances of a SEM are not friendly. Since the diode switch opens and closes at the LO rate, the RF and IF band-pass filters are alternately connected together and open-circuited. The input impedance of a SEM can change significantly over the period of the LO. We can use matching attenuators or other circuitry to alleviate this problem.

Conversion Loss

When a passive device is used to convert a RF signal into two equal-level IF signals, we will suffer an automatic 6 dB conversion loss. Half of the RF voltage goes to one sideband and half goes to the other sideband. In the single-ended mixer, the RF port is connected to the IF port only 50% of the time. The theoretical minimum conversion loss of a single-ended mixer is roughly 9 dB. In practice, realization effects cause the insertion loss to be 10 to 20 dB.

3.12 Single-Balanced Mixers (SBM)

The single-ended mixer is inexpensive and easy to realize, but has flaws such as poor interport isolation and many spurious signals. The *single-balanced mixer* (SBM) is an improvement over the SEM. Figure 3-44 shows the schematic diagram of one type of single-balanced mixer.

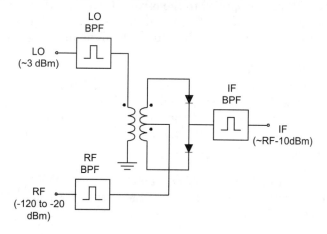

Figure 3-44 *Schematic diagram of a simple single-balanced mixer (SBM). Operation is similar to the SEM but the balun allows us to improve LO:IF and LO:RF isolation.*

This single-balanced mixer contains a balanced-to-unbalanced transformer, or *balun*. To understand the operation of the complete mixer, we first have to understand the operation of a balun.

Baluns

Baluns convert unbalanced signals to balanced signal and vice versa. An unbalanced signal is commonly carried on one wire and is referenced to a ground. The information in an unbalanced signal is derived from how the signal on the single wire varies with respect to ground.

A balanced signal is carried on two wires and is not referenced to ground. The information content of a balanced signal is derived from how the signal in one wire changes with respect to the other wire. To convert an unbalanced signal to a balanced signal, we first split the signal into two equal parts, then multiply one signal by -1 or $1 \angle 180°$. We put the positive signal on one wire and the negative signal on the other wire to form the balanced signal. To form an unbalanced signal from a balanced signal, we subtract the negative signal from the positive signal.

We will use the balun of Figure 3-44 to convert the unbalanced LO signal into a balanced signal. Since we now have the positive and negative versions of the LO together in the same system, we have the opportunity to add the two signals together. If we work everything right, we can completely cancel the LO signal out at the IF and RF ports of the SBM.

The Dot Convention

It is common practice to draw dots on the terminals of a transformer to indicate polarity. The dot convention shows how the transformer is wound and the polarity of output windings [see Figure 3-45(a)].

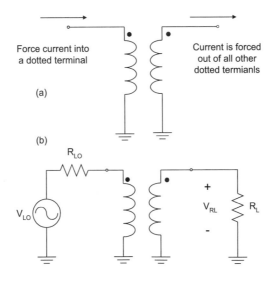

Figure 3-45 *Transformer (balun) dot convention.*

By definition, if we force a current into any dotted terminal, it will be forced out of all the other dotted terminals, as shown in Figure 3-45(a). Likewise, if we force a current into an undotted terminal, current will flow out of all the undotted terminals of the transformer. Of course, the magnitude of the current depends upon the turns ratio and upon the losses in the transformer.

Figure 3-45(b) shows the dot convention with respect to voltage. The voltage source V_{LO} forces a current into the primary's dotted terminal, so current flows out of the secondary's dotted terminal and through the load resistor. The voltage across the load resistor (V_{RL}), will be of the polarity shown in the figure. The dotted terminal will exhibit a positive voltage. It follows that if we apply a positive voltage to any dotted terminal, all of the other dotted output terminals will exhibit a positive voltage.

Balun Voltages

When speaking of mixers, the term *balun* usually refers to a balanced transformer with a center tap, as shown in Figure 3-46. The center tap can be grounded or connected to a signal source, or to a load.

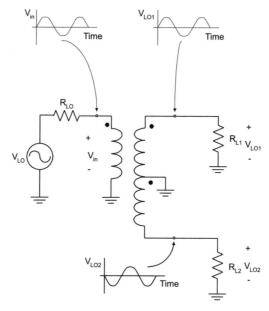

Figure 3-46 *Voltage generated in a simple balun (balanced-to-unbalanced transformer).*

If a signal is applied to the primary side of the balun (on the left of Figure 3-46), the secondary output voltages (V_{LO1} and V_{LO2}) appear with the polarities shown. In an ideal balun, the transformers are exactly matched and the voltage V_{LO1} exactly equals $-V_{LO2}$. In a non-ideal balun, the magnitudes of the two voltages are about equal and the two signals are approximately 180° out of phase (i.e., $V_{LO1} \cong -V_{LO2}$).

Figure 3-47 shows the circuit of Figure 3-46 with a slight modification. We have simply connected the two grounded ends of the load resistors together. If we assume that the resistors have exactly the same value (i.e., $R_{L1} = R_{L2} = R_L$) and that $V_{LO1} = -V_{LO2} = V_L$, then the current through the resistors is

$$I_{LO} = \frac{2V_L}{2R_L}$$

$$= \frac{V_L}{R_L}$$

3.70

Thus, current flows through the resistors R_{L1} and R_{L2} at a rate determined by V_{LO}.

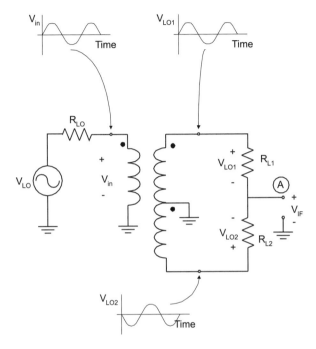

Figure 3-47 *Operation of a balanced-unbalanced transformer (BALUN). Transformer action produces equal and opposite voltages across R_{L1} and R_{L2}, canceling the LO voltage at the IF port.*

If the resistors have exactly the same value and the balun is perfectly balanced (V_{LO1} exactly equals $-V_{LO2}$), then the voltage V_{IF} at point A, sometimes referred to as a *virtual ground,* will be zero.

This effect does not apply only to sine waves. Any complex waveform can be applied to the transformer primary to achieve perfect cancellation at point A if the balance remains intact over the bandwidth of the signal. If the transformers are not exactly balanced or the two components are not exactly the same, then we will see a nonzero voltage at point A. The magnitude of the voltage at point A is a direct measure of the system's balance.

Center Tap Drive

Figure 3-48 shows another balun configuration. We are driving the center point of the balun with V_{RF}. We want to calculate what the voltages V_{LO} and V_{IF} will be. Again, we will assume the balun is perfectly balanced and $R_{L1} = R_{L2}$.

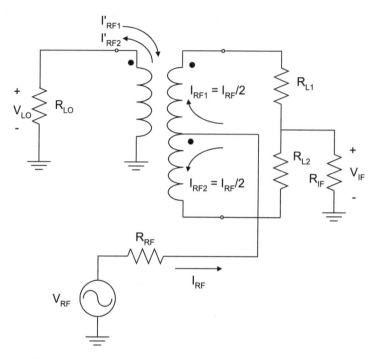

Figure 3-48 *Operation of a balanced-unbalanced transformer (Balun). Balun action produces equal and opposite currents at transformer action produces equal and opposite voltages across R_{L1} and R_{L2}, canceling the LO voltage at the IF port.*

Any current (I_{RF}) that is sourced by V_{RF} is split into I_{RF1} and I_{RF2} as it enters the transformer. If the balun is perfectly balanced and $R_{L1} = R_{L2}$, the current will split evenly as it enters the center tap of the balanced transformer and $I_{RF1} = I_{RF2} = I_{RF}/2$. Note that half of the RF current enters a dotted terminal and the other half enters an undotted terminal.

On the balun primary, I_{RF1} will generate I'_{RF1} in the direction shown; I_{RF2} will generate I'_{RF2} in the opposite direction. Since $I_{RF1} = I_{RF2}$, $I'_{RF1} = I'_{RF2}$ and the RF current exactly cancels at the LO port. By superposition, we can also say that any current sourced from the LO port will be cancelled at the RF port. In other words, the balun provides LO:RF isolation. The two RF currents I_{RF1} and I_{RF2} will recombine in-phase at the IF port, i.e., there is no isolation between the RF and IF ports of this mixer (see Figure 3-48).

Figure 3-49 sums up the analysis when we apply both a LO voltage and RF voltage. The balun will completely suppress the LO voltage at both the RF and IF ports of the mixer. The RF port is directly connected to the IF port.

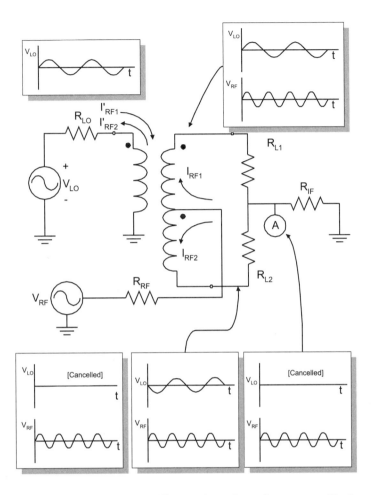

Figure 3-49 Balun operation, illustrating the voltage cancellation which is responsible for increased LO:IF and LO:RF isolation.

In Figure 3-50 we have replaced the resistors of Figure 3-49 with another type of component with no detrimental effect. A capacitor, inductor or any complex combination of the three (R, L or C) is possible. We can even replace the resistors with a nonlinear device such as a diode. However, in order to maintain a balance, we must insure that the two components are exactly identical. If the two components are inductors, for example, the inductors must have identical losses and internal capacitance. If we use diodes, the diodes must have the same dynamic impedance and impedances must track over temperature. Fortunately, diodes built on the same substrate under carefully controlled conditions track quite closely.

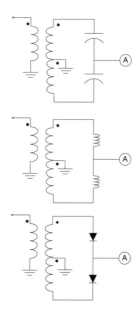

Figure 3-50 *A BALUN will provide isolation even when we replace the load resistors with complex, nonlinear loads. Each complex, nonlinear load must be as closely matched as possible to maintain balance and achieve high levels of cancellation.*

SBM Operation

The balun places the LO voltage across the two diodes, so the diodes will open and close at a rate determined by the LO (see Figure 3-51). If the balun is perfectly balanced and the diodes are exactly identical, the LO will be completely cancelled at both the RF and IF ports.

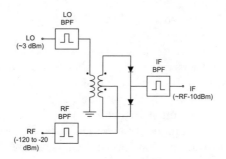

Figure 3-51 *Single balanced mixer (SBM) schematic. The two diodes must be carefully matched to provide high levels of LO:RF and LO:IF isolation.*

When the two diodes are both in their conducting or low-impedance states, the insertion loss between the RF port and the IF port of the mixer depends upon the forward resistance of the diodes. Since the resistance of the diodes is a strong function of the LO power, we can see that a high LO power leads to low mixer insertion loss. When the two diodes are in their nonconducting or high-resistance states, the RF port is disconnected from the IF port.

Figure 3-52 shows the conceptual operation of a single-balance mixer, which is identical to the conceptual operation of a single-ended mixer.

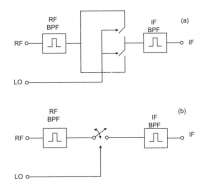

Figure 3-52 *Simplified single balanced mixer (SBM) operation. The diodes have been replaced by switches that are controlled by the LO.*

Figure 3-53 shows the applicable waveforms. The only difference between the waveform diagrams of the SEM and the SBM is that the LO must switch on two series-connected diodes rather than just one. This means that the SBM will usually require more LO power than the SEM.

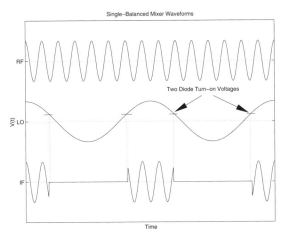

Figure 3-53 *Waveforms present in a single balanced mixer (SBM).*

Advantage and/Disadvantages

Cost

The SBM costs more than the SEM because of the extra diode and the addition of the balun. The SBM also costs more in terms of LO power because we now have to switch two series-connected diodes.

Interport Isolation

The addition of the balun and an additional diode causes the SBM to exhibit greater LO:RF and LO:IF isolation than the SEM. This is due to the diode match and balun balance. However, the SEM and the SBM offer similar RF:IF isolation. Since the LO is the strongest signal present in the receiver's conversion chain, keeping it under control is important. The SBM architecture addresses this most pressing mixer problem.

Spurious Performance

The spurious performance of the SBM is usually better than the SEM because the LO is not present at the IF port.

LO Power

Since a SBM must forward-bias two series-connected diodes, the SBM usually requires more LO power than the SEM (which relies on only one diode). The SBM requires at least 3 dBm of LO power.

Conversion Loss

Similar to the SEM, the RF port of the SBM is disconnected from the IF port for at least half the time, so the two mixers tend to exhibit similar insertion loss characteristics. Like the SEM, the SBM exhibits an insertion loss of 10 to 20 dB in practice.

Port Impedances

The input impedance of the RF and IF ports change significantly over a LO period. For half of the LO cycle, the two diodes are forward-biased and the RF port is connected to the IF port. For the other half of the LO cycle, the two diodes are in their high-impedance state; the impedances of the mixer's RF and IF ports are higher during this time period.

Effects of Poor Balance

The LO:IF and LO:RF isolation of the SBM are derived from two conditions. The two diodes in the mixer should be exactly the same and the balun should generate two perfect copies of the LO that differ only in their phase angle. The diodes are usually built on a common substrate to assure that they are made from the exact same material with the same doping levels. This structure also insures that the two diodes are at the same physical temperature. In practice, the SBM offers about 20 dB worth of LO suppression over the SEM at both the RF and IF ports (depending upon the bandwidth).

3.13 Double-Balanced Mixers (DBM)

The *double-balanced mixer* (DBM) offers improved performance over the SEM and SBM topologies. The DBM contains additional circuitry to improve the interport isolation over the single-ended and single-balanced designs. Furthermore, the design of the DBM acts to suppress the spurious products associated with the odd harmonics of both the LO and RF.

DBM Circuits

Figure 3-54 shows a common DBM topology. Both circuits are equivalent yet drawn differently. The diode topology shown is typically referred to as a diode ring. These diodes are manufactured on a single substrate to ensure that they are matched as closely as possible. For the sake of the initial analysis, we will assume that all of the diodes are identical and that the transformers are perfectly balanced.

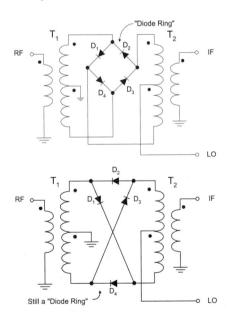

Figure 3-54 *Two equivalent circuit diagrams of a double-balanced mixer.*

RF, LO and IF are usually interchangeable. The LO can be forced into the IF port and the IF signal taken out of the LO port. The only practical difference is that each port may support a different frequency range. For example, the LO port in Figure 3-54 will extend to DC (or zero hertz) but the other two ports will not.

We will consider the diodes to operate as simple switches. Figure 3-55 shows an equivalent circuit for the DBM circuit of Figure 3-54 when the LO voltage is greater than the turn-on voltage of the diodes. Diodes D_2 and D_4 are in their conducting state; diodes D_1 and D_3 are in their nonconducting state. In this condition, the top of balun T_1 is connected to the top of balun T_2. The RF signal will pass through the mixer to the IF port with a 0° phase shift.

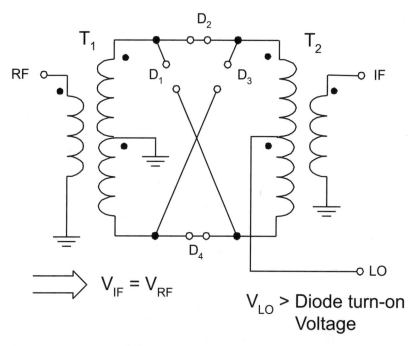

Figure 3-55 *Equivalent circuit of a double-balanced mixer (DBM) when the LO voltage is a diode drop above zero volts. The top of T_1 is connected to the top of T_2 and $V_{IF} = V_{RF}$.*

Figure 3-56 shows the DBM when the LO voltage is a diode drop below zero volts. Diodes D_1 and D_3 are conducting, and diodes D_2 and D_4 are nonconducting. The top of balun T_1 is connected to the bottom of balun T_2 so the RF signal will pass through the mixer to the IF port with a 180° phase shift.

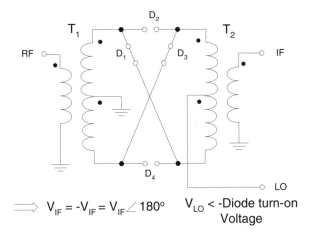

Figure 3-56 *Equivalent circuit of a double-balanced mixer (DBM) when the LO voltage is a diode drop below zero volts. The top of transformer T_1 is now connected to the bottom of transformer T_2 and $V_{IF} = -V_{RF}$.*

Figure 3-57 shows the waveforms over several cycles of the LO.

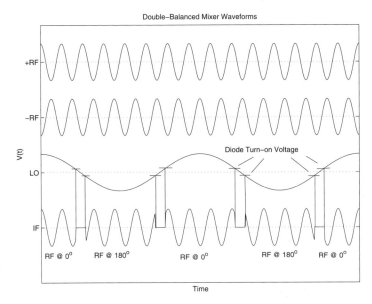

Figure 3-57 *RF, LO and IF voltages present in a double-balanced mixer.*

The RF is alternately connected to the IF port with a 0° phase shift, then with a 180° phase shift. In effect, we are multiplying the RF signal by

+1 when the LO voltage is greater than a forward diode drop and by −1 when the LO voltage is a diode drop below zero volts. This is a change from the SEM and SBM cases when we multiplied the RF by a +1/0 waveform. In equation form the multiplying waveform for the DBM is

$$W_{LO}(t) = \begin{cases} -1 \text{ when } V_{LO}(t) \leq \text{Diode Turn-on Voltage} \\ +1 \text{ when } V_{LO}(t) > \text{Diode Turn-on Voltage} \\ 0 \text{ Otherwise} \end{cases} \quad 3.71$$

Fourier analysis reveals an equivalent expression for $W_{LO(t)}$ is

$$W_{LO}(t) = \frac{4}{\pi} \sum_{k=0}^{\infty} \frac{1}{2k+1} \sin[(2k+1)\omega_{LO} t] \quad 3.72$$

There are two differences between Equation 3.72, which describes a DBM, and Equation 3.64, which describes SEMs and SBMs. The first difference is that Equation 3.72 does not include a constant or DC term. The DC term in Equation 3.64 is responsible for poor RF:IF isolation of the SEM and SBM. The second difference is that Equation 3.72 contains only odd harmonics of the LO.

Harmonic Suppression

Similar to the other two mixer topologies, the DBM tends to suppress the spurious products associated with the even-order LO harmonics. However, the DBM makes use of additional mechanisms to further suppress the even-order LO products as well as the even-order RF products. By noting certain symmetries in the time-domain, we can often comment on the odd or even Fourier components of the wave.

Fourier analysis reveals that if a time-domain function $f(t)$ satisfies the relationship.

$$f(t) = -f\left(t + \frac{T_0}{2}\right) \quad 3.73$$

for any T_0, then $f(t)$ contains only odd harmonics of $f_0 = 1/T_0$. This is a sufficient, but not necessary condition.

Figure 3-58 shows the equivalent LO-derived multiplying waveform generated by a double-balanced mixer. The six graphs represent the results for three values of LO power. In every case, the multiplying waveform satisfies Equation 3.73. It contains only the odd harmonics of the LO. The suppression of the even-order LO harmonics is a direct result of balun balance and diode match.

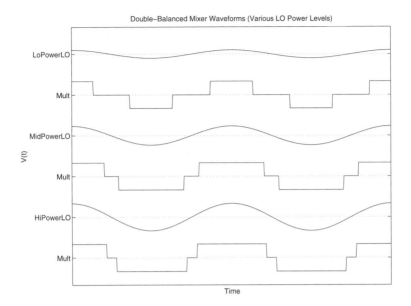

Figure 3-58 *Symmetrical waveforms generated in the DBM produce only odd LO harmonics. This effect helps to further suppress odd harmonic spurious products in the DBM.*

Port Swapping

Double-balanced mixer ports are generally interchangeable, i.e., the LO can be put into the IF port and the IF signal taken from the RF port. The mixer will still function although there might be a performance penalty. As an exercise, the interested reader might swap two of the ports and figure out how the DBM still performs the time-domain multiply necessary for mixer action. The frequency range of the three mixer ports may be different, however. For example, the port connected to the center tap of the two baluns usually extends down to DC; the other two ports (which are connected to baluns) do not operate at DC.

Advantages and Disadvantages

Through the use of baluns, the DBM exhibits a high isolation between all of its ports. The DBM also suppresses even-order spurious products due to diode and balun matching. The DBM makes more effective use of the available RF power than either the SEM or the SBM. The SEM and SBM gate the RF signal on and off. In other words, the IF port is connected to the RF port only a fraction of the time. Consequently, the IF port receives only half of the available RF power.

The DBM makes full use of the RF power by connecting the IF and RF ports, in some way, 100% of the time. This causes the conversion loss of the DBM to be less than the conversion loss of the SBM and SEM. The theoretical single-sideband insertion loss is about 4 dB (We see normally see 7 to 10 dB in practice).

Since the DBM must drive two series-connected diodes at a time, the LO drive power requirements of the DBM are higher than either the SEM or the SBM. Typically, we need at least +7 dBm of LO power to drive a DBM adequately. Some designs use special diodes to achieve a lower drive level, but as a result their linearity usually suffers.

3.14 Further Mixer Characteristics

LO Power and Conversion Loss

In any of the mixer types we have discussed, the LO turns the mixer's diodes on and off, which causes the series impedance of the mixer diodes to change from low impedance (when the diode is conducting) to high impedance (when the diode is nonconducting).

An analysis of semiconductor physics reveals that the small-signal resistance of a diode decreases with increasing forward voltage. In other words, the series resistance seen by the RF signal as it passes from the RF port to the IF port directly depends on the instantaneous voltage across the diode. A high-power LO will decrease this resistance and tend to decrease the insertion loss of the mixer. In other words, the mixer's insertion loss is a strong function of the LO power.

Figure 3-58 shows the LO-derived multiplying waveform $W_{LO}(t)$ for several levels of LO power. In each case, there is a period of time when $W_{LO}(t)$ is zero. During the off-time, the RF port is not connected to the IF port, and RF power is wasted. The conversion loss of the mixer is a direct function of this off-time and the length of the off-time period depends upon the LO power. A large LO power produces a short off time and thus a lower insertion loss. A small LO power produces a long off time and a higher insertion loss.

LO Power and Linearity

The LO alone controls when the diodes switch, which is approximately true when the LO power is very much larger than the RF power (for example by 20 dB). As the RF power approaches the power of the LO, the RF signal begins to affect the exact instant the diodes switch between their conducting and nonconducting states. This generates spurious signal power.

Applying a strong RF signal to a mixer can be compared to applying a strong RF signal to an amplifier. In an amplifier, the RF power begins to affect the DC bias levels of the transistors inside the amplifier. In a mixer, the RF signal causes the diodes to switch at the wrong moments. In both cases, the device is operated in its nonlinear region and the results are rather unpredictable. Generally, the LO power should be kept at least 20 dB higher than the maximum RF signal to confine the mixer to its weakly nonlinear range.

LO Noise

In a receiver, we use the zero crossings of the LO to switch the mixer diodes. For a first-order approximation, the mixer's diodes switch at the exact instant that the LO makes a zero crossing. Oscillator phase noise manifests itself as random changes in the zero-crossings of the LO waveform. Instead of a pure cosine wave, we can write the equation for a noisy oscillator as

$$V_{LO}(t) = A_{LO}(t)\cos[\omega_{LO}t + \phi_{LO}(t)] \qquad 3.74$$

where $A_{LO}(t)$ is the amplitude noise of the oscillator and $f_{LO}(t)$ represents the phase noise of the oscillator. If this noisy LO is mixed with a pure cosine wave, we obtain

$$V_{IF}(t) = \{A_{LO}(t)\cos[\omega_{LO}t + \phi_{LO}(t)]\}\{\cos(\omega_{RF}t)\}$$
$$= \frac{A_{LO}(t)}{2}\{\cos[(\omega_{LO} + \omega_{RF})t + \phi_{LO}(t)] + \cos[(\omega_{LO} - \omega_{RF})t + \phi_{LO}(t)]\} \qquad 3.75$$

Noise present on the LO is transferred to both sidebands. This is true for both amplitude and phase noise but, in practice, we usually find that the oscillator's phase noise is a greater problem than its amplitude noise.

Generally, it can be assumed that any signal converted from one frequency to another inherits the phase noise of the local oscillator. A noisy oscillator can contaminate a signal so badly that it cannot be demodulated.

Effects of Impedance Mismatch

Like most other RF components, mixers are characterized assuming they will be operated in a Z_0 environment. However, a mixer is is a device that demands that the outside world present a Z_0-load to it.

For optimum performance (including low spurious output and good inter-port isolation and such), we must design the systems to present the mixer with a wideband Z_0 termination on all of its ports. This presents a problem when we need to filter the RF or IF ports of the mixer. As we learned in Chapter 2, the input impedance of a filter in its stopband is usually not Z_0. The non-Z_0 filter impedance usually causes the mixer's spurious output power to increase, although it can also cause increased insertion loss and other effects.

The technical literature indicates that it is very important to terminate the mixer's IF port at the sum and the difference frequencies and at the LO frequency. For example, if we are making use of the difference product from the mixer's IF port, the difference frequency will see a Z_0 termination. We may have to take extra care to ensure a good match at the sum and LO frequencies.

There are several solutions to this problem. We can follow the mixer with an amplifier whose frequency range includes the sum, difference and other frequencies. We can usually count on an amplifier presenting the outside world with a Z_0 termination over its operating frequency range. However, the amplifier must exhibit sufficient dynamic range to process the large range of signals present on the mixer's IF port. The LO power leaving the mixer's IF port, for example, can be very strong (0 dBm or greater). The amplifier must be able to process the weak RF signals in the presence of the strong LO signal. We can also build special filters that are matched throughout their passband and their stopband.

A third solution is to place an attenuator directly on the IF port. The worst-case return loss of an attenuator is twice the value of the attenuation. For example, the maximum return loss of a 6 dB attenuator is 12 dB, even when we present the output of the attenuator with an open- or short-circuit. In this case, though, we must endure the signal loss through the attenuator.

Mixer SOI/TOI

A mixer has a second-order intercept point as well as a third-order intercept point. These numbers are a strong function of the LO power. As the LO power decreases, so do the second- and third-order intercepts of the mixer.

A third-order intercept test on a mixer is performed in much the same way we performed the test on an amplifier, except we have to consider the frequency conversion. Figure 3-59 shows the test technique.

We place two tones at the RF port (at $f_{RF,1}$ and $f_{RF,2}$). The IF port of the mixer will show two sets of four tones (each close to f_{SUM} and f_{DIFF}). The sum and the difference frequencies will each exhibit their own set of third-order distortion products. We normally select either the sum or the difference frequencies with an IF band-pass filter to examine the third-order distortion products at the desired output frequency.

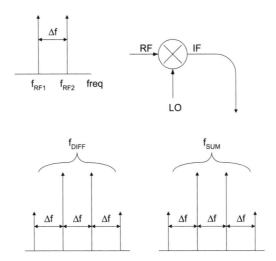

Figure 3-59 *Measuring mixer TOI. The third-order spurious signals appear at both the sum and difference frequencies.*

We measure the mixer's second-order intercept in the same way. We apply a single tone to the RF port of the mixer and note that the sum and the difference products at the mixer's IF port each shows a second-order output product. We normally filter either the sum or the difference frequency for use in the system and make the measurements at the selected frequency. Note that, if the IF band-pass filter is less than an octave in bandwidth, the IF band-pass filter will attenuate the second-order signal.

A mixer's intercept performance can be affected by how well the mixer's ports are terminated. A 2 dB improvement in the TOI of a mixer can be realized by properly terminating its ports.

3.15 Other Uses of Mixers

RF Switch

Figure 3-60 shows a double-balanced mixer used as a RF switch. We apply a switching waveform V_{Switch} to the mixer's LO port. Resistor R_{Limit} limits the current through the diodes. When the switching voltage is greater than a diode drop (about 0.65 volts), diodes D_2 and D_4 conduct. This connects balun T_1 directly to T_2. The RF signal passes through to the IF port. When the LO voltage is less than a diode drop (near zero volts), all of the diodes in the mixer are in their high impedance states and the RF signal is severely attenuated before arriving at the mixer's IF port. We can often achieve 25 or 30 dB on-to-off ratio with this simple technique.

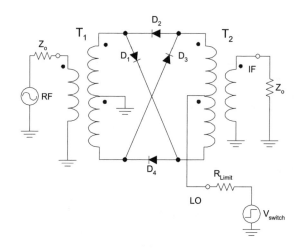

Figure 3-60 *A double-balanced mixer used as an RF switch. When the LO voltage is positive, diodes D_2 and D_4 turn on and pass the RF signal to the IF port. When the LO voltage is zero, all the diodes are off and the RF port is isolated from the IF port.*

Experimentally, we note that some of the RF signal leaks through the mixer when the LO voltage is slightly above zero volts, but well below the 0.65 volts required for full diode conduction. When the LO voltage is slightly above zero volts, the mixer passes the RF signal to the IF port with a 0° phase shift. When the LO voltage is slightly below zero volts, the mixer passes the RF signal to the IF port with a 180° phase shift. The amount of attenuation suffered between the RF port and the IF port depends upon the magnitude of the LO voltage.

When we use a DBM as an RF switch, we can often adjust the turn-off voltage so that the RF signal leaking through the mixer is cancelled by the RF signal we purposefully let through via the nonzero off voltage. This is one technique of increasing on/off ratio of the RF switch.

The ports of a mixer have frequency limits. Normally, the IF port will extend down to the lowest frequency, for example, often down to DC. This is the port we use to apply the control voltage to the RF switch. We can usually switch the ports of a mixer at will, as long as we pay attention to the frequency response of each port.

Voltage Variable Attenuator

The diodes of Figure 3-60 do not switch immediately when the LO voltage rises above 0.65 volts. There is a range of LO voltage, covering perhaps two-tenths of a volt, when the diodes are in transition between their conducting and nonconducting states. If we apply a LO voltage in this range, we can adjust the amount of RF power present at the IF port of the mixer. In this way, we can use a DBM as a voltage variable attenuator.

However, diodes are very temperature-sensitive devices. If we apply a constant control voltage to the mixer and measure the output power, we will find that the output power changes radically for even small changes in temperature. However, we can often compensate for this effect using automatic gain control or AGC. In an AGC system, the signal power is measured at some point in the system, then the gain of the system is adjusted based upon the power level we are measuring. This technique compensates for widely varying temperature ranges.

BPSK Modulator

When the LO voltage is in one polarity (above 0.65 volts, for example), the mixer passes the RF signal to the IF port with a 0° phase shift. When the LO voltage is in the other polarity (below −0.65 volts), the mixer passes the RF to the IF port with a 180° phase shift. In effect, we can use a signal on the LO port to alternate the phase of signal present on the IF port. If the LO is a digital data signal, we can use a DBM as a *binary phase shift keying* (BPSK) modulator.

Phase Detector

A mixer multiplies two signals together in the time domain.

$$V_{IF}(t) = [\cos(\omega_{RF}t)][\cos(\omega_{LO}t)]$$
$$= \frac{1}{2}\cos[(\omega_{RF} - \omega_{LO})t] + \frac{1}{2}\cos[(\omega_{RF} + \omega_{LO})t] \qquad 3.76$$

If the LO frequency equals the RF frequency (i.e., $\omega_{RF} = \omega_{LO}$) but they are offset by some phase θ, then we can write

$$V_{IF}(t) = [\cos(\omega t + \theta)][\cos(\omega t)]$$
$$= \frac{1}{2}\cos(\theta) + \frac{1}{2}\cos(2\omega t + \theta) \qquad 3.77$$

Finally, if we remove the signal at 2ω with a LPF, we are left with

$$V_{IF} = \frac{1}{2}\cos(\theta) \qquad 3.78$$

The output is proportional to the cosine of the phase difference between the RF and the LO signals. Figure 3-61 shows a graph of this function with respect to θ.

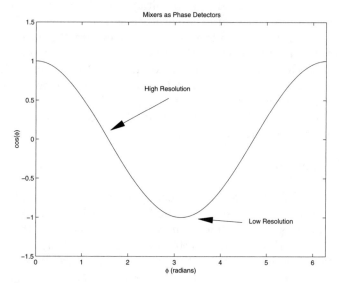

Figure 3-61 *Mixer output voltage as a function of the phase difference between the LO and RF signals. The LO and RF signals are at the same frequency.*

Since

$$\cos\left(\theta - \frac{\pi}{2}\right) = \sin(\theta) \qquad 3.79$$
$$\approx \theta \quad \text{for small } \theta$$

we can achieve fairly high resolution when the phase shift is in the neighborhood of 90°. In other words, we can use a DBM as a phase detector with high resolution as long as the phase difference between the two signals is approximately 90°.

3.16 Mixer Design Summary

The Mathematical Basis of Mixing

We assume the two input signals are

MIXERS | 341

$$V_{mod}(t) = A(t)\cos[\omega_1 t + \phi(t)]$$
$$V_{unmod}(t) = \cos(\omega_2 t)$$

where
  $A(t)$ = AM information waveform,
  $\phi(t)$ = PM or FM information waveform,
  $\omega_1 = 2\pi f_1$ = carrier frequency of the modulated wave,
  $\omega_2 = 2\pi f_2$ = carrier frequency of the unmodulated wave.

Second-Order Transfer Function

We can use a second-order process to perform the mixing process. The second-order operation is

$$V_{out,2}(t) = k_2 V_{in}^2(t)$$

The second-order output is

$$\begin{aligned}
V_{out,2}(t) &= k_2 [V_{mod}(t) + V_{unmod}(t)]^2 \\
&= k_2 \{A(t)\cos[\omega_1 t + \phi(t)] + \cos(\omega_2 t)\}^2 \\
&= \frac{k_2[1 + A^2(t)]}{2} + \frac{k_2 A^2(t)}{2}\cos[2\omega_1 t + 2\phi(t)] \\
&\quad + \frac{k_2}{2}\cos(2\omega_2 t) \\
&\quad + k_2 A(t)\cos[(\omega_1 + \omega_2)t + \phi(t)] \\
&\quad + k_2 A(t)\cos[(\omega_1 - \omega_2)t + \phi(t)]
\end{aligned}$$

We then filter to remove the unwanted terms.

$$V_{out,2,SUM}(t) = k_2 A(t)\cos[(\omega_1 + \omega_2)t + \phi(t)]$$
$$V_{out,2,DIFF}(t) = k_2 A(t)\cos[(\omega_1 - \omega_2)t + \phi(t)]$$

Time Domain Multiply

A time-domain multiply is the preferred way to perform the mix. This method produces only the sum and difference products.

$$V_{mult}(t) = V_{mod}(t)V_{unmod}(t)$$
$$= A(t)\cos[\omega_1 t + \phi(t)]\cos(\omega_2 t)$$
$$= \frac{A(t)}{2}\cos[(\omega_1 + \omega_2)t + \phi(t)]$$
$$+ \frac{A(t)}{2}\cos[(\omega_1 - \omega_2)t + \phi(t)]$$

The $A(t)/2$ term in this equation guarantees a minimum theoretical conversion loss of 6 dB. Despite our best efforts, we are generating higher-order distortion products than what is listed above.

Frequency Translation Equation

The output frequencies of a general nonlinear process are

$$f_{out} = |nf_1 \pm mf_2|$$

where
$n = 0, 1, 2, ...,$
$m = 0, 1, 2, ...,$
$n + m =$ the order of the response.

Three Forms

When the above equation is applied to mixers, we obtain three equations depending upon which quantities we know and which quantities we want to find.

$$f_{IF} = f_{LO} \pm f_{RF}$$
$$f_{RF} = f_{LO} \pm f_{IF}$$
$$f_{LO} = f_{RF} \pm f_{IF}$$

High-Side and Low-Side Injection

High-side and low-side LO frequencies are defined by

$$\text{Low-Side LO} \Rightarrow f_{LO} < f_{RF}$$
$$\text{High-Side LO} \Rightarrow f_{LO} > f_{RF}$$

Frequency Inversion

A frequency inversion occurs only when we use a high-side LO and select the lower sideband (at $f_{IF} = f_{LO} - f_{RF}$). In equation form,

$$\text{Spectrum Inversion when} \begin{cases} \text{Using HSLO} \\ \text{and} \\ \text{Selecting the Lower Sideband} \end{cases}$$

LSLO Image Frequencies
The image frequencies obtained when using a LSLO are

$$f_{LSIM} = f_{RF} - 2f_{IF}$$

HSLO Image Frequencies
The expression for the HSLO image frequencies is

$$f_{HSIM} = f_{RF} + 2f_{IF}$$

Suppression by a DBM

In order of highest suppression to lowest suppression, a DBM will suppress the following products:

- Even-by-Even,
- Even-by-Odd,
- Odd-by-Even,
- Odd-by-Odd,

where the order of the product is

$$\text{Order} = m + n$$

3.17 References

1. Hayward, W. A. *Introduction to Radio Frequency Design.* Englewood-Cliffs, NJ: Prentice-Hall, 1982.

2. Henderson, Bert C. *RF and Microwave Designers Handbook.* Watkins-Johnson Company, 1988/1989, 752, 759.

3. Peter, Will. "Reactive Loads: The Big Mixer Menace." *Microwaves*, no. 4 (1971).

4

Oscillators

> Purity lives and derives its life solely from the spirit of God.
> —David Hare (1917-1992)

4.1 Introduction

Oscillators play an important part in the ultimate performance of receivers. In this chapter, we will examine oscillator phase noise, frequency accuracy, drift and their cumulative effects on receiver performance.

Ideal Oscillator

Figure 4-1 shows some of the characteristics of an ideal oscillator. The time-domain plot of an ideal oscillator is a mathematically perfect sine wave [see Figure 4-1(a)]. There is no noise present on the signal. The period, defined as the time between zero crossings, is $1/f_0$ where f_0 is the frequency.

Figure 4-1(b) shows the zero crossings of the waveform are exactly deterministic: we know precisely when the waveform will cross zero.

Figure 4-1(c) shows that the single-sided Fourier spectrum of a perfect oscillator is a single, discrete impulse function with zero width. The spectrum contains no other discrete components such as harmonics and sampling sidebands. The signal has an infinite signal-to-noise ratio. The ideal oscillator is always exactly on the desired frequency over time, temperature, power supply variations, and so on.

Figure 4-1(d) shows the phasor representation of an ideal oscillator. A single phasor rotates counterclockwise about the origin at an angular frequency of ω_0.

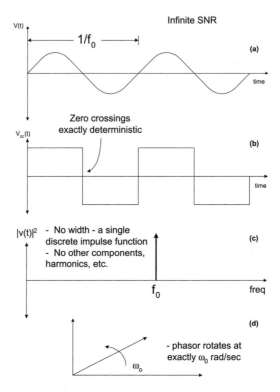

Figure 4-1 *Various representations of an ideal oscillator: (a) time domain, (b) zero crossings, (c) frequency domain and (d) phasor domain.*

Practical Oscillators

Actual oscillators are not ideal. Figure 4-2 shows the less-than-ideal properties of one such oscillator. The oscillator is contaminated by noise, as shown in Figure 4-2(a). The waveform now exhibits random amplitude and frequency variations.

Figure 4-2(b) indicates that the zero crossings have also taken on a random, nondeterministic character. We can no longer state exactly when a zero crossing can occur; we can only comment on the statistics.

Figure 4-2(c) reveals frequency domain characteristics of a less-than-ideal oscillator. The spectrum is no longer an impulse function with zero width. The exact shape of the spectrum depends on the details of upon the construction details of the oscillator. Figure 4-2(c) also shows oscillator drift. The frequency of a non-ideal oscillator will change with time, temperature and other environmental factors.

Changes in the oscillator's frequency over time periods which are greater than approximately one second are referred to as *drift*. Frequency changes over time periods less than a second or so are called *phase noise*. Both drift and phase noise represent the idea that the oscillator is not exactly at its nominal frequency.

Figure 4-2(d) shows the phasor representation of a noisy oscillator. The phasor still rotates about the origin at ω_0, but exhibits random variation of the amplitude and phase.

We represent the noise present on the oscillator by adding a second noise phasor to the phasor of the ideal oscillator. The magnitude and phase of the noise phasor are statistical quantities. It is frequently necessary to break up the noise phasor into its in-phase and quadrature components. The in-phase components represent the amplitude or AM oscillator noise, and the quadrature components are the phase or PM components of the noise.

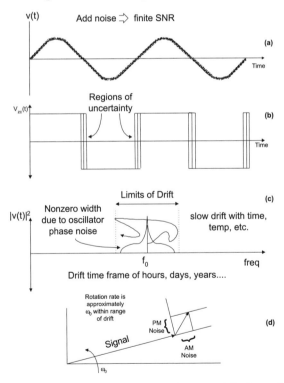

Figure 4-2 *Various representations of a nonideal oscillator: (a) time domain, (b) zero crossings, (c) frequency domain and (d) phasor domain.*

Example of a Poor Frequency Synthesizer

The spectrum of an imperfect oscillator usually contains discrete unwanted signals. Most commonly, the spectrum contains harmonics of the fundamental (or desired) signal. The output spectrum often contains other discrete signals such as subharmonics and frequencies that are remnants of the oscillator realization. The spectrum will also show the oscillator's phase noise.

Figures 4-3 through 4-8 show the spectral plots of a PLL synthesizer built by one of the authors. The principal requirements for this particular project were small physical size and low power consumption rather than low phase noise and low harmonic content. Consequently, the spectrum exhibits most of the problems discussed in the previous section. The synthesizer was tuned to 120 MHz.

Figure 4-3 shows the synthesizer's spectrum from 0 hertz to 1 GHz. Harmonics of varying strength are present at the oscillator's output. Normally, the second harmonic is strongest, followed by the third, fourth, and so on. However, in this particular design, the second harmonic is below the third harmonic, which is the strongest at 28 dBc. Note that the harmonics are observable to 840 MHz.

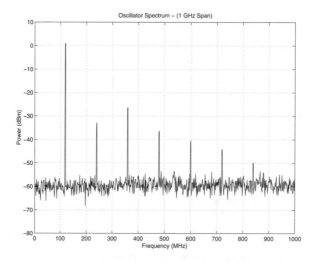

Figure 4-3 *Wideband spectrum of a phase-locked loop (PLL) frequency synthesizer. The fundamental is at 120 MHz. Note the various harmonics.*

Figure 4-4 shows the oscillator's spectrum over a 5 MHz bandwidth. The output spectrum suggests close-in phase noise which could be confirmed with a narrow scan. The spectrum also exhibits discrete, spurious signals approximately 1.935 MHz away and 50 dBc down from the carrier. These spurious products can probably be traced to some waveform present in the synthesizer circuitry.

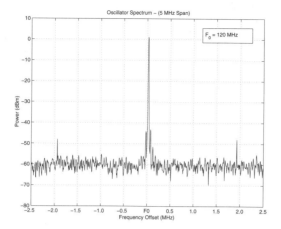

Figure 4-4 *The output of the PLL frequency synthesizer over a 5 MHz span. Note the discrete, nonharmonically related spurious signals at ±1.935 MHz.*

Figure 4-5 shows the synthesizer output across a 500 kHz span. The plot reveals that the close-in phase noise observed in Figure 4-4 contains several discrete tones approximately 50 kHz apart. The step size of this synthesizer is 50 kHz. These spurious signals are commonly referred to as *sampling sidebands*.

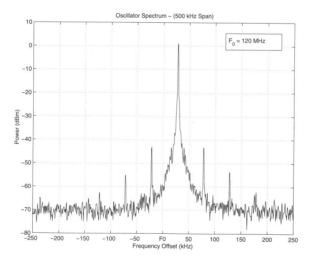

Figure 4-5 *The output of the PLL frequency synthesizer over a 500 kHz span reveals the oscillator's phase noise and several discrete components at 50 kHz intervals. The output is not exactly at its 120 MHz design frequency.*

Figure 4-6 shows that the oscillator is not exactly on frequency. The synthesizer frequency is about 28 kHz too high. This 230 part-per-million (ppm) or 0.023% error is quite large. We have centered the spectrum in Figure 4-6. The oscillator's close-in phase noise is now apparent over the 100 kHz span.

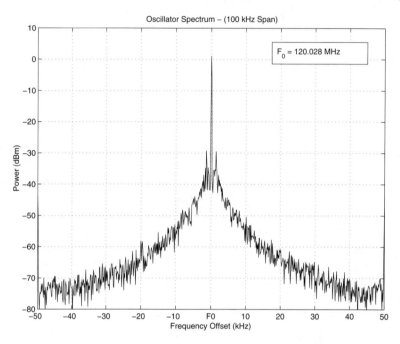

Figure 4-6 *The frequency synthesizer's output over a 100 kHz span. The oscillator's phase noise is quite apparent.*

Figure 4-7 shows the behavior of the oscillator's noise floor over a 10 kHz frequency span. The oscillator phase noise exhibits a change in slope as we approach the carrier. As we move toward the carrier, the noise increases, then levels off about 1200 hertz from the carrier. The noise then drops as we continue to approach the carrier. This noise profile is common in PLL frequency synthesizers. This synthesizer is particularly noisy due to a noisy *voltage-controlled oscillator* (VCO). We can also detect numerous spurious signals about 600 and 1200 hertz from the carrier. Figure 4-8 shows the oscillator's output over a 500-hertz span.

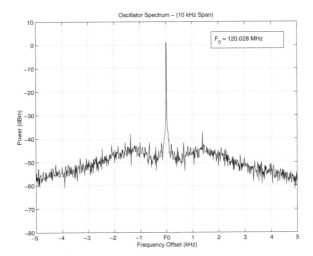

Figure 4-7 *The output of the frequency synthesizer over a 10 kHz span. The change in slope of the noise is due to the control loop nature of the design.*

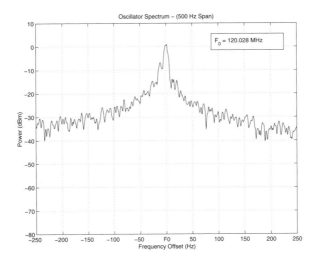

Figure 4-8 *The output of the frequency synthesizer over a 500-hertz span. The noise constantly increases as we approach the carrier.*

Receivers and Oscillators

In Chapter 3, we saw that any signal that is converted from one frequency to another by a local oscillator inherits the faults of that local oscillator. If the LO has significant phase noise, then signals converted by that LO will have significant phase noise. If the LO drifts with time and temperature, every signal converted by that LO will drift accordingly with time and temperature.

4.2 Phase Noise

Phase noise is a subtle characteristic of oscillators with far-reaching effects. Oscillator phase noise can limit the ultimate signal-to-noise ratio of any signal processed by a receiver. It can cause unwanted signals to mask wanted signals. In radar systems, receiver phase noise helps clutter mask the targets we want to see. Phase noise can limit the ultimate *bit error rate* (BER) of a system.

Representations of a Noisy Oscillator

Phase noise is defined as the short-term variation in an oscillator's phase or frequency. Generally, phase noise represents changes in the oscillator's phase or frequency observed over time periods of about a second or less. The term drift describes changes over periods of a second or greater.

As Figure 4-9(a) shows, we can represent a perfect oscillator by an impulse function in the frequency domain. The impulse is at the oscillator's center frequency f_0. We can show a noisy oscillator as a more complex function that remains centered about the oscillator's nominal frequency.

We can think of a nonideal oscillator as an ideal oscillator plus a series of sinusoids, each set off from the carrier by Δf [see Figure 4-9(b)]. Figure 4-10 shows the phasor diagram of the noisy oscillator, showing the carrier (at f_0) and one of the noise sidebands (at $f_0 + \Delta f$).

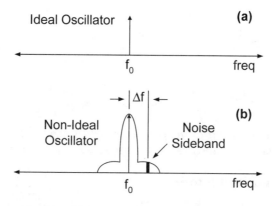

Figure 4-9 *A nonideal oscillator can be modeled as the arithmetic sum of ideal oscillators. Each noise sideband at offset Δf has a statistically described amplitude and phase.*

We can divide the noise vector into in-phase and quadrature components. When the magnitude of the carrier is much greater than the magnitude of the noise, we can label the in-phase component of the noise as "AM noise" because it changes only the amplitude of the carrier. We can label the quadrature component of the noise as *FM* or *PM noise* because this component can change only the phase of the carrier.

The in-phase and quadrature components of the noise vector are functions of time and they possess separate and unique Fourier spectra. There is no fundamental mechanism that forces the AM spectrum to be the same as the PM spectrum. In some oscillators, the two spectra are vastly different.

Figure 4-10 shows the oscillator output at one instant of time. The in-phase and quadrature vectors are both random processes. To describe their behavior over time, we have to use statistics. Figure 4-11 shows snapshots of the phasor taken over several intervals. Since we always sample the oscillator's output at $1/f_0$, the coherent or deterministic portion of the oscillator does not move. However, the noise vector will exhibit random amplitude and phase. The circles might represent the 1σ, 2σ and 3σ characteristics of the noise vector if the statistics are Gaussian.

The statistics for the in-phase and quadrature components can be different. For example, the standard deviation of the in-phase component could be three times the standard deviation of the quadrature component. In that case, the circles of Figure 4-11 would become ellipses.

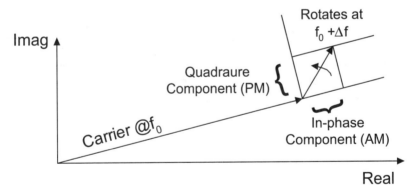

Figure 4-10 *The phasor representing the oscillator noise component at Δf is attached to the head of the fundamental oscillator phasor. We can break the noise phasor into in-phase and quadrature components.*

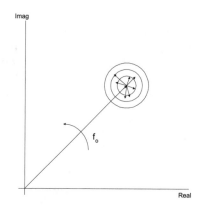

Figure 4-11 *If we sample the noise phasor over many intervals, we can develop circles that represent the statistics of the noise as a way to describe the phase and frequency of the oscillator over time.*

Amplitude Noise and Phase Noise

The phasor representation of Figure 4-11 shows snapshots of the oscillator's output taken at different times. The real part of this phasor is the projection onto the real or horizontal axis. Remember that the real part is the voltage we would see on an oscilloscope.

Figure 4-12 shows the phasor when it is just passing through the real axis. The real part of the vector and the instantaneous output voltage is maximum. At this instant, the quadrature component of the noise (the phase modulation) does not have much of an effect on the real part of the resultant vector. The uncertainty of the output is dominated by the in-phase or AM component of the noise.

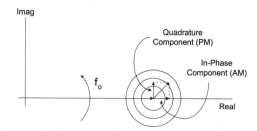

Figure 4-12 *The phasor, which represents the oscillator output, just as it passes through the real axis. The quadrature component of the noise does not affect the oscillator output at this time. The output uncertainty is dominated by the in-phase component of the noise.*

Figure 4-13 shows the phasor diagram when the oscillator's output waveform is just passing through zero. The projection of the in-phase component onto the real axis is zero. The uncertainty in the zero-crossings is entirely controlled by the quadrature component of the noise. Accordingly, the AM noise does not affect the zero crossings of the oscillator whereas the PM noise does.

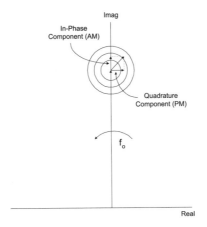

Figure 4-13 *The oscillator's output at zero-crossing time. The phasor, which represents the oscillator output, just as it passes through the imaginary axis. At this instant in time, the quadrature component of the noise dominates the oscillator's uncertainty. The in-phase component does not affect the oscillator's output.*

Phase Noise and Mixers

Mixers convert signals from one frequency to another. The conversion is usually performed by a time-domain multiply. A noisy LO will affect the signals being converted differently, depending upon how the mixer is realized. In most mixers, we use the local oscillator to switch diodes on and off. The diodes act like switches, alternately connecting and disconnecting the RF port to the IF port. When the LO is much stronger than the RF, the diodes switch synchronously with the LOs zero crossings. Since the AM noise does not affect the zero crossings of the oscillator's time-domain waveform, the AM noise is not passed onto the RF signal. In effect, any saturated switch mixer suppresses the AM noise of the LO.

If the LO is not strong enough to switch the diodes completely on and off, some AM noise will be impressed onto the RF signal as it passes through the mixer. Figure 4-14 shows a noisy local oscillator combined with a pure sine wave in a mixer. In almost all cases, the quadrature (or phase) noise component of the oscillator will be transferred to the IF signal. In some cases, the in-phase (or amplitude) noise will also be transferred to the IF signal, i.e., some of the modulation present on the LO will be passed onto the RF signal.

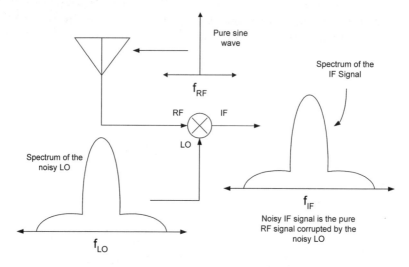

Figure 4-14 *The mixing process using a noisy local oscillator (LO). Noise present on the LO is transferred to every signal which is converted to a new frequency by the mixer.*

Effects of Phase Noise

Adjacent Channel Masking

In a channelized system, for example in the commercial FM broadcast band, a receiver is likely to collect many different signals at the same time. The signals will be at different frequencies and their power levels will vary dramatically. One of the most difficult receiver problems is the reception of a weak, desired signal in the presence of a strong, adjacent signal. The strong signal taxes the receiver's linearity, and the weak signal taxes the receiver's noise performance. LO phase noise makes this situation even more problematic.

In Figure 4-15, a typical channelized spectrum is applied to the RF port of a mixer. The channel spacing is f_{ch}. As the mixer converts the RF spectrum to the IF, it impresses the phase noise of the LO onto each signal. The result is the IF spectrum shown in Figure 4-15. The strong signal at f_2 combined with the phase noise of the LO has degraded the SNR of the signal at f_1. The signal at f_3 has been completely enveloped in noise. This effect is called *adjacent channel masking* or *receiver desensitization*.

OSCILLATORS | 357

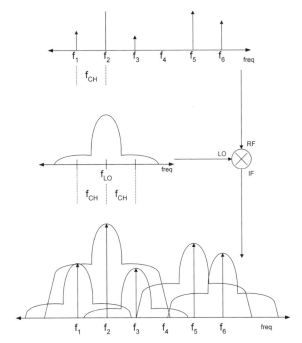

Figure 4-15 *Oscillator phase noise can cause large signals to mask weaker signals in channelized systems.*

In channelized systems, we are particularly interested in the levels of LO phase noise which are f_{ch} away from the carrier. As channelized signals on the air get closer together, phase noise specifications become more difficult. Ideally, we would like the phase noise of the local oscillator to have significantly receded by the time we have moved one channel spacing away from the LO's center frequency.

Radar

We use Doppler radar to measure moving targets (such as the radar guns used by the police). Doppler radar works by transmitting a signal at the target, then searching for the return. The change in frequency of the returned signal is a measure of the speed of the target. The transmitter frequency is usually several GHz, and the frequency shift caused by the Doppler shift is usually less than 10 kHz.

Figure 4-16 shows a typical return. The signal at f_1 is the transmitter feed through (at approximately 10 GHz). The signal at f_2 is the return from the target (at perhaps 10 GHz + 10 kHz). In this receiver, the phase noise of the transmitter is insufficient at a 10 kHz offset and the return signal is completely masked.

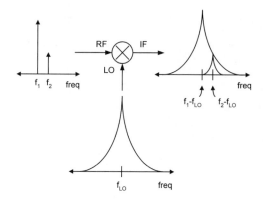

Figure 4-16 *Oscillator phase noise can cause clutter to mask the desired signal in a RADAR system.*

Ultimate SNR or Ultimate BER

Mathematical analysis reveals that LO phase noise can limit the ultimate *signal-to-noise ratio* (SNR) or the ultimate *bit error rate* (BER) of a receiver.

Phase Noise Description: $\mathcal{L}(f_m)$

Figure 4-17 shows the spectrum of a noisy oscillator as it might appear on a spectrum analyzer. Instead of a single, narrow impulse function, the spectrum has width.

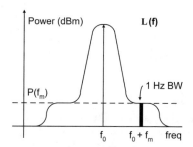

Figure 4-17 *Single-sideband phase noise is the oscillator power present at some offset f_m from the carrier. A measurement bandwidth of one hertz is assumed.*

One of the most common measures of phase noise is $\mathcal{L}(f_m)$, the single sideband phase noise. $\mathcal{L}(f_m)$ is the power present at some offset (f_m) from the carrier to the total signal power and is commonly expressed in dBc/hertz. When measuring a noise-like waveform, we must specify a measurement bandwidth. As usual, we will assume a 1-hertz measurement bandwidth.

$$\mathcal{L}(f_m) = \frac{\text{Power at offset } f_m \text{ from the carrier(1- hertz bandwidth)}}{\text{Total signal power}} \qquad 4.1$$

$$= \frac{P(f_m)_{1\text{ hertz}}}{\text{Total signal power}}$$

As Figure 4-17 indicates, $\mathcal{L}(f_m)$ refers to the signal power on just one side of the carrier (single sideband phase noise). When comparing two oscillators at the same frequency, the oscillator with the lower value of $\mathcal{L}(f_m)$, at a particular f_m, is the oscillator with the lower phase noise.

Figure 4-18 shows the output spectra of two oscillators operating at the same frequency. Their phase noise performance is markedly different: at some values of f_m oscillator #1 has lower phase noise while at other values of f_m oscillator #2 is lower. Note that noise spectra are not necessarily symmetrical about f_0.

$\mathcal{L}(f_m)$ is the most commonly used expression of phase noise. It is easy to measure with a spectrum analyzer and useful for comparing two oscillators which operate at the same frequency. However, $\mathcal{L}(f_m)$ is not very effective from a mathematical point of view.

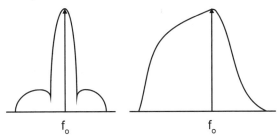

Figure 4-18 The spectra of two oscillators. Note that oscillator spectra can be nonsymmetrical about the carrier frequency f_0.

Measuring $\mathcal{L}(f_m)$

The phase-noise of the test equipment may influence the measurement. Quartz crystal oscillators usually exhibit very low phase noise (even a poorly designed crystal oscillator can exhibit excellent phase noise). Let us consider what might happen if we tried to measure $\mathcal{L}(f_m)$ of a crystal oscillator using a spectrum analyzer.

The spectrum analyzer, like any other receiver, uses an internal local oscillator to convert the input signal to some IF. The spectrum analyzer's LO is tunable over a wide range and its phase noise performance is likely to be poor compared to the crystal oscillator. Signals converted from one frequency to another by a LO will pick up the phase noise of the LO. Since the phase noise of the spectrum analyzer's LO is worse than the phase noise of the crystal oscillator, the phase noise of the LO will most likely be represented. Thus, measurements performed at the IF measures only the test equipment.

Sources of Phase Noise: The Leeson Model

Receiver designers and engineers need to know which oscillator technology will provide the lowest phase noise and how to design oscillators with low phase noise. The Leeson oscillator model, for example, is simple and provides answers to these problems. Figure 4-19 shows the Leeson model for an oscillator.

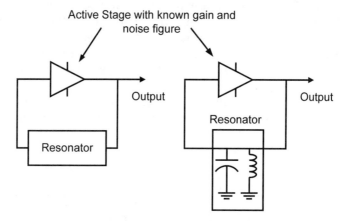

Figure 4-19 The Leeson model of an oscillator consists of a gain stage with a known gain and noise figure. The resonator determines the frequency of oscillation.

To build an oscillator, we feed the output of the active device through a resonant network. We then apply the output of the resonant network to the input of the active device, creating a feedback loop. Oscillation will occur at the frequencies where the power gain around the loop is greater than unity, and the phase shift around the loop is a multiple of 360°.

The oscillation frequency is a strong function of the phase response of the resonant network and the active device. The active device is wideband and exhibits a low phase slope. Therefore, the resonant network primarily determines the oscillation frequency. For illustration, a parallel LC configuration as the resonant network is shown in Figure 4-19, but other resonators such as crystals, transmission lines, and surface-acoustic wave devices are commonly used. Regardless of the resonator type, the following analysis is universal.

Flicker Noise

Figure 4-20 shows the noise spectrum of a typical active device. Above the corner frequency f_c the device exhibits a flat noise spectrum. Below the corner frequency the noise rises as f^{-1}. The corner frequency (f_c) in Figure 4-20 is about 150 hertz.

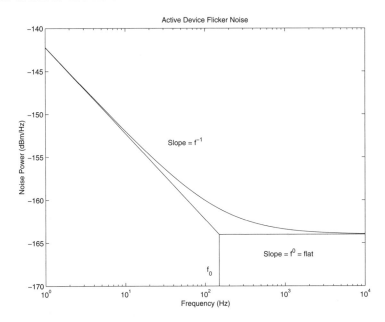

Figure 4-20 *The noise spectrum of a typical active device. Below some cutoff frequency f_c, the noise power increases as $1/f$.*

All active devices exhibit this flicker noise. The spectrum of the flicker noise affects the oscillator's output signal. As this noisy signal passes through the resonant circuit in the oscillator's feedback path, the f^{-1} and f^0 noise components are converted into f^{-3} and f^{-2} slopes, respectively. In conclusion we find that, by the time the signal leaves the oscillator, it has noise components with slopes of f^0, f^{-1}, f^{-2} and f^{-3}. Figure 4-21 shows the idealized spectrum of the oscillator's output.

The *additive white Gaussian noise* (AWGN) generated by the active device causes the phase noise in Figure 4-21. The AWGN causes both AM and PM noise to appear on the oscillator's output. It is desirable to use low-noise active devices to build low-noise oscillators.

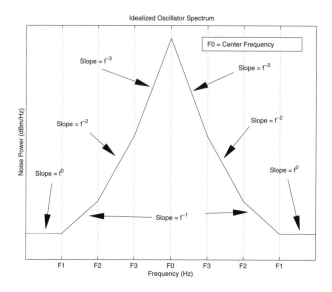

Figure 4-21 *The noise spectrum of a typical oscillator. The slopes of the noise power and the cutoff frequencies are derivable from the noise spectrum of the active device and the characteristics of the resonator used in the oscillator.*

Oscillator Limiting

When DC power is first applied to an oscillator, the internally generated noise of the active device passes through the resonant network and back into the active device. If the power gain and phase shift around the loop are sufficient, a signal will build up and oscillation occurs. The signal passing around the loop will continue to build in strength until limiting occurs in the active device which restricts the magnitude of the output waveform.

Phase Noise Expression

An analysis of Leeson's model is beyond the scope of this book. The result of the analysis is an approximation for the single-sided phase noise of an oscillator.

$$\mathcal{L}(f_m) = \frac{1}{2}\frac{FkT}{P_{avg}}\left[1 + \left(\frac{1}{f_m}\frac{f_0}{2Q_L}\right)^2\right] \qquad 4.2$$

where

$\mathcal{L}(f_m)$ = the single-sideband phase noise of the oscillator,
k = Boltzmann's constant = 1.38 E-23 watt-sec,

F = the noise factor of the active device (in linear terms),
T = the physical temperature in K,
P_{avg} = the average power taken from the oscillator,
f_0 = the carrier or center frequency of the oscillator,
f_m = the offset from the carrier frequency,
Q_L = the loaded Q of the oscillator's resonator.

Figure 4-22 shows some of these variables graphically.

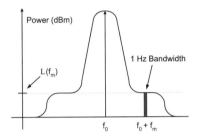

Figure 4-22 *Oscillator spectrum illustrating the variables in Equation 4.2.*

Minimizing Phase Noise

Resonator Q

Resonator Q is defined as

$$Q = 2\pi \frac{\text{Maximum energy stored}}{\text{Total energy lost}} \text{ per cycle} \qquad 4.3$$

Loaded Q is a measure of the energy stored by a resonant circuit when the resonator is placed in a circuit. Different resonator technologies have different Q's and certain technologies are more suited to particular applications. For example, quartz crystals have very high Q's but are difficult to tune. LC networks are easily tuned yet their Q's are lower.

Equation 4.2 indicates that phase noise and the loaded resonator Q are inversely proportional. If the resonator Q is increased, the phase noise of the oscillator decreases. To maximize the loaded Q of the resonator, high unloaded Q is required. Unloaded Q is a function of the type of oscillator the resonator used. Common choices include

- Crystal: Q range is 20k to 200k,
- Surface Acoustic Wave (SAW) Resonators : Q's of 2k to 12k are common,
- Dielectric Resonators : 500 to 5000,
- LC: Q's of 20 to 300.

This list is in the order of highest Q to the lowest unloaded Q. Equation 4.2 indicates that this list is also in the order of increasing oscillator phase noise.

With regard to phase noise, even the poorest crystal oscillator will outperform the best VCO because of the very large Q's of crystal oscillator resonators. From an ability-to-tune perspective, the VCO is the better choice. Experience has shown that we can change the frequency of a crystal oscillator perhaps 1000 ppm of the center frequency. Voltage-controlled oscillators can tune an octave or more, provided we accept poor phase noise.

Oscillator Center Frequency

Equation 4.2 shows that the phase noise increases with the oscillator center frequency. Consequently, all other things being equal, oscillators at lower frequencies are quieter than oscillators at higher frequencies. Note that poor resonator Q can spoil this relationship. A RC oscillator, which has a very poor Q, running at 5 kHz may be much noisier than a LC oscillator running at 100 MHz.

Offset Frequency

Equation 4.2 reveals that the phase noise increases as we approach the carrier. In other words, the value of $\mathcal{L}(f_m)$ increases with decreasing f_m.

4.3 Phase Modulation Review: Sinusoidal Modulation

Time Domain

For a sinusoidal modulating wave, the instantaneous phase of a PM signal is

$$\phi_{PM}(t) = 2\pi f_0 t + \Delta\phi_{pk} \cos(2\pi f_m t) \qquad 4.4$$

where
$\Delta\phi_{pk}$ = the peak phase shift,
f_m = the modulating frequency,
f_0 = the oscillator's carrier or RF frequency,
$\omega_m = 2\pi f_m$ = the modulating angular frequency,
$\omega_0 = 2\pi f_0$ = the oscillator's angular frequency.

The time-domain expression for a sinusoidally phase-modulated wave is

$$\begin{aligned}V_{PM}(t) &= V_{Pk} \cos\left[2\pi f_0 t + \Delta\phi_{Pk} \cos(2\pi f_m t)\right] \\ &= V_{Pk} \cos\left[\omega_0 t + \Delta\phi_{Pk} \cos(\omega_m t)\right]\end{aligned} \qquad 4.5$$

Frequency Domain

Fourier analysis reveals that Equation 4.5 can be rewritten as

$$V_{Pk}(t) = V_{Pk}J_0(\Delta\phi_{Pk})\cos(\omega_c t)$$
$$+V_{Pk}J_1(\Delta\phi_{Pk})\{\sin[(\omega_0+\omega_m)t]+\sin[(\omega_0-\omega_m)t]\}$$
$$-V_{Pk}J_2(\Delta\phi_{Pk})\{\cos[(\omega_0+2\omega_m)t]+\cos[(\omega_0-2\omega_m)t]\}$$
$$-V_{Pk}J_3(\Delta\phi_{Pk})\{\sin[(\omega_0+3\omega_m)t]+\sin[(\omega_0-3\omega_m)t]\}$$
$$+V_{Pk}J_4(\Delta\phi_{Pk})\{\cos[(\omega_0+4\omega_m)t]+\cos[(\omega_0-4\omega_m)t]\}$$
$$+L$$

4.6

where
$J_n(\Delta\phi_{pk})$ = the Bessel function of the first kind, order n, with an argument of $\Delta\phi_{pk}$.

Equation 4.6 takes the same form for a FM modulated wave. With some slight changes (for example, a sign, or changing a sine to a cosine in a few places), the spectrum of a PM wave is identical to the spectrum for an FM wave. Figure 4-23 shows the spectrum of a sinusoidally modulated PM waveform.

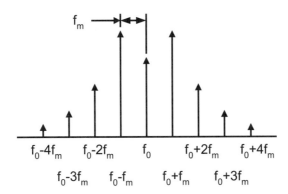

Figure 4-23 *The spectrum of a phase-modulated carrier under sinusoidal modulation.*

Bessel Functions

Equation 4.6 relates the magnitudes of the spectral components of a PM waveform with the Bessel functions of the first kind. Figures 4-24 and 4-25 show the first 6 Bessel functions, J_0 through J_5.

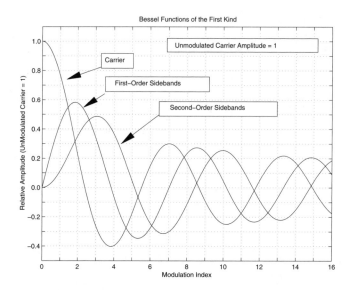

Figure 4-24 *The first three Bessel functions, J_0 through J_2.*

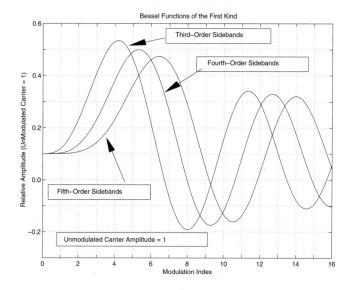

Figure 4-25 *Bessel functions J_3, J_4, and J_5.*

The 0th Bessel function describes the amplitude of the carrier (at f_0). The 1st Bessel function describes the amplitude of the two sidebands at $f_0 \pm f_m$, the 2nd Bessel function describes the amplitude of the sidebands at $f_0 \pm 2f_m$, the 3rd Bessel function describes the sidebands at $f_0 \pm f_m$, and so on. In general, the nth Bessel function describes the amplitude of the sideband at $f_0 \pm nf_m$. Normalized to 1 ohm, the power level of the unmodulated carrier is

$$P_{PM,\text{Unmodulated Carrier}} = \left(\frac{V_{Pk}}{\sqrt{2}}\right)^2 \qquad 4.7$$

The power level of the spectral component at $f_0 \pm nf_m$ is

$$P_{PM}(f_0 + nf_m) = P_{PM}(f_0 - nf_m) \qquad 4.8$$
$$= \left[\frac{V_{Pk}}{\sqrt{2}} J_n(\Delta\phi_{pk})\right]^2$$

The ratio of the power level any single spectral component to the power in the unmodulated carrier is

$$\frac{P_{PM}(f_0 + nf_m)}{P_{PM,\text{Unmodulated Carrier}}} = \frac{P_{PM}(f_0 - nf_m)}{P_{PM,\text{Unmodulated Carrier}}}$$
$$= \left[\frac{V_{Pk}}{\sqrt{2}} J_n(\Delta\phi_{pk})\right]^2 \bigg/ \left[\frac{V_{Pk}}{\sqrt{2}}\right]^2 \qquad 4.9$$
$$= J_n^2(\Delta\phi_{pk})$$
$$= 20\log|J_n(\Delta\phi_{pk})| \text{ in dB}$$

Small β Approximations

When the phase perturbation, $\Delta\phi_{pk}$, is small as it is with noise then several useful approximations may be used. These approximations are called the *small β approximations*. Rather arbitrarily, we say the small β approximations are valid when

$$\Delta\phi_{pk} \leq 0.2 \text{ radians} \qquad 4.10$$

For small values of $\Delta\phi_{pk}$,

1. The value of $J_0(\Delta\phi_{pk})$ is very close to unity or

$$J_0(\Delta\phi_{pk}) \approx 1 \text{ for } \Delta\phi_{pk} \leq 0.2 \qquad 4.11$$

2. The value of $J_1(\Delta\phi_{pk})$ is

$$J_1(\Delta\phi_{pk}) \approx \frac{\Delta\phi_{pk}}{2} \quad for\ \Delta\phi_{pk} \leq 0.2 \qquad 4.12$$

3. The values of the rest of the Bessel functions $J_2(\Delta\phi_{pk})$ through $J_n(\Delta\phi_{pk})$ are zero.

$$J_n(\Delta\phi_{pk}) \approx 0 \quad for\ \begin{cases} \Delta\phi_{pk} \leq 0.2 \\ n = 2,3,4,\ldots \end{cases} \qquad 4.13$$

Applying these approximations to Equation 4.6, the Fourier spectrum for a PM modulated wave under small β conditions is

$$\begin{aligned} V_{Pk}(t) &= V_{Pk} J_0(\Delta\phi_{Pk}) \cos(\omega_0 t) \\ &\quad + V_{Pk} J_1(\Delta\phi_{Pk}) \{\sin[(\omega_0 + \omega_m)t] + \sin[(\omega_0 - \omega_m)t]\} \\ &= V_{Pk} \cos(\omega_0 t) \\ &\quad + V_{Pk} \frac{\Delta\phi_{pk}}{2} \{\sin[(\omega_0 + \omega_m)t] + \sin[(\omega_0 - \omega_m)t]\} \end{aligned} \qquad 4.14$$

Frequency Domain (Small β)

We can make some approximations for the power spectrum of a PM wave under small β conditions.

- The power in the component at f_0 is

$$\frac{P_{PM}(f_0)}{P_{PM,\text{Unmodulated Carrier}}(f_0)} = J_0^2(\Delta\phi_{Pk}) \qquad 4.15$$
$$\approx 1$$

The power in the modulated carrier approximately equals the power in the unmodulated carrier.

- The power in each of the components at $f_0 \pm f_m$ is

$$\frac{P_{PM}(f_0 + f_m)}{P_{PM,\text{Unmodulated Carrier}}(f_0)} = \frac{P_{PM}(f_0 - f_m)}{P_{PM,\text{Unmodulated Carrier}}(f_0)}$$
$$= J_1^2(\Delta\phi_{Pk}) \qquad 4.16$$
$$\approx \frac{\Delta\phi_{Pk}^2}{4}$$

Since $\Delta\phi_{pk}$ must be less than 0.2 radians for the small β conditions to be valid, the sidebands will always be less than

$$\frac{P_{PM}(f_0 + f_m)}{P_{PM,\text{Unmodulated Carrier}}(f_0)} = \frac{P_{PM}(f_0 - f_m)}{P_{PM,\text{Unmodulated Carrier}}(f_0)} \qquad 4.17$$

$$\approx \frac{\Delta\phi_{Pk}^2}{4}$$

$$\leq \frac{(0.2)^2}{4} = 0.010 = -20 \text{ dB}$$

The sidebands of a phase modulated signal under small β conditions will always be less than 20 dB below the carrier. Figure 4-26 shows the spectrum of a phase modulated waveform under small β conditions.

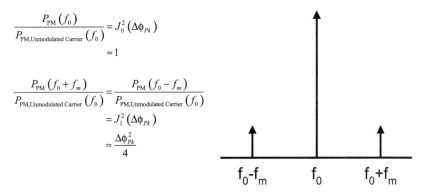

$$\frac{P_{PM}(f_0)}{P_{PM,\text{Unmodulated Carrier}}(f_0)} = J_0^2(\Delta\phi_{Pk})$$

$$\approx 1$$

$$\frac{P_{PM}(f_0 + f_m)}{P_{PM,\text{Unmodulated Carrier}}(f_0)} = \frac{P_{PM}(f_0 - f_m)}{P_{PM,\text{Unmodulated Carrier}}(f_0)}$$

$$= J_1^2(\Delta\phi_{Pk})$$

$$\approx \frac{\Delta\phi_{Pk}^2}{4}$$

Figure 4-26 *Spectrum of a sinusoidally modulated PM waveform under small β conditions. Only the carrier (at f_0) and the two sidebands (at $f_0 \pm f_m$) contain significant energy.*

Phasor (Small β)

Figure 4-27 shows a phasor diagram of a phase modulated wave under small β conditions. The only nonzero Fourier components are the carrier (at f_0) and the two sidebands (at $f_0 \pm f_m$). Figure 4-28 shows the phasor diagram when the two sidebands at $f_0 \pm f_m$ add together to produce the maximum phase deviation on the carrier.

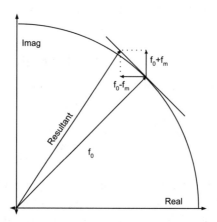

Figure 4-27 *Phasor diagram of a phase-modulated signal under small β conditions.*

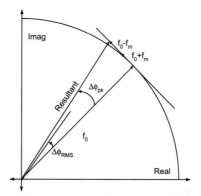

Figure 4-28 *Phasor diagram of a phase-modulated signal under small β conditions. The two sidebands at $f_0 \pm f_m$ are colinear at this instant in time. The extent of these sidebands produces a peak- and RMS-phase deviation of the carrier.*

The peak phase deviation $\Delta\phi_{pk}$ is given by

$$\tan(\Delta\phi_{Pk}) = \frac{2J_1(\Delta\phi_{Pk})}{J_0(\Delta\phi_{Pk})} \qquad 4.18$$

It is convenient to talk about a RMS phase deviation $\Delta\phi_{RMS}$ where

$$\Delta\phi_{RMS} = \frac{\Delta\phi_{Pk}}{\sqrt{2}} \qquad 4.19$$

Phase Modulation: General Modulation

For the following discussion, consider a perfect oscillator.

$$V_{\text{Perfect}}(t) = V_{Pk} \cos(\omega_0 t) \qquad 4.20$$

This oscillator has the desirable properties discussed earlier: an infinite SNR, a zero-width frequency spectrum and deterministic zero crossings. To model an oscillator with phase noise, we introduce a nondeterministic or noise-like waveform $\phi(t)$. The analysis is easier if we assume $\phi(t)$ is a Gaussian process (although that assumption does not have to be true). The frequency spectrum of (t) is $\phi(f_m)$. The Fourier transform of $\phi(t)$ is

$$\phi(f_m) = \Im[\phi(t)] \qquad 4.21$$

We also say that $\phi(t)$, being a Gaussian random process, has a RMS value ϕ_{RMS}. The mathematical model for an oscillator corrupted by phase noise is

$$V_{PN}(t) = V_{Pk} \cos[\omega_0 t + \phi(t)] \qquad 4.22$$

The addition of the phase noise term reduces the SNR of the oscillator, widens up its frequency spectrum and causes the zero crossings to be statistical, rather than deterministic. The characteristics of the noisy oscillator, its frequency spectrum and the statistics of its zero crossings, depend very strongly on the characteristics of $\phi(t)$ and $\phi(f_m)$.

Time Domain

For a general modulating waveform $\phi(t)$, the time domain equation for a phase-modulated wave is

$$V_{PN}(t) = V_{Pk} \cos[\omega_c t + \phi(t)]$$

where
$\phi(t)$ = the modulating waveform.

If we build an oscillator and want to measure its phase noise, we have to measure $\phi(t)$ given $V_{PN}(t)$.

Measuring $\phi(f_m)$

Figure 4-29 shows a phase detector consisting of a mixer and a *low pass filter* (LPF). Assuming that the LO has much lower phase noise than the oscillator under test, the voltage present at the IF port of the mixer is

$$V_{IF}(t) = k_{mix}\{V_{RF}(t)\}\{V_{LO}(t)\}$$
$$= k_{mix}\{\cos[\omega_0 t + \phi(t)]\}\left\{\cos\left(\omega_0 t - \frac{\pi}{2}\right)\right\} \quad 4.24$$
$$= \frac{k_{mix}}{2}\cos\left[\phi(t) - \frac{\pi}{2}\right] + \frac{k_{mix}}{2}\cos\left[2\omega_0 t + \phi(t) - \frac{\pi}{2}\right]$$

where
  k_{mix} is a constant related to the mixer's conversion loss.

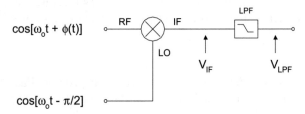

Figure 4-29 *Using a phase demodulator to measure the phase noise of an oscillator.*

With identical RF and LO frequencies and the cutoff frequency of the LPF set to remove the sum term from the output of the mixer and pass the difference term, we find

$$V_{LPF}(t) = \frac{k_{mix}}{2}\cos\left[\phi(t) - \frac{\pi}{2}\right] \quad 4.25$$

Since

$$\sin(\omega t) = \cos\left(\omega t - \frac{\pi}{2}\right) \quad 4.26$$

we can write for small values of $\phi(t)$,

$$\cos\left[\phi(t) - \frac{\pi}{2}\right] = \sin[\phi(t)] \quad 4.27$$
$$\approx \phi(t)$$

The output of the LPF is

$$V_{LPF}(t) \approx \frac{k_{mix}}{2}\phi(t) \quad 4.28$$

V_{LPF} may be applied to a suitable spectrum analyzer to view $\phi(f_m)$.

Measurement Difficulties

Let us examine some of the problems we might experience in measuring $\phi(t)$ as it was described. One obstacle is deriving the local oscillator. The LO should have much lower phase noise than the device under test, otherwise we measure the phase noise of the local oscillator.

The LO must be at the same frequency as the device under test. If the frequency difference is even one hertz, the results of Equations 4.24 through 4.28 will be inaccurate.

We can phase-lock the LO and the signal being measured. This produces a second signal whose frequency is exactly the same as the signal of interest but also introduces further measurement uncertainties.

Frequency Domain

The spectral analysis of a PM modulated wave under general conditions is quite complex. If we assume the small β conditions though, we can make various simplifying assumptions.

$\mathcal{L}(f_m)$

$\mathcal{L}(f_m)$ is the power spectrum of an oscillator as it would appear on a suitable spectrum analyzer when taking into account all of the measurement caveats we discussed earlier and normalizing the measurement bandwidth to 1 hertz.

Phasor

Figure 4-30 shows the phasor diagram of a carrier that has been small β phase-modulated by a random waveform. The *RMS* length of the unmodulated carrier is

$$\frac{V_{Pk}}{\sqrt{2}} \qquad 4.29$$

The tip of the vector representing the random waveform will gradually trace a path similar to that shown in Figure 4-30. The amplitude of the resultant will change slightly but this change can be ignored because mixers usually suppress small amplitude variations. The angle labeled $\phi(t)$ is the same $\phi(t)$ that appears in Equations 4.21 through 4.23.

374 | RADIO RECEIVER DESIGN

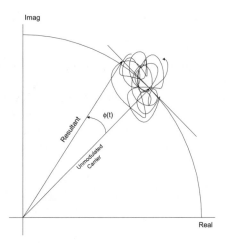

Figure 4-30 *Phasor diagram of an oscillator exhibiting phase noise. We model the noise component as a random vector placed at the end of the carrier vector, which creates a resultant with phase and amplitude variations.*

Figure 4-31 shows how $\mathcal{L}(f_m)$ is related to $\phi(t)$ and ϕ_{RMS} for two offset frequencies. We divide the spectral plot of the oscillator into narrow slices and consider the effect of each slice on the phasor description. The slice chosen in Figure 4-31(a) at $f_0 + f_{m1}$ contains only a small amount of power. The small amount of noise causes only a small phase ambiguity $\phi_{m1}(t)$, the larger noise power of Figure 4-31(b) causes a larger $\phi_{m2}(t)$.

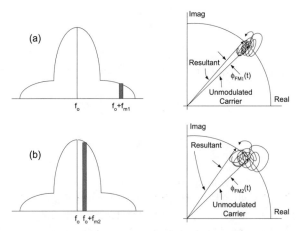

Figure 4-31 *The phase deviation in the oscillator at a particular offset is caused by the amount of noise power present at that frequency offset. Higher power causes a wider phase deviation.*

4.4 Relating $\phi(f_m)$ to $\mathcal{L}(f_m)$

Figure 4-32 shows the conceptual model we will use for this analysis. The box labeled "Phase Modulator" is a perfect oscillator whose phase can be controlled by an input waveform. The input waveform is $\phi(t)$ which has a frequency spectrum described by $\phi(f_m)$. The output waveform is $V_{PM}(t)$ and its frequency spectrum is $\mathcal{L}(f_m)$.

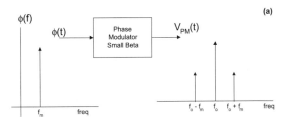

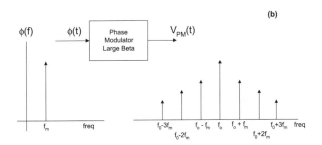

Figure 4-32 *Conceptual model used to analyze a noisy oscillator. Figure 4-32(a) shows the modulator under small β conditions. Figure 4-32(b) shows the modulator under large β conditions.*

Small β, Large β and Sinusoidal Modulation

Figure 4-32(a) shows the phase modulator with a sinusoidal input signal under small β conditions. One of the consequences of the small β conditions was that if we apply a sinusoidal input signal whose frequency is f_m to a phase modulator, the output spectrum will contain only three tones: $f_0 - f_m$, f_0 and $f_0 + f_m$. In practice, the spectra on the right side of Figure 4-32 represent the oscillator as we observe it on a spectrum analyzer. If we observe that the output of our oscillator exhibits equal-amplitude sidebands at $f_0 - f_m$ and $f_0 + f_m$, and we assume that the small β conditions apply, then we can model the oscillator as a small β phase modulator whose input signal is a sinusoid at frequency f_m. Figure 4-32(b) shows the same experiment under large β conditions. A sinusoidal input at frequency f_m produces an output spectrum with components at $f_0 - nf_m$, f_0 and $f_0 + nf_m$ where n runs from one to infinity. A large β output spectrum may not be used to draw conclusions about the input spectrum. For example, let us look at the signals at $f_0 - 3f_m$ and $f_0 + 3f_m$ in Figure 4-32(b). Under large β conditions, these output components could have been produced by input signals at f_m or $3f_m$.

Small β and Complex Modulation

Figure 4-33(a) shows a complex modulating waveform. We can model any complex waveform as a collection of sine waves. Under small β conditions, the signals present in the output spectrum at $f_0 \pm f_m$ are caused only by the components of the input signal at f_m.

In Figure 4-33(b), a frequency f_m is missing from the input waveform so the components at $f_0 \pm f_m$ will be missing from the output waveform. Similarly, if an input frequency component f_1 is particularly strong, the output components at $f_0 \pm f_1$ will be strong as well.

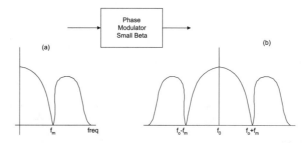

Figure 4-33 The relationship between the input and output spectra of a phase modulator. This representation is valid only under small β conditions.

The Magnitudes of $\phi(f_m)$ and $\mathcal{L}(f_m)$

Next we will find the power relationship between the input and output spectra of Figure 4-33. If we know the shape of the output spectrum $\mathcal{L}(f_m)$, what does that tell us about the shape of the input power spectrum $\phi(f_m)$? Under small β conditions, we know that $J_0(\Delta\phi_{pk}) = 1$, $J_1(\Delta\phi_{pk}) = \Delta\phi_{pk}/2$ and $J_n(\Delta\phi_{pk}) = 0$ for n equals two through infinity.

The first approximation [$J_0(\Delta\phi_{pk}) = 1$] results in equal modulated and unmodulated carrier magnitude. If the carrier is at 10 dBm when the phase modulator input waveform is zero volts [i.e., when $\phi(t) = 0$], the carrier will still be 10 dBm when $\phi(t)$ is not zero. The last approximation [$J_n(\Delta\phi_{pk}) = 0$

for $n = 2$ to infinity] indicates that the components at $f_0 \pm n \cdot f_m$ are negligible for $n \geq 2$. The middle approximation tells us about the relationship between the output and input spectra. The output spectrum, $\mathcal{L}(f_m)$ is given by

$$\mathcal{L}(f_m) = \frac{\text{Power in a 1-hertz bandwidth measured } f_m \text{ from the carrier}}{\text{Total signal power}} \qquad 4.30$$

$$= \frac{P(f_m)_{1 \text{ hertz}}}{\text{Total signal power}}$$

Under small β conditions, we can use Equations 4.15 and 4.16 to simplify the equation above to

$$\mathcal{L}(f_m) = \frac{\phi_{Pk}^2(f_m)}{4} \qquad 4.31$$

Since $\phi(t)$ is a nondeterministic, statistically described waveform, we must describe it using RMS rather than peak values. Since much of this analysis was done assuming sinusoidal modulation and we modeled the complex modulating waveform as a combination of sinusoids, we can write

$$\Delta\phi_{RMS} = \frac{\Delta\phi_{Pk}}{\sqrt{2}} \qquad 4.32$$

and

$$\mathcal{L}(f_m) = \frac{\phi_{RMS}^2(f_m)}{2} \qquad 4.33$$

Converting this equation to decibels produces

$$\mathcal{L}(f_m)_{dB} = 10 \log \left[\frac{\phi_{RMS}^2(f_m)}{2} \right] \qquad 4.34$$

$$= 20 \log [\phi_{RMS}(f_m)] - 3 \text{ dB}$$

If we add 3 dB to $\mathcal{L}(f_m)$, we will arrive at $\phi_{RMS}(f_m)$ as long as the small β approximations hold.

4.5 $S_\phi(f_m)$

Up to this point, we have specified a 1-hertz measurement bandwidth for both $\mathcal{L}(f_m)$ and $\phi(f_m)$. The quantity $S_\phi(f_m)$ is simply the power of $\phi(f_m)$ measured in a 1-hertz bandwidth or

$$S_\phi(f_m) = \frac{\phi^2_{RMS}(f_m)}{1 \text{ Hertz}} \quad (\text{rad})^2 \qquad 4.35$$

In other words, $S_\phi(f_m)$ is the spectral density of the phase fluctuations of an oscillator. This is the power of a phase discriminator's output measured in a 1-hertz bandwidth. It may be measured with an audio spectrum analyzer.

4.6 When the Small β Conditions are Valid

Previous derivations required that the oscillator meet the small β conditions. Given an oscillator, how can we tell from its output spectrum if it meets the small β criteria? The –10 dB/decade line drawn on Figure 4-34 represents an RMS phase deviation of approximately 0.2 radians integrated over any one decade of offset frequency. At 0.2 radians, the power in the higher-order sidebands of the phase modulation is still small compared to the power in the first sideband, which guarantees the small β criterion.

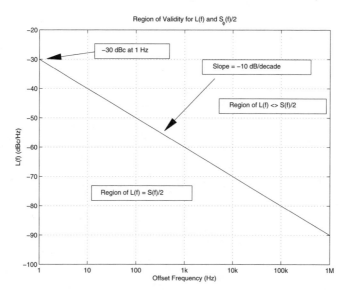

Figure 4-34 Regions of validity for small β conditions.

4.7 Phase Noise and Multipliers

Most frequency synthesizer designs involve multiplying and/or dividing a stable reference oscillator. For example, to generate a 1234.567 MHz LO, we might start with a 1 MHz crystal oscillator, divide it by 1000 to obtain a 1 kHz oscillator then multiply the 1 kHz oscillator by 1,234,567 to obtain the frequency needed.

Figure 4-35 shows the multiplication. We begin with a fixed reference oscillator (at f_{REF}). To generate a particular frequency, we multiply the oscillator by some number (n) to produce a new signal at nf_{REF}. Given the phase noise characteristics of the reference oscillator, what are the phase noise characteristics of the output of the multiplier? We will use the expression for a noisy oscillator to describe the reference oscillator

$$V_{REF}(t) = \cos[2\pi f_0 t + \phi(t)] \qquad 4.36$$

where
$\phi(t)$ is a random waveform with some RMS value ϕ_{RMS}.

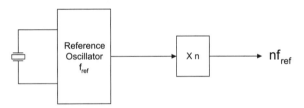

Figure 4-35 *A frequency multiplier.*

Passing the reference oscillator waveform through the frequency multiplier produces

$$V_{REF,mult}(t) = \cos[n\omega_0 t + n\phi(t)] \qquad 4.37$$

We have multiplied both the frequency and $\phi(t)$ by n. The quantity $n\phi(t)$ now has a RMS value of $n\phi_{RMS}$. Before the multiplication, the single-sideband phase noise of the reference oscillator at some arbitrary offset f_m is

$$\mathcal{L}_{REF}(f_m) = \frac{\phi_{RMS}^2(f_m)}{2} \qquad 4.38$$

After the multiplication, the single-sided phase noise at the same f_m is

$$\mathcal{L}_{\text{REF,mult}}(f_m) = \frac{[n\phi_{RMS}(f_m)]^2}{2} \qquad 4.39$$

$$= n^2 \frac{\phi_{RMS}^2(f_m)}{2}$$

The ratio of the single-sided phase noise power of the multiplied oscillator to the unmultiplied oscillator is

$$\frac{\mathcal{L}_{\text{REF,mult}}(f_m)}{\mathcal{L}_{\text{REF}}(f_m)} = n^2 \qquad 4.40$$

$$= 20\log(n) \quad \text{in dB}$$

If $n > 1$ (i.e., for frequency multipliers), then the $\mathcal{L}(f_m)$ of the multiplied oscillator is increased. This analysis assumes the small β conditions apply for both the unmultiplied and for the multiplied carrier at the f_m of interest.

Example 4.1 — Multiplied Oscillator Phase Noise

We measure the $\mathcal{L}(f_m)$ of an oscillator at $f_m = 100$ hertz and find that $\mathcal{L}(100 \text{ hertz}) = -80$ dBc. After multiplying the oscillator by 100, and assuming the small β conditions apply, what is $\mathcal{L}(f_m)$ of the multiplied oscillator at
a. $f_m = 100$ hertz,
b. $f_m = 10,000$ hertz?

Solution —

Using 4.40, we find

$$\left[\frac{\mathcal{L}_{\text{REF},1000x}(f_m)}{\mathcal{L}_{\text{REF}}(f_m)}\right]_{dB} = 20\log(n) \qquad 4.41$$

$$= 20\log(100)$$

$$= 40 \text{ dB}$$

a. The phase noise of the unmultiplied oscillator at 100 hertz away from the multiplied carrier is $-80 + 40 = -40$ dBc. Multiplying increases the phase noise.

b. We do not have enough information to answer this part of the problem. We need to know $\mathcal{L}(10,000 \text{ hertz})$ of the unmultiplied oscillator to solve the problem.

Equation 4.40 shows that each doubling of the carrier increases the phase noise measured at f_m by 6 dB. Likewise, each halving of the carrier results in a decrease in the phase noise measured at f_m by 6 dB.

The multiplier can be a nonlinear device utilizing some high-order distortion component or it can be a phase-locked loop. Also, Equation 4.40 shows what the minimum phase noise of the multiplier will be. The multiplier system can introduce excess phase noise into the system, and the output of the multiplier will be noisier than Equation 4.40 indicates.

4.8 Phase Noise and Dividers

Equation 4.40 is also true for frequency division. If we apply a signal to a digital flop-flop, for example, the output frequency equals one-half of the input frequency. We can divide the input signal by any number we desire using the appropriate digital techniques.

Example 4.2 — Frequency Division and Phase Noise
Given is an oscillator whose $\mathcal{L}(f_m)$ is –110 dBc when f_m = 10 kHz. What is $\mathcal{L}(10,000\text{ hertz})$ of the oscillator after frequency division by 12?

Solution —
Using Equation 4.40, we find that

$$\left[\frac{\mathcal{L}_{REF,\div 12}(f_m)}{\mathcal{L}_{REF}(f_m)}\right]_{dB} = 20\log(n)$$

$$= 20\log\left(\frac{1}{12}\right) \quad\quad 4.42$$

$$= -21.6 \text{ dB}$$

$\mathcal{L}(f_m)$ of the new oscillator at f_m = 10,000 hertz is –110 + (–21.6) = –131.6 dBc. Frequency division ($n < 1$) lowers $\mathcal{L}(f_m)$. We can think of frequency division as an averaging process (see Figure 4-36). We assume that the zero crossings of the input oscillator exhibit a certain amount of jitter or uncertainty that is related to the phase noise of the input oscillator.

Each rising edge of the input waveform exhibits some uncertainty, which contributes to the input's phase noise spectrum. When dividing by 8, the output changes only on every fourth rising edge of the input. The phase noise contributions of input transitions 1 through 5 are ignored because they do not affect the counter's output. As far as the output is concerned, the net effect is to spread the uncertainty of one zero crossing over the time required for eight input cycles, thus reducing the phase noise of the output waveform.

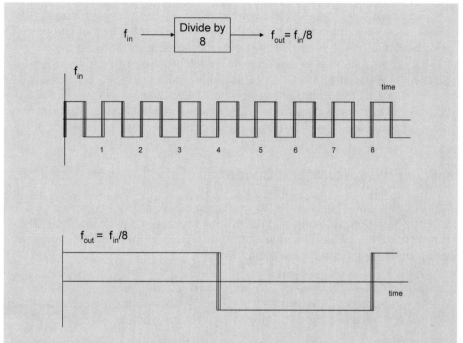

Figure 4-36 *The effects of a divider on oscillator phase noise. The uncertainty in one zero crossing of the input signal is spread across a longer time in the output signal.*

4.9 Incidental Phase Modulation

Oscillator phase noise represents an uncertainty in an oscillator's phase or frequency. As we have seen, any signal converted from one frequency to another by a mixer will inherit the phase noise of the LO used to perform the conversion. The LO phase noise can mask the information present on an information-bearing signal.

We can encode digital data into the phase of a sine wave. We may assign a binary 0 to 0° and a binary 1 to 180°. Anything that masks the true phase of the signal such as poor SNR or excess LO phase noise, can cause the phase of the signal to be misread.

Phase Demodulation

A phase demodulator produces an output voltage which is directly proportional to the phase difference between a reference signal (internally generated in the demodulator) and a data-bearing signal. In Figure 4-37, the information-bearing signal is

$$V_I(t) = \cos[\omega_0 t + \phi_I(t)] \qquad 4.43$$

The locally generated reference frequency is

$$V_{REF}(t) = \cos[\omega_0 t + \phi_{REF}(t)] \qquad 4.44$$

where
  $\phi_I(t)$ carries the information encoded as carrier phase,
  $\phi_{REF}(t)$ represents the phase noise on the locally generated reference signal.

The output of the phase demodulator is

$$\phi_I(t) - \phi_{REF}(t) \qquad 4.45$$

If the locally generated reference is quiet and $\phi_I(t) >> \phi_{REF,RMS}$, then the phase demodulator output represents the information-bearing $\phi_I(t)$. If $\phi_{REF,RMS}$ is significant with respect to $\phi_I(t)$, the phase demodulator output is corrupted by the local oscillator's phase noise. Let us assume a phase demodulator will produce an output of 1 volt when $\phi(t)$ is 180°. When ϕ_{IRMS} is 0°, the demodulator produces 0 volts. The phase demodulator is described by

$$\begin{aligned} k_\phi &= \frac{\Delta \text{Voltage}}{\Delta \text{Phase}} \\ &= \frac{1 \text{ Volt}}{180°} \\ &= 5.6 \frac{\text{mV}}{\text{Degree}} \end{aligned} \qquad 4.46$$

The LO phase noise introduces noise on the demodulator's output. This noise represents a nonreducible uncertainty and limits the demodulator's accuracy. Continuing the example, suppose we measure 8.2 mV$_{RMS}$ of noise on the phase demodulator's output when we apply a quiet carrier to the input of the phase demodulator in Figure 4-37 [i.e., $\phi_I(t) = 0$]. The equivalent phase noise of the reference oscillator is

$$\begin{aligned} \Delta \text{Phase}_{RMS} &= \frac{\Delta \text{Voltage}_{RMS}}{k_\phi} \\ &= \frac{8.2 \text{ mV}_{RMS}}{5.6 \text{ mV/degree}} \\ &= 1.5 \text{ degrees}_{RMS} \end{aligned} \qquad 4.47$$

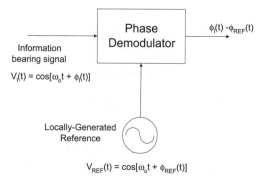

Figure 4-37 *A phase demodulator. The information is encoded in the phase of the received signal as $\phi_I(t)$. The LO phase noise is described by $\phi_{REF}(t)$ and $\phi_{REF,RMS}$.*

We can say the LO has 1.5 degrees$_{RMS}$ of incidental phase modulation.

IPM Definition

The incidental phase modulation (IPM) of an oscillator is

$$IPM = \sqrt{\int_{fa}^{fb} S_\phi(f_m) df_m} \quad (\text{radians}_{RMS}) \qquad 4.48$$

For small β, we can write

$$IPM = \sqrt{2\int_{f_a}^{f_b} \mathcal{L}_\phi(f_m) df_m} \quad (\text{radians}_{RMS}) \qquad 4.49$$

where

f_a and f_b represent the lower and upper frequency boundaries of the demodulated signal.

These two equations represent the amount of phase jitter present on an oscillator given its frequency spectrum $\mathcal{L}(f_m)$ or the spectral density of the phase fluctuations $S_\phi(f_m)$.

Conceptually, we measure the incidental phase modulation of an oscillator by assuming a phase demodulator constant k_ϕ. We apply a known phase-modulated signal and a noiseless reference signal to the demodulator and calculate the output voltage. We then calculate the output voltage when the demodulator input is a noiseless reference plus the noisy reference $V_{REF}(t)$ containing $\phi_{REF}(t)$. The ratio of the two output voltages is the IPM of the oscillator.

Example 4.3 — Incidental Phase Modulation

Figure 4-38 shows a local oscillator used in a 2.048 Mbps QPSK receiver. Find the IPM of this oscillator.

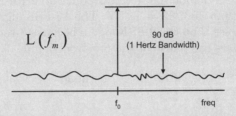

Figure 4-38 *The spectrum analyzer plot of a noisy oscillator (1-hertz resolution bandwidth).*

Solution —

We integrate Equation 4.49 from a lower frequency f_a to an upper frequency f_b. Rather arbitrarily, we will use a lower frequency limit that is equivalent to 100 symbols. In other words, occasionally, we might expect to receive 100 of the same symbols in a row.

The lower frequency limit is $2.048 \cdot 10^6/200 = 10.24$ kHz. The peak-to-null bandwidth of a 2.048 Mbps QPSK signal is 1.024 MHz. Assuming a little excess bandwidth, we use 1.5 MHz as the upper frequency. Equation 4.49 produces

$$IPM = \sqrt{2 \int_{10.24E3}^{1.5E6} (10^{-9}) df_m} \quad (\text{radians}_{RMS})$$

$$= \sqrt{(2)(10^{-9})(1.5 \cdot 10^6 - 10{,}240)} \qquad 4.50$$

$$= 0.055 \text{ radians}_{RMS}$$

$$= 3.1 \text{ degrees}_{RMS}$$

The output of this demodulator will produce a noise level equivalent to 3.1 degrees$_{RMS}$ of modulation. In other words, the IPM of this oscillator will cause a 3.1 degrees$_{RMS}$ uncertainty in the phase measurements of this demodulation.

IPM and SNR

The IPM of the receiver's LO will affect the ultimate SNR of a phase demodulated signal. We also have to consider the characteristics of the information-bearing signal. Imagine the RMS phase noise of a receiver is IPM$_{RMS}$. Figure 4-39 shows the constellation diagrams for an 8-state and a 16-state PSK signal. The phase noise of the receiver causes a phase measurement uncertainty of IPM$_{RMS}$, which is drawn on the diagram. The 8-state PSK demodulator will make a symbol error if the IPM of the local oscillator causes a phase error of more than 360°/16 = 22.5°. Since we are dealing with noise-like signals, we want the IPM to be much smaller than 22.5°.

The 16-state PSK demodulator will produce an error if the input signal deviates by more than 360°/32 = 11.25°. The 16-PSK signal is more sensitive to LO phase noise than the 8-PSK signal because it requires more accuracy when it recovers the phase of the received signal.

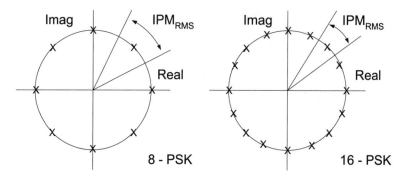

Figure 4-39 Incidental phase modulation (IPM) requirements of 8-PSK and 16-PSK.

Measuring IPM

Figure 4-40 diagrams a measurement setup. A phase demodulator produces a voltage proportional to the phase difference between an externally applied signal and an internally generated reference signal. The receiver generates the reference signal from the received signal.

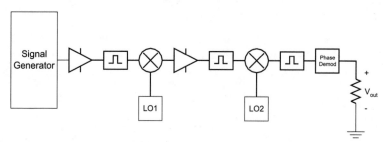

Figure 4-40 Measuring the incidental phase modulation (IPM) of a receiver.

The signal generator is phase-modulated at some arbitrary rate (for example, 1 kHz) and at some known phase deviation ($\Delta\phi_{pk}$), which corresponds to a RMS phase deviation ($\Delta\phi_{RMS}$) (see Figure 4-41).

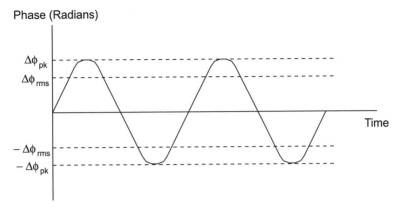

Figure 4-41 *Definitions of peak and RMS phase deviation.*

When the receiver demodulates the phase-modulated signal from the generator, the RMS voltage present on the output of the demodulator will be

$$V_{out,PM,RMS} = k_\phi \Delta\phi_{RMS} \qquad 4.51$$

where
k_ϕ is a constant associated with the phase demodulator.

Next, we remove the phase modulation from the signal generator and provide a quiet, noiseless carrier to the receiver. If the IPM of the signal generator is much smaller than the IPM of the receiver, the output of the phase demodulator will change to

$$V_{out,quiet,RMS} = k_\phi (IPM_{RMS}) \qquad 4.52$$

The ratio of these two voltages is

$$\frac{V_{out,PM,RMS}}{V_{out,quiet,RMS}} = \frac{k_\phi \Delta\phi_{RMS}}{k_\phi (IPM_{RMS})} \qquad 4.53$$
$$= \frac{\Delta\phi_{RMS}}{IPM_{RMS}}$$

Solving for IPM_{RMS}, we find

$$IPM_{RMS} = (\Delta\phi_{RMS}) \frac{V_{out,quiet,RMS}}{V_{out,PM,RMS}} \qquad 4.54$$

SNR and IPM

If the LOs in a receiver were perfect, the smallest voltage we would ever measure from the receiver's PM demodulator port would be zero volts. This occurs when the input signal is unmodulated (quiet carrier).

Under quiet carrier conditions the receiver produces noise due to the IPM of the receiver's LOs. This is the ultimate noise floor of the demodulator. It will never release a signal smaller than that voltage. If IPM is the limiting factor in a receiver, the ultimate SNR at the output will be

$$\begin{aligned}SNR_{IPM} &= \left(\frac{V_{out,PM,RMS}}{V_{out,quiet,RMS}}\right)^2 \\ &= \left(\frac{k_\phi \Delta\phi_{RMS}}{k_\phi IPM_{RMS}}\right)^2 \\ &= \left(\frac{\Delta\phi_{RMS}}{IPM_{RMS}}\right)^2 \\ &= 20\log\left(\frac{\Delta\phi_{RMS}}{IPM_{RMS}}\right)\end{aligned} \qquad 4.55$$

where

$\Delta\phi_{RMS}$ is the RMS phase deviation of the signal to be received.

Signals with small RMS phase deviations are more sensitive to IPM than signals with large RMS phase deviations.

Example 4.4 — Incidental Phase Modulation and SNR

An analog signal is broadcast using phase modulation with peak phase shifts of 0° and 180°. The IPM of the receiver is 3 degrees$_{RMS}$. Find the ultimate SNR present at the output of the phase detector.

Solution —

Assuming the analog signal is a sine wave, the RMS phase shift is

$$\begin{aligned}\Delta\phi_{RMS} &= \frac{\Delta\phi_{Pk}}{\sqrt{2}} \\ &= \frac{180°/2}{\sqrt{2}} \\ &= 63.6 \text{ degrees}_{RMS}\end{aligned} \qquad 4.56$$

The ultimate SNR present at the receiver's output is

$$SNR_{IPM} = 20\log\left(\frac{\Delta\phi_{RMS}}{IPM_{RMS}}\right)$$

$$= 20\log\left(\frac{63.6}{3}\right)$$

$$= 27 \text{ dB}$$

4.57

Incidental phase modulation is also known as β_ϕ, or *phase jitter*.

4.10 Incidental Frequency Modulation

Incidental frequency modulation (IFM) is a measure of the effects of a receiver's LO on frequency-modulated signals. IFM applies to frequency modulation in the same way that IPM applies to phase modulation.

Frequency Demodulation

A FM demodulator (or FM discriminator) produces a voltage whose instantaneous value is a direct function of the instantaneous frequency of a signal. For example, if we perform the FM demodulation at a 21.4 MHz IF, the discriminator produces a voltage proportional to the difference between the input signal's instantaneous frequency and 21.4 MHz.

In a receiver, the phase noise of the local oscillators causes measurement uncertainty. During frequency conversion, the signal acquires the phase noise of the local oscillators. Consequently, even a pure sine wave input acquires phase noise. Noise is always observed on the output of the FM demodulator as a result of the receiver's phase noise.

IFM Definition

Incidental frequency modulation (IFM) is defined as

$$IFM = \sqrt{\int_{fa}^{fb} f_m^2 S_\phi(f_m) df_m} \quad (\text{hertz}_{RMS})$$

4.58

Under small β conditions, we can write

$$IFM = \sqrt{2\int_{fa}^{fb} f_m^2 \mathcal{L}(f_m) df_m} \quad (\text{hertz}_{RMS})$$

4.59

where
f_a and f_b represent the lower and upper frequency boundaries of the demodulated waveform.

Given the frequency spectrum of an oscillator, these two equations represent the amount of variation in the instantaneous frequency of the oscillator.

Example 4.5 — Incidental Frequency Modulation
Figure 4-38 shows a local oscillator to be used in a cellular telephone receiving system. Find the IFM of this oscillator.

Solution —
We will evaluate Equation 4.59 from a lower frequency of 300 hertz to an upper frequency of 3000 hertz since this is the bandwidth of a common telephone channel. Equation 4.59 produces

$$IFM = \sqrt{2\int_{fa}^{fb} f_m^2 \mathcal{L}(f_m)df_m} \quad (\text{hertz}_{RMS})$$

$$= \sqrt{2\int_{300}^{3000} f_m^2 (10^{-9})df_m} \qquad 4.60$$

$$= \sqrt{(2\cdot 10^{-9})\left(\frac{1}{3}\right)(3000^3 - 300^3)}$$

$$= 4.2 \text{ hz}_{RMS}$$

Because of the phase noise on the local oscillator, it will appear as though the input signal has an uncertainty of 4.2 Hz_{RMS}.

IFM and SNR

The IFM of a receiver's LO will affect the maximum output SNR. Consider one FM signal that carries a printer signal at 75 baud which deviates 75 hertz and a second one that is a digital cellular telephone signal that deviates 20 kHz. In order to decide if the printer is sending a binary 0 or binary 1, we must measure the frequency with an accuracy of 37.5 hertz. To decode the cellular data properly, we have to measure the frequency of the signal with an accuracy of 10 kHz.

In the last example, the LO had an IFM of 4.2 Hz_{RMS} and, similar to IPM, IFM is a statistical quantity. The instantaneous frequency of the oscillator can be larger or smaller than 4.2 hertz. The printer signal (which deviates a mere 75 hertz) will be more sensitive to oscillator phase noise than the cellular telephone signal (which deviates 20 kHz).

Measuring IFM

Figure 4-42 is a measurement setup for the IFM of a receiver. It is very similar to the IPM setup except that FM demodulation and modulation are employed.

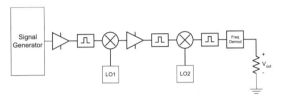

Figure 4-42 *Measuring the incidental frequency modulation (IFM) of a receiver.*

First, we frequency-modulate the signal generator at some arbitrary rate (for example, 1 kHz) and at some known peak frequency deviation Δf_{pk}. This corresponds to some RMS frequency deviation $\Delta \phi_{RMS}$ (see Figures 4-43 and 4-44). When we demodulate the FM signal from the generator, the RMS voltage from the receiver is

$$V_{out,FM,RMS} = k_{FM} \Delta f_{RMS} \qquad 4.61$$

where
  k_{FM} is a constant associated with the frequency demodulator,
  Δf_{RMS} is the RMS frequency deviation of the test signal (see Figure 4-44).

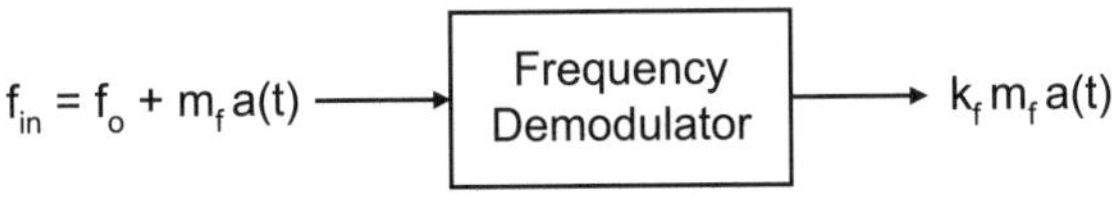

$$f_{in} = f_o + m_f a(t) \longrightarrow \boxed{\text{Frequency Demodulator}} \longrightarrow k_f m_f a(t)$$

Figure 4-43 *A frequency demodulator. The information is encoded in the frequency of the received signal. The LO phase noise is described by $\phi_{REF}(t)$ and $\phi_{REF,RMS}$.*

Next, we remove the frequency modulation from the signal generator and provide a quiet carrier to the receiver. If the IFM of the signal generator is much smaller than the IFM of the receiver, the output of the frequency demodulator is

$$V_{out,quiet,RMS} = k_{FM} IPM_{RMS} \qquad 4.62$$

The ratio of these two voltages is

$$\frac{V_{out,FM,RMS}}{V_{out,quiet,RMS}} = \frac{k_{FM}\Delta f_{RMS}}{k_{FM} IFM_{RMS}} \qquad 4.63$$

$$= \frac{\Delta f_{RMS}}{IPM_{RMS}}$$

Solving for IFM$_{RMS}$, we find

$$IPM_{RMS} = (\Delta f_{RMS})\frac{V_{out,quiet,RMS}}{V_{out,FM,RMS}} \qquad 4.64$$

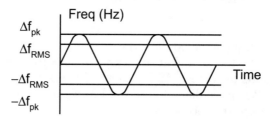

Figure 4-44 *Definitions of peak and RMS frequency deviation.*

SNR and IFM

If the receiver LOs were perfect, a quiet carrier would produce zero volts at the demodulated output. We have seen, however, that under quiet carrier input conditions, the receiver produces a noise voltage due to the IFM of the receiver's LOs. We can think of this as the ultimate noise floor of the demodulator. If IFM is the limiting factor in a receiver, the ultimate SNR present at the output of a receiver will be

$$SNR_{IFM} = \left(\frac{V_{out,FM,RMS}}{V_{out,quiet,RMS}}\right)^2$$

$$= \left(\frac{\Delta f_{RMS}}{IPM_{RMS}}\right)^2 \qquad 4.65$$

$$= 20\log\left(\frac{\Delta f_{RMS}}{IPM_{RMS}}\right)$$

where
Δf_{RMS} is the RMS frequency deviation of the signal of interest.

OSCILLATORS | 393

Signals with small RMS frequency deviations will exhibit a smaller IFM-limited SNR than signals with large RMS frequency deviations.

Example 4.6 — Incidental Frequency Modulation and SNR

A cellular telephone receiver processes an FM signal with a 20 kHz bandwidth. The signal is a voice-grade channel covering 300 to 3000 hertz. The IFM of the receiver is 5 Hz$_{RMS}$. Find the ultimate SNR present at the output of the receiver.

Solution —
For this system, the RMS frequency shift is

$$\Delta f_{RMS} = \frac{\Delta f_{PkPk}/2}{\sqrt{2}}$$
$$= \frac{20k/2}{\sqrt{2}}$$
$$= 7.1 \text{ kHz}_{RMS}$$

4.66

The ultimate SNR present at the receiver's output is

$$SNR = 20\log\left(\frac{\Delta f_{RMS}}{IFM_{RMS}}\right)$$
$$= 20\log\left(\frac{7.1 \text{ kHz}_{RMS}}{5 \text{ Hz}_{RMS}}\right)$$
$$= 32 \text{ dB}$$

4.67

4.11 Comparison of IPM and IFM

Since phase and frequency are closely related, we expect the IFM and IPM of an oscillator to be related. The equations describing IPM and IFM are

$$IPM_{RMS} = \sqrt{\int_{fa}^{fb} S_\phi(f_m)df_m} \quad (\text{radians}_{RMS})$$

4.68

and

$$IPM_{RMS} = \sqrt{\int_{fa}^{fb} f_m^2 S_\phi(f_m)df_m} \quad (\text{hertz}_{RMS})$$

4.69

Equations 4.68 and 4.69 are very similar except that the IFM equation contains an f_m^2 term. In practice, we observe that if we apply white noise (which has a flat frequency spectrum) to a FM demodulator, the power in the output spectrum increases parabolically with frequency. If we apply white noise to a phase demodulator, the output spectrum is flat, an observation that is validated by the equations.

4.12 Other Phase Noise Specifications

We characterize phase noise in many different ways. The following list contains the most common phase noise descriptions.

Spectral Density of the Frequency Fluctuations

This is the power spectral density of a frequency discriminator's output. We can measure it by passing the signal of interest through a frequency discriminator and then connecting a spectrum analyzer to the discriminator's output. The symbol commonly used is $S_{\delta f}(f_m)$, the units are Hz²/hertz. Note that $S_{\delta f}(f_m)$, is a function of f_m and not a single number.

Spectral Density of the Phase Fluctuations

This is the power spectral density of a phase discriminator's output. We can measure this quantity by phase-demodulating the noisy oscillator and examining the output on a spectrum analyzer. The symbol is $S_{\delta\phi}(f_m)$, the units are rad²/hertz. This is a function of f_m and not a single number.

Spectral Density of the Fractional Frequency Fluctuations

This quantity allows us to compare the phase noise characteristics of oscillators operating at different center frequencies. For example, if we pass an oscillator through a multiplier, we can use this quantity to tell us if the multiplier is adding excess phase noise. The symbol for the spectral density of the fractional frequency fluctuations is $S_y(f_m)$ and

$$S_y(f_m) = \frac{S_{\delta f}(f_m)}{f_0^2} \qquad 4.70$$

The units are 1/hertz.

Two-Point Allan Variance

Allan variance is a time-domain measure of oscillator phase noise. We determine the two-point Allan variance using a frequency counter to measure the oscillator frequency repetitively over a time period t. The Allan variance is the expected value of the RMS change in frequency with each sample normalized by the oscillator frequency. The symbol for Allan variance is $\sigma_y(t)$ and the quantity is unitless.

4.13 Spurious Considerations

Spurious signals are coherent signals generated by an oscillator other than the desired tone at f_0. There are two categories: harmonically related spurious signals and nonharmonically related spurious signals.

Harmonically Related Spurious Signals

Harmonically related spurious signals (or simply *harmonic spurious*) are coherent signals present at the output of an oscillator whose frequencies are integer multiples of the desired output signal, i.e., their frequencies are $2f_0$, $3f_0$, and so on. When the oscillator drives a mixer, we have already taken the LO harmonics into account because they are generated internally in the mixer's nonlinear elements. When the oscillator drives a transmitter, regulatory bodies are concerned about the radiated harmonic components. FCC part 15 requirements dictate strict guidelines for harmonic suppression. It is often necessary to filter in order to meet these requirements.

Figure 4-45 shows the harmonically related spurious of an oscillator. Each of the harmonic components exhibit the multiplied phase noise of the fundamental frequency (via Equation 4.40).

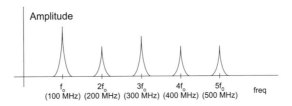

Figure 4-45 *Harmonically related spurious signals present at the output of an oscillator.*

Nonharmonically Related Spurious Signals

Nonharmonically related spurious signals (nonharmonic-spurious) are coherent signals present at the output of an oscillator whose frequencies are not integer multiples of the desired output signal. These signals are usually caused by realization effects within the oscillator. Nonharmonically related spurious signals also include signals that are subharmonics of the desired output signal.

Figure 4-46 shows the close-in spurious outputs of an oscillator. *Close-in* is a rather arbitrary term meaning that the spurious signal falls close to the fundamental component (usually within 1% of the oscillator's fundamental frequency). The close-in spurious signals in Figure 4-46 are typical of a phase-locked loop frequency synthesizer. Figure 4-47 shows the far-out spurious products that may also be present at the output of an oscillator. These signals include subharmonics of the fundamental output frequency as well as frequencies whose relationship to the desired output frequency is not apparent.

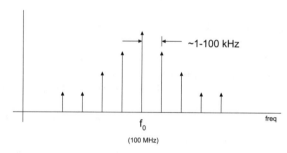

Figure 4-46 *Close-in spurious signals present at an oscillator's output. "Close-in spurious signals" are coherent signals whose frequencies differ from the fundamental signal by no more than 1%.*

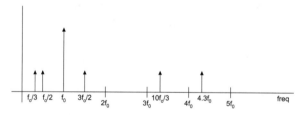

Figure 4-47 *Far-out spurious signal present at an oscillator's output. "Far-out spurious signals" are coherent signals whose frequencies differ from the fundamental signal by more than 1%.*

Effects of Oscillator Spurious Products

Harmonically related spurious products are manageable because they may be filtered. Moreover, these products are likely to be generated in any system due to the system's inherent nonlinearities. We expect harmonics to be present in this system and attempt to control only their level.

Nonharmonic spurious outputs are often more problematic. Figure 4-48 shows a receiving system with two input signals. The strong, undesired signal is at frequency f_1; the weak, desired signal is at frequency f_2. The receiver is tuned to f_2.

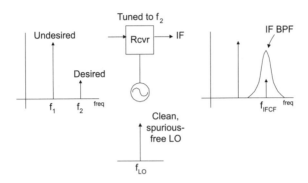

Figure 4-48 *The conversion process using a clean, spurious-free LO. The strong, undesired signal is at frequency f_1 and the weak, desired signal is at frequency f_2.*

The action of the receiver is to move the desired signal from f_2 to the receiver's IF center frequency. We then use the IF filter to pass only the desired signal. The IF filter also attenuates the other signals which were converted with the desired signal. Figure 4-48 shows how the system would behave if it had spurious-free local oscillators. Figure 4-49 shows how the receiver could behave if one of its local oscillators had a spurious output signal at an inconvenient frequency.

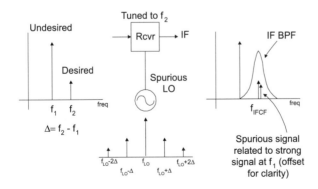

Figure 4-49 *The conversion process using an LO with close-in spurious components. The strong, undesired signal at f_2 combines with the LO spurious components to mask the weak, desired signal at f_1.*

Signals that are converted from one frequency to another by a mixer acquire the characteristics of the LO. If the LO has phase noise, the output signal will have phase noise. If the LO has spurious signals present in its output, then signals converted by the LO will contain spurious signals. If the LO exhibits spurious products which are Δf away from the desired output, the signals present at f_1 and f_2 both acquire these spurious products as they pass through the mixer. The spurious products, which are now present on the undesired signal, can mask the desired signal. For example, if a receiver's LO contains a spurious product whose frequency equals the IF, the signal may bleed through from the mixer's LO port to its IF port. If that occurs, every desired signal successfully converted to the IF by the mixer has to compete with this spurious product. In many cases, if the spurious signal at the receiver's IF is even slightly stronger than the desired signal, the receiver will demodulate the stronger signal. In this case, the oscillator's spurious output signal has degraded the receiver's ability to process weak signals.

Channelized Systems

Figure 4-50 depicts the spectral diagrams associated with channelized systems such as commercial broadcast FM radio. The channel spacing, i.e., the distance between adjacent stations, is 200 kHz.

The output of a phase-locked loop frequency synthesizer often contains spurious signals which are multiples of the channel spacing. As Figure 4-51 shows, when we use that oscillator to convert signals in a channelized system, the spurious output products cause adjacent channels to convert to the same IF. Both the desired and undesired signals must compete for the attention of the demodulator.

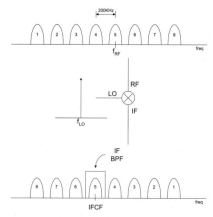

Figure 4-50 *The conversion process in a channelized system. We convert the desired RF signal to the center frequency of the IF filter.*

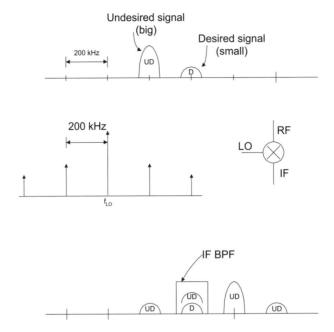

Figure 4-51 *The conversion process in a channelized system using an LO with spurious components. The undesired signal (UD) combines with the LO spurious signals to mask the desired signal (D).*

4.14 Frequency Accuracy

Frequency changes which occur over periods longer than a second are called oscillator *drift*. An oscillator may not produce the desired frequency when power is first applied. Over time, the oscillator's frequency will change slightly depending upon the temperature, power supply voltage and component aging.

Quantifying Drift

Oscillator drift is usually specified in parts per million (or ppm) of the center frequency. A typical specification might read 10 MHz ± 3 ppm. If the drift is specified as $f_0 \pm \Delta$ ppm, the frequency range of the oscillator is

$$f_{Range} = f_0 \pm \left[\frac{\Delta \text{ in ppm}}{10^6} f_0 \right] \qquad 4.71$$

Example 4.7 — Oscillator Drift in ppm

An oscillator is specified to be 100 MHz ± 2.5 ppm over some conditions of time and temperature. What is the range of output frequencies we can expect from this oscillator?

Solution —
Using Equation 4.71, we can write

$$f_{Range} = f_0 \pm \left[\frac{\Delta \text{ in ppm}}{10^6} f_0\right]$$

$$= 100 \text{ MHz} \pm \left[\frac{2.5}{10^6} 100 \text{ MHz}\right] \qquad 4.72$$

$$= 100 \text{ MHz} \pm 250 \text{ Hz}$$

Another common drift specification is *percent of center frequency*. A specification might read 40 MHz ± 0.001%. If the drift is specified as $f_0 \pm \Delta\%$, the oscillator's frequency will fall in the range of

$$f_{Range} = f_0 \pm \left[\frac{\Delta \text{ in \%}}{100} f_0\right] \qquad 4.73$$

Example 4.8 – Oscillator Drift in %

An oscillator has a drift specification of 25 MHz ± 0.006%. What is the range of output frequencies that an be expected from this oscillator?

Solution —
Equation 4.73 yields

$$f_{Range} = f_0 \pm \left[\frac{\Delta \text{ in \%}}{100} f_0\right]$$

$$= 25 \text{ MHz} \pm \left[\frac{0.006}{100} 25 \text{ MHz}\right] \qquad 4.74$$

$$= 25 \text{ MHz} \pm 1500 \text{ Hz}$$

Oscillator drift in ppm and percent are related by

$$(\Delta \text{ in \%}) = \frac{(\Delta \text{ in ppm})}{10{,}000} \qquad 4.75$$

Nomenclature

Initial Accuracy

Initial accuracy is generally defined as the difference between the oscillator output frequency and the specified frequency at 25°C at the time the oscillator leaves the manufacturer. If the oscillator includes a tuning feature, the initial accuracy is specified as a range.

Temperature Stability

Oscillator frequency varies with temperature. The specification might read "±5 ppm over a –30°C to +70°C temperature range." This means that the oscillator will be within 5 ppm of its initial frequency setting over the specified temperature range. This does not mean that the oscillator's frequency will be a linear function of temperature.

Figure 4-52 shows the frequency vs. temperature curves for several 1 MHz crystal oscillators. The stability specifications on the oscillators are X-ppm over a –30°C to +70°C temperature range, where "X" is ±10 ppm, ±5 ppm, ±2 ppm and ±1 ppm.

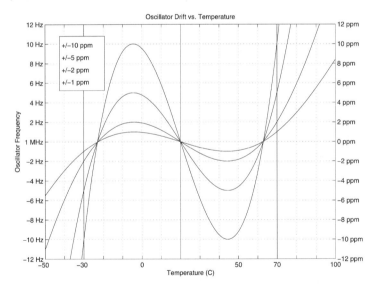

Figure 4-52 Frequency vs. temperature curves for several AT-cut crystal oscillators. Note that the frequency is not a linear function of temperature. Measured curves may not be as "clean" as these calculated curves.

The curves are roughly 5th order polynomials. At –30°C and +70°C, the oscillators are at extreme ends of their tolerances. They also reach the extremes at about –5°C and +45°C. Finally, actual oscillators exhibit hysteresis; they do not quite trace the same curve when the oscillator heats up as when it cools down, which is one of the fundamental limits on oscillator temperature stability.

Long-Term Stability

Over a long period of time, the frequency of an oscillator will change even if all the external parameters such as temperature and power supply voltage are held constant. This process is called *oscillator aging*. Figure 4-53 shows a typical oscillator aging curve. When the oscillator is initially turned on, the crystal ages quickly and stability improves rapidly with time. Most oscillators achieve their lowest aging rates within several months after turn-on. One cause of drift is the stresses generated in the crystal by the machining process when the crystal was made. These stresses affect the resonant frequency of the crystal and relax over time.

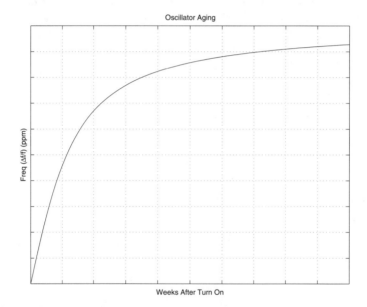

Figure 4-53 *Long-term oscillator aging. The frequency changes quickly at first, then more slowly as the stresses in the crystal relax.*

Example 4.9 — Cellular Telephones and Drift

Figure 4-54 shows a conversion scheme for a cellular telephone receiver. Assuming that we want to receive a signal at 884.460 MHz and given the accuracy specifications shown, find the range of possible output frequencies at the final IF (nominal frequency is 10.7 MHz).

OSCILLATORS | 403

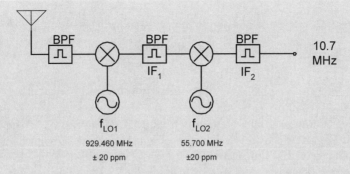

Figure 4-54 *A cellular telephone conversion scheme, with oscillator drift specifications.*

Solution —
Looking at the range of LO_1

$$f_{LO1} = 929.460 \text{ MHz} \pm 20 \text{ ppm}$$
$$= 929.460 \text{ MHz} \pm 18{,}589.2 \text{ Hz} \qquad 4.76$$
$$= 929.441\,410 \text{ MHz to } 929.478\,589 \text{ MHz}$$
$$= f_{LO1,L} \text{ to } f_{LO1,H}$$

The first conversion equation is $f_{IF} = f_{LO} - f_{RF}$, i.e., the RF signal will fall in the range of $f_{IF1,L}$ to $f_{IF1,H}$ where

$$f_{IF1,L} = f_{LO1,L} - f_{RF}$$
$$= 44.981\,410 \text{ MHz} \qquad 4.77$$
$$f_{IF1,H} = f_{LO1,H} - f_{RF}$$
$$= 45.018\,589 \text{ MHz}$$

Next, we calculate the possible ranges of LO_2.

$$f_{LO2} = 55.700 \text{ MHz} \pm 20 \text{ ppm}$$
$$= 55.700 \text{ MHz} \pm 1114 \text{ Hz} \qquad 4.78$$
$$= 55.698\,886 \text{ MHz to } 55.701\,114 \text{ MHz}$$
$$= f_{LO2,L} \text{ to } f_{LO2,H}$$

The second conversion equation is also $f_{IF} = f_{LO} - f_{RF}$, so the RF signal will fall in the range of $f_{IF2,L}$ to $f_{IF2,H}$ where

$$f_{IF2,L} = f_{LO2,L} - f_{IF1,H}$$
$$= 10.680\ 297 \text{ MHz}$$
$$f_{IF2,H} = f_{LO2,H} - f_{IF1,L}$$
$$= 10.719\ 703 \text{ MHz}$$

4.79

The RF signal might be as much as ±19,703 hertz off at the final 10.7 MHz IF. We know that the width of a cellular telephone signal is about 20 kHz. We would prefer that the entire 20 kHz signal falls inside the final IF bandwidth.

Figure 4-55 shows the situation graphically. We have calculated that the center frequency of the signal will fall between 10.680 297 and 10.719 703 MHz. Taking the 20 kHz signal bandwidth into account means that the final IF bandwidth must accept 10.670 297 to 10.729 703 MHz. Oscillator drift has increased the required IF bandwidth from approximately 20 kHz to 40 kHz. Since the signals are on 30 kHz centers, we may end up with two signals in the IF bandwidth.

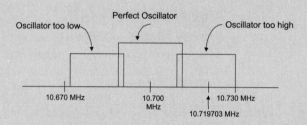

Figure 4-55 *The situation at the 10.7 MHz IF filter of Figure 4-54. The LO frequency accuracy can cause the desired signal to fall outside of the IF filter.*

Frequency Multiplication and Stability

If we use a multiplier to increase the frequency of an oscillator that drifts, the absolute drift (in hertz) increases directly with n, the multiplier. The drift, in percentage or ppm, remains constant. For example, if we multiply a 1 MHz ± 1 hertz reference oscillator by 100, the output signal will be at 100 MHz ± 100 hertz. In both cases, however, the accuracy of the oscillator is 1 ppm or 0.0001 %. Similarly, we can see that frequency division will decrease the amount of oscillator drift (in hertz) by the division factor n.

Example 4.10 — Frequency Multiplication and Drift
Three oscillators have stability specifications of ±4 ppm. The center frequency of the first oscillator is 1 MHz, the center frequency of the second oscillator is 250 MHz and the center frequency of the third oscillator is 1 GHz. Find the drift of these two oscillators in hertz.

Solution —
For the 1 MHz oscillator, we can write

$$1 \text{ MHz} \pm 4 \text{ ppm} = 1 \text{ MHz} \pm 4 \text{ Hz} \qquad 4.80$$

For the 250 MHz oscillator, we can write

$$250 \text{ MHz} \pm 4 \text{ ppm} = 250 \text{ MHz} \pm 1000 \text{ Hz} \qquad 4.81$$

For the 1 GHz oscillator, we can write

$$1 \text{ GHz} \pm 4 \text{ ppm} = 1 \text{ GHz} \pm 4000 \text{ Hz} \qquad 4.82$$

For a given value of temperature stability, higher frequency oscillators exhibit greater frequency drift than lower frequency oscillators. The high frequency oscillators will drift by a greater number of hertz than the low frequency oscillators. However, if we multiply or divide the frequency of an oscillator, the amount of drift expressed in ppm or percentage does not change; only the drift expressed in hertz changes.

Example 4.11 — Stability in ppm, hertz and %
A 1 MHz ± 0.5 ppm reference oscillator is multiplied by 1/5, 25 and 50. Express the drift of the input and output signals in terms of hertz, ppm and percentage.

Solution —
We will use Equations 4.71, 4.73 and 4.75.
- *Input Oscillator.* The frequency of the unmultiplied oscillator falls within the range of

$$1 \text{ MHz} \pm \left[\frac{0.5}{10^6} 1 \text{ MHz} \right] = 1 \text{ MHz} \pm 0.5 \text{ Hz}$$
$$= 1 \text{ MHz} \pm 0.00005\% \qquad 4.83$$

- *1/5x.* The output frequency range of the divider is

$$\frac{1}{5}(1 \text{ MHz} \pm 0.5 \text{ Hz}) = 200 \text{ kHz} \pm 0.1 \text{ Hz} \qquad 4.84$$

Equation 4.71 indicates

$$0.1 \text{ Hz} = \frac{(\Delta \text{ in ppm})}{10^6} 200 \text{ kHz}$$

$$\Rightarrow (\Delta \text{ in ppm}) = 0.1 \frac{10^6}{200 \cdot 10^3}$$

$$= 0.5 \text{ ppm}$$

4.85

- *25x*. The output frequency range of the 25x multiplier is

$$25(1 \text{ MHz} \pm 0.5 \text{ Hz}) = 25 \text{ MHz} \pm 12.5 \text{ Hz} \quad\quad 4.86$$

Equation 4.71 reveals

$$12.5 \text{ Hz} = \frac{(\Delta \text{ in ppm})}{10^6} 25 \text{ MHz}$$

$$\Rightarrow (\Delta \text{ in ppm}) = 12.5 \frac{10^6}{25 \cdot 10^6}$$

$$= 0.5 \text{ ppm}$$

4.87

- *50x*. The output frequency range is

$$50(1 \text{ MHz} \pm 0.5 \text{ Hz}) = 50 \text{ MHz} \pm 25 \text{ Hz} \quad\quad 4.88$$

Equation 4.71 shows

$$25 \text{ Hz} = \frac{(\Delta \text{ in ppm})}{10^6} 50 \text{ MHz}$$

$$\Rightarrow (\Delta \text{ in ppm}) = 25 \frac{10^6}{50 \cdot 10^6}$$

$$= 0.5 \text{ ppm}$$

4.89

Frequency multiplication and division both change the oscillator's deviation from its design value when the deviation is expressed in hertz. However, the deviation in ppm or percent is fixed, regardless of the multiplier or divider.

Timing Accuracy

Occasionally, we want to record elapsed time with nonideal oscillators. Both aging and initial offset cause errors in the perceived time lapse. The tables below summarize those errors. Table 4-1(a) assumes we start exactly on frequency, then the reference oscillator ages. Table 4-1(b) assumes that the reference oscillator is off by a fixed amount.

Table 4-1 (a) Oscillator aging and timing inaccuracies.

Aging/Day	1 day	1 week	1 month	1 year
10^{-7}	4.30 ms	210 msec	3.90 sec	580 sec
10^{-8}	430 μsec	21.0 msec	390 msec	58.0 sec
10^{-9}	43.0 μsec	2.10 msec	39.0 msec	5.80 sec
10^{-10}	4.30 μsec	210 μsec	3.90 msec	580 msec

Table 4-1 (b) Oscillator aging and timing inaccuracies.

Offset	1 day	1 week	1 month	1 year
10^{-5}	860 msec	6.00 sec	26.0 sec	320 sec
10^{-6}	86.0 msec	600 msec	2.60 sec	32.0 sec
10^{-7}	8.60 msec	60.0 msec	260 msec	3.20 sec
10^{-8}	860 μsec	6.00 msec	26.0 msec	320 msec
10^{-9}	86.0 μsec	600 μsec	2.60 msec	32.0 sec

4.15 Other Considerations

Tuning Speed

Tuning speed is the time required for a receiver to tune from frequency to frequency. After a tune command, how long is it before the receiver is settled at the new frequency and demodulating signals?

Tuning speed specifications vary. It can be defined as the time it takes the frequency synthesizer to tune to a new frequency, or the time it takes the converted waveform to settle at the IF port of the receiver, or when the synthesizers are within 1% of their final frequency. Tuning speed depends on realization. Some configurations tune quickly, others tune slowly. Other factors such as phase noise may dictate solutions which are slower.

Automatic Frequency Control

Due to oscillator drift with temperature, aging or other factors, the signals may not be exactly centered in a receiver's final IF filter. Each LO in the receiver gets to add its inaccuracies to the signal as the frequency conversion process proceeds.

Automatic frequency control (AFC) was developed to center a signal in the IF filter. As long as the desired signal is the largest signal in the IF passband, the receiver uses this signal to compensate for drift in the system. The user initially disables the AFC and tunes the receiver to the desired signal. The operator then turns the AFC on and the receiver centers the signal in the IF passband. This feature functions whether the desired signal or the receiver drifts. Provided the signal does not change too quickly, the receiver keeps it centered.

4.16 Oscillator Realizations

Receivers require locally generated oscillators to translate communications signals from frequency to frequency. To make intelligent decisions about frequency conversions schemes, we need to know how frequency synthesizers are built.

Phase-Locked Loops

The *phase-locked loop* (PLL) frequency synthesizer is one of the most common frequency synthesizers in use today. Several companies manufacture integrated circuits designed especially for these synthesizers. Designing PLL frequency synthesizers requires an understanding of control theory.

Figure 4-56 shows a block diagram for a simple PLL frequency synthesizer. The PLL is a control system used to impress the good characteristics of a crystal oscillator (frequency stability and low phase noise) onto a *voltage-controlled oscillator* (VCO) characterized by poor stability and poor phase noise. The result is a digitally tuned oscillator exhibiting good phase noise and stability.

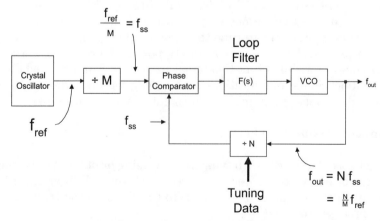

Figure 4-56 *Block diagram of a phase-locked loop (PLL) frequency synthesizer.*

OSCILLATORS | 409

When we design the conversion scheme, we will obtain requirements for the frequency synthesizers. The basic requirement is that the synthesizer must tune from a lower frequency to an upper frequency with a channel step size. For example, a PLL with a 20 kHz step size can tune in frequency increments of 20 kHz, but no finer. However, the ultimate step size of the receiver might be finer than the step size of an individual PLL synthesizer. There are often several frequency synthesizers present in a receiver that may be realized with different techniques. For example, it is common to have a PLL and a *numerically controlled oscillator* (NCO) in one receiver.

PLL Nomenclature

Locked vs. Unlocked

When the PLL is locked, the output has settled to the desired frequency and has been there for some time. The phase noise and accuracy of the reference are conveyed to the VCO. When locked, the PLL is in a steady-state condition.

The PLL is in its unlocked state for some period after a command to tune to a new frequency. The output has not yet settled to the desired frequency and the entire system is in a transient condition. The VCO phase-noise and stability are uncorrected by the reference.

Loop Bandwidth

A PLL control system compares the output of a system to some input, and then acts to change the output to match the characteristics of the input. In a PLL frequency synthesizer, the input is a stable, low-phase noise crystal oscillator. The output is a noisy, unstable VCO.

Two distinct regions characterize the phase noise of a PLL synthesizer: phase noise close to the carrier and phase noise far from the carrier. Figure 4-7 shows the transition. The action of the PLL corrects the noise close to the carrier to that of crystal oscillator. Far from the carrier, the phase noise resembles the uncontrolled phase noise of the VCO.

The PLL loop bandwidth is approximately equal to the offset from the carrier (in hertz) where the phase noise contribution from the VCO and reference oscillator are similar. The loop controls the output over a limited range of frequencies equal to the loop bandwidth.

Components of a PLL Frequency Synthesizer

Figure 4-56 shows a block diagram for a PLL frequency synthesizer. The functions of each block are explained below.

Reference Oscillator

The reference oscillator is the frequency reference of the system. The output of the PLL is locked to this oscillator and determines the stability and phase noise performance of the system. The crystal reference oscillator is often in the 1 to 10 MHz range. Since this oscillator is the system reference, the accuracy and stability of this oscillator should be excellent. If the entire receiver requires a frequency accuracy of ± 2 ppm, the stability of the reference oscillator must be at least ±2 ppm.

The PLL impresses the phase noise of the reference on the VCO so the phase noise of the output is a strong function of the phase noise of the reference. Reference oscillators are fixed-frequency, and high Q quartz crystals may be used for excellent phase noise.

If we ignore the components of the PLL and consider only the input and output ports, we can see that the PLL is essentially a frequency multiplier. As such, the phase noise of the output signal will be at least 20log(n) times worse than the phase noise of the input (lower limit). In typical PLL synthesizers other loop components contribute significant noise.

÷M

Crystal oscillators are easier, smaller and cheaper in the low MHz range. However, inside the PLL, we need a signal whose frequency equals the step size of the receiver. The ÷M counter divides the reference oscillator's output frequency down to the step size frequency. For example, a synthesizer with a 1 MHz reference oscillator and a 20 kHz step size would require a ÷M value of

$$\frac{1 \text{ MHz}}{20 \text{ kHz}} = 50 \qquad 4.90$$

A counter with at least 6 bits is required to accomplish this division. A ÷M counter needs at least $log_2(M)$ bits or flip-flops so the PLL synthesizer IC must have the capacity to perform the required frequency division. The ÷M counter is usually set at power up to a single value of m.

Phase Comparator

The phase comparator is one of the most subtle components of the synthesizer. Its function is to compare the phase of the divided-down reference oscillator (at $f_{SS} = f_{REF}/M$) with the phase of the divided-down output frequency (at f_{VCO}/N).

When locked, the phase comparator output contains information regarding the phase noise and stability of the VCO compared to the reference oscillator. If the VCO's phase (or frequency) is correct, the loop uses the phase comparator error signal to correct the problem. From the perspective of a control system, the output of the phase comparator is an error signal and the action of a control system is to drive the error signal to zero.

Frequency Acquisition

The phase comparator often performs a second function: frequency acquisition. When the synthesizer is tuned to a new frequency, new data is written to the ÷N counter and the PLL is unlocked. Initially, the VCO remains at the original frequency. Since the two signals applied to the phase comparator ($f_{SS} = f_{REF}/M$ and f_{VCO}/N) are different frequencies, measuring the phase difference between them is meaningless.

In this situation, the data from the phase comparator will be high enough in frequency so that it will not pass through the PLL's loop filter. The feedback path is effectively broken and the VCO will drift. In these situations, we need a method of gross frequency acquisition. The control signal should tell the VCO to either tune up or tune down in frequency. The digital phase detectors present in modern PLL frequency synthesizer ICs contain an automatic frequency acquisition feature. When the difference between the input frequencies is large, the error signal steers the VCO in the correct direction.

These digital phase detectors are referred to as *phase/frequency detectors*. They consist of a series of flip-flops that keep track of the edges of the input waveforms. When significantly more edges are derived on one input, the phase detector recognizes different input frequencies and acts to drive the VCO in the proper direction. Although digital phase detectors handle acquisition well, they are the major source of the close-in spurious products on the frequency synthesizer's output signal. These *sampling sidebands* are discrete signals present on the output of the PLL synthesizer. They fall at

$$f_0 \pm n f_{SS} \qquad 4.93$$

where f_0 is the desired output frequency of the synthesizer and f_{SS} is the step size of the PLL synthesizer.

The phase/frequency detectors produce digital waveforms that are filtered by the loop integrator to provide the control voltage for the VCO. Some of the energy from the digital waveform passes through the synthesizer's loop filter and FM modulates the VCO, producing the sampling sidebands. Figure 4-5 shows the sampling sidebands of a PLL frequency synthesizer. This oscillator has a tuning step of 50 kHz.

Loop Filter

Designers set loop parameters via the loop filter. The loop filter determines the tuning speed, the settling time, the spurious suppression and the phase noise performance of the synthesizer. However, all of these parameters are not independently adjustable. The filter is simply a LPF with carefully controlled gain and phase characteristics but the design involves many tradeoffs (see references 10 through 13).

Voltage-Controlled Oscillator

Voltage-controlled oscillators (VCOs) are tunable oscillators whose frequency is controlled by an applied voltage. The VOC determines the tuning range of the synthesizer and sets the phase noise and spurious characteristics of the output.

Tuning Range

In general, a VCO that tunes over a wide frequency range has a higher phase noise than a narrow-range VCO. To explain this effect, we first need to define VCO tuning constant gain, which is the VCO's change in frequency divided by the corresponding change in input voltage or,

$$k_{VCO} = \frac{\Delta f_{VCO}}{\Delta V_{VCO}} \qquad 4.92$$

Example 4.12— VCO Tuning Constant or VCO Gain
If the frequency of a VCO changes from 100 MHz to 200 MHz as the input voltage changes from 10 volts to 30 volts, find the VCO gain.

Solution —
Using Equation 4.92, we find

$$\begin{aligned} k_{VCO} &= \frac{\Delta f_{VCO}}{\Delta V_{VCO}} \\ &= \frac{200-100}{30-10} \\ &= 5 \,\frac{\text{MHz}}{\text{Volt}} \end{aligned} \qquad 4.93$$

Let us assume we have two VCOs, which both require an input range of 5 to 20 volts to tune over their entire frequency range. The first oscillator, tuning over a 10 MHz range, has a tuning constant of

$$k_{VCO1} = \frac{\Delta f_{VCO}}{\Delta V_{VCO}}$$

$$= \frac{10 \text{ MHz}}{15 \text{ Volts}} \qquad 4.94$$

$$= 0.67 \frac{\text{MHz}}{\text{Volt}}$$

The second oscillator (100 MHz tuning range) has a tuning constant of

$$k_{VCO2} = \frac{\Delta f_{VCO}}{\Delta V_{VCO}}$$

$$= \frac{100 \text{ MHz}}{15 \text{ Volts}} \qquad 4.95$$

$$= 6.67 \frac{\text{MHz}}{\text{Volt}}$$

Imagine that there is a small amount of noise present on the VCO's input line. This noise can be thermal noise or it can be an unwanted, discrete signal. Let us assume the undesired input voltage has an RMS value of 10 μV_{RMS}. The 10 μV_{RMS} of noise will cause a frequency deviation of

$$f_{dev1} = (10 \, \mu \text{Volt}_{RMS}) k_{VCO1}$$

$$= (10 \, \mu \text{Volt}_{RMS}) \left(0.67 \frac{\text{MHz}}{\text{Volt}} \right) \qquad 4.96$$

$$= 6.7 \text{ Hz}_{RMS}$$

on the narrow-tuning VCO. That same 10 μV_{RMS} of noise will cause a frequency deviation of

$$f_{dev2} = (10 \, \mu \text{Volt}_{RMS}) k_{VCO1}$$

$$= (10 \, \mu \text{Volt}_{RMS}) \left(6.67 \frac{\text{MHz}}{\text{Volt}} \right) \qquad 4.97$$

$$= 66.7 \text{ Hz}_{RMS}$$

The wider the VCO tunes, the more susceptible it is to undesired signals on its tuning line. If the undesired signal is noise, then the oscillator exhibits excess phase noise. If the undesired signal is a discrete waveform, the oscillator will exhibit discrete sidebands. Although it helps to decrease the amount of thermal noise present on the control line of a VCO, some of the noise is due to the components internal to the VCO itself.

Phase Noise

A PLL frequency synthesizer is a control system that is capable of controlling the phase noise of the output only within the loop bandwidth. The loop bandwidth covers

$$f_0 - f_{LBW} \quad \text{to} \quad f_0 + f_{LBW} \qquad 4.98$$

where

f_0 is the current output frequency of the synthesizer,
f_{LBW} is the loop bandwidth of the PLL synthesizer.

Inside the loop bandwidth, the PLL disciplines the VCO phase noise with the phase noise of the reference oscillator. Consequently, inside the loop bandwidth, the phase noise of the VCO is suppressed. At frequencies well outside of the loop bandwidth, the control loop does not affect the VCO phase noise. The phase noise at these offsets approaches the phase noise of the basic VCO.

The PLL phase noise in the neighborhood of the loop bandwidth is a composite of the multiplied reference noise and the undisciplined VCO noise. It we want low phase noise outside of the loop bandwidth, we must choose a VCO with low phase noise at the offsets of interest.

÷N

The ÷N counter in concert with the reference oscillator and the ÷M counter controls the output frequency of the loop. In normal operation, both the reference and the ÷M counter are fixed. The ÷N counter is changed to alter the output frequency of the synthesizer. To shift the output frequency, a new number is written to the ÷N counter. When the ÷N counter is changed, the loop unlocks for a period of time and then settles to the new output frequency. This is the tuning speed of the synthesizer.

Spectrum

Figures 4-3 through 4-8 show various views of the spectrum of a PLL frequency synthesizer.

PLL Design Tradeoffs

The design of a PLL synthesizer involves many trade-offs. The following discussion focuses on how to define and quantify the various tradeoffs and their interaction.

Rules of Thumb

The list below contains generalized design guidelines for typical PLL frequency synthesizers.

- A VCO with a large tuning range is noisier than a similar VCO with a small tuning range.

- A PLL frequency synthesizer containing a VCO with a large tuning range will tend to have stronger close-in spurious signals than a synthesizer containing a VCO with a small tuning range.

- A PLL frequency synthesizer with a large loop bandwidth will tune faster than a synthesizer with a small loop bandwidth.

- Synthesizers whose loop bandwidths are a large percentage of the step size will exhibit higher close-in spurious signals than an equivalent synthesizer with a small percentage bandwidth.

- Because of stability concerns and other realization factors, the loop bandwidth is typically less than 10% of the synthesizer's step size.

- The phase noise of the output is determined by the VCO for offsets well outside of the loop bandwidth. The phase noise of the output is determined by the reference oscillator for offsets well inside the loop bandwidth. Near the loop bandwidth, the phase noise is a combination of both.

- Synthesizers with small step sizes must have small loop bandwidths. Therefore, these synthesizers will tend to tune slowly and their VCO's must have good close-in phase noise.

Phase Noise

In order to realize a low-phase noise synthesizer, we need either

- a wide loop bandwidth, which will let us discipline the phase noise of the VCO to large offset frequencies, or

- a narrow loop bandwidth and a VCO whose phase noise is acceptable for offsets greater than the loop bandwidth.

A wide loop bandwidth implies a large step size because the loop bandwidth is limited to 10% of the step size. Therefore, PLL frequency synthesizers with small step sizes, low phase noise and a wide tuning range are difficult to build. One of the following configurations is much easier to realize.

- Small step size and low phase noise, but narrow-tuning range.
- Small step size and wide tuning range, but with poor phase noise.
- Low phase noise and wide tuning range, with large step size.

Well-designed communications receivers recognize these limits in their conversion schemes. The first LO of a wideband receiver will often tune over a very large range but with a 1 MHz step size. This is the *coarse-step* synthesizer. The large step size allows for a large loop bandwidth and good phase noise performance. The coarse-step synthesizer places the signal of interest somewhere in the first IF filter. The first IF filter must be wide enough to handle this situation.

The second LO has a small step size (perhaps the ultimate step size of the receiver) but it tunes over only 1 MHz. The small step size forces a narrow loop bandwidth. The phase noise of the raw VCO will be low since the VCO is narrow-tuning. This synthesizer is often called a *fine-step synthesizer*.

Spurious Signals

The magnitude of the sampling sidebands is determined by the digital phase/frequency comparator, the loop filter and the VCO tuning characteristics. Commonly used digital phase/frequency detectors produce discrete signals at their output ports even when the synthesizer is locked. These signals contain strong components at the step size frequency and its harmonics. When the synthesizer has a step size of 1 kHz, the signals present at the phase/frequency detector outputs have strong components at 1 kHz, 2 kHz, 3 kHz, and so on. We have discussed the dangers associated with large VCO gains. The larger the VCO gain, the more sensitive is the VCO to noise or other unwanted signals on its control line.

Figure 4-56 shows that the loop filter lies between the digital phase/frequency detector and the VCO. We can think of the loop filter as a low-pass filter whose cut-off frequency is directly related to the loop bandwidth of the synthesizer. The loop filter can be used to suppress the digital waveforms from the phase/frequency detector. The lower the cut-off frequency of the loop filter, the more suppression the filter provides for the digital waveforms. This implies that synthesizers whose loop bandwidth is a small percentage of the step size will tend to have lower spurious output power than synthesizers with large percentage bandwidths. For example, if two PLL synthesizers are identical except that one has a 1 kHz loop bandwidth and the other has a 10 kHz loop bandwidth, we would expect to see a lower spurious output power from the synthesizer with the 1 kHz loop bandwidth.

On the one hand, a large loop bandwidth provides better phase noise. On the other hand, the narrow loop bandwidth offers lower close-in spurious signal power. Separate filters are often added to the loop whose sole purpose is to suppress the sampling sidebands. The phase and magnitude response of these sideband suppression filters must be carefully evaluated, lest they introduce instability.

Tuning Speed

Normally, the synthesizer with the widest loop bandwidth will tune the fastest. Control signals with a wide loop bandwidth travel quickly around the loop and the loop can react quickly. In receivers that contain several PLL frequency synthesizers, the tuning speed is determined by the PLL with the narrowest loop bandwidth and thus by the synthesizer with the smallest step size.

A second option for the fine-step synthesizer is a *numerically controlled oscillator* (NCO). As will be shown, a NCO offers a very fine step size and a very quick tuning speed though its spurious output is only fair and its frequency range is limited.

4.17 Numerically Controlled Oscillator (NCO)

A *numerically controlled oscillator* (NCO) generates sine waves using a digital counter, ROM tables and a digital-to-analog converter (DACs). The NCO offers very small step sizes (< 1 hertz, in some cases) and very fast tuning speeds. However, the output exhibits many nonharmonically related spurious signals at unusual (but predictable) frequencies. The maximum output frequency of NCO is limited but technology will advance the speed.

Figure 4-57 shows the basic configuration. The NCO uses an n-bit phase accumulator driven at a reference clock at f_{REF}. At each rising edge of the reference clock, the phase accumulator output increases by the phase increment, P_i. The output of the phase accumulator is fed into a read-only-memory (ROM) which converts the phase information into a digitized sine wave of b-bits.

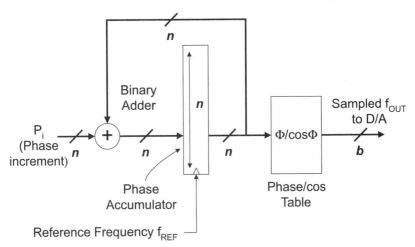

Figure 4-57 *Simplified block diagram of a numerically controlled oscillator (NCO). Practical NCOs have modifications that allow modulation.*

Operation

To provide a simple example, we will assume a 4-bit data path ($n = 4$). Typically, NCOs use a larger number of bits (some devices accommodate a 32-bit data path). The frequency resolution of an NCO is related to the number of bits in its data path.

The output of the binary adder is the modulo-16 sum of the two inputs. The user applies the phase increment at the P_i port and, as will be shown, this value controls the output frequency f_{out}. The $\phi/\cos\phi$ table is a ROM containing a digitized sine wave. The output of the phase accumulator is zero to 2^{n-1} which corresponds to 0 to 359.999... degrees of the sine wave. The number of bits we apply to the $\phi/\cos\phi$ table is the *phase resolution* of the NCO. In the example, the NCO's phase resolution is four bits.

Table 4-2 NCO phase accumulator.

T (μsec)	Phase Increment P_i				
	1	2	3	4	5
0	0	0	0	0	0
1	1	2	3	4	5
2	2	4	6	8	10
3	3	6	7	12	15
4	4	8	12	0	20=4
5	5	10	15	4	9
6	6	12	18=2	8	14
7	7	14	5	12	19=3
8	8	16=0	8	0	8
9	9	2	11	4	13
10	10	4	14	8	18=2
11	11	6	17=1	12	7
12	12	8	4	16=0	12
13	13	10	7	4	17=1
14	14	12	10	8	6
15	15	14	13	12	11
16	16=0	16=0	16=0	16=0	16=0

The value of P_i determines the output frequency. The higher the value of P_i, the higher the output frequency. Assume that the phase accumulator initially contains all zeroes and that P_i is 01_H (= 0001_B). At every rising edge of f_{REF}, the phase accumulator copies the data present on its input port to its output. The phase accumulator holds this data on its output port until f_{REF} produces another rising edge.

On successive rising edges of the clock line, the phase accumulator output becomes 02_H (0010_B), 03_H (0011_B), 04_H (0100_B), 05_H (0101_B), 06_H (0110_B), 07_H (0111_B), ... , $0F_H$ (1111_B), 00_H (0000_B) ... The phase accumulator output is incrementing by 1 every time we get a rising edge on the f_{REF} line. When the phase accumulator gets to 15 ($0F_H = 1111_B$), the output rolls over and starts at 0 again. In effect, we have built a modulo-16 counter that increments by 1 on each cycle of the clock.

When P_i is 02_H (0010_B), the modulo-16 counter increments by 2s. When P_i is 3, we have a counter that increments by 3s, and so on. In effect, we have built a user-programmable frequency divider. If we assume that the clock line provides a rising edge every micro-second (i.e., a 1 MHz clock rate), then the Table 4-2 shows the output of the phase accumulator over time for various values of P_i. Figure 4-58 shows this data graphically.

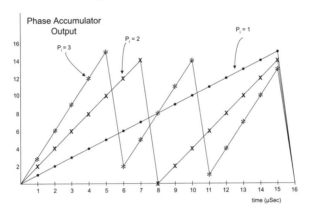

Figure 4-58 *The output of the phase accumulator of Figure 4-57 for different values of phase increment P_i. This is the data from Table 4-2.*

Frequency of Output

Using the 1 MHz clock rate, the output frequency is

$$f_{out} = \frac{P_i}{16}(1 \text{ MHz}) \qquad 4.99$$

For any P_i and f_{REF}, the output frequency is

$$\text{Spur Level (due to PM)} = -6p + 5.17 \quad (\text{dB}) \qquad 4.100$$

where
 n = the number of bits carried by the phase accumulator and binary adder,

f_{REF} = the reference frequency,
P_i = the phase increment supplied by the user.

More Bits

We can increase the number of discrete frequencies available from the NCO by increasing the number of bits carried through the NCO. The number of discrete frequencies generated by an NCO is given by

$$\text{Spur Level (due to AM)} = -6.02b + 1.76 \quad (\text{dB}) \qquad 4.101$$

The number of frequencies is not 2^n because of sampling effects in the DAC and rollover effects in the phase accumulator. The practical number of frequencies produced is on the order of 80% of the 2^{n-1} figure.

Example 4.13 — NCO Rollover
Examine the behavior of the 4-bit NCO example when $P_i = 0\ F_H$.

Solution —
Assuming the clock occurs at a 1 MHz rate and P_i is $0F_H$, the NCO will produce the following output stream:

Table 4-3 NCO phase accumulator rollover.

T (μsec)	Phase Acc Output	T (μsec)	Phase Acc Output
0	15 (0F$_H$)	10	5 (05$_H$)
1	14 (0E$_H$)	11	4 (04$_H$)
2	13 (0D$_H$)	12	3 (03$_H$)
3	12 (0C$_H$)	13	2 (02$_H$)
4	11 (0B$_H$)	14	1 (01$_H$)
5	10 (0A$_H$)	15	0 (00$_H$)
6	9 (09$_H$)	16	15 (0F$_H$)
7	8 (08$_H$)	17	14 (0E$_H$)
8	7 (07$_H$)	18	13 (0D$_H$)
9	6 (06$_H$)	19	12 (0C$_H$)

The frequency of the output waveform is the same as if the phase increment were 01_H. This is an effect of the modulo-16 adder and of the Nyquist theorem. Nyquist states that a waveform must be sampled at least twice as fast as its bandwidth. In the NCO case, we have to provide at least two output samples at the highest frequency we generate. This limits the theoretical maximum frequency of the NCO to $f_{CLK}/2$. We are limited to a maximum output frequency of about 40% of f_{CLK} after we address filtering and other realization concerns.

NCO Phase Noise

Since the output frequency of a NCO is less than the input frequency, the output can have better phase noise performance than the input by the $20 \log(n)$ rule.

Tuning Speed

As long as the phase increment (P_i) is constant, the NCO output frequency will be constant. When we change P_i, the frequency of the NCO changes. Frequency changes are instantaneous and phase-continuous. In other words, there is no abrupt transient waveform present at the output of the D/A converter. The time-domain waveform is smooth and continuous. If we examined the output of a PLL frequency synthesizer as it changes frequency, we would see a complex waveform full of transients. At the instant we change the ÷N counter in the PLL IC, we generate a complex transient in the loop, which is followed by the VCO sweeping to the new frequency.

Modulation

The digital nature of the NCO provides excellent control of the output waveform. The frequency changes are phase-continuous. A phase continuous waveform has definite advantages when we use complex waveforms to transmit digital data. Phase-continuous waveforms, for example, require less frequency spectrum than other waveforms. With some minor changes to the architecture of Figure 4-57, we can accommodate phase, amplitude modulation and frequency modulation. We can also configure the NCO to produce both sine and cosine waves. The two signals, always 90° apart, are useful in basebanding circuitry.

Practical Aspects of NCOs

NCOs generate spurious signals along with the desired output tone. The source of these signals lies in the realization details of Figure 4-57. Figure 4-59 shows the n-bit phase accumulator divided into two parts: one part is p bits wide, the other is $q = n - p$ bits wide. A typical commercial NCO IC contains a 32-bit phase accumulator. If we fed all 32 bits into the phase/cos table, we would need over four billion addresses in the phase/cosine table.

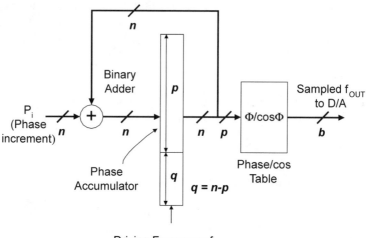

Figure 4-59 A more practical block diagram of a NCO. Only the most significant bits of the phase accumulator are passed to the phase/cos lookup table.

We carry the full n bits in the phase accumulator and binary adder, which produces a very small frequency step size. Only the *most significant bits* (MSBs) of the phase accumulator will be presented to the phase/cos table. The value p is often in the range of eight to twelve bits. This phase truncation is one source of spurious NCO signals.

Phase Truncation

Looking at Figure 4-59, let us assume a seven-bit phase accumulator ($n = 7$) and feed the four MSBs into the phase/cos table ($p = 4$). The phase accumulator "knows" the output phase to seven bits but it is not passing that knowledge on to the phase/cos table lookup ROM. Consequently, the ROM always looks up a slightly wrong phase value and the DAC output is always slightly incorrect. Let us assume P_i is 3. For each clock cycle, the state of the NCO is shown in Table 4-4.

The first two cycles (clocks 0 through 10 and clocks 11 through 21) are both 11 cycles long. The third cycle (clocks 22 through 31) is 10 cycles long. At clock #32, the state of the NCO is identical to its state at clock #0, and this cycle will repeat. The NCO will perpetually produce this string of [11,11,10,11,11,10,11,11,10 ...] long cycles (two long cycles, then a short cycle).

The average frequency is correct but the instantaneous frequency is always a little off. Thus, the NCO output always exhibits a little frequency modulation, causing output spurs. The exact level of the spurs and their frequencies depend on the number of bits carried through the phase accumulator and on how many bits we use in the sine lookup table. Some numbers produce no FM at all because they do not affect the bits not presented to the sine lookup table. As an exercise, work though Figure 4-59 and Table 4-4 when P_i is 4.

Table 4-4 NCO Phase Truncation.

Clock #	Phase Accumulator	Output to the phase/cos table Lookup ROM (Top 3 MSBs)	Clock #	Phase Accumulator	Output to the phase/cos table Lookup ROM (Top 3 MSBs)
0	0	0	16	16	4
1	3	0	17	19	4
2	6	1	18	22	5
3	9	2	19	25	6
4	12	3	20	28	7
5	15	3	21	31	7
6	18	4	22	2	0
7	21	5	23	5	1
8	24	6	24	8	2
9	27	6	25	11	2
10	30	7	26	14	3
11	1	0	27	17	4
12	4	1	28	20	5
13	7	1	29	23	5
14	10	2	30	26	6
15	13	3	31	29	7
			32	0	0

Figures 4-60 through 4-63 show NCO spectra generated under the following conditions: Ten-bit phase accumulator (n = 10) and a five-bit $\phi/\cos\phi$ table lookup (p = 5). The clock, f_{REF}, runs at 100 MHz

- *Figure 4-60*. Phase increment P_i set to 32. (The frequency is 3,125,000 hertz.) This is a "perfect" P_i number which generates a spectrum with no FM.

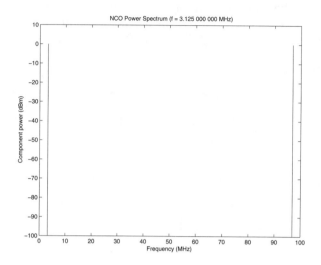

Figure 4-60 *NCO output for $P_i = 32$. This is a "perfect" P_i which generates no unwanted spurious signals.*

- ***Figure 4-61***. Ten-bit phase increment is set to 33. (The frequency is 3,222,656.25 hertz.) This is one step size away from the perfect P_i setting. A one-bit change in the phase increment has changed the output spectrum from almost no spurs to spurs that are only 30 dB below the desired output. This is a very common effect in NCOs.

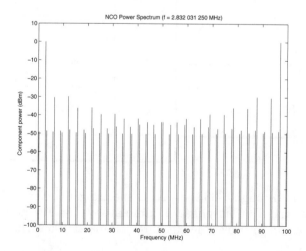

Figure 4-61 *NCO output for $P_i = 33$. Although we are only one step size away from the "perfect" P_i of 32, the spurious output has changed significantly.*

OSCILLATORS | 425

- **Figure 4-62**. P_i is set to 38. (The frequency is 3,710,937.5 hertz.) Spurious products are still only 30 dB below the desired signal but their frequencies have changed.

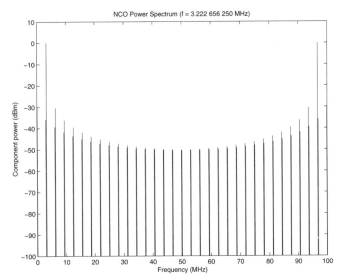

Figure 4-62 NCO output for $P_i = 29$. The spectrum is similar to Figure 4-61 but the frequencies and amplitudes of many of the spurious signals have changed.

- **Figure 4-63**. P_i is set to 29 (The frequency is 2,832031.25 hertz).

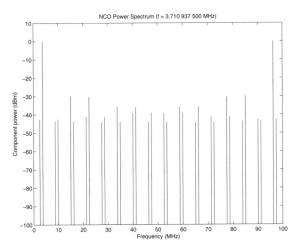

Figure 4-63 NCO output for $P_i = 38$. Again, the amplitudes and frequencies of the spurious signals are radically different from Figures 4-62 and 4-63.

Figures 4-64 and 4-65 show the effect of increasing p, the number of bits sent to the $\phi/cos\phi$ lookup table. In both cases, P_i is ten bits and the value is 17. The f_{REF} clock is 100 MHz. This configuration produces an output signal at 1.660 156 250 MHz. We are sending 5 bits phase resolution to the sine lookup table in Figure 4-64.

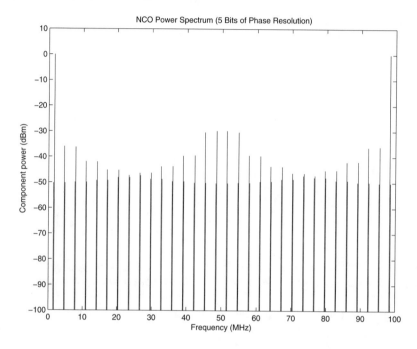

Figure 6-64 *NCO output when sending 5 bits of phase resolution to the $\phi/cos\phi$ lookup table. $P_i = 17$, $f_{Ref} = 100$ MHz and the output frequency is 1.660 156 250 MHz. The worst-case spurious products are 30 dBc.*

In Figure 4-65, we are using 8 bits of phase resolution. More bits of phase resolution passed to the lookup tables generates fewer spurious signals. A rough approximation of the NCO spur level due to phase quantization is

$$\text{Spur Level (due to PM)} = -6p + 5.17 \quad (\text{dB}) \qquad 4.102$$

where p = the number of bits sent to the $\phi/cos\phi$ table.

OSCILLATORS | 427

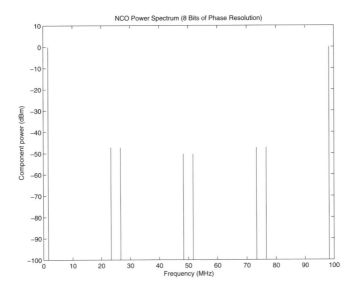

Figure 4-65 *The NCO from Figure 4-64 except we are sending 8 bits of phase resolution to the $\Phi/\cos\Phi$ lookup table. The three extra bits have significantly decreased the number of spurious products in the NCO output and have also decreased their magnitude to 50 dBc.*

Phase Dithering

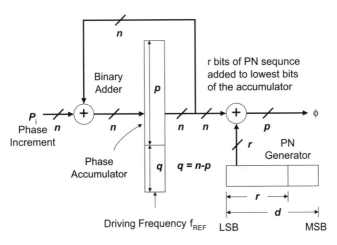

Figure 4-66 *One method of randomly dithering the phase of an NCO using a pseudo-random PN Sequence.*

The spurious signals present on an NCO's output are due to repetitive events occurring in the NCO. If we can force those events to become more random, the coherent spurious products are changed into a more noiselike phenomenon. Figure 4-66 shows one method of randomizing the output of the phase accumulator.

The PN Generator is a pseudorandom sequence generator that produces semirandom numbers. We add this random noise to the output of the phase accumulator, then pass the randomized data on to Phase/Cos table. Figures 4-67 and 4-68 show the output of the NCO with and without phase dithering.

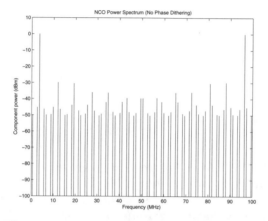

Figure 4-67 The output of a NCO without the phase dithering mechanism of Figure 4-66. The spurious signals are coherent tones which are 30 dBc.

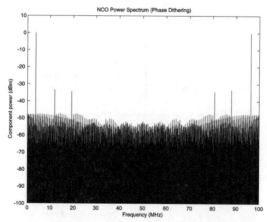

Figure 4-68 The output of a NCO when we dither the oscillator's phase. The spurious signals have been converted into pseudo-random noise and the worst-case spurious signal is now 35 dBc.

We have reduced most of the spurs below 50 dBc. The strength of the two strongest spurs are reduced about 10 dB, which raised the noise floor over the entire output bandwidth. We did not remove spurious energy; we simply spread it.

Figure 4-69 shows the architecture of a commercial NCO. This IC features a 32-bit data path, an eight bit digital-to-analog converter (DAC) path, and both cosine (in-phase) and sine (quadrature) output ports.

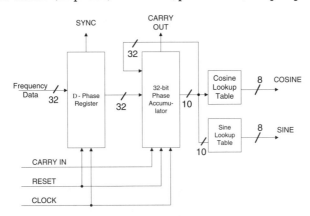

Figure 4-69 The architecture of a commercial NCO IC.

D/A Converters (DACs)

The distortion in the figures shown thus far is due only to phase truncation. The NCO output signals are represented to full floating-point precision, which means we have not considered DAC distortion.

The NCO of Figure 4-69 contains a 10-bit Digital-to-Analog converter (DAC) to transform the digital output of the sine ROM into an analog waveform. Because of its digital nature, the NCO can express the instantaneous value of a sine wave only with finite precision. The output waveform is quantized, which means that there is a nonzero error between an ideal sine wave and the output waveform of the NCO.

There are two nonlinear effects at work in an actual DAC. One is the distortion the DAC introduces because it represents an analog waveform in discrete steps. The second source of distortion arises from DAC nonlinearities. DACs exhibit missing codes, non-uniform step sizes and a host of other parameters that generate nonlinear distortion. Currently, these parameters are the ultimate limit to NCO spurious performance.

Amplitude Quantization

Let us ignore phase truncation effects and concentrate on the effects of amplitude quantization. Amplitude quantization behaves similar to phase truncation. The frequency and amplitude of the nonlinear products are strong functions of P_i, the phase increment, and of b, the number of bits fed to the DAC. If the DACs are perfect, larger values of b lower the noise on the NCOs' output. Figures 4-70 and 4-71 show two identical NCOs: one uses an 8-bit DAC, the other uses a 12-bit DAC. The nonlinear products shown in these figures are a result of DAC nonlinearities.

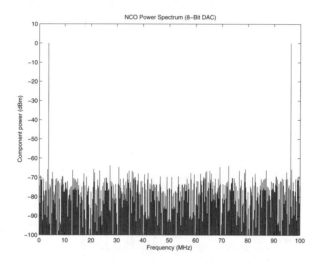

Figure 4-70 *NCO output for an ideal 8-bit DAC. The spurious products are about 70 dBc.*

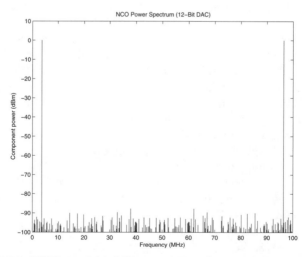

Figure 4-71 *NCO output of Figure 4-70 when we increase the number of DAC bits to 12. The spurious products are now about 90 dBc.*

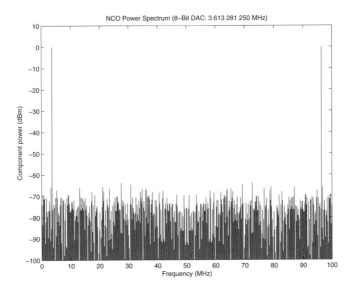

Figure 4-72 *NCO output tuned to one frequency. Note the quantity and character of the spurious signals.*

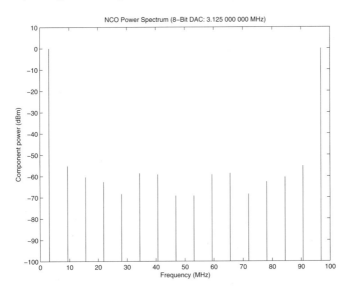

Figure 4-73 *The NCO from Figure 4-72 tuned to a slightly different frequency. The quantity and strength of the spurious products of a NCO can change dramatically with slight changes in frequency.*

Compare Figures 4-72 and 4-73. Both NCOs use 8-bit DACs but are tuned to slightly different frequencies. The character of the spurious signals is very different. Again, the distortion presented here is caused entirely by the DAC nonlinearities. We can separate the NCO's output waveform into two components: the ideal sine wave and an error signal. The NCO output signal is the time-domain sum of the ideal sine wave and the error signal. Likewise, the frequency spectrum of the NCO's output is the sum of the frequency-domain plot of the ideal sine wave plus the frequency-domain plot of the error function.

The nature of the error function depends upon the exact frequency the NCO is generating and the number of bits in the DAC converter. For example, at some frequency settings, the NCO will generate a sine wave whose exact values just happen to coincide with the values that the DAC converter can produce. The error function is zero and the output spectrum is a pure, ideal sine wave.

With other frequency settings, the exact values of the ideal sine wave and the NCO hardly ever correspond and the result is a complex, nonzero error function with a complex, nonzero frequency spectrum. The error functions are repetitive and any repetitive signal present in the error waveform translates into unwanted spurious products in the frequency domain.

The repetition rate of the error function will be related to the desired NCO output signal. That does not mean the repetition rate of the error function is the same as the desired output signal. There is no simple way to determine the spurious output frequencies given the architecture of an NCO and the desired output frequency.

The NCO spur level due to DAC quantization is roughly

$$\text{Spur Level (due to AM)} = -6.02b + 1.76 \quad \text{(dB)} \qquad 4.103$$

where b = the number of amplitude bits in the DAC.

Nonideal DACs

Nonideal D/A converters also contribute to the spurious output signals present in a NCO. In an ideal DAC converter, each output state differs from its immediate neighbor by the same amount (Δ). For example, if we apply some binary number N to the input of a DAC converter, the output is V_N. If we apply $N-1$ or $N+1$ to the input of the DAC, the outputs should be $V-\Delta$ or $V+\Delta$, respectively. This is a linearity specification. Since actual DAC converters are nonlinear, this effect causes the spurious output spectrum to increase. As of this writing, non-ideal DAC performance is the limiting factor on NCO spurious performance.

4.18 Crystal Reference Oscillators

An uncompensated crystal oscillator can provide at least ±100 ppm of stability over a 0°C to 70°C temperature range. At high frequencies and small channel spacings this accuracy is not sufficient.

Oscillator manufacturers have developed several techniques to improve the accuracy and temperature stability of a basic crystal oscillator. The choice of a particular type of compensation is determined by the required stability. The next most significant specifications are usually power consumption and physical size.

Ovenized Crystal Oscillators (OCXO)

One way to prevent an oscillator from drifting with temperature is to maintain the resonant element at a constant temperature, which is the basis for the OCXO. In an OCXO, the crystal and its associated electronics are mounted inside a temperature-controlled module. This keeps the temperature of the frequency-determining elements constant. The oscillator is designed to work at a temperature that is above the highest expected ambient (usually around 80°C).

The advantage of an OCXO is its stability, which is unequalled by other oscillator types. Typical fractional stabilities are ±1 ppm to ±100 ppb for a temperature range of −55°C to +85°C. Oscillator manufacturers can obtain better stabilities over narrower temperature ranges.

The disadvantages of OCXOs are power, size, warm-up time and cost. The oven power is determined mainly by the amount of insulation used and the temperature differential between the oven and the external environment. Increasing the amount of insulation to reduce heat loss requires an increase in size and results in a trade-off between power and size.

Warm-up is defined as the time required for the oven to reach operating temperature and for the frequency to stabilize. It depends largely on the available power and the thermal mass of the oven. Typical warm-up times are from 2 to 10 minutes.

OCXOs require power for the oscillator, the oven and the oscillator's output stages. Oscillator manufacturers can design their units to run off a single power supply but dual supplies are helpful. The power supply feeding the electronics has to be well-regulated and low noise (typical values are 5 to 24 volts with 12 volts being a happy medium). The oven supply can be noisy and not well-regulated but it must be able to supply from 5 to 20 watts at turn on and 1 to 3 watts after the oven is up to temperature.

Most OCXOs are tunable over a small frequency range, either with an adjustment screw internal to the device or with a voltage control (VCOCXO). The control feature will affect both the stability and the phase noise of the unit. A typical fractional tuning range is ±1 ppm.

Temperature-Compensated Crystal Oscillators (TCXO)

Temperature-compensated crystal oscillators (TCXOs) are less stable in frequency than OCXO's. However, the warm-up time of the TCXO is much shorter (100 msec is typical) and their power consumption is much less (15 to 150 mW). TCXO's typically cost only a fraction of an OCXO.

A TCXO uses a temperature-compensation network to discipline the basic crystal oscillator. Temperature sensitive components in the compensation circuitry tune the oscillator to offset the basic oscillator's temperature drift. With the usual compensation techniques, frequency stabilities of ±1 ppm over a –55°C to +85°C temperature range are realizable. Designers can achieve better stabilities over narrower temperature ranges.

Most TCXOs are tunable over a small frequency range to account for aging. The typical tuning range is ±5 ppm, although ranges of up to ±50 ppm can be realized. As with the OCXO, a large tuning range generally degrades the phase noise of the oscillator. The phase noise of a TCXO is typically a little worse than an OCXO because of the temperature compensation network. TCXOs require power for the oscillator, the compensation network and the output circuitry. They can usually function with just one power supply.

Digital Temperature-Compensated Crystal Oscillators (DTCXO)

The *digital temperature-compensated crystal oscillator* (DTCXO) uses a digital temperature compensation network or a microprocessor to discipline the basic crystal oscillator over temperature. A microprocessor inside the DTCXO monitors the temperature of the resonator and other critical elements, then adjusts a temperature-compensating network to accommodate for the oscillator's frequency shift. The microprocessor lets the designer generate complex tuning curves based on current temperature, supply voltage and the recent temperature history of the resonant element. Disadvantages include large physical size and increased complexity. The compensation network slightly increases the phase noise of the oscillator.

Uncompensated Crystal Oscillators (XO)

Uncompensated crystal oscillators are simple crystal oscillators with no temperature compensation. They are used in low precision applications such as digital system clocks. Typical frequency stability is in the ±10 to ±1000 ppm range, depending upon the cost and operating temperature range. Typically, hybrid assembly techniques are used to achieve small physical size and large production volume. XOs are usually smaller in size and less expensive than TCXOs.

4.19 Oscillator Design Summary

Single Sideband Phase Noise

The *single sideband phase noise* of an oscillator is the oscillator's power spectrum as measured on a spectrum analyzer (in a 1-hertz bandwidth). We are also assuming the phase noise of the spectrum analyzer is invariably better than the phase noise of the device under test. In equation form, the single sideband phase noise is

$$\mathcal{L}(f_m) = \frac{\text{Power in a 1-hertz bandwidth measured } f_m \text{ from the carrier}}{\text{Total signal power}}$$

$$= \frac{P(f_m)_{1\,\text{hertz}}}{\text{Total signal power}}$$

Leeson Model of a Noisy Oscillator

An approximation for the single-sideband phase noise of a generic oscillator is

$$\mathcal{L}(f_m) = \frac{1}{2}\frac{FkT}{P_{avg}}\left[1+\left(\frac{1}{f_m}\frac{f_0}{2Q_L}\right)^2\right]$$

where
$\mathcal{L}(f_m)$ = the single-sideband phase noise of the oscillator,
k = Boltzmann's constant = 1.38 E-23 watt-sec,
F = the noise factor of the active device (in linear terms),
T = the physical temperature in K,
P_{avg} = the average power taken from the oscillator,
f_0 = the carrier or center frequency of the oscillator,
f_m = the offset from the carrier frequency,
Q_L = the loaded Q of the oscillator's resonator.

Phase Modulation

The time-domain expression for a sinusoidally phase-modulated wave is

$$V_{PM}(t) = V_{Pk}\cos\left[2\pi f_0 t + \Delta\phi_{Pk}\cos(2\pi f_m t)\right]$$
$$= V_{Pk}\cos\left[\omega_0 t + \Delta\phi_{Pk}\cos(\omega_m t)\right]$$

where

$\Delta\phi_{pk}$ = the peak phase shift,
f_m = the modulating frequency,
f_0 = the oscillator's carrier or RF frequency,
$\omega_m = 2\pi f_m$ = the modulating angular frequency,
$\omega_0 = 2\pi f_0$ = the oscillator's angular frequency.

According to Fourier analysis this equation can be rewritten as

$$V_{Pk}(t) = V_{Pk}J_0(\Delta\phi_{Pk})\cos(\omega_c t)$$
$$+V_{Pk}J_1(\Delta\phi_{Pk})\{\sin[(\omega_0+\omega_m)t]+\sin[(\omega_0-\omega_m)t]\}$$
$$-V_{Pk}J_2(\Delta\phi_{Pk})\{\cos[(\omega_0+2\omega_m)t]+\cos[(\omega_0-2\omega_m)t]\}$$
$$-V_{Pk}J_3(\Delta\phi_{Pk})\{\sin[(\omega_0+3\omega_m)t]+\sin[(\omega_0-3\omega_m)t]\}$$
$$+V_{Pk}J_4(\Delta\phi_{Pk})\{\cos[(\omega_0+4\omega_m)t]+\cos[(\omega_0-4\omega_m)t]\}$$
$$+L$$

where

$J_n(\Delta\phi_{pk})$ = the Bessel function of the first kind, order n, with an argument of $\Delta\phi_{pk}$.

Small β Conditions

Rather arbitrarily, we say the small β approximations are valid when

$$\Delta\phi_{pk} \leq 0.2 \ radians$$

For small values of $\Delta\phi_{pk}$, we note the following relationships:

1. The value of $J_n(\Delta\phi_{pk})$ is very close to unity or

$$J_0(\Delta\phi_{pk}) \approx 1 \ for \ \Delta\phi_{pk} \leq 0.2$$

The power in the component at f_0 is

$$\frac{P_{PM}(f_0)}{P_{PM,Unmodulated\ Carrier}(f_0)} = J_0^2(\Delta\phi_{Pk})$$

The power in the modulated carrier approximately equals the power in the unmodulated carrier.

2. The value of $J_1(\Delta\phi_{pk})$ is

$$J_1(\Delta\phi_{pk}) \approx \frac{\Delta\phi_{pk}}{2} \quad \text{for } \Delta\phi_{pk} \leq 0.2$$

The power in each of the components at $f_o \pm f_m$ is

$$\frac{P_{PM}(f_0+f_m)}{P_{PM,\text{Unmodulated Carrier}}(f_0)} = \frac{P_{PM}(f_0-f_m)}{P_{PM,\text{Unmodulated Carrier}}(f_0)}$$

$$\approx \frac{\Delta\phi_{Pk}^2}{4}$$

$$\leq \frac{(0.2)^2}{4} = 0.010 = -20 \text{ dB}$$

The sidebands of a phase modulated signal under small β conditions will always be less than 20 dB below the carrier.

3. The values of the rest of the Bessel functions $J_1(\Delta\phi_{pk})$ through $J_n(\Delta\phi_{pk})$ are zero.

$$J_n(\Delta\phi_{pk}) \approx 0 \quad \text{for } \begin{cases} \Delta\phi_{pk} \leq 0.2 \\ n = 2,3,4,\ldots \end{cases}$$

Applying these approximations, the Fourier spectrum for a PM modulated wave under small β conditions is

$$V_{Pk}(t) = V_{Pk}J_0(\Delta\phi_{Pk})\cos(\omega_0 t)$$
$$+ V_{Pk}J_1(\Delta\phi_{Pk})\{\sin[(\omega_0+\omega_m)t]+\sin[(\omega_0-\omega_m)t]\}$$
$$= V_{Pk}\cos(\omega_0 t)$$
$$+ V_{Pk}\frac{\Delta\phi_{pk}}{2}\{\sin[(\omega_0+\omega_m)t]+\sin[(\omega_0-\omega_m)t]\}$$

Small β Conditions and Single-Sideband Phase Noise

Under small β conditions, $\mathcal{L}(f_m)$ and $\phi^2_{RMS}(f_m)$ are related by

$$\mathcal{L}(f_m) = \frac{\phi^2_{RMS}(f_m)}{2}$$

In decibels,

$$\mathcal{L}(f_m)_{dB} = 10\log\left[\frac{\phi^2_{RMS}(f_m)}{2}\right]$$

$$= 20\log[\phi_{RMS}(f_m)] - 3 \text{ dB}$$

When the small β approximations apply, we add 3 dB to $\mathcal{L}(f_m)$ to find $\phi_{RMS}(f_m)$.

Phase Noise and Multipliers

Phase noise increases with multiplication as

$$\frac{\mathcal{L}_{\text{REF,mult}}(f_m)}{\mathcal{L}_{\text{REF}}(f_m)} = n^2$$

$$= 20\log(n) \quad \text{in dB}$$

This equation indicates that each doubling of the carrier increases the phase noise measured at f_m by 6 dB. Likewise, each halving of the carrier results in a decrease in the phase noise measured at f_m by 6 dB.

Incidental Phase Modulation

The incidental phase modulation (IPM) of an oscillator is

$$\text{IPM} = \sqrt{\int_{f_a}^{f_b} S_\phi(f_m) df_m} \quad (\text{radians}_{\text{RMS}})$$

For small β, we can write

$$\text{IPM} = \sqrt{2\int_{f_a}^{f_b} \mathcal{L}_\phi(f_m) df_m} \quad (\text{radians}_{\text{RMS}})$$

where f_a and f_b represent the lower and upper frequency boundaries of the demodulated signal.

If IPM is the limiting factor in a receiver, the ultimate SNR at the output will be

$$SNR_{IPM} = \left(\frac{\Delta\phi_{RMS}}{IPM}\right)^2$$

$$= 20\log\left(\frac{\Delta\phi_{RMS}}{IPM}\right)$$

where $\Delta\phi_{RMS}$ is the RMS phase deviation of the signal we want to receive.

Incidental Frequency Modulation

Incidental frequency modulation (IFM) is defined as

$$IFM = \sqrt{\int_{fa}^{fb} f_m^2 S_\phi(f_m) df_m} \quad (\text{hertz}_{RMS})$$

For small β, we can write

$$IFM = \sqrt{2\int_{fa}^{fb} f_m^2 \mathcal{L}(f_m) df_m} \quad (\text{hertz}_{RMS})$$

where f_a and f_b represent the lower and upper frequency boundaries of the demodulated waveform.

If IFM is the limiting factor in a receiver, the ultimate SNR present at the output of a receiver will be

$$SNR_{IFM} = \left(\frac{\Delta f_{RMS}}{IFM}\right)^2$$

$$= 20\log\left(\frac{\Delta f_{RMS}}{IFM}\right)$$

where $\Delta\phi_{RMS}$ is the RMS frequency deviation of the signal of interest.

NCO Spur Levels

According to a rough approximation, the NCO spur level due to phase quantization is

$$\text{Spur Level (due to PM)} = -6p + 5.17 \quad (\text{dB})$$

where p = is the number of bits sent to the $\phi\cos\phi$ generator.

The NCO spur level due to DAC quantization is approximately

$$\text{Spur Level (due to AM)} = -6.02b + 1.76 \quad (\text{dB})$$

where b is the number of amplitude bit in the DAC.

Frequency Accuracy

If the drift of an oscillator is specified as $f_0 \pm \Delta$ ppm, the frequency range will be

$$f_{Range} = f_0 \pm \left[\frac{\Delta \text{ in ppm}}{10^6} f_0 \right]$$

If the drift is specified as $f_0 \pm \Delta\%$, the oscillator's frequency will fall in the range of

$$f_{Range} = f_0 \pm \left[\frac{\Delta \text{ in \%}}{100} f_0 \right]$$

4.20 References

1. Cercas, Francisco A. B., M. Tomlinson, and A. A. Albuquerque. "Designing with Digital Frequency Synthesizers." *Proceedings of the RF Expo East*, 1990.

2. *Crystal Oscillators Catalog*. Vectron Laboratories, Inc. (1990-1991).

3. Engleson, Morris and Ron Breaker. "Interpreting Incidental FM Specifications." *Frequency Technology Magazine*, no. 2 (1969): 13.

4. *Frequency Control Products: Handbook for the Design and Component Engineer*. Orlando: Piezo Technology, 1990.

5. Grebenkemper, C. John. "Local Oscillator Phase Noise and Its Effect on Receiver Performance." *Watkins-Johnson Technical Note* 8, no. 6 (1981).

6. Leeson, D. B. "A Simple Model of Feedback Oscillator Noise Spectrum." *Proceedings of the IEEE* 54, no. 2 (1966): 329-330.

7. "Low Phase Noise Applications of the HP-8662A and 8663A Synthesized Signal Generators." *Hewlett-Packard Application Note* 283-2, no. 12 (1986).

8. McCune, Earl W., Jr. "Control of Spurious Signals in Direct Digital Synthesizers." *Digital RF Solutions Application Note* AN1004, (1990).

9. Payne, John B. "Measure and Interpret Short-Term Stability." *Microwaves*, no.7 (1976):

10. Przedpelski, Andrzej. "Analyze, Do not Estimate, Phase-Lock-Loop," *Electronic Design*, 10 May 1978.

11. ---. "Optimize Phase-Lock Loops to Meet Your Needs, or Determine Why You Can't." *Electronic Design*, 13 September 1978.

12. ---. "Suppress Phase-Lock-Loop Sidebands Without Introducing Instability." *Electronic Design*, 13 September (1978): 134-137.

13. ---. "Programmable Calculator Computes PLL Noise, Stability." *Electronic Design*, 31 March 1981.

14. Scherer, Dieter. "Design Principles and Test Methods for Low Phase Noise RF and Microwave Sources." *RF and Microwave Measurement Symposium and Exhibition*, October 1987.

15. ---. "Today's Lesson: Learn about Low-Noise Design." Parts I and II. *Microwaves* 4 (1979): 116-122, 5 (1979): 72-77.

16. "RF and Microwave Phase Noise Measurement Seminar." *Hewlett-Packard*, March 1988.

17. "Signal Generator Spectral Purity." *Hewlett-Packard Application Note* 388, (1990).

18. Terman, Frederick E. *Radio Engineering*. 2nd ed. New York: McGraw-Hill, 1937.

19. Upham, Art. "Low-Noise Oscillator Paper." Hewlett-Packard Inc.

20. Zavrel, Robert J., Jr. "Digital Modulation Using the NCMO." *RF Design Magazine*, no. 4 (1988): 27.

5

Amplifiers and Noise

Noise is the most impertinent of all forms of interruptions. It is not only an interruption, but also a disruption of thought.
— Arthur Schopenhauer

He that loves noise must buy a pig.
— John Ray

The noise is so great, one cannot hear God thunder.
— R. C. Trench

5.1 Introduction

Many noise properties are statistical in nature. Thermal noise, also known as white or Gaussian noise, is the fundamental limit on the smallest signal a receiving system can process. In this chapter, we will examine noise from a mathematical point of view and relate this knowledge to receiving systems.

5.2 Equivalent Model for a RF Device

Figure 5-1 shows the general noise model that will be used throughout this chapter. The signal source can be a signal generator, antenna or another RF amplifier. We will model the signal source as a voltage source V_S in series with a resistor R_S. The resistor R_S represents the output impedance of the signal source. R_S can also be used to account for the noise present at the output port of the signal source. We will assume this resistor is at some noise temperature (T_S). T_S is not the physical temperature of the signal source but accounts for the noise power present at the output of the signal source. Eventually, we will replace the source notation (R_S and T_S) with an antenna notation (R_{ant} and T_{ant}) to emphasize the receiving aspects of the discussion.

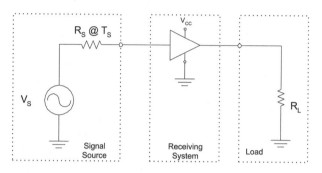

Figure 5-1 *Noise model of a receiving system.*

The receiving system can be a single RF amplifier (as shown in Figure 5-1) or it can be a complex system of amplifiers, mixers, attenuators and oscillators. At this point, we will only look at its input and output ports. Figure 5-2 shows the model we will be using to describe an average receiving system.

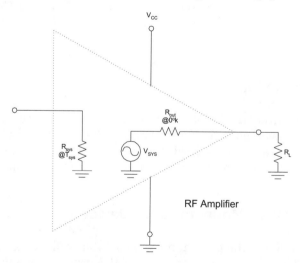

Figure 5-2 *Model of a receiving system emphasizing the noise aspects of the device.*

The receiver accepts power from the external source. The system measures the power delivered to R_{sys} and adjusts its internal voltage source V_{sys} so that a fixed multiple of the input power is delivered to R_L, i.e.,

$$P_{RL} = G_P P_{in} \qquad 5.1$$

where
 P_{in} = the power delivered to the system's input resistor by some external source (in linear units such as watts or milliwatts).
 P_{RL} = the power delivered to the load resistor R_L by the system (in linear units).
 G_P = the power gain of the system (usually > 1, in linear units).

We can write Equation 5-1 in a logarithmic format

$$P_{RL,dBm} = G_{p,dB} + P_{in,dBm} \qquad 5.2$$

where
 $P_{in,dBm}$ = the power delivered to the receiving system's input resistor by some external source (in dBm).
 $P_{RL,dBm}$ = the power delivered to the load resistor R_L by the receiving system in dBm.
 $G_{P,dB}$ = the power gain of the system (usually > 0 dB in dB).

We can also write

$$P_{RL,dBW} = G_{p,dB} + P_{in,dBW} \qquad 5.3$$

where
 $P_{in,dBW}$ = the power delivered to the receiving system's input resistor by some external source in dBW.
 $P_{RL,dBW}$ = the power delivered to the load resistor R_L by the receiving system (in dB).

In other words, the power gain of a system does not change with how the power is described (whether it is in dBW, dBm, or dBf).

Mental Model

A convenient mental model of the RF system in Figures 5-1 and 5-2 can be established as follows.

- The system measures the input power dissipated in R_{sys}. It multiplies this power by some power gain G_P and delivers the multiplied power to the load resistor R_L.

- The waveform developed across the output resistor (R_L) is a duplicate of the waveform developed across R_{sys}.

- The input power dissipated in R_{sys} includes both signal and noise power.

Power Gain and Noise

The receiving system does not register the difference between signal power and noise power. The external source, an antenna, for example, will deliver both signal power and noise power to R_{sys}. The system will add noise of its own to the input signal, then amplify the total package by the power gain.

Noise Temperature

The temperature (T_{sys}) of the input resistor R_{sys} is the noise temperature of the system. T_{sys} is a direct measure of the noise added by the system. Although there are many sources of noise inside the receiving system, we are going to "blame" the temperature of the input resistor. We universally assume the noise temperature of the output resistor (R_{out} of Figure 5-2) is 0 K. In other words, the output resistor is noiseless.

Power Supply

The receiving systems of Figures 5-1 and 5-2 have connections to both a power supply (V_{cc}) and ground. Extraneous noise can leak in through these connections if the power supply is not adequately filtered.

Matching

Most of the analysis in this chapter assumes that the various components are matched, i.e., that $R_S = R_{sys} = R_{out} = R_L$. The actual system will deviate from this ideal. (We will discuss the effects of mismatching later.)

5.3 Noise Fundamentals

If we measure the AC voltage present across the terminals of a physical resistor, we will find that there is a voltage present. This is *thermal noise*. The effect responsible for the noise involves the physical temperature of the resistor. Since the resistor exists at some temperature above absolute zero (0 K), the atoms and electrons inside the material are in constant, random motion. The hotter the resistor, the faster the electrons jiggle. This jiggling represents a random current and, since we have a random current moving through a resistive material, there will be a random voltage present across the material. Thermal noise places a limit on the smallest signal we can process. Other sources of noise are independent of the thermal effect.

Thermal Noise

Figure 5-3 shows a resistor of resistance R Ω at a physical temperature of T K. This resistor is connected to a lossless and noiseless band-pass filter whose noise bandwidth is B_n.

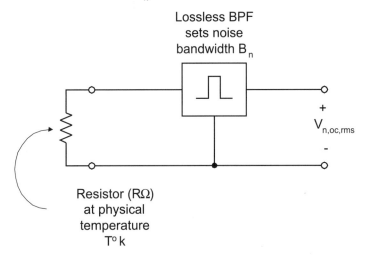

Figure 5-3 *Model of a receiving system emphasizing the noise aspects of the device.*

Figure 5-4 shows the noise voltage we will see across the terminals of the resistor. Thermal noise is completely random; there is no deterministic relationship between the voltage at one instant in time and the voltage in the next instant.

However, the noise voltage can be described statistically. If we examine a large number of instantaneous voltage samples, we find that thermal noise has a Gaussian amplitude distribution that is characterized by the familiar bell-shaped curve. The Gaussian bell curve is shown to the right of the time-domain plot of random noise in Figure 5-4. The shaded area represents the percentage of the time the instantaneous noise voltage lies between voltage e1 and voltage e2.

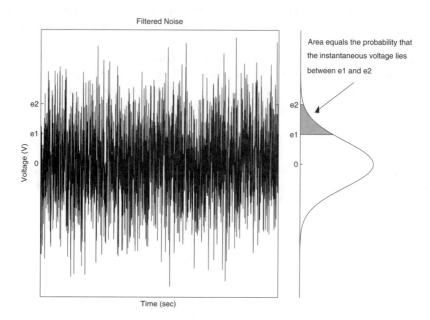

Figure 5-4 *The statistical description of Gaussian noise. The RMS voltage of the noise is the standard deviation of the Gaussian bell curve.*

RMS Value

Since we can describe noise only in statistical terms, we cannot state that the noise voltage will always be less than a certain value. After a long enough waiting period, the noise will eventually exceed any arbitrary voltage. When discussing noise, it is more meaningful to speak of noise in terms of its RMS or *standard deviation value*. We then can apply the properties of the Gaussian statistics to predict how often the noise voltage is likely to exceed any particular value.

Statistical Description

We almost universally assume that noise is a Gaussian-distributed phenomenon. We need both the mean and standard deviation to describe a Gaussian process completely. Since the RF world is an AC-coupled arena, the mean value of noise is almost universally taken to be 0 volts. In other words, noise has no DC component and only the standard deviation is needed to describe the distribution. If we measure the open-circuit AC voltage across the output terminals of the band-pass filter (see Figure 5-3), we will find that the RMS voltage is

AMPLIFIERS AND NOISE | 449

$$V_{n,oc,RMS} = \sqrt{4kTB_n R} \qquad 5.4$$

where

$V_{n,oc,RMS}$ = the open-circuit RMS noise voltage produced by the resistor. This noise voltage ($V_{n,oc,RMS}$) is the standard deviation value we will use in calculations involving Gaussian statistics.

k = Boltzmann's constant = $1.381 \cdot 10^{-23}$ joules/K (or $1.381 \cdot 10^{-23}$ watt-sec/K). This number represents the conversion factor between two forms of energy. It gives the average mechanical energy per particle which can be coupled out electrically per K (another useful expression for Boltzmann's constant is $1.381 \cdot 10^{-20}$ milliwatt-sec/K).

T = temperature in K. The hotter the resistor, the faster the atoms and electrons jiggle. With increasing temperature, each particle contains more energy on the average and the noise voltage across the resistor rises. We often use room temperature (about 290 K) for this number. We know

$$T_K = T_{°C} + 273 \qquad 5.5$$

and

$$T_{°C} = \frac{5}{9}(T_{°F} - 32) \qquad 5.6$$

B_n = the noise bandwidth of the measurement in hertz. The noise power is spread evenly across the frequency spectrum, i.e., the wider the measurement bandwidth, the more noise voltage we will measure. Figure 5-14 shows the effects of changing the noise bandwidth. As the noise bandwidth narrows, the noise voltage becomes smaller (see Equation 5.4).

R = the value of the resistor in ohms. The larger the resistance the noise current must flow through, the more voltage will be generated.

Voltage Source Model

Figure 5-5 shows a model for a noisy resistor. We will replace the noisy resistor by an ideal, noiseless resistor in series with a voltage source. The voltage source's value is $V_{n,oc,rms}$. This noise generator is not one of the usual sinusoidal voltage sources often found in circuit theory. This voltage source generates random Gaussian noise and has to be treated as such. All noise phenomena discussed in this chapter are based upon this model.

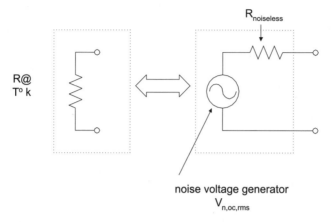

Figure 5-5 *The voltage source model for a noisy resistor. The noisy resistor is replaced by a noiseless resistor in series with a Gaussian noise voltage source.*

Example 5.1 — Resistor Thermal Noise Voltage
Find the open-circuit, RMS noise voltage present across the following resistors:
a. a 1-megaohm resistor at room temperature (about 290 K) measured in a 200 kHz noise bandwidth.
b. a 50-ohm resistor at 1000 K measured in a 10 MHz noise bandwidth.

Solution —
a. Substituting $T = 290$ K, $B_n = 200$ kHz and $R = 1$ MΩ into Equation 5.4, we find $V_{n,oc,rms} = 56.6$ μV_{RMS}.
b. Substituting $T = 1000$ K, $B_n = 10$ MHz and $R = 50$ Ω into Equation 5.4, we find $V_{n,oc,rms} = 5.25$ μV_{RMS}.

Current Source Model
We can also model the noisy resistor of Figure 5-5 as a current source in parallel with an ideal noiseless resistor. Figure 5-6 shows this model.

The value of the current source is

$$I_{n,sc,RMS} = \sqrt{\frac{4kTB_n}{R}} \qquad 5.7$$

where

$I_{n,sc,RMS}$ = the short-circuit RMS noise current produced by the resistor.

$I_{n,sc,RMS}$ is the standard deviation of the noise current generated by the current source of Figure 5-6. All of the other variables are defined above.

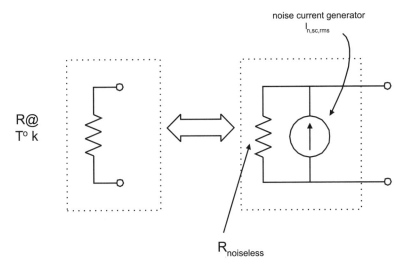

Figure 5-6 *The current source model for a noisy resistor. The noisy resistor is replaced by a noiseless resistor in parallel with a Gaussian noise current source.*

Example 5.2 — Resistor Thermal Noise Current
Find the RMS noise current present in the following short-circuited resistors:
a. a 1-megaohm resistor at room temperature (about 290 K) measured in a 200 kHz noise bandwidth.
b. a 50-ohm resistor at 1000 K measured in a 10 MHz noise bandwidth.
These are the same resistors, noise bandwidths, and temperatures we used in Example 5.1.

Solution—
a. Substituting $T = 290$ K, $B_n = 200$ kHz and $R = 1$ MΩ into Equation 5.7, we find $I_{n,sc,RMS} = 56.6$ pA$_{RMS}$.
b. Substituting $T = 1000$ K, $B_n = 10$ MHz and $R = 50$ Ω into Equation 5.7 we find $I_{n,sc,rms} = 105$ nA$_{RMS}$.

Example 5.3 — Equivalency of the Voltage and Current Noise Models
Show that the voltage and current noise models of Figures 5-5 and 5-6 are equivalent models.

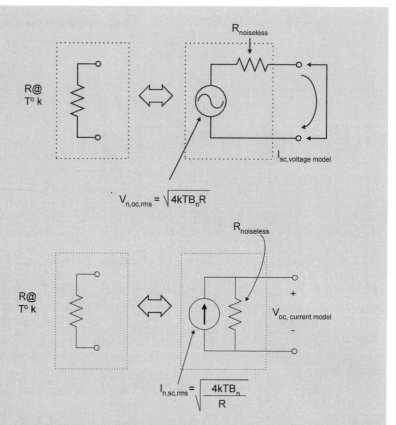

Figure 5-7 *Example 5.3. Equivalency of the voltage and current noise models.*

Solution —

Looking at Figure 5-7, we short the terminals of the noisy resistor and find the current that flows.

$$I_{sc,\text{Voltage Model}} = \frac{V_{n,oc,RMS}}{R}$$
$$= \frac{\sqrt{4kTB_nR}}{R}$$
$$= \sqrt{\frac{4kTB_n}{R}}$$
$$= I_{n,sc,RMS}$$

5.8

The voltage and current models deliver the same short-circuit current. Let us look at the open-circuit voltage of the current-source model (see Figure 5-7).

$$V_{oc,\text{Current Model}} = I_{n,sc,RMS}R$$
$$= \left(\sqrt{\frac{4kTB_n}{R}}\right)R \quad\quad 5.9$$
$$\sqrt{4kTB_nR}$$
$$= V_{n,oc,RMS}$$

Since the short-circuit current and the open-circuit voltages are the same, Thevanin shows that the circuits are equivalent. Figure 5-8 summarizes the two resistor models.

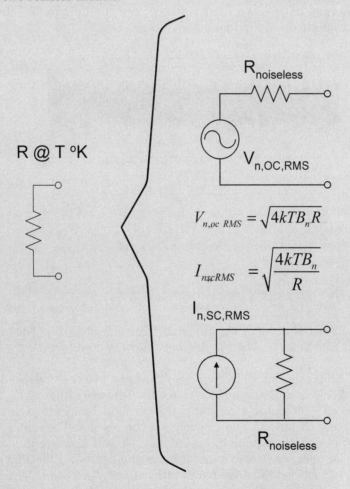

Figure 5-8 *Example 5.3. Voltage and current source models for a noisy resistor.*

5.4 Thermal Noise Properties

The thermal resistor noise we have described in the previous section is also known as Gaussian noise, white noise, Johnson noise, or Nyquist noise. We will refer to thermal resistor noise simply as thermal noise.

Flat Frequency Spectrum

The most common type of noise is *additive white Gaussian noise* (AWGN). "Additive" refers to the fact that the noise is arithmetically added to the signal of interest. The "white" refers to the shape of the frequency spectrum. Thermal noise has a flat frequency spectrum. Figure 5-9 shows the frequency spectrum of thermal noise voltage.

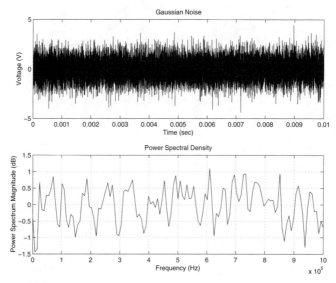

Figure 5-9 Thermal noise in the time and frequency domains. The noise power present in a given bandwidth is independent of center frequency.

Thermal noise contains all frequencies. The noise power is independent of frequency. Looking at Equation 5.4, we see that the magnitude of $V_{n,oc,rms}$ depends only on Boltzmann's constant, the physical temperature of the resistor, the noise bandwidth and the value of the resistor. The center frequency is not part of the equation. Accordingly, we will measure the same amount of noise energy between 1 and 2 hertz and between 1,000,000 and 1,000,001 hertz provided the noise bandwidths of both filters are identical.

In reality, the bandwidth of thermal noise is limited. A close examination of the physics reveals that the noise spectrum does exhibit an upper frequency limit. The 3 dB bandwidth of the thermal noise spectrum is about 10^{12} hertz (or one terahertz).

Random

Thermal noise is random, and its amplitude distribution is Gaussian, i.e., if we measure the instantaneous amplitudes of noise voltage over a period of time and plot a histogram, we will find that the histogram resembles a Gaussian bell curve.

Coherence Time

The random nature of noise does not necessarily mean that the noise voltage at one instant in time is completely uncorrelated with the noise voltage at the next instant in time. For practical reasons, we always view noise through a filter. The filter will not let the noise voltage change too rapidly from one instant to the next. For example, suppose we are observing a noise source after it has been passed through a 10 hertz lowpass filter. This filter will react very slowly to changing inputs, and we would not expect to see the noise voltage on the filter's output to change faster than the filter would allow. This property is called the *coherence time* of the noise and is related to the bandwidth of the filter through which we observe noise. Coherence time is roughly the amount of time that passes between consecutive samples of the noise so that the new sample will be uncorrelated to the previous sample. The coherence time causes the autocorrelation function of the noise to widen. The autocorrelation function will be shallower and fatter than the impulse response we expect from wideband noise.

Statistical Description

The noise voltage can be described statistically. When we take a large number of samples, thermal noise has a Gaussian amplitude distribution, characterized by the bell-shaped curve. Figure 5-10 shows a time-domain plot of thermal noise superimposed over the Gaussian bell curve. If we assume the mean value of the noise is zero ($\mu = 0$ or no DC component), the RMS voltage will be equal to the standard deviation of the Gaussian voltage distribution.

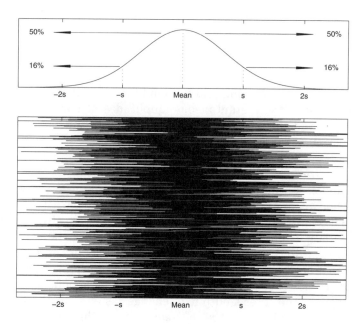

Figure 5-10 *The instantaneous noise voltage follows a Gaussian probability distribution.*

Example 5.4 — Expected Values of Resistor Noise
We are given a 50-ohm resistor at room temperature (290 K). We measure the noise voltage through a 100 MHz lowpass filter and assume the filter has a noise bandwidth of 102 MHz.
a. How often will the magnitude of the noise voltage be greater than 16 µV?
b. How often will the magnitude of the noise voltage exceed 16 µV if we heat the resistor to 600 K?

Solution —
a. Use Equation 5.4 to find $V_{n,oc,RMS}$ for T = 290 K, B_n = 102 MHz and R = 50 Ω.

$$V_{n,oc,RMS} = \sqrt{4kTB_nR}$$
$$= \sqrt{4(1.38 \cdot 10^{-23})(290)(102 \cdot 10^6)(50)} \quad 5.10$$
$$= 9.04 \ \mu V_{RMS}$$

The standard deviation of the noise is 9.04 µV, which is also the standard deviation of the Gaussian bell curve. The mean is 0 volts. The normalizing equation is

$$z = \frac{x - \mu}{\sigma} \qquad 5.11$$

We can write

$$z_{16\mu V} = \frac{16 - 0}{9.04} = 1.77 \qquad 5.12$$

We now know that 16 µV is 1.77 standard deviations out on a Gaussian bell curve whose standard deviation is 9.04.

Figure 9-3 in Chapter 9 shows the area of the Gaussian curve lying above z = 1.77 is 0.0384, which means that the noise voltage will be greater than 16 µV 3.84% of the time. Using symmetry, we also know that the noise voltage will be less than −16 µV for 3.84% of the time. The magnitude of the noise voltage will be greater than 16 µV for 7.68% of the time.

b. Use Equation 5.4 to find $V_{n,oc,RMS}$, with T = 600 K, B_n = 102 MHz and R = 50 ohms.

$$\begin{aligned} V_{n,oc,RMS} &= \sqrt{4kTB_nR} \\ &= \sqrt{4(1.38 \cdot 10^{-23})(600)(102 \cdot 10^6)(50)} \\ &= 13.0 \; \mu V_{RMS} \end{aligned} \qquad 5.13$$

Again, we use Equation 5.11 to normalize 16 µV to a Gaussian curve with a standard deviation of 13 µV.

$$z_{16\mu V} = \frac{16 - 0}{13.0} = 1.23 \qquad 5.14$$

16 µV is 1.23 standard deviations out on the Gaussian curve. Note that heating the resistor increased the RMS noise voltage from the resistor. This effectively spreads out the Gaussian curve. Since the bell curve is flatter, 16 µV does not represent as many standard deviations as it did in part a.

Figure 9-3 in Chapter 9 shows that the magnitude of the noise voltage will be greater than 16 µV 10.93% of the time. The instantaneous noise voltage will be less than −16 µV for 10.93% of the time. The magnitude of the noise will be greater than 16 µV for 21.9% of the time.

Noise Rules of Thumb

The following rules of thumb are derived from analyzing the Gaussian bell curve. The rules assume we are dealing with Gaussian-distributed noise and the value of the noise voltage is in RMS units. We also assume the mean of the noise voltage is 0 volts.

- The magnitude of the peaks of Gaussian noise will not exceed three times the RMS value of the noise signal (in reality, we will exceed this value only 0.26% of the time).

- Given a noise voltage of $V_{n,RMS}$, we can construct the following table using Gaussian statistics.

Table 5-1 Peak values of Gaussian noise.

Threshold Value	% of time the magnitude of the noise peaks will exceed the threshold value	% of time the noise peaks will exceed the threshold value
$0V_{n,RMS}$	100%	50%
$1V_{n,RMS}$	31.8%	15.9%
$2V_{n,RMS}$	4.56%	2.28%
$3V_{n,RMS}$	0.26%	0.13%
$3.3V_{n,RMS}$	0.097%	0.048%
$4V_{n,RMS}$	63 ppm	32 ppm
$5V_{n,RMS}$	0.57 ppm	0.29 ppm
$6V_{n,RMS}$	$1.98 \cdot 0^{-3}$ ppm	$0.990 \cdot 10^{-3}$ ppm
$7V_{n,RMS}$	$2.6 \cdot 10^{-6}$ ppm	$1.3 \cdot 10^{-6}$ ppm

- Some specifications speak of a peak noise voltage. Although we should inquire further to be certain, this specification frequently means 3.3 times the RMS value. From the table above, the magnitude of the noise peaks will exceed 3.3 times the RMS value no more than 0.1% of the time.

- For band-limited Gaussian noise viewed through a band-pass filter, the statistically expected number of maxima per second is

$$\frac{\text{Maxima}}{\text{Second}} = \sqrt{\frac{3(f_U^5 - f_L^5)}{5(f_U^3 - f_L^3)}} \quad \text{for a BPF} \qquad 5.15$$

where
f_L = the lower 3 dB frequency of the band-pass filter,
f_U = the upper 3 dB frequency of the band-pass filter.

For narrow band-pass filters (i.e., for bandwidths of less than 10% of the center frequency), $f_U \cong f_L$ and we obtain

$$\frac{\text{Maxima}}{\text{Second}} = \frac{f_U + f_L}{2} \approx \text{the Filter's Center Frequency} \qquad 5.16$$

For a lowpass filter ($f_L = 0$) the expected number of maxima per second is

$$\frac{\text{Maxima}}{\text{Second}} = 0.775 f_U \text{ for a LPF} \qquad 5.17$$

Equations 5.16 and 5.17 describe only the number of maxima per second. To find the number of maxima and minima per second, multiply by 2. The average number of zero crossings per second at the output of a narrow-band-pass filter of rectangular shape when the input is a sine wave in Gaussian noise is

$$N_{\text{Zero Crossings}} = 2f_0 \sqrt{\frac{SNR + 1 + \left(\frac{B_n^2}{12 f_0^2}\right)}{SNR + 1}} \qquad 5.18$$

where
f_0 = the center frequency of the filter,
B_n = the filter's noise bandwidth,
SNR = the signal-to-noise power ratio in linear terms.

For the case of Gaussian noise only, we specify the SNR = 0 and the equation becomes

$$N_{\text{Zero Crossings}} = 2f_0 \sqrt{1 + \left(\frac{B_n^2}{12 f_0^2}\right)} \qquad 5.19$$

If half of the zero-crossings are positive-going and half are negative-going, a frequency counter connected to this system would read a frequency of $N_{\text{Zero-Crossings}}/2$, assuming there was enough signal to drive the counter adequately. For random white noise with frequency limits f_U and f_L, the expected number per second of either the positive- or negative-going sense is

$$\frac{\text{Zero Crossings}}{\text{Second}} = \sqrt{\frac{f_U^3 - f_L^3}{3(f_U - f_L)}} \qquad 5.20$$

$$= \sqrt{\frac{1}{3}(f_U^2 + f_U f_L + f_L^2)}$$

For a lowpass filter, if $f_L = 0$,

$$\frac{\text{Zero Crossings}}{\text{Second}} = 0.577 f_U \qquad 5.21$$

For narrow band-pass filters with bandwidths of less than 10% of the center frequency ($f_U \cong f_L$), we find

$$\frac{\text{Zero Crossings}}{\text{Second}} = \frac{f_U + f_L}{2} \qquad 5.22$$

When we make measurements with average-responding meters, we are interested in the following ratios:

- For Gaussian noise,

$$\frac{\text{RMS Value}}{\text{Average Value}} = \sqrt{\frac{\pi}{2}} \qquad 5.23$$
$$= 1.25$$
$$= 1.96 \text{ dB}$$

- For a sine wave,

$$\frac{\text{RMS Value}}{\text{Average Value}} = \frac{\pi}{2\sqrt{2}} \qquad 5.24$$
$$= 1.11$$
$$= 0.91 \text{ dB}$$

- For a square wave,

$$\frac{\text{RMS Value}}{\text{Average Value}} = 1 \qquad 5.25$$
$$= 0 \text{ dB}$$

The average value above is the full-wave rectified average.

Combining Independent Noise Sources

There are many important cases where two different noise sources are present in a system at the same time. Common examples are transistor amplifiers, operational amplifiers and receiving systems. Figure 5-11 shows a basic example. Resistors R_1 through R_n are noisy resistors; R_L is noiseless. We are interested in the noise voltage present across R_L due to the noise sources in R_1 through R_n.

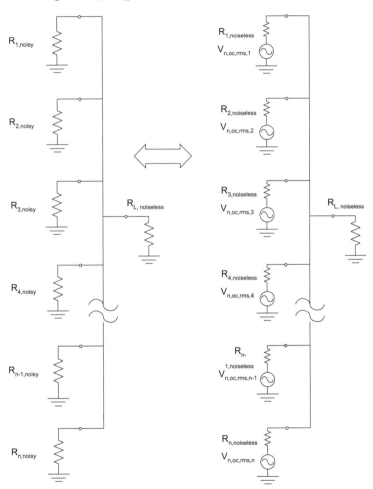

Figure 5-11 *Independent noise sources present in a system. Adding the RMS noise voltages together produces an incorrect result. We calculate the noise power delivered to R_L by each noise source, then add the powers together to produce the correct result.*

Since the noise voltages in resistors R_1 through R_n are random, we cannot make any assumptions about their voltages at any particular instant in time. Although their RMS voltages cannot be added together, we can add their powers. To solve problems like the one shown in Figure 5-11, we first short out all of the noise sources but one. We then find the power dissipated in R_L due to the active noise source. In Figure 5-11 for example, we would first short out $V_{n,oc,rms,2}$ through $V_{n,oc,rms,n}$ and leave $V_{n,oc,rms,1}$ in the circuit. We then find the noise power dissipated in R_L as a result of $V_{n,oc,rms,1}$.

Next, we would short out all of the noise sources except for $V_{n,oc,rms,2}$ and find the power dissipated in R_L due to the noise generated by R_2. We continue by examining the power delivered to R_L by each resistor in succession until we have found the power delivered to R_L by each noise source in the circuit. Finally, we add all of the powers together to find the total noise power dissipated in R_L. We can find the total noise voltage across R_L using

$$P_{noise,RL} = \frac{V_{n,oc,RMS,RL}^2}{R_L} \qquad 5.26$$

Example 5.5 — Combining Noise Sources

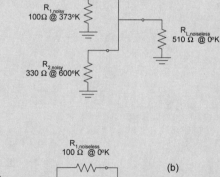

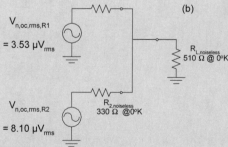

Figure 5-12 *Example 5.5. Find the noise voltage across the noiseless load resistor R_L.*

Using Figure 5-12(a), assume R_1 is a 100-ohm resistor placed in boiling water (100°C = 373 K), R_2 is a 330-ohm resistor at 600 K and R_L is a 510-ohm noiseless resistor. Find $V_{n,RMS,RL}$, the RMS noise voltage across R_L. Assume B_n = 6 MHz.

Solution —
In Figure 5-12(b), the two noisy resistors R_1 and R_2 are separated into two noiseless resistors $R_{1,noiseless}$ and $R_{2,noiseless}$ in series with their equivalent noise voltage sources. Using Equation 5.4, we find $V_{n,oc,rms,R1}$ = 3.53 μV_{rms} and $V_{n,oc,rms,R2}$ = 8.10 μV_{rms}.

First, we short-circuit $V_{n,oc,rms,R2}$ producing the circuit in Figure 5-13(a). Using voltage division, we find the voltage across R_L due to $V_{n,oc,rms,R1}$ is 2.38 μV_{rms}.

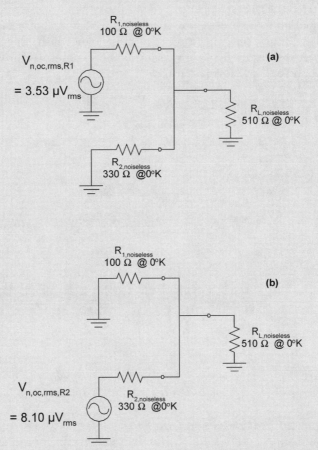

Figure 5-13 *Example 5.5. Calculate the noise delivered to the load resistor by each source in turn to find the total noise power delivered to the load resistor.*

Using Equation 5.26, the noise power delivered to R_L by $R_{1,\text{noisy}}$ is $(2.38 \mu V_{rms})^2/510 = 11.1$ fW. Similarly, we find the noise power delivered to R_L by $V_{n,oc,rms,R2}$ when $V_{n,oc,rms,R1}$ is shorted [Figure 5-13(b)]. Voltage division reveals the voltage across R_L due to $V_{n,oc,rms,R2}$ is 1.64 μV_{rm}. The noise power delivered to R_L by $R_{2,\text{noisy}}$ is $(1.64 \mu V_{rms})^2/510 = 5.26$ fW.

Finally, we add the noise powers together to find the total noise power delivered to R_L: $P_{noise,RL} = 11.1 + 5.26 = 16.4$ fW. Using Equation 5.26, the RMS noise voltage across R_L is 2.89 μV_{rms}. Note that adding the two noise voltages above produces the incorrect answer: $V_{n,rms,RL}$ 2.38 + 1.64 = 4.02 μV_{rms}.

This is exactly the way many circuit-analysis computer programs such as SPICE analyze the noise performance of complicated circuits — one noise source at a time. Remember that noise *voltages* do not add, and noise *powers* add.

Bandwidth Effects

Equation 5.4 expresses the relationship between noise bandwidth B_n and $V_{n,oc,RMS}$. Simply put, the wider the noise bandwidth, the larger the RMS noise voltage. Figure 5-14 shows thermal noise observed through a 50 kHz, a 10 kHz and a 2 kHz lowpass filter.

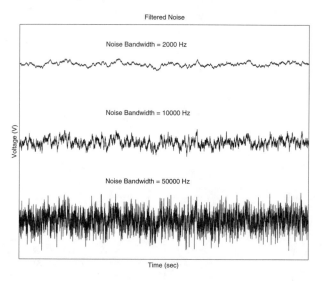

Figure 5-14 *Noise observed through filters with varying noise bandwidths. The narrower the filter bandwidth, the lower the RMS noise voltage.*

Response of Voltmeters to Gaussian Noise

Average responding meter. Random noise with the same RMS value as a sine wave has an average value of 1.05 dB less than the average value of the sine wave (1.96 − 0.91 = 1.05 dB). An average responding meter calibrated to read the RMS value of a sine wave will therefore read 1.05 dB too low compared to the RMS value of a random noise input.

Average meter responding with both sine wave and noise applied. An average responding meter with equal RMS levels of a random-noise signal and a sine wave signal applied reads 2.18 dB higher than it would for the random noise alone, or 1.13 dB higher than it would for the sine wave alone (2.18 − 1.13 = 1.05 dB).

Miscellaneous Comments

Although some of the following points may seem obvious, we have often found it necessary to include them.

- Filters will filter noise. Noise is just like any other signal.

- Attenuators will attenuate noise.

- Noise can and does exist at levels below kT_0B_n, or −174 dBm/Hz.

Noise behaves just like any other signal a system processes. The difference is that the received signal is wanted and the noise is unwanted. The choices available we have to minimize the amount of noise we are receiving are limited.

5.5 Noise Power

So far we have been discussing noise voltages and currents. Because receiving systems deal with energy and power, we will examine thermal resistor noise in terms of noise power. First we will define the terms source and load.

A *source* is an energy source. It contains some type of signal generator, which we will model as a voltage source in series with a source resistor. We will universally label the source's series resistor as R_S, which can be either noisy or noiseless, depending upon the problem. A source can be an antenna, a signal generator, a noisy resistor, an RF amplifier or a receiver's output port.

A *load* is an energy sink. We model it as a resistor (either noisy or noiseless, depending upon the problem) and label it R_L. We are interested in the noise and/or signal power delivered to R_L.

5.6 One Noisy Resistor

Figure 5-15(a) shows a noisy source resistor ($R_{S,noisy}$) connected to a noiseless load resistor ($R_{L,noiseless}$). In a receiving system, we almost universally match the value of the load resistor to the source resistor., i.e., R_S equals R_L.

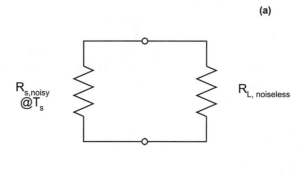

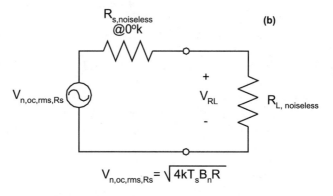

Figure 5-15 *Calculating the noise delivered by a noisy resistor to a noiseless resistor.*

Figure 5-15(b) shows the noisy source resistor replaced by its noise voltage source $V_{n,oc,rms,RS}$ and a noiseless series resistor $R_{S,noiseless}$. The noise power delivered to the load resistor R_L is

$$P_{RL} = \frac{V_{RL}^2}{R_L} \qquad 5.27$$

Since $R_L = R_S$,

$$V_{RL} = \frac{1}{2} V_{n,oc,RMS,RS} \qquad 5.28$$
$$= \frac{1}{2}\sqrt{4kT_S B_n R_S}$$

and

$$P_{RL} = \frac{\left(V_{n,oc,RMS,RS}/2\right)^2}{R_L}$$

$$= \frac{\left(\sqrt{4kT_S B_n R_S}/2\right)^2}{R_L} \qquad 5.29$$

$$= \frac{kT_S B_n R_S}{R_L}$$

Since $R_S = R_L$, we can write

$$P_{RL} = kT_s B_n \qquad 5.30$$

In a matched system, the amount of noise that a noisy resistor passes to a noiseless resistor is $kT_S B_n$. This is true regardless of the values of the two resistors. Every resistor, whether it is a piece of wire with milliohms of resistance or a 10-megohm resistor, will deliver $kT_S B_n$ worth of noise power to its matched load resistor.

In a receiving system, the source and load resistances are matched to each other to allow the load resistor to receive the maximum amount of signal power from the signal source. By matching the system, we will also receive the maximum amount of noise power as well.

Example 5.7 — $kT_S B_n$

How much noise power does a 50-ohm source resistor deliver to a matched load if the source resistor is at 100 K? Use a noise bandwidth of 4 kHz.

Solution —

Using Equation 5.30 with T = 100 K and B_n = 4 kHz, we find that P_{RL} = 5.52·10⁻¹⁸ W. Note that we did not really need to know the value of the source resistor.

Room Temperature (T_0)

As a standard, radio engineers, particularly those dealing with terrestrial radio links, often assume the temperature of the source resistor is room temperature. We use T_0 for room temperature.

$$\begin{aligned}\text{Room Temperature} &= T_0 \\ &= 290\ K \\ &= 17\,°C \\ &= 62\,°F\end{aligned} \qquad 5.31$$

Note that T_0 is actually a little colder than a comfortable room temperature.

N_0

The noise power delivered to a matched load resistor by a source resistor (at room temperature) is N_0.

$$N_0 = kT_0 B_n \qquad 5.32$$

Also, note that

$$\begin{aligned} kT_0 &= (1.38 \cdot 10^{-23})(290) \\ &= 4 \cdot 10^{-21} \,\frac{\text{W}}{\text{Hz}} \\ &= -174 \,\frac{\text{dBm}}{\text{Hz}} \end{aligned} \qquad 5.33$$

Equation 5.33 states that when we are dealing with a 1-hertz noise bandwidth, a source resistor at room temperature will deliver –174 dBm of power to a matched, noiseless load resistor. In reality, we rarely deal with systems with 1-hertz noise bandwidths.

We will use Equation 5.33 as a starting point for more realistic calculations. If we are analyzing a system with a noise bandwidth of B_n, the noise power delivered by a room temperature source resistor to its matched source resistor is

$$N_{0,dBm} = -174 \,\frac{\text{dBm}}{\text{Hz}} + 10\log(B_n) \qquad 5.34$$

Equations 5.33 and 5.34 are classic examples of unit misuses. Although logarithmic quantities have no units associated with them, we frequently include bogus units with these equations for clarity or as a memory jogger.

Example 5.7 — $kT_0 B_n$

How much noise energy does a matched source resistor at room temperature deliver to its load if we observe the noise through a filter with a 200 kHz noise bandwidth? Use Equation 5.32.

Solution —

Using Equation 5.32 with B_n = 200 kHz and T_0 = 290 K produces N_0 = 801·10⁻¹⁸ W.

Example 5.8 —174 dBm/Hz

Use Equation 5.34 to solve the example above.

Solution —
With $B_n = 200$ kHz, Equation 5.34 produces

$$N_{0,dBm} = -174\,\frac{\text{dBm}}{\text{Hz}} + 10\log(B_n)$$
$$= -174 + 10\log(200{,}000) \qquad 5.35$$
$$= -174 + 53$$
$$= -121\,\text{dBm}$$

Note that

$$-121\,\text{dBm} = 10\log(N_{0,mW}) \qquad 5.36$$
$$\Rightarrow N_0 = 794 \cdot 10^{-18}\,\text{W}$$

Considering round-off errors, this is the same answer as in the previous example.

5.7 System Model — Two Noisy Resistors

We can now examine a realistic model for a receiving system. Figure 5-16(a) shows an antenna connected to a receiver that is drawn as a simple RF amplifier. Figure 5-16(b) shows the model in more detail. Source V_{ant} represents the signal collected by the receiving antenna. The resistor R_{ant} represents the output impedance of the antenna. It also models the noise collected by the antenna via its noise temperature T_{ant}. The more noise collected by the antenna, the higher T_{ant} becomes.

The input impedance of the receiving system is modeled by R_{sys}. T_{sys} accounts for all of the internally generated noise of the receiving system. (The noiseless resistors R_{out} and R_L are not important for the noise analysis.) The receiving system measures the power dissipated in R_{sys} and delivers a copy of this signal to the load resistor R_L.

The only noisy components in Figure 5-16(b) are the two resistors, R_{ant} at T_{ant} and R_{sys} at T_{sys}. The other components in the figure are noiseless. By definition, all the noise in the system has to come from either R_{ant} or R_{sys}.

Figure 5-17(a) shows the receiving system stripped down to its bare noise essentials. Figure 5-17(b) shows the same circuit, but the two resistors have been replaced with their noise models. Since the two noise sources are uncorrelated, we have to examine the noise performance of the circuit one noise generator at a time.

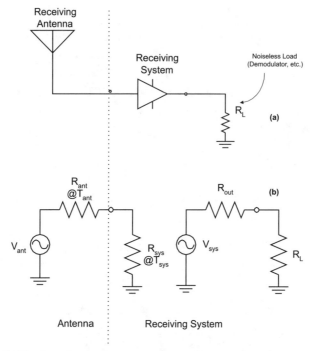

Figure 5-16 *A noisy antenna and a noisy receiver in a receiving system configuration. The noise present at the input of the system comes from the antenna and from the noise generated internally by the receiver.*

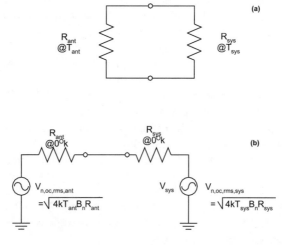

Figure 5-17 *The receiving system of Figure 5-16 emphasizing the fundamental noise contributors.*

Case 1. First, we short $V_{n,oc,rms,sys}$ and leave $V_{n,oc,rms,ant}$ in the circuit [Figure 5-18(a)]. Using voltage division, we find that the noise voltage across R_{sys} is half of the generator voltage $V_{n,oc,rms,ant}$ because $R_{ant} = R_{sys}$. Furthermore, we can say that the voltage across R_{ant} is also one-half of the generator voltage $V_{n,oc,rms,ant}$. In other words,

$$V_{n,Rsys,Case1} = V_{n,Rant,Case1}$$
$$= \frac{1}{2} V_{n,oc,RMS,ant} \qquad 5.37$$

where
$V_{n,Rant,Case\ 1}$ = the noise voltage across R_{ant},
$V_{n,Rsys,Case\ 1}$ = the noise voltage across R_{sys}.

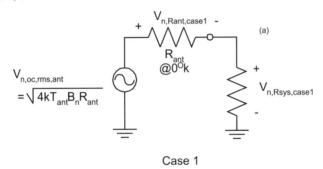

Case 1

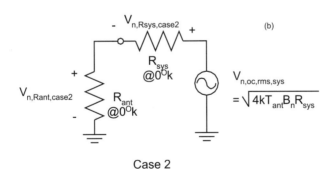

Case 2

Figure 5-18 *The receiving system noise model of Figure 5-17. Analyze the effects of one noise source at a time.*

Since the $V_{n,Rant,Case\ 1} = V_{n,Rsys,Case\ 1}$, the noise power dissipated by each resistor is the same and equal to

$$P_{n,\text{Rant,Case1}} = P_{n,\text{Rsys,Case2}}$$
$$= \frac{V_n^2}{2}$$
$$= \frac{\left(\sqrt{4kT_S B_n R}/2\right)^2}{R} \quad\quad 5.38$$
$$= kT_{Ant} B_n$$

The noise power delivered to R_{sys} equals the noise power delivered to R_{ant}. The noise power delivered to each resistor is $kT_{ant}B_n$. The source of the noise power is the thermal noise generator inside of R_{ant}.

Case 2. Figure 5-18(b) shows the second circuit where $V_{n,oc,rms,ant}$ is shorted and $V_{n,oc,rms,sys}$ is left in the circuit. Using the same techniques, we find the noise power dissipated in each resistor is the same and equal to

$$P_{n,\text{Rant,Case2}} = P_{n,\text{Rsys,Case2}} \quad\quad 5.39$$
$$= kT_{Sys} B_n$$

As with **Case 1**, the noise power delivered to R_{sys} equals the noise power delivered to R_{ant}. The noise power delivered to each resistor is $kT_{sys}B_n$ and the source of the noise power is the thermal noise generator inside of R_{sys}. The total noise power dissipated by each resistor is the sum of the noise powers from **Case 1** and **Case 2**. The total noise power dissipated by each resistor is

$$P_{n,\text{Rant}} = P_{n,\text{Rsys}}$$
$$= k\left(T_{ant} + T_{sys}\right)B_n \quad\quad 5.40$$

Equation 5.40 presents a simple yet far-reaching relationship. We have shown that all the noise present in the system depends only on the noise temperature of the antenna (T_{ant}) and on the noise temperature of the system (T_{sys}).

Example 5.9 — Power Transfer in Resistors of Unequal Temperatures

Figure 5-19 shows two matched resistors at unequal temperatures. One resistor (R_{Hot}) is submersed in boiling water at 100°C = 373 K. The second resistor (R_{Cold}) is submersed in a dry-ice bath at –79°C = 194 K. Assume that the band-pass filter has a noise bandwidth of 25 MHz and find
a. the power delivered to R_{Cold} from R_{Hot},

b. the power delivered to R_{Hot} from R_{Cold},
c. the net power flow and direction

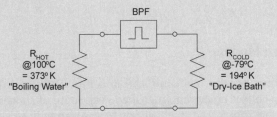

Figure 5-19 *Example 5.9. Resistors at different temperatures.*

Solution —
Since the two resistors are matched (i.e., their values are equal), we can use Equations 5.38 and 5.39. We find

a. The power delivered to R_{Cold} from R_{Hot} is

$$P_{\text{RCold from RHot}} = kT_{Hot}B_n$$
$$= (1.38 \cdot 10^{-23})(373)(25 \cdot 10^6) \qquad 5.41$$
$$= 129 \text{ fW}$$

The noise source present in R_{Hot} delivers 129 femtowatts to both R_{Hot} and R_{Cold}.

b. As in part a., the power delivered to R_{Hot} from R_{Cold} is

$$P_{\text{RHot from RCold}} = kT_{Cold}B_n$$
$$= (1.38 \cdot 10^{-23})(194)(25 \cdot 10^6) \qquad 5.42$$
$$= 66.9 \text{ fW}$$

The noise source present in R_{Cold} delivers 66.9 femtowatts to both R_{Hot} and R_{Cold}.

c. Since 129 femtowatts flows from R_{Hot} to R_{Cold} and 66.9 femtowatts flows from R_{Cold} to R_{Hot}, there is a net power flow of 62.1 femtowatts from R_{Hot} to R_{Cold}.

In theory, the power flowing from R_{Hot} to R_{Cold} would eventually cause R_{Hot} to cool down and R_{Cold} to heat up until the temperatures of the two resistors were equal. There would be no net power flow from one resistor to another. In reality, the outside environment supplies the energy necessary to sustain this heat transfer. The boiling water forces R_{Hot} to stay at 100°C, and the dry-ice bath forces the temperature of R_{Cold} to stay at $-79°C$. For the general case, the net power flow from a hot resistor to a cooler one is

$$P_{\text{Net Flow}} = k(T_{Hot} - T_{Cold})B_n \qquad 5.43$$

Antenna Noise Models

In a receiving system, we built the antenna and positioned it in the environment to collect electromagnetic waves. Some of those waves will be the signals we are interested in receiving and some will be Gaussian noise at the same frequency of the received signal. The noise is at the same frequency as the signal; we cannot filter it. If the noise and signal were at different frequencies, they could be separated by filtering.

To quantify the noise performance of the antenna, we use the model shown in Figure 5-20. The voltage source (V_{ant}) represents the ability of the antenna to collect signals. The larger the signal the antenna receives, the larger the value of V_{ant}. The value of the series resistor R_{ant} represents the output impedance of the antenna (typically 50, 75, or 300 ohms). We have to be careful when matching the antenna to the characteristic impedance of the system. There is no physical resistor inside the antenna — R_{ant} comes from matching the characteristic impedance of free space to the system characteristic impedance.

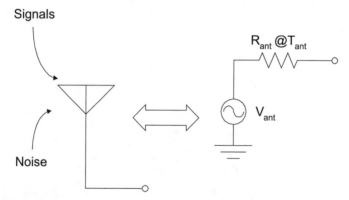

Figure 5-20 *Signal and noise models of a receiving antenna.*

Antenna Noise Temperature

Figure 5-20 indicates that R_{ant} exists at some noise temperature (T_{ant}). The larger the noise power we measure at the output of the antenna, the larger the value of T_{ant}. The physical temperature of the antenna does not influence the value of T_{ant}. The noise the antenna produces is simply the noise it collects from its environment. Figure 5-21 emphasizes this point.

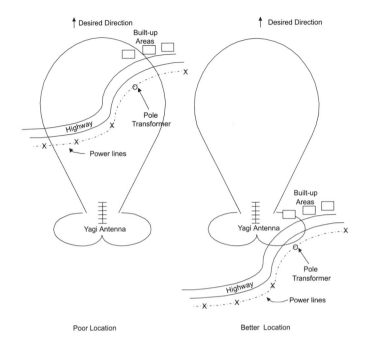

Figure 5-21 Since an antenna collects noise from the environment, the same antenna can have a widely varying noise temperature, depending upon what the antenna views. The noise temperature of the leftmost antenna will higher than the noise temperature of the rightmost antenna.

The noise temperature of the antenna can be reduced by repositioning it with respect to sources of external noise.

Example 5.10 — Antenna Noise Temperature

When we use Figure 5-22 as a guide, the signal power we can measure out of a FM broadcast antenna is −94 dBm. The characteristic impedance of the system is 50 ohms and the noise bandwidth is 210 kHz.

a. Find V_{ant}.
b. At 2:30 AM, we measure −125 dBm of noise power from the antenna. Find T_{ant} assuming $T_{sys} << T_{ant}$.
c. At 9:00 AM, as people turn on their computers (generating EMI), start their cars and generate ignition noise, the noise power measured at the terminals of the antenna increases to −114 dBm. Find T_{ant} assuming $T_{sys} << T_{ant}$.

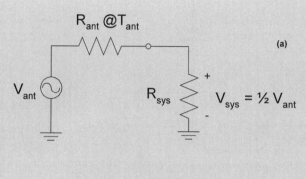

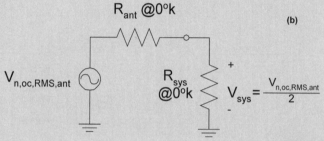

Figure 5-22 Example 5.10. Signal and noise models of an FM broadcast band antenna.

Solution —
a. Since this is a matched system with a characteristic impedance of 50 ohm, we know $R_{sys} = R_{ant} = 50\ \Omega$ and, since $R_{sys} = R_{ant}$, we also know the voltage across the load resistor $V_{sys} = 1/2\ V_{ant}$. Using

$$P_{dBm} = 10 \log\left(\frac{P_{Watts}}{0.001}\right) \quad 5.44$$

we know

$$-94\ \text{dBm} = 10 \log\left(\frac{P_{Sys,watts}}{0.001}\right) \quad 5.45$$
$$\Rightarrow P_{Sys} = 398\ \text{fW}$$

Using

$$P = \frac{V^2}{R_L} \quad 5.46$$

with $R_{sys} = 50\ \Omega$ and $P_{sys} = 398$ fW, we find

$$398 \cdot 10^{-15} = \frac{V_{Sys}^2}{50}$$
$$\Rightarrow V_{Sys} = 4.46 \ \mu V_{RMS}$$
5.47

Since $V_{ant} = 2 \ V_{Sys}$, we know $V_{ant} = 8.92 \ \mu V_{RMS}$.

b. If $T_{sys} << T_{ant}$, we know all the noise present in R_{sys} is due to T_{ant}. Equation 5.30 indicates the power delivered to a noiseless load resistor from a source resistor at a given temperature. Applying Equation 5.30 to the problem, we have $P_{sys} = -125$ dBm $= 316 \cdot 10^{-18}$ W and $B_n = 210$ kHz.

$$P_{Rsys} = kT_{ant}B_n$$
$$\Rightarrow 316 \cdot 10^{-18} \text{ Watts} = (1.38 \cdot 10^{-23})T_{ant}(210{,}000)$$
5.48
$$\Rightarrow T_{ant} = 109 \ K = -164 \ ^\circ C$$

Note that the noise temperature of the antenna is very cold (109 K = –164°C).

c. Since $T_{sys} << T_{ant}$, we know all the noise present in R_{sys} is due to T_{ant}. Following the procedure outlined in part b. above, we find $P_{sys} = -114$ dBm $= 3.98$ fW. With $B_n = 210$ kHz, Equation 5.30 produces

$$P_{Rsys} = kT_{ant}B_n$$
$$3.98 \text{ fW} = (1.38 \cdot 10^{-23})T_{ant}(210{,}000)$$
5.49
$$\Rightarrow T_{ant} = 1374 \ K = 1101 \ ^\circ C$$

The noise temperature of the antenna changed even though we have not changed the system or the physical temperature of antenna. The noise temperature of the antenna is almost entirely due to the noise the antenna receives. Since the noise power received by the antenna can change throughout the day, so can the noise temperature of the antenna.

Example 5.11 — Satellite Antenna Noise Temperature

For about 20 days a year, geosynchronous satellites pass between the earth and sun for several minutes a day (see Figure 5-23). When the geometry is right, the satellite is in front of the sun for approximately 70 minutes. During the time the satellite is positioned between the earth and sun, the noise temperature of the ground station receive antenna increases dramatically. This occurs because the antenna "sees" the electrical noise radiated by the sun as well as the normal background noise.

478 | RADIO RECEIVER DESIGN

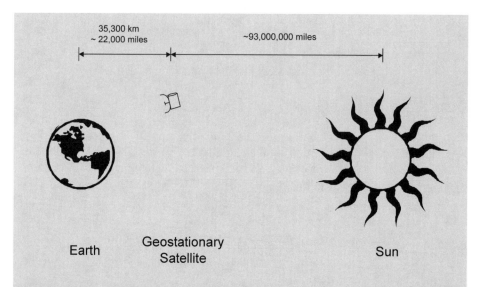

Figure 5-23 Example 5.11. *The geometry of the earth, sun and satellite causes an increase in the earth station receiving antenna because the earth station antenna is forced to look at the sun as well as the satellite.*

This geometry causes another problem. For another 20 days a year, the earth is between the geostationary satellite and the sun. Although this does not cause the antenna noise temperature to increase, it forces the satellite to run off batteries for the time the sun is eclipsed. Some of the communications satellites require approximately 2000 watts, which presents a battery problem given the environmental and reliability problems of spaceborne equipment.

Characteristics of Antenna Noise Temperature

Antenna noise temperature changes with frequency. Figure 5-24 shows the temperature of an antenna pointed at a 5° elevation off of the horizon. Note that the noise temperature is different at different frequencies. Antenna noise temperature changes with direction. If the antenna is pointed at one section of the environment, we might see a very small noise temperature. Changing the antenna's orientation even slightly can have a significant effect on the noise power.

AMPLIFIERS AND NOISE | 479

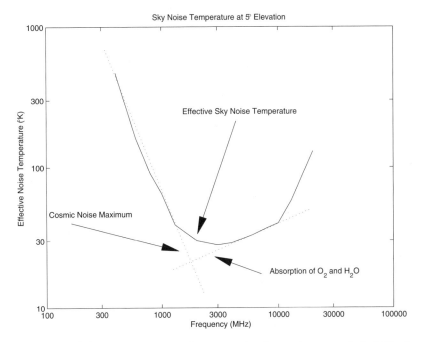

Figure 5-24 *Noise temperature of one particular antenna pointed 5° above the horizon.*

5.8   Internally Generated System Noise Model

Internally generated system noise adds noise to the signal due to several effects.

Physical Temperature

The receiving system is physically at some temperature above absolute zero (0 K). The thermal noise voltage generated by the resistors inside of the system develops Gaussian noise.

Recombination Noise

Recombination noise is the random generation and recombination of holes and electrons inside the active devices due to thermal effects. When a hole and electron combine, they create a small current spike. Since there are many recombination events occurring at once, these small spikes combine to form Gaussian noise.

Partition Noise

Partition noise is the result of the random division of emitter current between the base and collector in bipolar transistors. This random division manifests itself as thermal noise. Partition noise is absent in FET's because there is no chance for current splitting.

Shot Noise

The DC current used to bias the transistors inside the system's amplifiers contributes noise. Current flowing across the barriers of transistors, diodes and vacuum tubes is not smooth and continuous but a flow of discrete electric charges. Since electrical current is the flow of electrons, a continuous current is the sum of pulses of current caused by each electron as it crosses the barrier. Watching the number of electrons crossing the barrier over a period of time, we will find that sometimes a little more than the average number of electrons cross the barrier, sometimes a little less than average cross. This variation is called *shot noise*. The RMS value of shot noise is

$$I_{\text{Shot Noise,RMS}} = \sqrt{2qI_{DC}B_n} \qquad 5.50$$

where
q = the electron charge = $1.6022 \cdot 10^{-19}$ coulomb,
I_{DC} = the direct current in amperes,
B_n = the noise bandwidth in hertz.

Shot noise is Gaussian and has a flat frequency spectrum.

Example 5.12 — Shot Noise

Using a 1 MHz noise bandwidth, find the shot noise current ($I_{ShotNoise,RM}$) for the following conditions:
a. I_{DC} = 1 Amp,
b. I_{DC} = 1 mA,
c. I_{DC} = 1 pA.

Solution —

Using Equation 5.50 with B_n = 1 MHz, we find
a. Using I_{DC} = 1 Amp, we find $I_{ShotNoise,RMS}$ = 566 nA. Note that $I_{ShotNoise,RMS}$ is about 0.000057% of the 1 Amp nominal current. The noise is about

$$20\log\left(\frac{566 \cdot 10^{-9}}{1}\right) \qquad 5.51$$

or 125 dB below the 1 Amp DC level.

b. Using $I_{DC} = 1$ mA, we find $I_{ShotNoise,RMS} = 566$ pA. Note that $I_{ShotNoise,RMS}$ is about 0.057% of the 1 mA nominal current. The noise is about

$$20 \log\left(\frac{566 \cdot 10^{-12}}{10^{-6}}\right) \qquad 5.52$$

or 65 dB below the 1 μA DC level.

c. For $I_{DC} = 1$ pA, the shot noise current $I_{ShotNoise,RMS} = 566$ fA. $I_{ShotNoise,RMS}$ is about 57% of the 1 pA current. The noise is about

$$20 \log\left(\frac{566 \cdot 10^{-15}}{10^{-12}}\right) \qquad 5.53$$

or 5 dB below the 1 pA DC level.

Note that as the DC current becomes smaller, the shot noise problem worsens.

Equation 5.50 assumes the charges passing through some barrier act independently of one another. This is true for transistors, diodes and vacuum tubes where the current flows by diffusion, but not for metallic conductors (i.e., a simple piece of wire). The movement of each electron strongly depends upon the movement of its neighbors. A simple resistive circuit will exhibit far less shot noise than Equation 5.50 indicates.

Amplifier Noise Model

Figure 5-16 shows the models used in this chapter to describe and analyze the noise performance of a receiving system. As a mathematical tool, we will use the two noisy resistor models we discussed in Section 5.7.

Equivalent Input Noise Power

Referring to Equation 5.40 and to Figure 5-16, the total noise power dissipated by R_{sys} is

$$N_{in} = k\left(T_{ant} + T_{sys}\right)B_n \qquad 5.54$$
$$= kT_{ant}B_n + kT_{sys}B_n$$

where
N_{in} = the total noise power dissipated in R_{sys} due to the sum of the external noise generated by R_{ant} and the internal noise generated by R_{sys} (in linear units, for example, watts or milliwatts),
T_{ant} = the noise temperature of the antenna,
T_{sys} = the noise temperature of the system.

N_{in} is the *equivalent input noise power*, or the *system noise floor*. The equivalent input noise power is the total noise power dissipated by R_{sys}. This is the sum of any noise power delivered to the amplifier from external sources (like the antenna) plus any noise power generated internally within the amplifier.

Converting Equation 5.54 into more convenient units produces

$$N_{in,dBW} = 10\log(T_{ant} + T_{sys}) + 10\log(B_n) - 228.6 \qquad 5.55$$

and

$$N_{in,dBm} = 10\log(T_{ant} + T_{sys}) + 10\log(B_n) - 198.6 \qquad 5.56$$

Equations 5.54 through 5.56 show the amount of noise power present on the input of an amplifier given the temperature of the input source resistor and the noise performance of the amplifier. Equation 5.54 states that the portion of input noise generated by the system is directly related to T_{sys}. The lower T_{sys}, the lower the receiver's internally generated noise power. In other words, the noise characteristics of the receiver are completely defined by T_{sys}.

Example 5.14 — System Noise Floor (General Case)
a. Find the noise floor for a system with an antenna noise temperature of 200 K, an amplifier noise temperature of 75 K and a noise bandwidth of 35 MHz.
b. What percentage of the total system noise power comes from the antenna noise temperature?
c. What percentage of the total system noise power comes from the amplifier noise temperature?

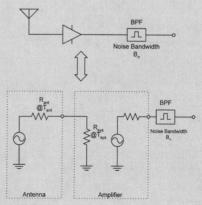

Figure 5-25 *Example 5.13.*

Solution—
In Figure 5-25, $T_{ant} = 200$ K, $T_{sys} = 75$ K and $B_n = 35$ MHz.
a. Using Equation 5.54 produces

$$N_{in} = k(T_{ant} + T_{sys})B_n$$
$$= (1.38 \cdot 10^{-23})(200 + 75)(35 \cdot 10^6)$$
$$= 133 \cdot 10^{-15} \text{ watts}$$
$$= -99 \text{ dBm}$$

5.57

or, using Equation 5.56, we can write

$$N_{in,dBm} = 10\log(T_{ant} + T_{sys}) + 10\log(B_n) - 198.6$$
$$= 10\log(275) + 10\log(35 \cdot 10^6) - 198.6$$
$$= -99 \text{ dBm}$$

5.58

b. The noise power contributed by the antenna is due to the antenna noise temperature, i.e.,

$$N_{in,ant} = kT_{ant}B_n$$
$$= (1.38 \cdot 10^{-23})(200)(35 \cdot 10^6)$$
$$= 96.6 \cdot 10^{-15} \text{ watts}$$

5.59

The percentage of the total input noise power contributed by the antenna is

$$N_{in,\%} = \frac{96.6}{133}$$
$$= 73\%$$

5.60

c. The noise power contributed by the amplifier is

$$N_{in,sys} = kT_{sys}B_n$$
$$= (1.38 \cdot 10^{-23})(75)(35 \cdot 10^6)$$
$$= 36.2 \cdot 10^{-15} \text{ watts}$$

5.61

The percentage of the total input noise power contributed by the amplifier is

$$N_{in,\%} = \frac{36.2}{133}$$
$$= 27\%$$

5.62

Equivalent Output Noise Power

The amplifier applies its power gain (G_P) to the equivalent input noise power N_{in} and produces

$$N_{out} = G_p k (T_{ant} + T_{sys}) B_n$$
$$= G_p k T_{ant} B_n + G_p k T_{sys} B_n \qquad 5.63$$

where

N_{out} = noise power delivered to R_L by the RF amplifier of Figure 5-16.

Converting Equation 5.63 into more convenient units produces

$$N_{out,dBW} = G_{p,dB} + 10\log(T_{ant} + T_{sys}) + 10\log(B_n) - 228.6 \qquad 5.64$$

and

$$N_{out,dBm} = G_{p,dB} + 10\log(T_{ant} + T_{sys}) + 10\log(B_n) - 198.6 \qquad 5.65$$

System Noise Temperature

As we saw above, there are many sources of noise inside of a receiving system. Figure 5-16 shows a model of a receiving system. For this model, we are going to "blame" all of the receiving system's internally generated noise on the noise temperature (T_{sys}) of the resistor (R_{sys}).

5.9 Signal-to-Noise Ratio (SNR)

In receiving systems, we are interested in the ratio of signal power to noise power or *signal-to-noise ratio* (SNR or S/N). The SNR indicates how well we can interpret the received signal; it directly affects the signal's quality. In the digital domain, SNR relates directly to the expected number of bits we will receive in error, i.e., the *bit error rate* (BER). Usually, the SNR is measured at the output of a receiver where the demodulation is performed, but it is equally valid to measure or calculate it at the receiver's input.

Definition

The signal-to-noise ratio is defined as

$$SNR = \frac{\text{Total Signal Power}}{\text{Total Noise Power}} \text{ in a given bandwidth} \qquad 5.66$$

Note that some bandwidth is implied. The signal plus noise waveform will always be filtered and the filter will restrict the amount of noise power present. The filter will also restrict the amount of signal power present if we are not careful.

A high number for a SNR is desirable because it means that the waveform contains much more signal power than it does noise power. This implies that the characteristics of the signal coming out of the receiver are almost entirely controlled by the received signal and not by the received noise. A low SNR means that the waveform's signal power is not much higher than the waveform's noise power.

Example 5.14 — Input Noise Power

Given a system with a 6.2 MHz noise bandwidth, an antenna temperature of 120 K and a system noise temperature of 220 K, find the output signal-to-noise power ratio for an input signal of –77 dBm.

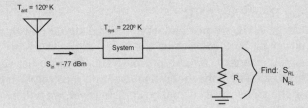

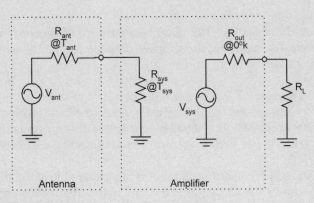

Figure 5-26 Example 5.14.

Solution —

Using Figure 5-26 as a guide, we know that $T_{ant} = 120$ K and $T_{sys} = 220$ K. We also know the signal power delivered to R_{sys} is –77 dBm. Using Equation 5.56, we find that total noise power dissipated in R_{sys} is

$$\begin{aligned}N_{Rsys,dBm} &= 10\log(T_{ant}+T_{sys})+10\log(B_n)-198.6\\ &= 10\log(120+220)+10\log(6.2\cdot 10^6)-198.6\\ &= -105.4\text{ dBm}\end{aligned}$$ 5.67

The noise power dissipated in R_{sys} is –105.3 dBm; the signal power dissipated in the same resistor is –77 dBm. The signal-to-noise ratio is –77 + (–105.3) = 28.3 dB.

Since the system applies its gain equally to both the signal and the noise, the output SNR is the same as the input SNR. For example, if the gain of the amplifier is 20 dB, we would measure –57 dBm of signal power and –85.3 dBm of noise power at the output of the amplifier. The output SNR would be –57 –(–85.3) = 28.3 dB.

Increasing Received SNR

For a large SNR, we need a large received signal power, a small amount of noise power or both present at the input of the receiver. To increase the received signal strength, we can decrease the distance between the receiver and transmitter, equip both the transmitter and receiver with better antennas or increase the transmitted signal power. However, we will eventually reach a point where the received signal power cannot be increased any further and we have to decrease the amount of noise power that is being received.

Equation 5.54 offers three options to reducing the amount of noise we receive.

- We can decrease T_{sys}, the noise temperature of the system. Decreasing T_{sys} involves selecting the proper components for the system and connecting them appropriately.

- We can decrease T_{ant}, the noise temperature of the receiving antenna. To some extent, we can do this by using a narrow-bandwidth antenna. Since the antenna looks at a smaller amount of the universe, it will tend to receive less noise. Reducing the antenna side and back lobes can also help.

- We can reduce the antenna noise temperature by operating only during quiet times such as the middle of the night. We can also try to re-orient the link so that the receiving antenna points to a quiet section of real estate. In general, trying to reduce the received noise power through reducing T_{ant} is difficult and results are uncertain at best.

- Since the received noise power is directly proportional to the noise bandwidth of the system, we can reduce the noise power by reducing the noise bandwidth (B_n) of the system. This is one of the primary reasons we use the narrowest filter bandwidth possible. The minimum filter bandwidth is usually set by the bandwidth of the signal to be processed. We make the filter as narrow as possible but just wide enough to accept all of the signal power without significant distortion.

Example 5.15 — Signal-to-Noise Objectives

Most systems need at least 10 dB of SNR to function effectively. Received signal characteristics such as bit error rate and understandability drop off rapidly below approximately 15 dB.

Some of the SNR objectives that are most favorable include

50 dB : Telephony
45 dB or higher : Television
70 dB : Compact-disk quality audio
< 10^{-6} : Data link bit error rates

Television SNR and picture quality
55 dB : "Very good, noise barely perceptible"
45 dB : "Acceptable, some noise visible"
35 dB : "Not acceptable, poor picture"

For a 10^{-8} bit error rate, we need
10.6 dB SNR for coherent PSK
14.5 dB SNR for non-coherent FSK
15.0 dB SNR for non-coherent ASK

5.10 Noise Factor and Noise Figure

Determining the noise temperature of an antenna can be quite difficult. In carefully controlled situations, such as microwave telephone links or satellite communications systems, this seems less of a problem because we are typically spending a considerable amount of money on the link and can afford to take the time to actually measure the antenna's noise temperature as well as take steps to minimize it. Also, we have a narrow antenna beamwidth and a relatively clear, fixed transmission path.

However, there are a great many situations when it is uncertain what the antenna noise temperature will be. Several situations which exhibit poorly characterized noise performance are cellular telephone equipment, ham radio gear, commercial broadcasting (AM, FM and television) and hand-held ("walkie-talkie") situations. In these cases, we cannot characterize the noise performance of the antenna. Yet answers to questions such as "How far away can we get and still talk?" have to be found. Since we cannot perform an accurate noise analysis of these situations, we have to rely on guess work.

N_0

It is assumed that the antenna noise temperature will be $T_0 = 290$ K. By definition, a signal source with a noise temperature of T_0 will deliver a noise power of N_0 to a matched load resistor where

$$\begin{aligned} N_{0,\text{Watts}} &= kT_0 B_n \\ &= \left(1.38 \cdot 10^{-23}\right)(290) B_n \\ &= \left(4.00 \cdot 10^{-21}\right) B_n \end{aligned} \qquad 5.68$$

Also

$$\begin{aligned} N_{0,dBW} &= 10\log(B_n) - 204 \\ N_{0,dBm} &= 10\log(B_n) - 174 \end{aligned} \qquad 5.69$$

For the sake of analysis, we will assume we are operating in a system with a 1-hertz noise bandwidth. This assumption is obviously impractical, but we use it as a starting place for computations. For a noise bandwidth of 1 hertz,

$$\begin{aligned} N_{0,\text{dBW},1\,\text{Hz}} &= -204 \; dBW \\ N_{0,\text{dBm},1\,\text{Hz}} &= -174 \; dBm \end{aligned} \qquad 5.70$$

Note that N_0 describes only the noise power the antenna delivers to a matched load resistor. N_0 is independent of the noise internally generated in the receiving system.

Definitions

The noise temperature concept we have developed in this chapter describes noise and its effects on receiving systems in the most general terms. These equations will always work and are always applicable. Once we have defined a standard antenna noise temperature of $T_0 = 290$ K, we can define terms to make the calculations easier.

When the noise temperature of the receiving antenna is T_0, we use *noise factor* to quantify the amount of noise power the system adds to a signal. *Noise factor* and *noise temperature* of a system both describe how much noise the system adds to a signal. Noise factor is most useful when the input noise power is N_0. The definition of noise factor is

$$F = \frac{\text{Noise power delivered by a noisy component}}{\text{Noise power delivered by a noiseless component}} \qquad 5.71$$

when the component's input noise power is $N_0 = kT_0 B_n$

Occasionally, other definitions of noise factor can be found; all are derivations of Equation 5.71. The noise power delivered by a noisy receiver is the equivalent input noise of the receiver times the receiver's power gain or $G_p k(T_0 + T_{amp})B_n$. The noise power delivered by an ideal noiseless receiver is $G_p k T_0 B_n$. Combining these two expressions with Equation 5.71 produces

$$F = \frac{G_p k(T_0 + T_{amp})B_n}{G_p k T_0 B_n} \qquad 5.72$$

The simplified version reads

$$F = \frac{T_0 + T_{amp}}{T_0} \quad F \geq 1 \qquad 5.73$$

Noise Factor and Noise Figure

To convert between noise factor (F, a linear term) and noise figure (F_{dB}, a term in decibels), we use

$$F_{dB} = 10\log(F) \qquad 5.74$$

Knowing the difference between noise factor and noise figure is very important from a mechanical point of view. Many mistakes result from plugging noise figure into equations that need noise factor.

Noise Factor and Noise Temperature Relationships

Rearranging Equation 5.73 produces

$$T_{amp} = T_0(F_{amp} - 1) \qquad 5.75$$

Both T_{sys} and F describe exactly the noise performance of a system. However, noise figure is a little handier to calculate when the temperature of the input noise is 290 K.

490 | RADIO RECEIVER DESIGN

Example 5.16 — Noise Figure/Noise Factor/Noise Temperature Conversions

a. Given an amplifier with a noise figure of 2 dB, find the noise factor and noise temperature of the amplifier.
b. Another amplifier has a noise temperature of 290 K, find the noise figure and noise factor.
c. A third amplifier has a noise factor of 100. Find the noise figure and noise temperature.

Solution —

a. Using Equation 5.74 with $F_{dB} = 2$ dB implies $F = 1.58$. Equation 5.75 with $F = 1.58$ produces $T_{sys} = 170$ K.
b. $T_{sys} = 290$ K. Using Equation 5.75 produces a noise factor of $F = 2$. Using $F = 2$ in Equation 5.74 yields a noise figure of $F_{dB} = 3$ dB.
c. Using Equations 5.74 and 5.75 with $F = 100$ gives $F_{dB} = 20$ dB and $T_{sys} = 28700$ K.

Example 5.17 — Table of Noise Figure/Noise Temperature

Table 5-2 Noise Figures and the equivalent noise temperatures.

F_{dB}	T_{sys} (K)	F_{dB}	T_{sys} (K)
0.0	0.0	6.0	865
0.1	6.8	7.0	1,160
0.2	13.7	8.0	1,540
0.3	20.7	9.0	2,010
0.5	35.4	10.0	2,610
1.0	75.1	15.0	8,880
1.5	120	20.0	28,700
2.0	170	30.0	290,000
3.0	290	40.0	2,900,000
4.0	438	50.0	29,000,000
5.0	627	100.0	2.9×10^{12}

Equivalent Input Noise Power

The equivalent input noise power of a component is the sum of the

AMPLIFIERS AND NOISE | 491

noise delivered to the component by the external source plus the noise generated internally in the component. We model the internally generated noise by assuming the input resistor is at a temperature T_{sys}. The equivalent input noise power, or noise floor of a system with a noise figure of F, a noise temperature of T_{sys} and an input noise temperature of T_0 is

$$N_{in,equ} = k(T_0 + T_{sys})B_n \qquad 5.76$$

Combining Equations 5.73 and 5.76 produces

$$N_{in,equ} = k[T_0 + (F-1)T_0]B_n \qquad 5.77$$
$$= FkT_0B_n$$

Converting Equation 5.77 into decibels produces

$$N_{in,equ,dBm} = F_{dB} + 10\log(kT_0) + 10\log(B_n) \qquad 5.78$$

We know

$$10\log(kT_0) = -174\frac{dBm}{Hz} \qquad 5.79$$

Accordingly, Equation 5.78 becomes

$$N_{in,equ,dBm} = F_{dB} - 174\frac{dBm}{Hz} + 10\log(B_n) \qquad 5.80$$

Equation 5.80 describes the equivalent input noise power for a system with a noise figure of F_{dB} and a noise bandwidth of B_n when the input noise power, i.e., the power delivered by the external source is kT_0B_n.

Let us examine the terms of Equation 5.80.

- -174 dBm/Hz. This term represents the noise temperature of the antenna (or other external source). When we assumed the antenna noise temperature to be 290 K, we effectively set this term to -174 dBm/Hz. The noise delivered to the system by the antenna is kT_0B_n. We assume a 1-hertz noise bandwidth and kT_0 = 174 dBm/Hz.

- F_{dB}. This term describes the noise generated internally by the system. Note that the higher the noise figure, the higher the equivalent input noise power. A low noise figure is desirable.

- $10\log(B_n)$. The noise is observed through a filter with some bandwidth.

This term takes the noise bandwidth into account.
Note that the larger the noise bandwidth, the larger the equivalent input noise power. In order to minimize the input noise power, we have to operate in the smallest bandwidth possible.

Example 5.18 — System Noise Floor with N_0 Input Noise Power

Figure 5-27 shows an antenna whose noise temperature is 290 K. The antenna is connected to an amplifier with a noise figure of 7 dB. The noise bandwidth is 6.7 MHz and the amplifier has a power gain of 16 dB. Find the equivalent input noise power and the output noise power using
a. Equation 5.80,
b. Equation 5.77,
c. Equation 5.54.

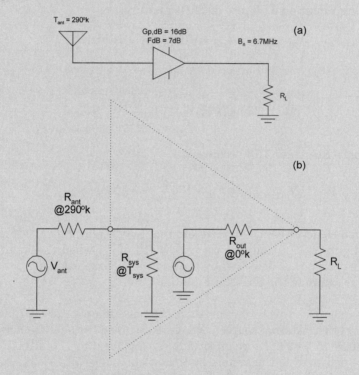

Figure 5-27 *Example 5.18.*

Solution —
a. Since the noise temperature of the antenna is T_0, we can use the noise figure of the amplifier directly with Equation 5.80 to find the equivalent input noise power.

$$N_{in,dBm} = F_{dB} - 174\frac{\text{dBm}}{\text{Hz}} + 10\log(B_n) \qquad 5.81$$
$$= 7 - 174\frac{\text{dBm}}{\text{Hz}} + 10\log(6.7 \cdot 10^6)$$
$$= -99 \text{ dBm}$$

The equivalent input noise power is −99 dBm. The amplifier applies its power gain to this noise power, so the output noise power is

$$-99 \text{ dBm} + G_{p,dB} = -99 + 16 = -83 \text{ dBm}.$$

b. We first convert the noise figure and gain into linear units.

$$7 \text{ dB} = 10\log(F) \qquad 16 \text{ dB} = 10\log(G_p)$$
$$\Rightarrow F = 5.01 \qquad \Rightarrow G_p = 39.8 \qquad 5.82$$

Applying Equation 5.77 produces

$$N_{in,equ} = FkT_0 B_n$$
$$= (5.01)(1.13 \cdot 10^{-23})(290)(6.7 \cdot 10^6) \qquad 5.83$$
$$= 134.4 \cdot 10^{-15} \text{ W}$$
$$= -99 \text{ dBm}$$

$$N_{out} = G_p N_{in}$$
$$= (39.8)(134.4 \cdot 10^{-15} \text{ W})$$
$$= 5.35 \cdot 10^{-12} \text{ W} \qquad 5.84$$
$$\approx -83 \text{ dBm}$$

As in part a., the equivalent input noise power is −99 dBm and the output power is about −83 dBm.

c. Equation 5.54 requires noise temperatures. Using Equation 5.75, we convert the 7 dB (or 5.01) amplifier noise figure into noise temperature using Equation 5.75.

$$T_{amp} = T_0(F - 1)$$
$$= 290(5.01 - 1) \qquad 5.85$$
$$= 1160 \text{ K}$$

We now can find the equivalent input noise power using Equation 5.54.

$$N_{in} = k(T_{ant} + T_{sys})B_n$$
$$= (1.38 \cdot 10^{-23})(1160 + 290)(6.7 \cdot 10^6)$$
$$= 134.2 \cdot 10^{-15} \text{ W}$$
$$= -99 \text{ dBm}$$

5.86

We arrive at the same answer as before.

Equivalent Output Noise Power

Once we know the equivalent input noise power being dissipated in $R_{amp,in}$, we can find the output noise power by multiplying by the power gain or

$$N_{out} = G_p F k T_0 B_n \qquad 5.87$$

In decibels, this is

$$N_{out,dBm} = G_{p,dB} + F_{dB} - 174 \frac{\text{dBm}}{\text{Hz}} + 10\log(B_n) \qquad 5.88$$

Signal-to-Noise Ratio and Noise Figure

Let us examine the signal-to-noise ratios at the input and output of an amplifier whose noise figure is F and whose power gain is G_p. The noise temperature of the external source is $T_0 = 290$ K. Using Figure 5-26 as a guide, let S equal the signal power dissipated in R_{sys}. The noise power delivered to R_{sys} by the external source is $N_0 = kT_0 B_n$. The input signal-to-noise ratio is

$$\left(\frac{S}{N}\right)_{in} = \left(\frac{S}{kT_0 B_n}\right) \qquad 5.89$$

The signal power delivered to R_L is the input signal power times the amplifier's power gain or

$$S_{out} = G_p S_{in} \qquad 5.90$$

Equation 5.87 states that the noise power delivered to R_L is

$$N_{out} = G_p F k T_0 B_n \qquad 5.91$$

The signal-to-noise ratio at the output of the amplifier is

$$\left(\frac{S}{N}\right)_{out} = \left(\frac{G_p S_{in}}{G_p F k T_0 B_n}\right) \qquad 5.92$$

Combining Equations 5.89 and 5.92 produces

$$\frac{(S/N)_{in}}{(S/N)_{out}} = \frac{S_{in}/(kT_0 B_n)}{(G_p S_{in})/(G_p F k T_0 B_n)} \qquad 5.93$$

Canceling out all of the like terms, we find

$$\frac{(S/N)_{in}}{(S/N)_{out}} = F \text{ when } N_{in} = N_0 \qquad 5.94$$

The noise factor of a system measures the degradation of the signal-to-noise ratio as the signal passes through the system. As with most calculations involving noise figure, Equation 5.94 is valid only when the input noise is N_0. Equation 5.94 is derived from Equation 5.71 and is commonly quoted as the definition for noise figure.

Example 5.19 — S/N Degradation

We have been given a system with an antenna noise temperature of T_0, a gain of 10 dB, a noise figure of 3 dB and a noise bandwidth of 30 kHz. We apply a signal of –110 dBm to the system.
a. Find the noise power delivered to the system by the antenna.
b. Find the equivalent input noise power.
c. Find the output noise power.
d. Find the output signal power.
e. Verify that the noise figure of the system equals the degradation in the signal-to-noise ratio as the signal passes through the system.

Solution —

Since the input noise power is N_0 (i.e., the input noise temperature is 290 K), we can use the equations for noise figure.
a. The noise delivered to the system by the external amplifier is given by Equation 5.34.

$$\begin{aligned} N_{0,dBm} &= -174 \frac{\text{dBm}}{\text{Hz}} + 10\log(B_n) \\ &= -174 + 10\log(30{,}000) \\ &= -129 \text{ dBm} \end{aligned} \qquad 5.95$$

496 | RADIO RECEIVER DESIGN

The antenna delivers −129 dBm of noise power to the system. This result does not include any noise the system adds.

b. Using Equation 5.80, we find the total noise dissipated in R_{sys} is

$$N_{in,dBm} = F_{dB} - 174\,\frac{\text{dBm}}{\text{Hz}} + 10\log(B_n)$$
$$= 3 - 174 + 10\log(30{,}000) \qquad 5.96$$
$$= -126\text{ dBm}$$

This is the same equation as in part a., but we have added the system's noise figure to account for the noise power added by the system.

c. The output noise power equals the input noise power (in dBm) plus the power gain of the amplifier (in dB).

$$N_{out,dBm} = N_{in,dBm} + G_{p,dB}$$
$$= -126 + 10 \qquad 5.97$$
$$= -116\text{ dBm}$$

d. The output signal power is the input signal power (in dBm) plus the amplifier's power gain.

$$S_{out,dBm} = S_{in,dBm} + G_{p,dB}$$
$$= -110 + 10 \qquad 5.98$$
$$= -100\text{ dBm}$$

e. The signal-to-noise ratio at the system's input is

$$\left(\frac{S}{N}\right)_{in} = -110 - (-129) \qquad 5.99$$
$$= 19\text{ dB}$$

The signal-to-noise ratio at the system's output is

$$\left(\frac{S}{N}\right)_{in} = -110 - (-116) \qquad 5.100$$
$$= 16\text{ dB}$$

The degradation of the signal-to-noise ratio through the system is

$$\frac{(S/N)_{in}}{(S/N)_{out}} = 19 - 16 \qquad 5.101$$
$$= 3 \text{ dB}$$

The system's noise figure is 3 dB.

5.11 Cascade Performance

The system in Figure 5-28(a) contains several amplifiers in cascade, i.e., the output of one amplifier feeds the input of the next. Given the power gain and the noise characteristics of all the devices in a cascade, we want to find the noise performance of the cascade. Let

$G_{p,1}$ = the power gain of the first amplifier in linear terms,
$G_{p,2}$ = the power gain of the second amplifier in linear terms,
$G_{p,n}$ = the power gain of the nth amplifier in linear terms,

T_1 = the noise temperature of the first amplifier,
T_2 = the noise temperature of the second amplifier,
T_n = the noise temperature of the nth amplifier,

F_1 = the noise factor of the first amplifier in linear terms,
F_2 = the noise factor of the second amplifier in linear terms,
F_n = the noise factor of the nth amplifier in linear terms,

$G_{p,n,dB}$ = the power gain of the nth amplifier in decibels,
$F_{n,dB}$ = the noise figure of the nth amplifier in decibels,

$G_{p,cas}$ = the power gain of the cascade in linear terms,
$F_{cas,dB}$ = the noise figure of the cascade in linear terms,
T_{cas} = the noise temperature of the cascade,
$G_{p,cas,dB}$ = the power gain of the cascade in decibels,
$F_{cas,dB}$ = the noise figure of the cascade in decibels.

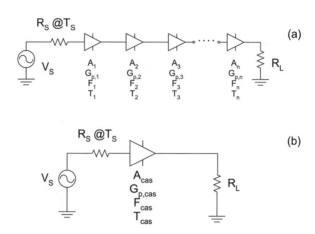

Figure 5-28 *An n-element cascade and its single-element equivalent.*

For a first-pass analysis, we will assume that the entire system is exactly matched to the system's characteristic impedance (Z_0). This means that $R_S = R_L = Z_0$, and that the input and output impedances of all the amplifiers are exactly Z_0. (This is not always true in practice but we will make corrections to the analysis later.) The goal is to describe the cascade as a single amplifier with a power gain of $G_{p,cas}$, a noise temperature of T_{cas} and a noise factor F_{cas} [see Figure 5-28(b)].

Gain Performance of a Cascade

Using Figure 5-28(a) as a guide, we would like to find the total power gain of the cascade ($G_{p,cas}$) given the power gains of the individual components which make up the cascade. First, we will present amplifier A_1 with an input signal power of $S_{in,A1}$. The signal power present at the output of A_1 is

$$S_{out,A1} = S_{in,A1} G_{p,1} \qquad 5.102$$

The output of amplifier A_1 is connected to the input of amplifier A_2. The input signal applied to A_2 is $S_{out,A1}$ and the signal present at the output of A_2 is

$$S_{out,A2} = S_{out,A1} G_{P,2} \qquad 5.103$$
$$= S_{in,A1} G_{P,1} G_{P,2}$$

Finally, amplifier A_3 sees an input signal of $S_{out,A2}$ to the input of amplifier A_3. The output of A_3 is

$$S_{out,A3} = S_{out,A2} G_{P,3}$$
$$= S_{in,A1} G_{P,1} G_{P,2} G_{P,3} \qquad 5.104$$

We continue this process until we reach the end of the cascade, i.e., the output of the nth amplifier. The signal out of the nth amplifier is

$$S_{out,n} = G_{p,1} G_{p,2} G_{p,3} \cdots G_{p,n} S_{in,A1} \qquad 5.105$$

The power gain of the cascade is

$$G_{p,cas} = G_{p,1} G_{p,2} G_{p,3} \cdots G_{p,n} \qquad 5.106$$

Converting Equation 5.106 to decibels, we find

$$G_{p,cas,dB} = G_{p,1,dB} + G_{p,2,dB} + G_{p,3,dB} + \ldots + G_{p,n,dB} \qquad 5.107$$

Example 5.20 — Cascade Gain Calculation
Find the gain of the cascade shown in Figure 5-29.

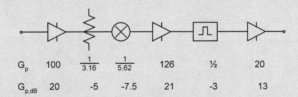

Figure 5-29 *Example 5.20.*

Solution —
Using Equation 5.106, we find

$$\begin{aligned}
G_{p,cas} &= G_{p,1} G_{p,2} G_{p,3} \cdots G_{p,n} \\
&= (100)\left(\frac{1}{3.16}\right)\left(\frac{1}{5.62}\right)(126)\left(\frac{1}{2}\right)(20) \\
&= 7.09
\end{aligned} \qquad 5.108$$

We can also find the solution using Equation 5.107.

$$\begin{aligned}
G_{p,cas,dB} &= G_{p,1,dB} + G_{p,2,dB} + G_{p,3,dB} + \ldots + G_{p,n,dB} \\
&= 20 + (-5) + (-7.5) + 21 + (-3) + 13 \\
&= 38.5 \text{ dB} \\
&= 7.09
\end{aligned} \qquad 5.109$$

Practical Effects

If we build up a cascade and measure its power gain, we will find that the actual results deviate slightly from Equations 5.106 and 5.107. There are several reasons for the deviation. One of the implicit assumptions we made as we derived the cascade gain equations was that all of the amplifiers had input and output impedances which were exactly matched to the system impedance. In other words, the terminal impedances of all the amplifiers were exactly equal to Z_0. If this is the case, then each amplifier accepts all of the signal power available from the amplifier preceding it. Equations 5.106 and 5.107 are exactly accurate when all of the amplifiers are matched.

In practice, however, amplifiers are not exactly matched to the system's characteristic impedance. This is especially true for wideband amplifiers, i.e., amplifiers whose frequency range is greater than one octave. For example, one vendor specifies a wideband amplifier with a maximum input/output VSWR of 2.5 over the entire band of operation (6 octaves). We know

$$R_L = Z_0(VSWR) \quad \text{or} \quad R_L = \frac{Z_0}{VSWR} \qquad 5.110$$

The R_L must be

$$R_L = Z_0(VSWR) \qquad R_L = \frac{Z_0}{VSWR}$$
$$= 50(2.5) \qquad \qquad = \frac{50}{2.5} \qquad 5.111$$
$$= 125\,\Omega \qquad \qquad = 20\,\Omega$$

The terminal impedance on this amplifier can vary from 20 to 125 ohms (in a 50-ohm system). The terminal impedances usually change with frequency. For example, the input impedance may be 120 ohms at one frequency, 75 ohms at another frequency and 25 ohms at a third frequency. Another factor we did not account for when we derived Equations 5.106 and 5.107 is that the gain of a typical amplifier will not always be the same over time, temperature and frequency. For example, one manufacturer sells an amplifier with a gain specification of 25 dB –1 dB. The –1 dB span addresses the change in amplifier gain with time, temperature and frequency. We normally run through equations 5.106 and 5.107 only once, using the nominal gain for each amplifier. The effect is that the cascade will, on average, exhibit the gain given by equations 5.106 and 5.107. However, the actual gain of the cascade may be higher or lower than expected, or may vary with frequency.

Noise Temperature of a Cascade

To determine the noise temperature of the cascade in Figure 5-28(b) we have to take into consideration the noise and gain characteristics of the cascade elements. Refer to Figure 5-30 as you follow along this derivation.

Let us apply an input noise power of $kT_{ant}B_n$ to the cascade. Using Equation 5.54, we find the equivalent input noise power to amplifier A_1 is

$$N_{in,A1} = k(T_{ant} + T_1)B_n \qquad 5.112$$

which includes noise supplied by the external source (T_{ant}) and noise generated internally by A_1 (T_1). Amplifier A_1 applied its power gain to the input noise; accordingly, the noise available at the output of A_1 is

$$\begin{aligned} N_{out,A1} &= G_{p,1}k(T_{ant} + T_1)B_n \\ &= G_{p,1}kT_{ant}B_n + G_{p,1}kT_1B_n \end{aligned} \qquad 5.113$$

Figure 5-30 shows this noise power schematically. The blocks are labeled "$G_{p,1}kT_1B_n$" and "$G_{p,1}kT_{ant}B_n$."

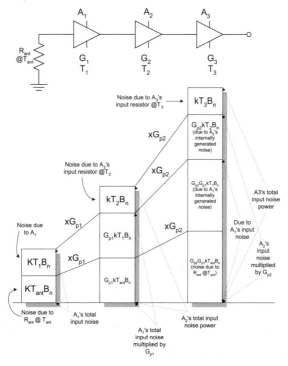

Figure 5-30 *Graphical description of the noise present in a three-element cascade.*

Since the output of A_1 is connected to the input of amplifier A_2, Equation 5.113 also shows the external noise applied to the input of A_2. We can rearrange Equation 5.113.

$$N_{in,A2} = k[G_{p,1}(T_{ant} + T_1)]B_n \qquad 5.114$$

We can think of the term $[G_{p,1}(T_{ant} + T_1)]$ as the temperature of the noise applied to A_2's input. Amplifier A_2 accepts this noise power from the external source and adds some noise of its own. The noise added by amplifier A_2 is due to T_2, the noise temperature of A_2's input resistor. In Figure 5-30, the noise power contributed by A_2 is labeled "kT_2B_n."

The total noise power dissipated in A_2's input resistor is

$$N_{in,A2} = k[G_{p,1}(T_{ant} + T_1) + T_2]B_n \qquad 5.115$$

where $N_{in,A2}$ is a result of the noise delivered by A_1 and the noise generated internally by A_2. A_2 applies its power gain to the input noise power. The noise power present at the output of A_2 is

$$N_{out,A2} = G_{p,2}k[G_{p,1}(T_{ant} + T_1) + T_2]B_n \qquad 5.116$$

Figure 5-30 shows $N_{out,A2}$. The blocks are labeled "$G_{p,2}kT_2B_n$," "$G_{p,2}G_{p,1}kT_1B_n$," and "$G_{p,2}G_{p,1}kT_{ant}B_n$." We can rewrite Equation 5.116.

$$N_{out,A2} = k\{G_{p,2}[G_{p,1}(T_{ant} + T_1) + T_2]\}B_n \qquad 5.117$$

The term $\{G_{p,2}[G_{p,1}(T_{ant}+T_1)+T_2]\}$ is the temperature of the noise available from A_2. Amplifier A_3 accepts the noise from A_2 and adds its own noise to the system (the block labeled "kT_3B_n" in Figure 5-30). The total noise power present at the input of A_3 is

$$N_{in,A3} = k[G_{p,2}G_{p,1}(T_{ant} + T_1) + G_{p,2}T_2 + T_3]B_n \qquad 5.118$$

The noise power available from the output of amplifier A_3 is

$$N_{out,A3} = k[G_{p,3}G_{p,2}G_{p,1}(T_{ant} + T_1) + G_{p,3}G_{p,2}T_2 + G_{p,3}T_3]B_n \qquad 5.119$$

When we carry this process through to the nth amplifier, we can write the noise power available from the output of the nth amplifier as

AMPLIFIERS AND NOISE | 503

$$\begin{aligned}N_{out,An} = k\big[&G_{p,n}\cdots G_{p,3}G_{p,2}G_{p,1}(T_{ant}+T_1)\\ +&G_{p,n}\cdots G_{p,3}G_{p,2}T_2\\ +&G_{p,n}\cdots G_{p,3}T_3\\ +&\ldots\\ +&G_{p,n}T_n\big]B_n\end{aligned} \qquad 5.120$$

Dividing the output noise power given by Equation 5.120, by the power gain of the cascade produces the equivalent input noise power to the cascade.

$$\begin{aligned}N_{in,cas} &= \frac{N_{out,cas}}{G_{cas}}\\ &= \frac{N_{out,cas}}{G_{p,n}\cdots G_{p,3}G_{p,2}G_{p,1}}\\ &= k\frac{\big[G_{p,n}\cdots G_{p,3}G_{p,2}G_{p,1}(T_{ant}+T_1)+G_{p,n}\cdots G_{p,3}G_{p,2}T_2+G_{p,n}\cdots G_{p,3}T_3+\cdots\big]}{G_{p,n}\cdots G_{p,3}G_{p,2}G_{p,1}}B_n\end{aligned} \qquad 5.121$$

Equation 5.121 simplified becomes

$$N_{in,cas} = k\left[(T_{ant}+T_1)+\frac{T_2}{G_{p,1}}+\frac{T_3}{G_{p,1}G_{p,2}}+\ldots+\frac{T_n}{G_{p,1}G_{p,2}G_{p,3}\cdots G_{p,n-1}}\right]B_n \qquad 5.122$$

Looking at the cascade as a single device with a noise temperature of T_{cas}, Equation 5.54 indicates that the equivalent input noise power of the cascade is

$$N_{in,cas} = k(T_{ant}+T_{cas})B_n \qquad 5.123$$

Equating 5.54 and 5.122 produces

$$k(T_{ant}+T_{cas})B_n = k\left[(T_{ant}+T_1)+\frac{T_2}{G_{p,1}}+\frac{T_3}{G_{p,1}G_{p,2}}+\ldots+\frac{T_n}{G_{p,1}G_{p,2}G_{p,3}\cdots G_{p,n-1}}\right]B_n \qquad 5.124$$

Canceling out like terms produces an expression for the noise temperature of the cascade.

$$T_{cas} = T_1+\frac{T_2}{G_{p,1}}+\frac{T_3}{G_{p,1}G_{p,2}}+\ldots+\frac{T_n}{G_{p,1}G_{p,2}G_{p,3}\cdots G_{p,n-1}} \qquad 5.125$$

Note that the power gains are linear, not logarithmic or decibel quantities.

Example 5.21 — Noise Temperature of a Cascade

Find the noise temperature of the cascade shown in Figure 5-31. Use Equation 5.125.

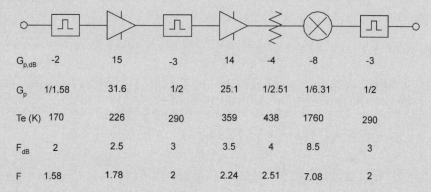

$G_{p,dB}$	-2	15	-3	14	-4	-8	-3
G_p	1/1.58	31.6	1/2	25.1	1/2.51	1/6.31	1/2
Te (K)	170	226	290	359	438	1760	290
F_{dB}	2	2.5	3	3.5	4	8.5	3
F	1.58	1.78	2	2.24	2.51	7.08	2

Figure 5-31 *Example 5.21 and Example 5.22.*

Solution —

Using Equation 5.125, we find

$$T_{cas} = T_1 + \frac{T_2}{G_{p,1}} + \frac{T_3}{G_{p,1}G_{p,2}} + K + \frac{T_n}{G_{p,1}G_{p,2}G_{p,3}\ldots G_{p,n-1}}$$

$$= 170 + \frac{226}{(1/1.58)} + \frac{290}{(1/1.58)(31.6)} + \frac{359}{(1/1.58)(31.6)(1/2)}$$

$$+ \frac{438}{(1/1.58)(31.6)(1/2)(25.1)} \qquad 5.126$$

$$+ \frac{1760}{(1/1.58)(31.6)(1/2)(25.1)(1/2.51)}$$

$$+ \frac{290}{(1/1.58)(31.6)(1/2)(25.1)(1/2.51)(1/6.31)}$$

$$= 170 + 357 + 14.5 + 35.9 + 1.8 + 17.6 + 18.3$$

$$= 615 \text{ K}$$

Noise Factor of a Cascade

Given the noise factors of all the elements making up the cascade, combining Equation 5.125 with Equation 5.75 produces an expression for the noise factor of a cascade

$$T_0(F_{cas} - 1) = T_0(F_1 - 1)$$
$$+ \frac{T_0(F_2 - 1)}{G_{p,1}}$$
$$+ \frac{T_0(F_3 - 1)}{G_{p,1}G_{p,2}}$$
$$+ \ldots$$
$$+ \frac{T_0(F_n - 1)}{G_{p,1}G_{p,2}G_{p,3}\ldots G_{p,n-1}} \qquad 5.127$$

Simplifying produces

$$F_{cas} = F_1 + \frac{F_2 - 1}{G_{p,1}} + \frac{F_3 - 1}{G_{p,1}G_{p,2}} + \ldots + \frac{F_n - 1}{G_{p,1}G_{p,2}\ldots G_{p,n-1}} \qquad 5.128$$

The power gains and noise factors of Equation 5.128 are the linear, not the logarithmic, quantities.

Example 5.22 — Noise Factor of a Cascade
Find the noise factor/noise figure of the cascade shown in Figure 5-31. Use Equation 5.128.

Solution —
Using Equation 5.128, we find

$$F_{cas} = 1.58 + \frac{1.78 - 1}{(1/1.58)} + \frac{2 - 1}{(1/1.58)(31.6)} + \frac{2.24 - 1}{(1/1.58)(31.6)(1/2)}$$
$$+ \frac{2.51 - 1}{(1/1.58)(31.6)(1/2)(25.1)}$$
$$+ \frac{7.08 - 1}{(1/1.58)(31.6)(1/2)(25.1)(1/2.51)} \qquad 5.129$$
$$+ \frac{2 - 1}{(1/1.58)(31.6)(1/2)(25.1)(1/2.51)(1/6.31)}$$
$$= 1.58 + 1.23 + 0.05 + 0.12 + 0.0060 + 0.071 + 0.63$$
$$= 3.12$$
$$= 4.95 \text{ dB}$$

A noise figure of 4.95 dB is equivalent to a 615 K noise temperature (from Equation 5.75). The results of Examples 5.21 and 5.22 are equivalent.

> Calculating the noise temperature or noise factor of a cascade by hand is tedious. If you will be performing the calculation frequently, it is best to find or write a computer program to perform the task.

In Practice

As with gain, the noise temperature and noise figure of an actual cascade are not always what Equations 5.125 and 5.128 dictate. The plot of the noise performance versus frequency of a real-world cascade will exhibit peaks and valleys just like the gain performance will. There are several reasons for this behavior.

- The terminal impedances and gain ripple of each amplifier in the cascade cause errors when we evaluate the cascade noise equations. Since the noise performance of the cascade is intimately connected to the cascade's gain, any error when evaluating the gain will result in error when evaluating the noise performance.

- The noise figure or noise temperature of a component is usually not constant with time, temperature and frequency. As the noise performance changes, the error in the calculated noise performance increases.

- The noise figure or noise temperature specification associated with an amplifier or mixer purchased from a vendor is valid only when the component has been terminated in its characteristic impedance on both its input and output. If the terminations are wrong, the component can exhibit widely different noise characteristics from what is specified by the manufacturer. Usually, but not always, the noise performance worsens as we move away from the characteristic impedance.

    Figure 5-32 shows two examples. In Figure 5-32(a), a noisy resistor connected to an amplifier through a band-pass filter. The amplifier has a power gain G_p and a noise figure of F_{dB}. Remember that the gain and noise figure were measured when the amplifier was terminated in a wideband Z_0.

    When we discussed filters, we found that the input and output impedance of a filter can change rapidly as we move away from the passband of the filter. This can radically alter the gain and noise characteristics of the amplifier at these frequencies. The amplifier can exhibit gain and noise "humps" caused by the non-Z_0 impedance present on its input. This effect is not too problematic because the system's behavior outside of the frequency band of interest is not relevant. It can be an obstacle, however, when we are performing frequency conversions using mixers and have an elevated noise level at the mixer's image frequency. Improper termination can also cause the amplifier to break into oscillation, which will distort any notion of the amplifier's gain or noise performance.

- Figure 5-32(b) shows three RF amplifiers in series. Since the input and output impedances of the amplifier can be different from the system's Z_0, each amplifier may not be terminated properly, and its noise performance may be compromised. Since the effect of Figure 5-32(b) can occur both in- and out-of-band, it can be more of a troublemaker than the effect illustrated in Figure 5-32(a).

All these effects combined produce the noise performance that will, on average, behave as Equations 5.125 and 5.128 dictate. However, like the gain, the noise performance of the cascade will have high spots (the cascade is noisier than expected) and low spots (the cascade is quieter than expected) on its noise versus frequency plot.

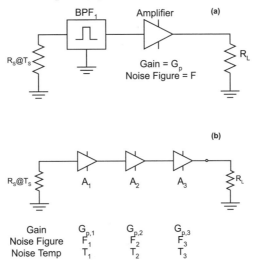

Figure 5-32 *Errors associated with the cascade noise equations.*

5.12 Examining the Cascade Equations

Equations 5.125 and 5.128 indicate the noise performance of a cascade when the characteristics of the pieces that make up the cascade are given. The two equations offer insight into the effects of the individual components on the cascade performance. Here we will only discuss Equation 5.125, which deals with noise temperature. The same conclusions apply to the noise figure Equation 5.128.

- It is easy to isolate the contribution each component makes to the noise performance of the cascade. For example, the only place where the noise temperature of amplifier A_3 appears in Equation 5.125 is the term

$$\frac{T_3}{G_{p,1} G_{p,2}} \qquad 5.130$$

which quantifies the contribution amplifier A_3 makes to the cascade's overall noise temperature.

- Since all of the terms in Equation 5.125 are positive, the noise temperature of the cascade will always be greater than T_1, the noise temperature of the first amplifier. T_1 is the minimum noise temperature of the cascade. For a low-noise cascade, we use an amplifier with a low noise temperature for the first amplifier (A_1).

- The amount of noise a particular amplifier contributes to the cascade depends on the amount of power gain preceding the amplifier. In Equation 5.125, the amount of noise that temperature A_3 adds to the cascade is diminished by the power gain of the first two amplifiers. This implies that if we want to build a system with a small noise figure, we should use as much gain as possible as early in the cascade as possible.

5.13 Minimum Detectable Signal (MDS)

Equation 5.54 indicates that when we connect a receiver whose noise temperature is T_{sys} to an antenna with a noise temperature of T_{ant}, the noise power present at the input of the receiver is

$$N_{in} = k\left(T_{ant} + T_{sys}\right) B_n \qquad 5.131$$

The equivalent noise power present on the input to a receiver sets a lower limit on the smallest signal the system can detect. The smallest signal we can detect is the *minimum detectable signal* (MDS). We arbitrarily assume a signal is detectable when the signal power equals the equivalent input noise power or

$$S_{MDS} = k\left(T_{ant} + T_{sys}\right) B_n \qquad 5.132$$

In other words, if we apply a signal to the system's input and the signal's power equals the system's MDS, then the signal-to-noise ratio will be unity or, equivalently, 0 dB. Note that, although the system may be able to detect a signal whose power is the MDS of the system, the system may not be able to process the signal. In other words, it may not produce a suitable bit error rate or output signal-to-noise ratio.

Input Noise (N_0)

If we assume that the antenna noise temperature is T_0, we can derive several equations that describe the MDS.

$$S_{MDS,T0} = F_{sys} k T_0 B_n \qquad 5.133$$

and

$$S_{MDS,T0,dBm} = F_{sys,dB} - 174 \frac{\text{dBm}}{\text{Hz}} + 10\log(B_n) \qquad 5.134$$

where
F_{sys} = the noise figure of the system,
T_{sys} = the noise temperature of the system,
B_n = the noise bandwidth.

5.14 Noise Temperature Measurement

After we have built an amplifier, we measure its noise performance. Normally, we perform this task with a noise figure meter. Figure 5-33 shows a rough schematic diagram of the HP-8970 noise figure meter.

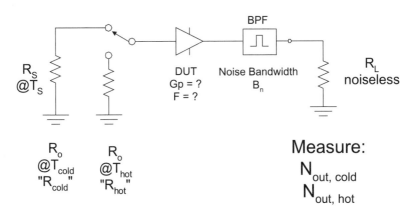

Figure 5-33 *Simplified block diagram of HP-8970 noise figure meter.*

At the left, we have two resistors, "R_{hot}" and "R_{cold}." Both have the value R_0, which is the characteristic impedance of the system. The two resistors are at different noise temperatures: one is at T_{hot} (usually about 10,000 K), and the other is at T_{cold} (usually about room temperature or 290 K). We know T_{hot} and T_{cold} with a high degree of accuracy. The noise figure meter alternately connects first the hot resistor to the device under test, then it connects the cold resistor to it. The output of the DUT goes to a band-pass filter that sets the noise bandwidth of the measurement. The noise bandwidth of the HP-8970 noise figure meter is about 4 MHz. (The actual bandwidth, however, does not matter because it cancels out in the equations. We have to keep in mind that this filter does not change during the measurement.)

Finally, we measure the noise energy coming out of the band-pass filter in a load resistor R_L. When we connect the hot resistor to the DUT, we will measure a noise power of $N_{out,hot}$. When we connect the cold resistor to the DUT, we will measure a quantity of noise power ($N_{out,cold}$). During the analysis, it will be convenient to form the ratio

$$Y = \frac{N_{out,hot}}{N_{out,cold}} \qquad 5.135$$

Noise Source

R_{hot} and R_{cold} used to be precision resistors at precisely controlled physical temperatures. Traditionally, the cold resistor was placed in a liquid nitrogen bath and the hot resistor was placed in either boiling water or ice water. Figure 5-34 shows the form the two-resistor noise source takes today. The noise figure meter alternately connects and disconnects a +28 VDC power supply to an avalanche noise diode. The avalanche diode generates a lot of the noise when the +28 VDC is applied. It generates very little noise when the DC power supply is disconnected.

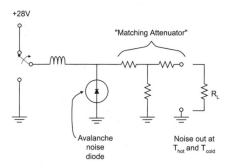

Figure 5-34 *Simplified block diagram of a noise generating pod. This device generates noise at well-calibrated temperatures.*

The three resistors in a T-formation on the right side of Figure 5-34 form a *matching attenuator*, or *matching pad*. The matching pad helps keep the input impedance of the noise generator constant when it is turned on and when it is turned off. The matching pad is also partly responsible for making sure T_{cold} is 290 K.

This method of generating noise is repeatable and accurate. It is less trouble than keeping a Dewar flask of liquid nitrogen around and boiling up some water every time we want to make a noise temperature measurement. The diode is reverse-biased into the avalanche region. Such noise sources have become very popular because of their small size and low power requirements. Recent design advances have produced noise sources that are stable with time, have a broad frequency range and a low reflection coefficient (i.e., they are well matched to Z_0).

Excess Noise Ratio (ENR)

The noise source is characterized by its *excess noise ratio* (ENR). The ENR is related to T_{hot} and T_{cold} by

$$ENR_{dB} = 10\log\left(\frac{T_{hot} - T_{cold}}{T_{cold}}\right) \qquad 5.136$$

We have shown the ENR data from one particular noise generator in Table 5-3 below. This particular set of data comes from a Hewlett Packard 346B noise source which covers a frequency range of 10 MHz to 18.0 GHz.

The excess noise ratios of well-matched devices are usually about 15 dB (i.e., $T_{hot} \approx 10,000$ K). Higher excess noise ratios are possible by giving up impedance match and flat frequency response. Noise generators are usually calibrated at the factory, and each noise generator is slightly different. The factory prints the data on the noise generator as frequency vs. ENR.

Table 5-3 — *ENR vs. frequency for one particular HP346B noise pod.*

Freq (GHz)	ENR (dB)	Freq (GHz)	ENR (dB)	Freq (GHz)	ENR (dB)
0.01	14.99	6.0	14.80	13.0	15.38
0.1	15.14	7.0	14.89	14.0	15.44
1.0	14.94	8.0	14.98	15.0	15.53
2.0	14.90	9.0	15.12	16.0	15.53
3.0	14.72	10.0	15.28	17.0	15.57
4.0	14.73	11.0	15.38	18.0	15.57
5.0	14.74	12.0	15.41	-	-

Example 5.23 — ENR, T_{cold} and T_{hot}
One particular noise pod has an ENR of 15.5 dB at 6.0 GHz. Assuming T_{cold} is 290 K, what noise temperature does this equal?

Solution—
Using Equation 5.136, we find

$$ENR_{dB} = 10\log\left(\frac{T_{hot} - T_{cold}}{T_{cold}}\right)$$
$$15.5 \text{ dB} = 10\log\left(\frac{T_{hot} - 290}{290}\right) \qquad 5.137$$
$$T_{hot} = 10{,}580 \text{ K}$$

Measuring Noise Temperature

We measure the noise temperature of an unknown unit by alternately applying the hot and cold noise power to the device. If we measure the output noise power in both cases, we have all the information we need to calculate the gain and noise temperature of the unknown device. When the input noise temperature is T_{hot} (see Figure 5-33), the noise power present on the output of the device under test is

$$N_{out,hot} = G_{p,sys} k (T_{hot} + T_{sys}) B_n \qquad 5.138$$

where
$G_{p,sys}$ = the power gain of the DUT,
T_{sys} = the noise temperature of the DUT.

When the input noise temperature is T_{cold}, the output noise power is

$$N_{out,cold} = G_{p,sys} k (T_{cold} + T_{sys}) B_n \qquad 5.139$$

If we take the ratio of Equations 5.138 and 5.139, we can isolate the noise temperature of the DUT.

$$Y = \frac{N_{out,hot}}{N_{out,cold}}$$
$$= \frac{G_{p,sys} k (T_{hot} + T_{sys}) B_n}{G_{p,sys} k (T_{cold} + T_{sys}) B_n} \qquad 5.140$$
$$= \frac{(T_{hot} + T_{sys})}{(T_{cold} + T_{sys})}$$

$$T_{sys} = \frac{T_{hot} - YT_{cold}}{Y-1}$$ 5.141

> **Example 5.24 — Measuring Noise Temperature**
>
> Radio astronomers use very large antennas coupled with very low-noise, cryogenically cooled amplifiers to observe astronomical events. Occasionally, astronomers like to verify the noise performance of their system. Because it is inconvenient and time consuming to disassemble the antenna, remove the amplifier and perform the test, another way was found.
>
> To perform the test, a calibrated source of hot noise and a calibrated source of cold noise is used. For the cold measurement, the astronomers simply point the antenna at an empty area of space. The temperature of this cold noise is about 6 K (after taking into account the atmosphere). A radio star supplies the hot noise. Radio stars emit large amounts of noise at a relatively constant temperature. Many such stars have been observed and calibrated. Since we can supply the system with two noise sources at different temperatures, we can verify the system's performance easily and accurately.

Measuring Gain

When we use a noise figure meter to measure the gain and noise figure of a device, we must first calibrate the meter. We perform the calibration before we make any measurements by connecting the output of the noise pod directly to the input of the noise figure meter. The noise figure meter then measures its own noise temperature and gain. After calibration, we attach the unknown device. The power gain of the DUT is

$$G_{p,sys} = \frac{N_{out,hot} - N_{out,cold}}{N_{out,hot,cal} - N_{out,cold,cal}}$$ 5.142

where

$N_{out,hot,cal}$ = the noise power measured by the noise figure meter when the input noise is at T_{hot} and the meter is in calibration mode (i.e., when the output of the noise pod is connected directly to the input of the noise figure meter).

$N_{out,cold,cal}$ = the noise power measured by the noise figure meter when the input noise is at T_{cold} and the meter is in calibration mode.

Note that we have made all the measurements without knowing the noise bandwidth of the noise figure meter. The only requirement is that B_n stay the same when we apply the cold noise, and when we apply the hot noise.

Example 5.25 — Noise Temperature Measurement

To measure the noise figure of a microwave receiver, we are using a noise source with an ENR of 15.3 dB at 4.5 GHz. The cold noise temperature of the noise source is T_0. When we present the receiver with the cold input noise, we measure −69.0 dBm of noise power on the receiver's output. When we feed the hot noise into the receiver, we measure −67.7 dBm. What is the receiver's gain and noise figure?

Solution —

The problem states $T_{cold} = T_0 = 290$ K. Using Equation 5.136, we can find T_{hot}.

$$ENR_{dB} = 10\log\left(\frac{T_{hot} - T_{cold}}{T_{cold}}\right)$$

$$15.3 \text{ dB} = 10\log\left(\frac{T_{hot} - 290}{290}\right) \quad\quad 5.143$$

$$T_{hot} = 10{,}120\,°K$$

Using Equation 5.135, the Y-factor for this noise source is

$$Y = \frac{N_{out,hot}}{N_{out,cold}}$$

$$= -67.7 \text{ dBm} - (-69 \text{ dBm})$$

$$= 1.3 \text{ dB} \quad\quad 5.144$$

$$= 1.35$$

Plugging these results into Equation 5.141 produces

$$T_{sys} = \frac{T_{hot} - YT_{cold}}{Y - 1}$$

$$= \frac{10{,}120 - (1.35)(290)}{1.35 - 1} \quad\quad 5.145$$

$$= 27{,}800\,K$$

The noise figure from Equation 5.73 is

$$F_{sys} = \frac{T_0 + T_{sys}}{T_0}$$

$$= \frac{290 + 27{,}800}{290} \quad\quad 5.146$$

$$= 96.9$$

$$= 19.9 \text{ dB}$$

Practical Effects

Noise figure meters can be tricky instruments. Unless we are careful, we can receive bad or misleading measurements. Some of the noise-figure measurement problems we have encountered over the years and their causes are listed here.

- Note that we have managed to measure both the gain and noise figure of the DUT without knowing the noise bandwidth of the measurement system. This is important from a practical point of view because it is difficult to mass-produce a filter with a precise noise bandwidth, especially over time and temperature. All that is required of the noise bandwidth is that it remain the same when we measure the hot noise as when we measure the cold noise.

- When we performed the derivations above, we assumed the gain of the device under test remained the same when we switched the input between the hot noise and the cold noise. This is not always a valid assumption. Many receivers, for example, incorporate an automatic gain control or AGC feature. The receiver varies its gain with varying input power levels. When measuring the noise figure of a receiver, the AGC has to be disabled to avoid unfavorable results. Leaving the AGC enabled when making noise figure readings will make the noise temperature appear worse than it really is.

- When performing a noise figure measurement, the noise power presented to the device under test is very small. In Example 5.26 above, the noise powers presented to the DUT are

$$N_{in,cold} = kT_{cold}B_n \quad N_{in,hot} = kT_{hot}B_n \qquad 5.147$$
$$= -104 \text{ dBm} \quad\quad = -88.5 \text{ dBm}$$

assuming a 10 MHz noise bandwidth. These are very small signal powers. Figure 5-35 shows a typical laboratory setup. The cable running between the noise figure meter and the noise pod carrier is either 28 V_{DC} or 0 V_{DC}. When the noise figure meter supplies 28 V_{DC} to the noise pod, the pod generates noise with a noise temperature of T_{hot}. When the noise figure meter supplies 0 V_{DC} to the noise pod, the pod produces noise with a temperature of T_{cold}.

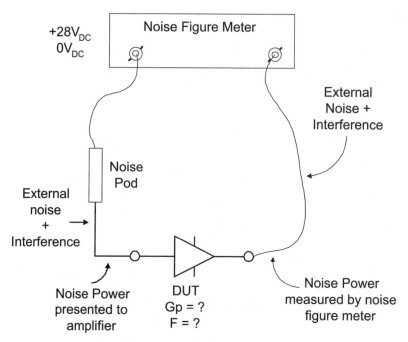

Figure 5-35 *Measuring the noise figure of a receiver with a wideband.*

Since two noise powers generated by the noise pod are at fairly small levels, the cable running between the noise pod and the DUT and the cable running between the DUT and the noise figure meter can cause problems. If the cables have poor shielding, they can allow external noise and interference into the system. Since the noise presented to the DUT and the noise measured by the noise figure meters are not what we expect them to be, we will receive incorrect results. In short, it is preferable to use double-shielded or even hard-line cable when you are making noise figure measurements.

When measuring a system that performs a frequency conversion, another problem can arise. Figure 5-36(a) shows a setup of a system we might use to measure the noise figure of a receiver. Figure 5-35(b) is a schematic representation of the receiver we are testing.

The tow oscillators might present a problem. If significant energy from either of these two oscillators is present on the output port of the receiver (and the frequency of the oscillator is right), the noise figure meter will measure the power in the oscillator. The meter will assume that the energy is due to the noise power presented to the receiver by the noise pod. Consequently, results will be poor, and the noise temperature of the DUT will look worse than it really is. To solve this problem, we place a band-pass filter between the receiver's output and the input port of the noise figure meter.

AMPLIFIERS AND NOISE | 517

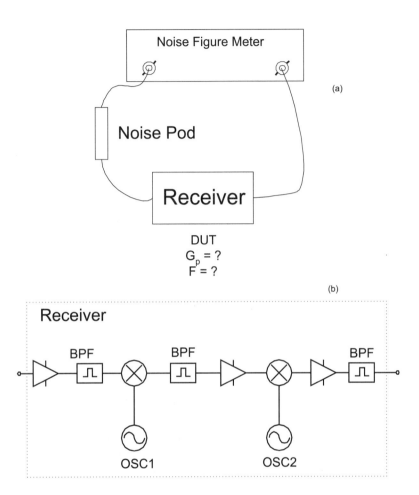

Figure 5-36 *Measuring the noise figure of a device that performs a frequency conversion.*

Figure 5-37 shows a block diagram of a receiver. We would like to measure its noise figure but, due to mechanical constraints, we can only connect to the receiver at point A. When we connect the noise figure meter, we find the noise figure is much higher than we expect it to be based upon the calculations. Since sensitivity measurements agree reasonably well with noise figure calculations, the problem must arise from the noise figure measurement.

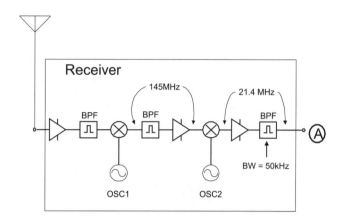

Figure 5-37 *Measuring the noise figure of a receiver with a wideband output device.*

The problem lies in the last two components. The last amplifier is a wideband unit (10 to 500 MHz) and has a fairly high noise figure (approximately 10 dB). As a consequence, the amplifier driving point A has a high level of noise present at its output at all times. Even if there is no input signal, the amplifier will provide

$$G_p k T_{amp} B_n \text{ watts} \qquad 5.148$$

to the outside world.

The band-pass filter preceding the amplifier is only 50 kHz wide. Any noise power generated by the noise pod of the noise figure meter has to pass through this 50 kHz filter to be measured. Most noise figure meters have about a 4 MHz noise bandwidth. The noise figure meter simply measures the noise power present when the noise pod is on and when it is off. It compares the two power readings to measure the gain and noise figure of the DUT.

When the noise pod is turned off, the noise figure meter measures the noise coming through the 50 kHz filter combined with the wideband noise generated by the amplifier. When the noise pod is turned off, the noise figure meter measures the noise passing through the 50 kHz filter (which has now changed because the noise pod is generating more noise power). The noise figure meter also measures the wideband noise generated by the amplifier. This noise never changes, because it was not generated in the noise pod. The noise figure meter sees essentially the same amount of noise power when the noise pod is off as when it is on. If you care to unfold the math of Equations 5.135 and 5.141, you will see that the noise figure meter interprets this condition as if it is measuring a device with a high noise figure.

The solution was to place a 10 kHz band-pass filter between point A and the noise figure meter. This eliminated all of the wideband noise from the final amplifier. Any noise measured by the noise figure meter is due to the noise that the noise pod presents to the input of the receiver.

Example 5.26 — Noise Figure Measurement Error

Use Equations 5.135 and 5.141 to show that if $N_{out,hot} \cong N_{out,cold}$, the noise temperature of the DUT is large.

Solution —

When $N_{out,hot} \cong N_{out,cold}$, Equation 5.135 produces

$$Y = \frac{N_{out,hot}}{N_{out,cold}}$$
$$\approx \frac{1}{1}$$
$$\approx 1$$

5.149

In reality, Y will be slightly greater than 1. When Y is slightly greater than 1, Equation 5.141 shows

$$T_{sys} = \frac{T_{hot} - YT_{cold}}{Y - 1}$$
$$\approx \frac{T_{hot} - YT_{cold}}{\text{Small Number}}$$
$$= \text{Big Number}$$

5.150

All of these measurement anomalies tend to make the DUT look noisier than it actually is.

5.15 Lossy Devices

Not all of the devices we use to design radio systems provide power gain; some cause power loss. In this section, we will look at the gain and noise characteristics of these lossy devices.

Gain Performance of Lossy Devices

The terms *gain* and *loss* are frequently used when discussing signals and systems. When a device is specified as having a 6 dB gain, it means the device will accept a signal, multiply its power by 4 times (or add 6 dB) and present that power to the outside world. If the input signal power is P_{in}, then the output signal power is

$$P_{out} = 4P_{in} \qquad 5.151$$

or, in decibels,

$$P_{out,dBm} = P_{in,dBm} + 6 \text{ dB} \qquad 5.152$$

A 6 dB loss means that the device accepts a signal, divides the signal power by 4 (or subtracts 6 dB from the input signal power), then presents that power to the outside world. Let the input signal power be P_{in}. The output signal power will be

$$P_{out} = \frac{P_{in}}{4} \qquad 5.153$$

or

$$P_{out,dBm} = P_{in,dBm} - 6 \text{ dB} \qquad 5.154$$

A loss of x dB is equivalent to a gain of –x dB. Mathematically, gains and losses are treated the same.

Example 5.27 — Gains and Losses in Cascades

Figure 5-38 shows a cascade made up of blocks with gain and losses. Find the cascade gain.

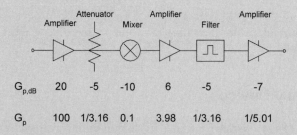

Figure 5-38 Example 5.27. A cascade with gains and losses.

Solution—

The cascade gain is

$$\begin{aligned} G_{p,cas,dB} &= 20 + (-5) + (-10) + 6 + (-5) + (-7) \\ &= -1 \text{ dB} \end{aligned} \qquad 5.155$$

The cascade has a gain of –1 dB or a loss of 1 dB. We can find the cascade gain using only the linear terms.

$$G_{p,cas} = (100)\left(\frac{1}{3.16}\right)\left(\frac{1}{10}\right)(3.98)\left(\frac{1}{3.16}\right)\left(\frac{1}{5.01}\right) \qquad 5.156$$

$$= 0.795$$

$$= -1.0 \text{ dB}$$

Noise Performance of Lossy Devices

When designing systems, we occasionally find it is necessary to introduce loss into a system at a particular point. One of the most commonly used devices for intentionally introducing loss is a resistive attenuator. Attenuators are often referred to as *pads*. A 6 dB attenuator is a 6 dB pad.

Characteristic Impedance

Attenuators are designed in view of the characteristic impedance of their operating system. Figure 5-39 shows two 6 dB attenuators, one for a 50-ohm system and one for a 300-ohm system, with differing resistor values.

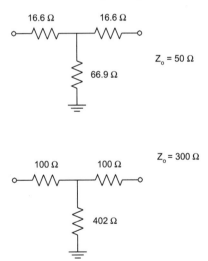

Figure 5-39 *Two 6 dB attenuators. The top attenuator is designed for a 50-ohm system. The bottom attenuator is designed for a 300-ohm system.*

The first design parameter of an attenuator is its characteristic impedance. If we terminate one end of the attenuator in its characteristic impedance and look into the other end, we will still see the characteristic impedance. Figure 5-40 shows the 300-ohm pad terminated in 300 ohm. The input impedance Z_{in} is

$$Z_{in} = 100 + \frac{(402)(400)}{402 + 400}$$
$$= 100 + 200$$
$$= 300 \, \Omega$$

5.157

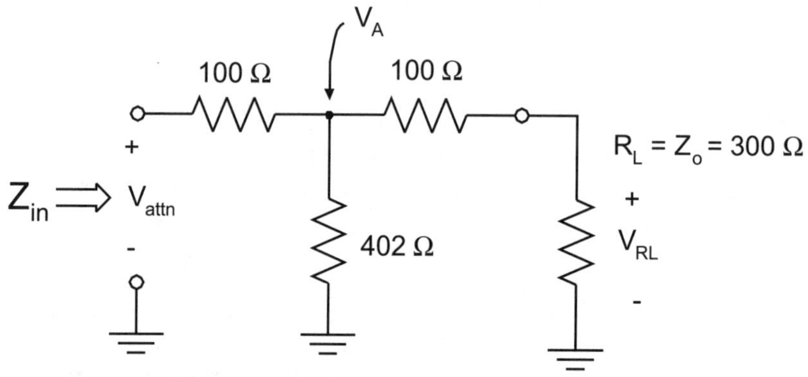

Figure 5-40 *Calculating the input impedance of an attenuator. The attenuator must be terminated in its characteristic impedance.*

Note that we will see the characteristic impedance only when the attenuator is terminated in its characteristic impedance. If we replaced the 300-ohm resistor with an open circuit, the input impedance would be 502 ohms. If we replaced the 300-ohm resistor with a short circuit, the input impedance would be 180 ohms.

Attenuation

The second characteristic of an attenuator is the designed-in attenuation, or loss value. In other words, we want the signal to suffer a precise, designed-in amount of loss as it passes through the attenuator. For example, we apply a sine wave to the attenuator in Figure 5-41. V_{Attn} is 0.548 V_{RMS} (which will dissipate 1 mW in 300 ohm). When the attenuator is terminated in 300 ohm, the input impedance of the attenuator is also 300 ohms. The power delivered to the attenuator is

$$P_{Attn} = \frac{V_{Attn,RMS}^2}{Z_0}$$
$$= \frac{(0.548)^2}{300}$$
$$= 1 \, \text{mW}$$
$$= 0 \, \text{dBm}$$

5.158

There are many ways to find $V_{RL,RMS}$. Here we will use voltage division. The voltage V_A (Figure 5-40) is

$$V_A = (0.548)\frac{200}{300}$$
$$= 0.365\ V_{RMS}$$
 5.159

The voltage across R_L is V_{RL} and

$$V_{RL} = (0.365)\frac{300}{400}$$
$$= 0.274\ V_{RMS}$$
 5.160

Finally, the power delivered to R_L is

$$P_{RL} = \frac{V_{RL,RMS}^2}{R_L}$$
$$= \frac{(0.274)^2}{300}$$
$$= 0.250\ mW$$
 5.161

The power gain of the attenuator is

$$G_p = \frac{P_{in}}{P_{out}}$$
$$= \frac{0.250}{1.00}$$
$$= 0.25$$
$$= -6\ dB$$
 5.162

The power gain of the attenuator is -6 dB. Equivalently, we can say that the power loss of the attenuator is 6 dB. In short, we design an attenuator to operate in some characteristic impedance. When we terminate the attenuator in its characteristic impedance Z_0, the input impedance of the attenuator will be Z_0 and the attenuator will reduce the signal power by the amount specified. When the attenuator is not terminated in its characteristic impedance on both the source and the load sides, the input and output impedances as well as the loss value of the attenuator will change.

System without Attenuator

Figure 5-41(a) shows a signal source (V_S in series with R_S) connected directly to a load resistor R_L. The system is matched, i.e., $R_S = R_L$.

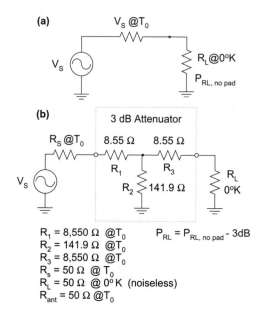

Figure 5-41 A system with and without a 3 dB attenuator in the signal path.

We will assume that the noise temperature of R_S is T_0 and that we measure a signal power of $P_{RL,\text{ no pad}}$ in R_L. Since R_S is at T_0 and the system is matched, R_L will dissipate a noise power of N_0 where

$$P_{\text{noise,RL,NoPad}} = N_0 \qquad 5.163$$
$$= kT_0 B_n$$

If we assume a 1-hertz noise bandwidth, then

$$P_{\text{noise,RL,NoPad}} = kT_0 B_n$$
$$= (1.38 \cdot 10^{-23})(290)(1) \qquad 5.164$$
$$= 4 \cdot 10^{-21} \text{ W}$$

System with Attenuator

Figure 5-41(b) shows the same system with a resistive attenuator between the signal source and the load resistor. By design, this attenuator has a power loss of 3 dB or a power gain of –3 dB. We are interested in the total noise power delivered to the load resistor when the attenuator is in place.

The noise power generated by the source resistor R_S is attenuated by 3 dB before it reaches the load resistor. However, the resistors that make up the attenuator will generate noise of their own since they are at some nonzero physical temperature. The noise generated by R_1, R_2 and R_3 will contribute to the total noise power dissipated in the load resistor.

To make the mathematics a easier, we will assume that the noise temperature of all the resistors, except R_L, is T_0 (290 K). We have to keep in mind that the noise temperature and physical temperatures of a resistor can be different, although there can be a strong relationship between the two. R_L is noiseless since its noise temperature is 0 K. We will assume that the characteristic impedance of the system is 50 ohms, so $R_S = R_L = 50$ ohms. We will use a noise bandwidth of 1 hertz.

First we find the values of the noise voltage sources in series with each resistor in the attenuator of Figure 5-42(b). Repeated application of Equation 5.4 produces the following table.

Table 5-4 Noise voltages generated by resistors in Figure 5-41.

Resistor Value (ohms)	$V_{n,oc,RMS}$ (V_{RMS})
$R_S = 50\ \Omega$	895 pV_{RMS}
$R_1 = 8.55\ \Omega$	370 pV_{RMS}
$R_2 = 141.9\ \Omega$	1507 pV_{RMS}
$R_3 = 8.55\ \Omega$	370 pV_{RMS}

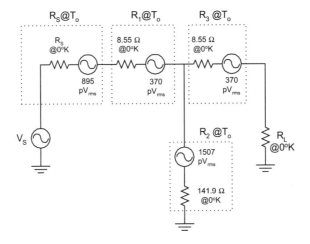

Figure 5-42 The system of Figure 5-41(b). The noise-producing resistors have been replaced by their equivalent circuits — a noise generator in series with a noiseless resistor.

Figure 5-42 shows the noise-equivalent circuit of Figure 5-41(b) with noiseless resistors and noise voltage generators. The noise generators in each resistor are uncorrelated, which means that we have to examine the noise power delivered to the load resistor by each noise generator separately and add up all the noise powers to produce the final result.

First, let us look at the effect of the noise generator in R_s. Figure 5-43 shows the circuit from Figure 5-42 with all the voltage sources removed except for the 895 pV_{rms} generator present in R_{st}. Analysis of the loop currents i_1 and i_2, as shown in Figure 5-43, produces

$$895 \text{ pV} = (200.5)i_1 + (-141.9)i_2 \qquad 5.165$$
$$0 = (-141.9)i_1 + (200.5)i_2$$

We can rewrite these equations in matrix format as

$$\begin{bmatrix} 895\text{p} \\ 0 \end{bmatrix} = \begin{bmatrix} 200.5 & -141.9 \\ -141.9 & 200.5 \end{bmatrix} \begin{bmatrix} i_1 \\ i_2 \end{bmatrix}$$

$$\begin{bmatrix} i_1 \\ i_2 \end{bmatrix} = \begin{bmatrix} 8.943 \text{ pA}_{\text{RMS}} \\ 6.33 \text{ pA}_{\text{RMS}} \end{bmatrix} \qquad 5.166$$

The noise power delivered to R_L by the noise generator present in R_s is

$$P_{\text{Noise in } R_L \text{ due to } R_S} = i_{2,RMS}^2 R_L$$
$$= (6.33 \cdot 10^{-12})^2 (50) \qquad 5.167$$
$$= 2.00 \cdot 10^{-21} \text{ W}$$

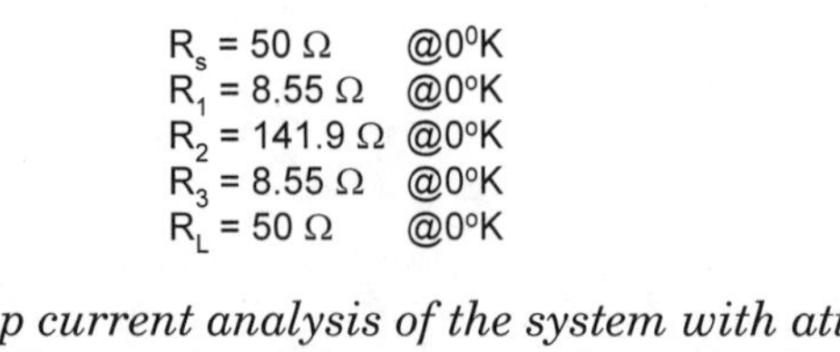

$R_s = 50 \ \Omega$ @0°K
$R_1 = 8.55 \ \Omega$ @0°K
$R_2 = 141.9 \ \Omega$ @0°K
$R_3 = 8.55 \ \Omega$ @0°K
$R_L = 50 \ \Omega$ @0°K

Figure 5-43 *Loop current analysis of the system with attenuator shown in Figure 5-41 (b). This analysis determines the noise delivered to the load resistor by R_s only.*

When the 3 dB attenuator is in the system, the noise power delivered to the load resistor is $2.00 \cdot 10^{-21}$ watts. When the 3 dB attenuator was not in the system, Equation 5.164 showed that the noise power delivered to the load resistor was $4 \cdot 10^{-21}$ watts. The 3 dB attenuator reduces the amount of power which reaches the load resistor by 1/2 or 3 dB. However, the noise power generated by R_1, R_2 and R_3 has yet to be accounted for. Figure 5-45 shows the circuit topology we will use to analyze the noise effect of R_1.

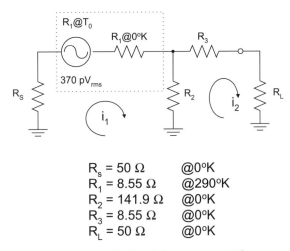

$R_s = 50\ \Omega$ @0°K
$R_1 = 8.55\ \Omega$ @290°K
$R_2 = 141.9\ \Omega$ @0°K
$R_3 = 8.55\ \Omega$ @0°K
$R_L = 50\ \Omega$ @0°K

Figure 5-44 *Loop current analysis of the system with attenuator shown in Figure 5-41 (b). This analysis determines the noise delivered to the load resistor by R_1 only.*

The current-loop matrix is

$$\begin{bmatrix} 370p \\ 0 \end{bmatrix} = \begin{bmatrix} 200.5 & -141.9 \\ -141.9 & 200.5 \end{bmatrix} \begin{bmatrix} i_1 \\ i_2 \end{bmatrix} \qquad 5.168$$

$$\begin{bmatrix} i_1 \\ i_2 \end{bmatrix} = \begin{bmatrix} 3.697\ \text{pA}_{RMS} \\ 2.62\ \text{pA}_{RMS} \end{bmatrix}$$

The noise present in R_L due to the noise generator in R_1 is

$$\begin{aligned} P_{\text{Noise in RL due to R1}} &= i_{2,RMS}^2 R_L \\ &= \left(2.62 \cdot 10^{-12}\right)^2 (50) \\ &= 343 \cdot 10^{-24}\ \text{watts} \end{aligned} \qquad 5.169$$

Applying the same analysis techniques to R_2 and R_3, we find

$$P_{\text{Noise in RL due to R2}} = i_{2,RMS}^2 R_L$$
$$= (4.40 \cdot 10^{-12})^2 (50) \qquad 5.170$$
$$= 969 \cdot 10^{-24} \text{ watts}$$

and

$$P_{\text{Noise in RL due to R3}} = i_{2,RMS}^2 R_L$$
$$= (3.70 \cdot 10^{-12})^2 (50) \qquad 5.171$$
$$= 684.5 \cdot 10^{-24} \text{ watts}$$

The total noise power delivered to the load resistor is the sum of the noise powers delivered to R_L by R_s, R_1, R_2 and R_3, and we can write

$$P_{\text{Total Noise in R}_L} = P_{\text{Noise from R}_S} + P_{\text{Noise from R}_1} + P_{\text{Noise from R}_2} + P_{\text{Noise from R}_3}$$
$$= (200 \cdot 10^{-21}) + (343 \cdot 10^{-24}) + (969 \cdot 10^{-24}) + (684 \cdot 10^{-24}) \qquad 5.172$$
$$= 4.00 \cdot 10^{-21} \text{ W}$$

The total noise power dissipated in R_L due to the noise generated by R_S, R_1, R_2 and R_3, is 4.00·100⁻²³ watts. This is the same noise power delivered to the load resistor of Figure 5-41(a) when the attenuator was not in the system. The 3 dB attenuator of Figure 5-41(b) attenuates the noise power generated by R_s by 3 dB. However, the resistors that make up the attenuator add just enough noise so that the noise power dissipated in R_L is back up to kT_0B_n. The attenuator does reduce the noise power but the thermal noise generated by the lossy elements of the attenuator returns the same amount of noise into the system.

Example 5.28 — Noise Factor of Attenuators at T_0

Figure 5-45 shows a system with a 3 dB attenuator in place between the source and load. The physical temperature of the attenuator is T_0. Find
a. the noise figure of the attenuator.
b. the noise temperature of the attenuator.

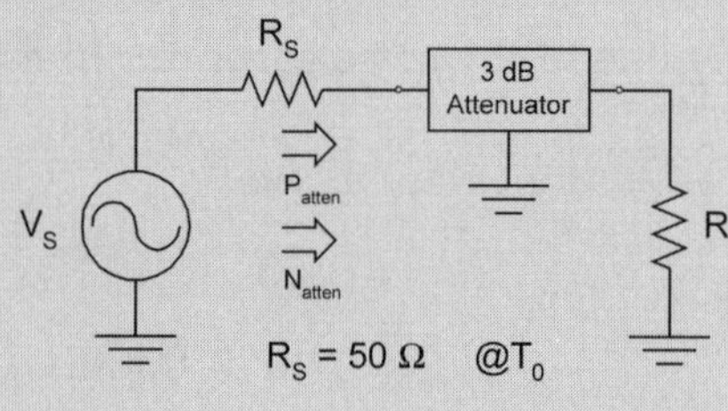

Figure 5-45 *Example 5.28. The noise factor of a resistive attenuator.*

Solution —
a. Let us assume the attenuator absorbs 1 mW of signal power (i.e., P_{ATTN} = 1 mW = 0 dBm). We have just shown that the source resistor will deliver a noise power of $N_0 = kT_0B_n$ = 4.00E-21 W = −174 dBm to a matched load, in a 1-hertz noise bandwidth. Since the attenuator presents a matched load to the signal source, we know N_{ATTN} = −174 dBm. The input signal-to-noise ratio is

$$\left(\frac{S}{N}\right)_{in} = 0 - (-174)$$
$$= 174 \text{ dB} \quad \quad 5.173$$

On the output side, the attenuator reduces the signal power by 3 dB so the signal power delivered to R_L is −3 dBm. We have just shown that a 3 dB attenuator at room temperature will deliver N_0 = −174 dBm of noise power to its load resistor. The output signal-to-noise ratio is

$$\left(\frac{S}{N}\right)_{out} = -3 - (-174)$$
$$= 171 \text{ dB} \quad \quad 5.174$$

Equation 5.94 shows that if the input noise to a system is N_0, the noise figure of a device is the ratio of the input SNR to the output SNR. The noise figure of the attenuator is

$$F = \frac{(S/N)_{in}}{(S/N)_{out}}$$
$$= 174 \text{ dBm} - 171 \text{ dBm} \quad \quad 5.175$$
$$= 3 \text{ dB}$$

The noise figure of the T_0 3 dB attenuator is 3 dB. This is always true for a resistive attenuator if the noise temperature of the attenuator is T_0.

b. Equation 5.75 expresses the relationship between noise factor and noise temperature. The noise figure of the attenuator is 3 dB, its noise factor is 2, and its noise temperature is

$$T_{amp} = T_0(F_{amp} - 1)$$
$$= 290(2 - 1) \quad \quad 5.176$$
$$= 290 \text{ } K$$

Noise Figures of Lossy Devices

Example 5.29 illustrates that if the physical temperature of a resistive attenuator is T_0, the noise figure of the attenuator (in decibels) equals the loss of the attenuator (in decibels), or

$$F_{\text{Attenuator,dB}} = \text{Attenuator's Loss}_{\text{dB}} \qquad 5.177$$

When the attenuator's physical temperature is T_0

assuming that the noise temperature of the attenuator equals the physical temperature.

Example 5.29 — Lossy Devices in Cascade

Figure 5-46(a) shows a resistive attenuator at T_0 followed by an amplifier. Show that the noise figure of the cascade is

$$F_{sys,dB} = F_{Attn,dB} + F_{Amp,dB} \qquad 5.178$$

when the noise figure of the attenuator equals the attenuator's loss or

$$F_{Attn} = \frac{1}{G_{p,Attn}} \qquad 5.179$$

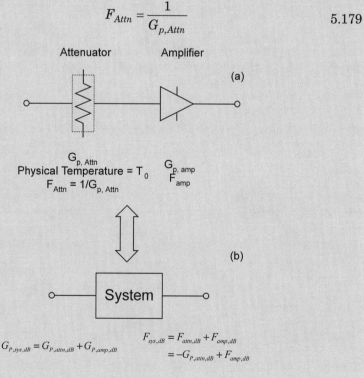

Figure 5-46 Example 5.29. Lossy devices in cascade.

AMPLIFIERS AND NOISE | 531

Solution —
The noise factor equation for a two-element cascade is

$$F_{sys} = F_1 + \frac{F_2 - 1}{G_{p,1}} \qquad 5.180$$

where F_{sys}, F_1, F_2 and $G_{p,1}$ are all in linear terms. The noise figure of an attenuator whose physical temperature is T_0 is

$$F_{Attn} = \frac{1}{G_{p,Attn}} \quad \text{or} \quad F_{Attn,dB} = -G_{p,Attn,dB} \qquad 5.181$$

Combining the two equations produces

$$\begin{aligned} F_{cas} &= F_{Attn} + (F_{amp} - 1)F_{Attn} \\ &= F_{Attn} + F_{Attn}F_{amp} - F_{Attn} \\ &= F_{Attn}F_{amp} \end{aligned} \qquad 5.182$$

Converting to decibels,

$$\begin{aligned} F_{cas,dB} &= 10\log(F_{attn}F_{amp}) \\ &= F_{attn,dB} + F_{amp,dB} \end{aligned} \qquad 5.183$$

Example 5.30 — Lossy Devices in Cascade II

Find the noise figure of the two-element cascade for the following conditions:
a. a 10 dB attenuator followed by an amplifier with a 4 dB noise figure,
b. a 3 dB attenuator followed by an amplifier with a 6 dB noise figure,
c. a 4 dB attenuator, a 6 dB attenuator and an amplifier with a 7 dB noise figure.

Solution—
Using Equation 5.183, we find
a. $F_{cas,dB} = 10 + 4 = 14$ dB,
b. $F_{cas,dB} = 3 + 6 = 9$ dB,
c. $F_{cas,dB} = 4 + 6 + 7 = 17$ dB.

Example 5.31 — Noise Factor of Attenuators at 0 K

Figure 5-47 shows a system with a 3 dB attenuator in place between the source and load. We have set the physical temperature of the attenuator to 0 K. Find

a. the noise figure of the attenuator,
b. the noise temperature of the attenuator.

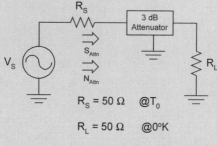

Figure 5-47 *Example 5.31. The noise factor of a resistive attenuator with a noise temperature of 0 K.*

Solution —
a. Let us assume the signal power delivered to the attenuator (S_{ATTN}) is 1 mW = 0 dBm. Assuming a 1-hertz noise bandwidth, we know the noise power delivered to the attenuator is $N_{ATTN} = N_0 = kT_0B_n = 4.00 \cdot 10^{-21}$ W = 0174 dBm. The input signal-to-noise ratio is

$$\left(\frac{S}{N}\right)_{in} = 0 - (-174)$$
$$= 174 \text{ dB}$$
5. 184

At the output, the signal power delivered to the load resistor is –3 dBm. However, since the attenuator is now at 0 K, it will not generate any noise of its own. The noiseless attenuator will decrease the input noise power as much as the input signal power. Consequently, the load resistor will dissipate 174 –3 = –177 dBm of noise power. The output signal-to-noise ratio is

$$\left(\frac{S}{N}\right)_{out} = -3 - (-177)$$
$$= 174 \text{ dB}$$
5.185

Using Equation 5.94, the noise figure of the cold attenuator is

$$F = \frac{(S/N)_{in}}{(S/N)_{out}}$$
$$= 174 \text{ dBm} - 174 \text{ dBm}$$
$$= 0 \text{ dB}$$
5.186

The noise figure of the 0 K 3 dB attenuator is 0 dB. When a resistive attenuator is at a physical temperature other than T_0, the noise figure does not necessarily equal the attenuator's power loss.

b. Using Equation 5.75, we find the noise temperature of the 0 K attenuator is

$$T_{amp} = T_0(F_{amp} - 1)$$
$$= 290(1-1)$$
$$= 0\,°K$$
       5.187

These examples illustrate that the noise figure of a resistive attenuator depends upon its physical temperature. When the attenuator is at room temperature, the noise figure of the attenuator equals the loss of the attenuator. When the attenuator is at absolute zero (0 K), the noise figure of the attenuator is 0 dB. The noise temperature of a lossy, resistive attenuator is related to its physical temperature by

$$T_{Attn,Noise} = \frac{T_{Attn,Physical}(1 - G_{p,Attn})}{G_{p,Attn}}$$
       5.188

where
  $T_{Attn,Noise}$ = the noise temperature of the attenuator in K,
  $T_{Attn,Physical}$ = the physical temperature of the attenuator in K,
  $G_{p,Attn}$ = the power gain of the attenuator in linear terms.

This relationship can be verified experimentally. We used an HP-8970 noise figure meter to measure the noise figure of a 6 dB attenuator at various temperatures and generated the following table:

Table 5-5 *Measured noise figure of an attenuator at various physical temperatures.*

Physical Temperature (°C)	Measured Noise Figure (dB)	Measured Noise Temperature (K)	Power Gain (dB)
-31.6	5.41	718	-6.05
-21.0	5.52	744	-6.05
+26.4	6.20	919	-6.05
+52.7	6.60	1036	-6.05
+78.0	6.75	1082	-6.05

Applying Equation 5.188 to Table 5-5 above produces

Table 5-6 *Measured and calculated noise figures of a resistive attenuator at various physical temperatures.*

Physical Temperature (°C)	Measured Noise Figure (dB)	Measured Noise Temperature (K)	Power Gain (dB)
-31.6	5.41	718	-6.05
-21.0	5.52	744	-6.05
+26.4	6.20	919	-6.05
+52.7	6.60	1036	-6.05
+78.0	6.75	1082	-6.05

Measured results agree well with theoretical results.

Example 5.32 — Noise Temperature of a Resistive Attenuator

Find the noise temperature and noise figure of a 10 dB attenuator whose physical temperature is
a. 300°C,
b. –100°C.

Solution —
The power gain of a 10 dB attenuator is –10 dB = 1/10 = 0.1. Using Equation 5.188, we find
a. 300°C = 573 K, so the noise temperature of the 10 dB pad is

$$T_{Attn,Noise} = \frac{T_{Attn,Physical}(1-G_{p,Attn})}{G_{p,Attn}} \quad \quad 5.189$$

$$= \frac{573(1-0.1)}{0.1}$$

$$= 5160 \ K$$

Equation 5.73 yields the noise figure.

$$F = \frac{T_0 + T_{amp}}{T_0}$$

$$= \frac{290 + 5160}{290} \quad \quad 5.190$$

$$= 18.8$$

$$= 12.7 \ dB$$

Note that an attenuator warmer than room temperature has a noise figure that is higher than its loss.

b. $-100°C = 173$ K. The noise temperature is

$$T_{Attn,Noise} = \frac{T_{Attn,Physical}(1 - G_{p,Attn})}{G_{p,Attn}} \qquad 5.191$$

$$= \frac{173(1 - 0.1)}{0.1}$$

$$= 1560\ K$$

Equation 5.73 provides the noise figure.

$$F = \frac{T_0 + T_{amp}}{T_0}$$

$$= \frac{290 + 1560}{290} \qquad 5.192$$

$$= 6.4$$

$$= 8.0\ \text{dB}$$

An attenuator cooler than room temperature has a noise figure that is lower than its loss.

In summary, the noise power added by the attenuator depends entirely upon the noise temperature of the attenuator. The noise temperature is almost directly related to the attenuator's physical temperature. For example, if we were to repeat the analysis of Figure 5-42 assuming the physical temperature of R_1, R_2 and R_3 was different from T_0, we would find a different amount of noise power is dissipated in R_L.

There are many topologies for resistive attenuators, including pi-, T-, bridged-T, balanced and distributed attenuators. Although we performed an analysis on only one topology, the analysis holds for any type of resistive attenuator and is also valid for lossy filters (i.e., filters with insertion loss) as long as they are matched to the system's characteristic impedance.

Signal Loss without Attenuators

Attenuators are used most frequently to produce a signal loss in a system. Other ways of introducing loss into a system include methods such as introducing an intentional mismatch into a system, reactive power splitting and losses via signal spreading causing path loss. Since these alternative do not involve resistive losses, they usually do not possess the noise penalties associated with resistive lossy attenuators.

5.16 Amplifier and Noise Summary Data

Open-Circuit Noise Voltage Across a Resistor

The open-circuit RMS noise voltage present across the terminals of a resistor is

$$V_{n,oc,RMS} = \sqrt{4kTB_n R}$$

where
$V_{n,oc,rms}$ = the open-circuit RMS noise voltage produced by the resistor,
k = Boltzmann's constant = $1.381 \cdot 10^{-23}$ joules/K (or $1.381 \cdot 10^{-23}$ watt-sec/K),
T = temperature in K.

Short-Circuit Noise Current Through a Resistor

The short-circuit RMS current present through a resistor is

$$I_{n,sc,RMS} = \sqrt{\frac{4kTB_n}{R}}$$

where
$I_{n,sc,rms}$ = the short-circuit RMS noise current produced by the resistor.

Combining Uncorrelated Noise Sources

When combining separate, uncorrelated noise sources, their powers must be added together. Adding their RMS voltages together, will produce the wrong number. One method of combining independent noise sources together is to remove all but one noise source, then calculate the power delivered to the load by the one source. Next, we remove everything but a second noise source, repeat the process and continue calculating the effect of single noise sources until you have run through them all. Finally, we add the powers.

Thermal Noise Delivered to a Receiver

The noise power delivered by a noisy resistor to a matched, noiseless resistor is

$$N = kT_s B_n$$

When the noisy resistor represents the noise available from an antenna, this equation represents the noise delivered by the antenna to a receiver. This noise power does not include the internal noise generated by the receiver.

The noise power delivered by a noisy resistor (at T_S) to a noisy load resistor (at T_L) is

$$N = k(T_S + T_L)B_n$$

where
N = the total noise power dissipated in the load resistor R_L by the source resistor R_S and by the load resistor R_L,
T_S = the noise temperature of the source,
T_L = the noise temperature of the load,
B_n = the noise bandwidth of the measurement.

If the source is an antenna whose noise temperature is T_{ant} and the load is a receiver whose noise temperature is T_{sys}, noise delivered to the system (or the system's input noise floor) is

$$N_{Floor} = k(T_{ant} + T_{sys})B_n$$

Expressed in dBm,

$$N_{Floor,dBm} = 10\log(T_{ant} + T_{sys}) + 10\log(B_n) - 198.6$$

N_0

It is commonly assumed that the noise temperature of an antenna is T_0 or 290 K. When this is true, the antenna delivers the noise power N_0 to a matched load where

$$\begin{aligned}N_{0,Watts} &= kT_0 B_n \\ &= (1.38 \cdot 10^{-23})(290)B_n \\ &= (4.00 \cdot 10^{-21})B_n\end{aligned}$$

We note

$$\begin{aligned}kT_0 &= (1.38 \cdot 10^{-23})(290) \\ &= 4 \cdot 10^{-21} \frac{\text{Watts}}{\text{Hz}} \\ &= -174 \frac{\text{dBm}}{\text{Hz}}\end{aligned}$$

Thus,

$$N_{0,dBm} = 10\log(B_n) - 174$$

Using Noise Figure

Noise figure, like noise temperature, accurately describes the noise behavior of a receiving system. The relationship between system noise figure and system noise temperature is

$$F_{sys} = \frac{T_0 + T_{sys}}{T_0} \quad F_{sys} \geq 1$$

and

$$T_{sys} = T_0(F_{sys} - 1)$$

If the antenna noise temperature is T_0, the receiver's noise floor is

$$N_{Floor} = k(T_0 + T_{sys})B_n$$

We can write

$$\left.\begin{array}{l} N_{Floor} = F_{sys}kT_0 B_n \\ N_{Floor,dBm} = F_{sys,dB} + 10\log(B_n) - 174 \end{array}\right\} \text{when antenna temperature is } T_0$$

Minimum Detectable Signal (MDS)

The criteria for minimum detectable signals is that the MDS signal power equals the input noise power into the receiver. When the antenna noise temperature is T_{ant}, the MDS is

$$S_{MDS} = k(T_{sys} + T_{ant})B_n$$

When the antenna noise temperature is T_0, the MDS is

$$S_{MDS} = k(T_{sys} + T_0)B_n$$

$$S_{MDS,dBm} = F_{sys,dB} - 174\frac{\text{dBm}}{\text{Hz}} + 10\log(B_n)$$

Cascade Equations

The cascade equations for noise temperature and noise figure are

$$T_{cas} = T_1 + \frac{T_2}{G_{p,1}} + \frac{T_3}{G_{p,1}G_{p,2}} + \ldots + \frac{T_n}{G_{p,1}G_{p,2}G_{p,3}\ldots G_{p,n-1}}$$

$$F_{cas} = F_1 + \frac{F_2 - 1}{G_{p,1}} + \frac{F_3 - 1}{G_{p,1}G_{p,2}} + \ldots + \frac{F_n - 1}{G_{p,1}G_{p,2}\ldots G_{p,n-1}}$$

Noise Figure and Resistive Attenuators

The noise figure of a resistive attenuator at room temperature is about equal to the attenuator's loss. When we place a resistive attenuator (whose noise temperature is T_0) at the input of second device, the gain and noise figure of the two-element cascade is

$$G_{cas,dB} = G_{Attn,dB} + G_{2,dB}$$
$$F_{cas,dB} = F_{Attn,dB} + F_{2,dB}$$

where
  $G_{Attn,dB}$ = the gain of the attenuator,
  $G_{2,dB}$ = the gain of the second device,
  $G_{Attn,dB}$ = the noise figure of the attenuator = $-G_{Attn,dB}$,
  $F_{2,dB}$ = the noise figure of the second device.

Add the loss of the resistive attenuator to the noise figure of the second device to produce the noise figure of the cascade.

5.17 References

1. Bryant, James. "Ask the Applications Engineer: 8 Op Amp Issues." *Analog Dialogue* 24, no. 3 (1990): 12-13.

2. Freeman, Roger L. *Reference Manual for Telecommunications Engineering*. New York: John Wiley and Sons, 1985.

3. "Fundamentals of RF and Microwave Noise Figure Measurement." *Hewlett-Packard Application Note* 57-1 (July 1983).

4. Gardner, Floyd M. *Phase-Lock Technique*. New York: John Wiley and Sons, 1979.

5. Ha, Tri T. *Solid-State Microwave Amplifier Design*. New York: John Wiley and Sons, 1981.

6. Horowitz, Paul and Winfield Hill. *The Art of Electronics*. Cambridge: Cambridge U P, 1980.

7. Jorden, Edward C. *Reference Data for Engineers: Radio, Electronics, Computer and Communications*. 7th ed. Indianapolis: Howard W. Sams and Co, 1985.

8. Motchenbacher, C. D. and F. C. Fitchen. *Low-Noise Electronic Design*. New York: John Wiley and Sons, 1973.

9. Perlow, Stewart M. "Basic Facts about Distortion and Gain Saturation." *Applied Microwaves Magazine*, no. 5 (1989): 107.

10. Ryan, Al and Tim Scranton. "DC Amplifier Noise Revisited." *Analog Dialogue* 18, no. 1 (1984): 3-10.

11. Sklar, Bernard. *Digital Communications: Fundamentals and Applications*. Englewood Cliffs, NJ: Prentice-Hall, 1988.

12. Skolnik, Merrill I. *Introduction to Radar Systems*. 2nd ed. New York: McGraw-Hill, 1980.

13. Williams, Richard A. *Communications Systems Analysis and Design: A Systems Approach*. Englewood Cliffs, NJ: Prentice-Hall, 1987.

6

Linearity

> Nothing is more noble, nothing more venerable than fidelity.
> —Cicero

6.1 Introduction

In Chapter 5, we found that thermal noise sets the limit on the smallest signal a receiver can process. Linearity, on the other hand, sets the limit on the largest signal that can be processed. A system will alter or distort any signal that is too large.

Definitions

Since there is some contention about describing the various harmonics of a signal, the following nomenclature is used throughout this book:

- Fundamental: The frequency is f_0.
- First Harmonic: The frequency is also f_0.
- Second Harmonic: The frequency is $2f_0$.
- Third Harmonic: The frequency is $3f_0$.
- n^{th} Harmonic: The frequency is nf_0.

6.2 Linear and Nonlinear Systems

Although every system is nonlinear to one extent or another, this discussion focuses on linear systems for the following reasons:

- Almost every nonlinear system can be modeled as a linear system over a narrow operating range. For example, RF amplifiers are nonlinear devices but, if the input signal power is kept small enough, the amplifier can be modeled as a linear device.

- Even simple filters, which consists of common inductors and capacitors, are nonlinear devices. At very high power levels, the dielectrics of the capacitors behave in a nonlinear fashion and the inductors can saturate. As a result, the value of the component changes with the voltage across it or the current through it. This is a nonlinear process.

- The mathematics describing linear systems is simple and well-understood. The equations are reasonably easy to program into a computer, which makes numerical simulation quick and uncomplicated. A system that is modeled as linear yields the most information because the system's nonlinearities can be described as deviations from its linear behavior.

Linear Systems

In Chapter 2, we discussed linear systems and their characteristics. We found that a linear system has the following characteristics:

- The only frequencies present in a linear system are directly traceable to some voltage or current source in the system. For example, if a linear system is excited with a 1 MHz sine wave, it will be impossible to find any voltage or current in the system at any other frequency (excluding internally generated noise).

- Superposition works. The system behaves the same regardless of the magnitude of the input signal power. It does not matter whether the system is driven with a 10-femtowatt signal (very small) or a 10- megawatt signal (very large). The terminal impedances and the various transfer functions all remain unchanged.

- A linear system does not change its characteristics when more than one signal at a time is applied. Increasing the number of sine waves or complex waveforms will not result in a system change.

Weakly Nonlinear Systems

Nonlinear effects are troublesome to describe mathematically and programming the equations into a computer is a difficult task. The numerical simulations tend to be slow, which makes repetitive calculations unwieldy and sluggish. To keep things simple, we will examine systems which are weakly nonlinear. We will add signals to the system which are just strong enough to cause nonlinear effects to occur, but not so strong that the system characteristics change dramatically.

Gain Compression

One common nonlinear effect is a change in a system's voltage transfer function with changing power levels. For example, a system may exhibit a voltage gain of G_V or, equivalently, a power gain of G_p when small signals are applied to its input. If the input signal power is increased, the voltage and power gains of the system change. Since the gain usually decreases, this is commonly referred to as *gain compression*.

Harmonic Distortion

Another common nonlinear effect is harmonic distortion. When a large signal is applied to a nonlinear system, signals at the output of the system can be observed that were generated internally inside the system. For example, if a signal whose frequency is f_0 is applied to a system, we will observe signals at DC (or 0 hertz), f_0, $2f_0$, $3f_0$, ... , at the output.

If two signals, at f_1 and f_2, are applied to the input of a nonlinear system, signals at $\pm nf_1 \pm mf_2$ will be observed at the system's output. The variables n and m are integers, including zero. These distortion products are always present at the output of any nonlinear system. However, if the input signals are weak enough, the distortion generated by the system may be too weak to notice. Distortion products are often ignored if their power levels are below the noise.

Mixers, frequency doublers and the like are purposely nonlinear. They are used because they develop signal frequencies at their outputs that are not present at their inputs. All systems are nonlinear to one degree or another. However, at regular input power levels, the nonlinearities are so small that they can be ignored. Further, if the input signal power is small enough, every system is a linear system.

6.3 Amplifier Transfer Curve

Figure 6-1 shows a circuit diagram of a Rf amplifier along with a source and load. The signal voltage at the input of the amplifier is V_{in} and the signal voltage across the load resistor is V_{out}. The power supply terminals are labeled V_{cc} and V_{bb}.

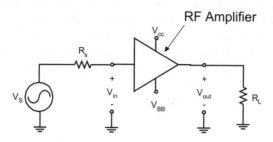

Figure 6-1 *Circuit diagram of a Rf amplifier with input and output voltages.*

Figure 6-2 shows the large signal voltage transfer curve of one amplifier. Note the flattening at the top and bottom of the plot and the sharp slope to the curve as the curve passes through (0,0). The flattening denotes saturation, and the slope about (0,0) is related to the voltage gain of the amplifier.

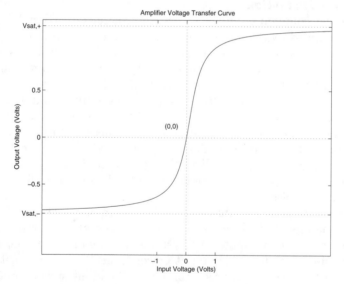

Figure 6-2 *Large signal voltage transfer curve of the amplifier shown in Figure 6-1.*

Large Signals

If we drive the amplifier with a large sine wave, we might observe the input and output voltages of Figure 6-3.

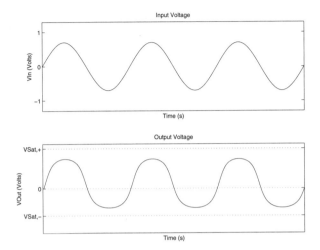

Figure 6-3 *Large signal input and output voltages of the RF amplifier of Figure 6-1.*

The output is a distorted version of the input. Saturation effects keep the output wave from exceeding $V_{sat}+$ or $V_{sat}-$. The output waveform is also asymmetrical. This is the most general case — other amplifiers may behave differently, and this amplifier might behave differently for a different input power level. The values of $V_{sat},+$ and $V_{sat}-$ are related to the power supply voltages and the bias points of the transistors inside the amplifier. Figure 6-4 shows the spectra of the input and output waveforms.

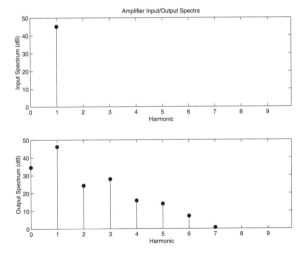

Figure 6-4 *Spectra of the input and output voltages of Figure 6-3.*

The input is a mathematically pure sine wave, and the output shows significant harmonic content. As mentioned earlier, one characteristic of a nonlinear system is that new frequencies are developed in the nonlinear process.

Small Signals

When the amplifier is driven with miniscule signals, it is operating under small signal conditions. Since the input voltage will be restricted to a very small range, we will be operating the amplifier very close to the (0,0) point shown in Figure 6-1. Figure 6-5 shows that the magnified neighborhood around (0,0) is very nearly a straight line. The slope of the linear approximation is related to the amplifier's power gain.

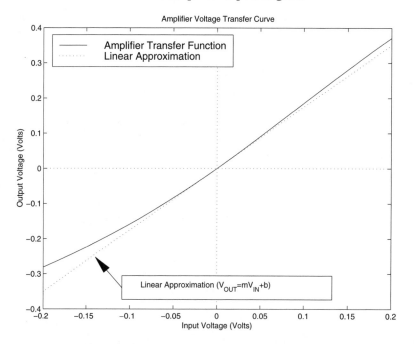

Figure 6-5 *Voltage transfer curve of Figure 6-2, magnified around (0,0). The linear approximation holds for small input signals.*

Example 6.1 — Amplifier Power Gain
Show that the slope of the voltage transfer function of an amplifier is directly related to its power gain. Assume the amplifier operates in a system with characteristic impedance Z_0.

Solution —

Use Figure 6-5 as a guide. We have drawn a straight line, i.e., a first-order curve fit through the (0,0) point and matched the slope of the line to the slope of the amplifier's transfer function at the origin. Let m equal the slope of the line. From geometry, we know

$$V_{out}(t) = mV_{in}(t) + b \qquad 6.1$$

where b = the y-intercept of the line. The y-intercept is zero in this case.

$$V_{out}(t) = mV_{in}(t)$$
$$\Rightarrow \frac{V_{out}(t)}{V_{in}(t)} = m \qquad 6.2$$

Referring to Figure 6-1, the power gain of the amplifier is

$$G_p = \frac{V_{out}^2(t)/R_L}{V_{in}^2(t)/R_{amp.in}} \qquad 6.3$$

Since the amplifier operates in a system with a Z_0 characteristic impedance, we know that $R_{amp,in} = R_L = Z_0$ and

$$G_p = \frac{V_{out}^2(t)}{V_{in}^2(t)} \qquad 6.4$$
$$= m^2$$

The slope of the straight-line approximation in Figure 6-5 is 1.75. The power gain of this amplifier is 1.75^2 or $3.06 = 4.9$ dB.

Summary

Figures 6-6 and 6-7 sum up the behavior of nonlinear devices. Figure 6-6 shows the amplifier under small-signal conditions. The signal level at the input is 1 μV_{pk}, which is –110 dBm in a 50-ohm system. The signal present at the output of the RF amplifier looks very much like the input signal in both the time and frequency domains. The only measurable difference is the increase in voltage and power. Since the input signal is –110 dBm and the output signal is –90 dBm, the small signal power gain of this amplifier is 20 dB.

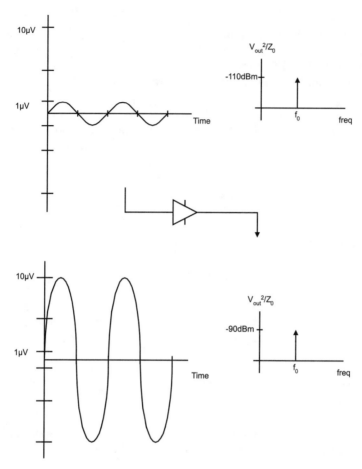

Figure 6-6 *A nonlinear amplifier under small signal conditions.*

Figure 6-7 shows the amplifier operating under large signal conditions. The input signal voltage is 1 V_{pk}, which is +10 dBm in a 50-ohm system. Except for the change in power level, the large input signal has the same appearance as the small signal in both the time and frequency domains.

However, the amplifier's output signal is now severely distorted. In the time domain, the output is clearly different from the input waveform. The frequency domain shows a series of sine waves harmonically related to the input sine wave. The power levels of the output harmonics are related to the input power level and to the shape of the amplifier's voltage transfer curve. These harmonics are clearly not present on the input and were generated inside the amplifier. These signals, generated in nonlinear devices, are often referred to as *intermodulation products* or simply as *intermods*.

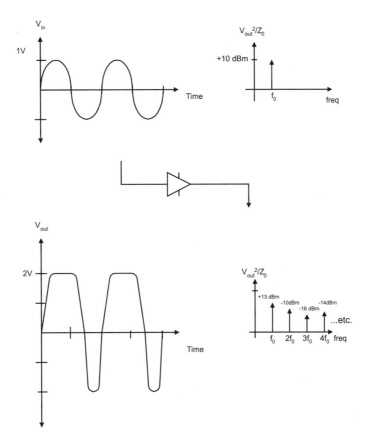

Figure 6-7 *A nonlinear amplifier under large signal conditions.*

Nonlinear Device

This amplifier meets several of the criteria for a nonlinear device.

- Its behavior changes when signals of different powers are applied.

- The amplifier distorts the waveform, producing frequencies not present at the input. This is especially noticeable at high power levels, but the effect occurs at all input power levels. For small signals, these distortion products are usually very small and can be ignored.

For single input tones at a frequency of f_0, the distortion takes the form of discrete signals at nf_0 where n extends from zero to infinity. For multiple input signals, the expression is more involved.

In general, every component discussed in this book is nonlinear to some extent. This includes devices such as amplifiers, mixers, filters, attenuators and transmission lines. However, some devices are more linear than others. For example, we can pass several thousand watts (60 dBm = 30 dBW = 1000 watts) of RF power through a properly designed transmission line or filter with no visible signs of nonlinear distortion. Yet, most of the amplifiers and mixers reviewed in this book show significant distortion at input power levels of just 1 mW (0 dBm = – 30 dBW = 1 mW). From a linearity perspective, the two most troublesome components are amplifiers and mixers. Diode switches and voltage-controlled attenuators can also be problematic.

Nonlinear Behavior

Nonlinear behavior limits the large signal handling capability of an amplifier just as noise limits the small-signal capability. For small signals, the distortion generated by a nonlinear device may be very low in power. The distortion products below the noise floor or the minimum detectable signal of the system can be ignored. As the input power is increased, the system will eventually be able to detect the distortion generated by the nonlinear device. At some input power level, the distortion power will become too great and the output waveform will become unacceptable.

6.4 Polynomial Approximations

How can devices be compared from a linearity point of view? When a signal is applied to a device, how much distortion will the device generate? Polynomials provide the solution. Since we can express any arbitrary curve as a polynomial, the error between the approximation and the curve will be as small as we like if we carry enough terms. A polynomial is an equation of the form

$$y = k_0 + k_1 x + k_2 x^2 + k_3 x^3 + \ldots + k_n x^n \qquad 6.5$$

where $k_n x^n$ is the nth-order term of the polynomial.

RF Amplifier

To model the voltage transfer curve of a RF amplifier, Equation 6.5 is modified to include more familiar variables.

$$V_{out}(t) = k_0 + k_1 V_{in}(t) + k_2 V_{in}^2(t) + k_3 V_{in}^3(t) + \ldots + k_n V_{in}^n(t) \qquad 6.6$$

Matching Derivatives

There are several numerical methods available for generating a polynomial to approximate a given curve. The method used here forces a match between the derivatives of the curve and the derivatives of the polynomial. Figure 6-8 shows the amplifier transfer function and its first-, second- and third-order polynomial approximations. The equations for the approximations are

First-Order or Linear Approximation:

$$V_{out}(t) = 1.752 \cdot V_{in}(t) \qquad 6.7$$

Second-Order Approximation:

$$V_{out}(t) = 1.752 \cdot V_{in}(t) + 1.446 \cdot V_{in}^2(t) \qquad 6.8$$

Third-Order Approximation:

$$V_{out}(t) = 1.752 \cdot V_{in}(t) + 1.446 \cdot V_{in}^2(t) - 3.229 \cdot V_{in}^3(t) \qquad 6.9$$

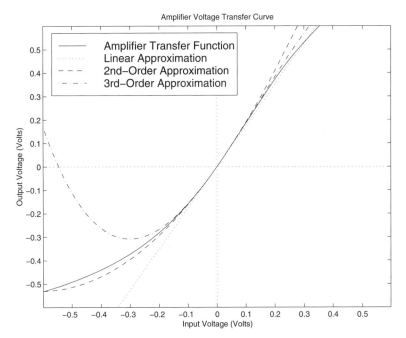

Figure 6-8 *Amplifier transfer function with its first, second and third order polynomial approximations.*

Observations

Close examination of Figure 6-8 reveals that the higher-order polynomials match the original curve better at the (0,0) point. Note that the polynomial approximations diverge rapidly from the original curve with increasing input signal. The higher-order curves diverge quicker than the lower-order curves.

Weakly Nonlinear

In the following analysis, the amplifier will be operating close to the (0,0) point on the voltage transfer curve. Under small signal conditions, the linear approximation holds. As the input signal increases in power, we must describe the transfer function with a second-order polynomial to account for the distortion observed. As the input signal level increases further, the amplifier should be described using the third-order approximation. It is not necessary to continue increasing the polynomial order as the input signal increases. For reasons we will discuss later, we can usually stop at the third-order polynomial.

The polynomial approximations are valid for small signals and near the (0,0) point of Figure 6-8. With increasing input level, the approximations no longer describe the amplifier adequately and the analysis breaks down. In other words, as long as we do not allow the input signal to become too strong, the following analysis is accurate and useful. If the input signals are strong enough that the amplifier's transfer curve cannot be accurately described by a matching polynomial, the analysis becomes obsolete.

6.5   Single-Tone Analysis

We can approximate the nonlinear voltage transfer function of a Rf amplifier with a polynomial of the form of Equation 6.6. Since we are interested in amplifying RF signals, let us look at the output voltage of the nonlinear device when the input voltage is a single cosine wave, i.e., when $V_{in} = \cos(\omega t)$.

The k_0 Term

The 0th term of the polynomial is independent of the input signal. This term represents the DC present at the output of the amplifier. Since we are usually interested in signals whose frequencies are greater than 0 hertz and since RF amplifiers are usually AC-coupled, this term is ignored. In most RF devices, this term is usually very small or equals zero.

The $k_1 V_{in}^1$ Term

The k_1 term represents the power gain of the amplifier. Large k_1's translate into large power gains; small k_1's translate into small power gains. Equation 6.4 relates the power gain of the amplifier to k_1 term. For a sinusoidal input voltage of

$$V_{in}(t) = A\cos(\omega t) \qquad 6.10$$

the $k_1 V_{in}^1$ term produces an output voltage of

$$V_{out,1}(t) = k_1 A\cos(\omega t) \qquad 6.11$$

This *linear* or *first-order term* is not responsible for any distortion in the amplifier's output. If a voltage V_{in} is applied to a linear device, the output voltage V_{out} will contain only the same frequency. As will be shown, this is not necessarily true for the higher-order polynomial terms.

The $k_2 V_{in}^2$ Term

The second-order $k_2 V_{in}^2$ term of the polynomial is responsible for some of the signal distortion present on the amplifier's output. For a sinusoidal input signal of

$$V_{in}(t) = A\cos(\omega t) \qquad 6.12$$

the $k_2 V_{in}^2$ term will generate an output voltage of

$$\begin{aligned} V_{out,2}(t) &= k_2 A^2 \cos^2(\omega t) \\ &= \frac{k_2 A^2}{2}\left[1 + \cos(2\omega t)\right] \\ &= \frac{k_2 A^2}{2} + \frac{k_2 A^2}{2}\cos(2\omega t) \end{aligned} \qquad 6.13$$

The second-order term produces energy at frequencies other than the input frequency. Equation 6.13 shows output energy at DC or 0 hertz (the $k_2 A^2/2$ term) and energy at twice the input frequency (the $k_2 A^2/2 \cdot \cos(2\omega t)$ term). Clearly, the second-order term is at least partially responsible for the nonlinear behavior of the amplifier since it produces frequencies at the amplifier's output that were not present at the input. We often use the term second-order distortion to describe the effects caused by the second-order term of the approximating polynomial.

The $k_3 V_{in}^3$ Term

This is the third-order term of the polynomial. For a sinusoidal input signal of

$$V_{in}(t) = A\cos(\omega t) \qquad 6.14$$

the third-order term produces

$$V_{out,3}(t) = k_3 A^3 \cos^3(\omega t)$$
$$= \frac{k_3 A^3}{4}\left[3\cos(\omega t) + \cos(3\omega t)\right] \qquad 6.15$$

Equation 6.15 shows energy at the fundamental frequency (the original input frequency or f_0) and at three times the input frequency (at $3f_0$). This is third-order distortion. Similar to the second-order term, the third-order term is responsible for some of the amplifier's nonlinear behavior.

The $k_4 V_{in}^4$ Term

For a sinusoidal input, the fourth-order term produces

$$V_{out,4}(t) = \frac{k_4 A^4}{8}\left[3 + 4\cos(2\omega t) + \cos(4\omega t)\right] \qquad 6.16$$

The fourth order term generates energy at DC, at the second harmonic of the input signal and at the fourth harmonic of the input signal. The second-order term produces energy at DC and at the second harmonic of the input waveform. The fourth-order term produces energy at DC and the second harmonic as well as the fourth harmonics of the input signal.

The $k_5 V_{in}^5$ Term

The fifth-order term produces

$$V_{out,5}(t) = \frac{k_5 A^5}{16}\left[10\cos(\omega t) + 5\cos(3\omega t) + \cos(5\omega t)\right] \qquad 6.17$$

which contains energy at the fundamental frequency, the third harmonic and the fifth harmonic. The fifth-order term of the polynomial spawns signals at the fundamental, the third and the fifth harmonics of the input signal, and the third-order term creates energy at only the fundamental and the third harmonics.

The $k_6 V_{in}^6$ Term

Under sinusoidal drive, the sixth-order term produces

$$V_{out,6}(t) = \frac{k_6 A^6}{32} \left[10 + 15\cos(2\omega t) + 6\cos(4\omega t) + \cos(6\omega t) \right] \qquad 6.18$$

The output has energy at DC, the second, fourth and sixth harmonics.

Observations

Several conclusions can be drawn from the data of the previous section. Although these conclusions can be mathematically proven, we will forgo a rigorous analysis and simply state the results.

Even n

For even n, the term $\cos^n(\omega t)$ will produce only even harmonics of the fundamental. The harmonics range from DC (which is the 0th harmonic) up to and including the nth harmonic. These are the *even-order responses* of the system.

Odd n

For odd n, $\cos^n(\omega t)$ will produce only odd harmonics of the fundamental. The harmonics begin with the fundamental or first harmonic up to and including the nth harmonic. Responses that are odd and greater than one are called the *odd-order responses* of the system.

n^{th}-Order Polynomial and n^{th}-Order Harmonics

For a given n, $\cos^n(\omega t)$ will produce harmonics up to and including the nth harmonic, but no higher. For example, $\cos^5(\omega t)$ will generate only odd harmonics up to and including the 5th harmonic.

Power of Each Harmonic

At the output of the amplifier, the strength of a given harmonic is determined by several terms of the polynomial. For example, given a 6th-order polynomial that describes an amplifier's voltage transfer function, the amplitude of the second harmonic will be determined by the second-, fourth- and sixth-order terms of the polynomial.

Power Decreases as n Increases

When we operate the amplifier in its weakly nonlinear input range, the amplitude of the nth harmonic will tend to decrease as n increases. For example, the third harmonic will be smaller than the second harmonic; the fourth harmonic will be smaller than the third harmonic, and so on. However, if the input signal power is increased to the point where we are operating in the amplifier's grossly nonlinear range, none of these approximations hold. Remember that the polynomial approximation is valid only when the input voltage is small. Devices such as mixers and frequency doublers are built to be purposefully nonlinear. Designers go out of their way to enhance particular nonlinearities and suppress the linear responses. The *decreasing amplitude with increasing n* rule does not apply to mixers, frequency doublers and the like.

6.6 Two-Tone Analysis

So far we have assumed that only one sine wave at a time was present in the system. In practice, however, this is normally not the case. Therefore, we have to analyze the system when it is driven by more than one sine wave.

Commercial FM Radio

In the United States, the commercial FM radio broadcast band inhabits the spectrum from 87.8 to 108.0 MHz. Each channel is centered on an odd multiple of 100 kHz: 87.9, 88.1, 883, 88.5, ... 107.7 and 107.9 MHz. There are 101 allocated channels and each channel can occupy up to 200 kHz of bandwidth. Figure 6-9 shows the FM broadcast band graphically. A typical commercial FM broadcast receiver in an urban environment might see 30 signals simultaneously at the output of the antenna.

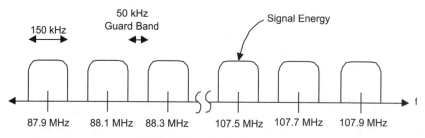

Figure 6-9 *The commercial FM broadcast band in the United States.*

Two-Tone Analysis

How will a nonlinear device behave when many different signals are simultaneously applied to its input? In practice, the signals will likely be at different frequencies and different power levels.

Proceeding directly to the general case is difficult. To make the initial analysis a little easier, we will analyze the behavior of a nonlinear device when only two sine waves are applied to the input and assume that both sine waves are the same power level. The input signal will be

$$V_{in}(t) = A_1 \cos(\omega_1 t) + A_2 \cos(\omega_2 t) \qquad 6.19$$
$$= A_1 \cos(2\pi f_1 t) + A_2 \cos(2\pi f_2 t)$$

where
$\omega_1 = 2\pi f_1,$
$\omega_2 = 2\pi f_2,$
f_1 = frequency of tone #1,
f_2 = frequency of tone #2,
A_1, A_2 = the amplitude of each cosine wave.

The k_0 Term

This term represents the DC present at the output of the amplifier. We will ignore it for the purpose of this discussion.

The $k_1 V_{in}^1$ Term

This term again represents the power gain of the amplifier. The larger the value of k_1, the larger the power gain. For the dual-tone input described by Equation 6.19, the output of the amplifier due to the $k_1 V_{in}^1$ term is

$$V_{out,1}(t) = k_1 [A_1 \cos(\omega_1 t) + A_2 \cos(\omega_2 t)] \qquad 6.20$$

As in the single-tone case, the linear term does not change or distort the input signal; it simply applies power gain.

The $k_2 V_{in}$ Term

When we apply the two-tone input signal described by Equation 6.19 to a nonlinear device, the second-order term produces an output voltage of

$$V_{out,2}(t) = k_2 [A_1 \cos(\omega_1 t) + A_2 \cos(\omega_2 t)]^2$$
$$= k_2 \left\{ \frac{A_1^2 + A_2^2}{2} + \frac{A_1^2 \cos(2\omega_1 t)}{2} + \frac{A_2^2 \cos(2\omega_2 t)}{2} \right. \qquad 6.21$$
$$\left. + A_1 A_2 \cos[(\omega_1 + \omega_2)t] + A_1 A_2 \cos[(\omega_1 - \omega_2)t] \right\}$$

As in the single-tone case, the second-order term produces energy at 0 hertz (or DC) and at the second harmonic of each input signal (i.e. at $2\omega_1$ and at

$2\omega_2$). Equation 6.21 contains two terms we have not previously observed: $\cos[(\omega_1 + \omega_2)t]$ and $\cos[(\omega_1 - \omega_2)t]$. We define

- the difference frequency as $\Delta\omega = |\omega_1 - \omega_2|$ or $\Delta f = |f_1 - f_2|$
- the sum frequency as $\Sigma\omega = |\omega_1 + \omega_2|$ or $\Sigma f = |f_1 + f_2|$

The spurious signals generated by the second-order term tend be the strongest of all the spurious signals. Figure 6-10 shows the portion of the output spectrum of a Rf amplifier which is due only to second-order distortion. The output tones are concentrated around DC and $f_1 + f_2$. There are no components at either f_1 or f_2.

When building systems, nonlinear distortions are often avoided at all costs although they can be used in a positive way. A mixer is a device built with enhanced second-order performance (see Chapter 3). We use the sum and difference products arising from the $k_2 V_{in}^2$ term to perform frequency translation, phase detection and other functions.

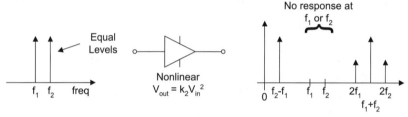

Figure 6-10 *Output spectrum of a nonlinear device due only to second-order distortion.*

Example 6.2 — Second-Order Distortion and the FM Broadcast Band

Find the second-order output frequencies for the following input conditions:
a. $f_1 = 87.9$ MHz, $f_2 = 88.1$ MHz
b. $f_1 = 107.7$ MHz, $f_2 = 107.9$ MHz
c $f_1 = 87.9$ MHz, $f_2 = 107.9$ MHz.

These frequencies represent the center and band edges of the US commercial FM broadcast band.

Solution —
Using Equation 6.21, we find
a. We have output frequencies at
DC,
$2f_1 = 175.8$ MHz,

$2f_2 = 176.2$ MHz,
$f_2 + f_1 = 88.1 + 87.9 = 176$ MHz,
$|f_2 - f_1| = |88.1 - 87.9| = 0.2$ MHz.

b. We have output frequencies at
DC,
$2f_1 = 215.4$ MHz,
$2f_2 = 215.8$ MHz,
$f_2 + f_1 = 107.9 + 107.7 = 215.6$ MHz,
$|f_2 - f_1| = |107.9 - 107.7| = 0.2$ MHz.

c. We have output frequencies at
DC,
$2f_1 = 175.8$ MHz,
$2f_2 = 215.8$ MHz,
$f_2 + f_1 = 107.9 + 87.9 = 195.8$ MHz,
$|f_2 - f_1| = |107.9 - 87.9| = 20.0$ MHz.

Figure 6-11 shows the output spectrum for these three cases with the appropriate levels marked. Considering this example, we can see that the second-order distortion frequencies possible in this scenario are DC, 0.2 to 20.0 MHz, 175.8 to 215.8 MHz. The second-order distortion of the amplifier will not produce any signals whose frequencies are outside this range.

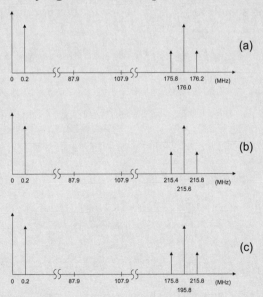

Figure 6-11 Output spectrum of the three cases discussed in Example 6.2.

The $k_3 V_{in}^3$ Term

When a nonlinear device is driven with two equal-level sine waves, the third-order term in the approximating polynomial produces an output voltage of

$$V_{out,3}(t) = k_3 \big[A_1 \cos(\omega_1 t) + A_2 \cos(\omega_2 t) \big]^3$$

$$= \left(\frac{3A_1^3}{4} + \frac{3A_1 A_2^2}{4} \right) \cos(\omega_1 t) + \left(\frac{3A_2^3}{4} + \frac{3A_1^2 A_2}{4} \right) \cos(\omega_2 t)$$

$$+ \frac{A_1^3}{4} \cos(3\omega_1 t) + \frac{A_2^3}{4} \cos(3\omega_2 t)$$

$$+ \left(\frac{3A_1^2 A_2}{4} \right) \cos\big[(2\omega_1 + \omega_2)t\big] + \left(\frac{3A_1^2 A_2}{4} \right) \cos\big[(2\omega_1 - \omega_2)t\big]$$

$$+ \left(\frac{3A_1 A_2^2}{4} \right) \cos\big[(2\omega_2 + \omega_1)t\big] + \left(\frac{3A_1 A_2^2}{4} \right) \cos\big[(2\omega_2 - \omega_1)t\big]$$

6.22

As might be expected from the single-tone analysis, the third-order term produces energy at the fundamental frequencies, f_1 and f_2, and at three times the input frequencies, $3f_1$ and $3f_2$. We also find several new products at $(2f_1 + f_2)$, $(2f_2 + f_1)$, $(2f_1 - f_2)$ and $(2f_2 - f_1)$.

In a nonlinear device with both linear and third-order components, most of the power at f_1 and f_2 will be generated by the $k_1 V_{in}$ term of the approximating polynomial. The relative power levels given above assume that all of the power in the signals at f_1 and f_2 is due to the third-order term.

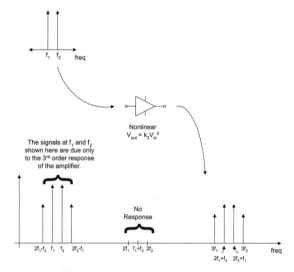

Figure 6-12 *Output spectrum of a nonlinear device due only to third-order distortion.*

Figure 6-12 shows the portion of the output spectrum of a RF amplifier which is due only to the third-order distortion.

Example 6.3 — Third-Order Distortion and the FM Broadcast Band
Find the third-order output frequencies for the following input conditions:
a. $f_1 = 87.9$ MHz, $f_2 = 88.1$ MHz,
b. $f_1 = 107.7$ MHz, $f_2 = 107.9$ MHz,
c. $f_1 = 87.9$ MHz, $f_2 = 107.9$ MHz,
d $f_1 = 97.9$ MHz, $f_2 = 98.1$ MHz.

Solution —
Using Equation 6. 22, we find
a. We have third-order output frequencies at
 $f_1 = 87.9$ MHz,
 $f_2 = 88.1$ MHz,
 $3f_1 = 263.7$ MHz,
 $3f_2 = 264.3$ MHz,
 $2f_2 + f_1 = 2(88.1) + 87.9 = 264.1$ MHz,
 $2f_1 + f_2 = 2(87.9) + 88.1 = 263.9$ MHz,
 $|2f_2 - f_1| = |2(88.1) - 87.9| = 88.3$ MHz,
 $|2f_1 - f_2| = |2(87.9) - 88.1| = 87.7$ MHz.

b. We have third-order output frequencies at
 $f_1 = 107.7$ MHz,
 $f_2 = 107.9$ MHz,
 $3f_1 = 323.1$ MHz,
 $3f_2 = 323.7$ MHz,
 $2f_2 + f_1 = 2(107.9) + 107.7 = 323.5$ MHz,
 $2f_1 + f_2 = 2(107.7) + 107.9 = 323.3$ MHz,
 $|2f_2 - f_1| = |2(107.9) - 107.7| = 108.1$ MHz,
 $|2f_1 - f_2| = |2(107.7) - 107.9| = 107.5$ MHz.

c. We have third-order output frequencies at
 $f_1 = 87.9$ MHz,
 $f_2 = 107.9$ MHz,
 $3f_1 = 263.7$ MHz,
 $3f_2 = 323.7$ MHz,
 $2f_2 + f_1 = 2(107.9) + 87.9 = 303.7$ MHz,
 $2f_1 + f_2 = 2(87.9) + 107.9 = 283.7$ MHz,
 $|2f_2 - f_1| = |2(107.9) - 87.9| = 127.9$ MHz,
 $|2f_1 - f_2| = |2(87.9) - 107.9| = 67.9$ MHz.

d. We have third-order output frequencies at

$f_1 = 97.9$ MHz,
$f_2 = 98.1$ MHz,
$3f_1 = 293.7$ MHz,
$3f_2 = 394.3$ MHz
$2f_2 + f_1 = 2(98.1) + 97.9 = 294.1$ MHz,
$2f_1 + f_2 = 2(97.9) + 98.1 = 293.9$ MHz,
$|2f_2 - f_1| = |2(98.1) - 97.9| = 98.3$ MHz,
$|2f_1 - f_2| = |2(97.9) - 98.1| = 97.7$ MHz.

Figure 6-13 shows the output spectrum for these four cases.

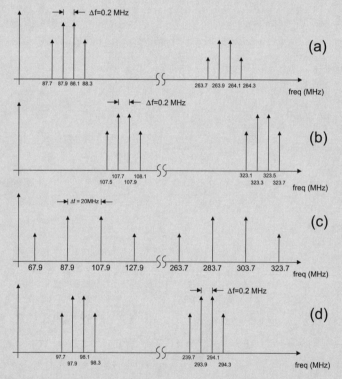

Figure 6-13 *Output spectrum of the four cases described in Example 6.3.*

In 6-13(a), 6-13(b) and 6-13(d), the third-order difference products produce signals that lie over other possible FM stations. In part 6-13(d), for example, the two third-order products at 97.7 MHz and 98.3 MHz both lie directly on top of valid FM broadcast frequencies. If we wanted to listen to a weak station at 97.7 MHz, the two strong signals at 97.9 and 98.1 MHz could combine to mask the 97.7 MHz. Third-order distortion presents a fundamental limitation on the linearity of most systems once we consider filtering effects.

LINEARITY | 563

The higher-order terms of the approximating polynomial will be important to later discussions. We will briefly describe the behavior of the higher-order terms under two-tone sinusoidal drive, giving only the frequencies that result.

The $k_4 V_{in}^4$ Term

Under two-tone sinusoidal drive, the fourth-order term produces signals at DC (or 0 Hz),
$2f_1, 2f_2,$
$4f_1, 4f_2,$
$f_1 \pm f_2$
$2f_1 \pm 2f_2,$
$3f_1 \pm f_2, 3f_2 \pm f_1.$

The fourth-order term produces energy at DC, the second harmonics of f_1 and f_2 and the fourth harmonics of f_1 and f_2. We also observe distortion products at several sum and difference combinations.

The $k_5 V_{in}^5$ Term

The fifth-order term of the approximating polynomial produces signals at:
DC (or 0 Hz),
$f_1, f_2,$
$2f_1 \pm f_2, 2f_2 \pm f_1$
$3f_1, 3f_2,$
$3f_1 \pm 2f_2, 3f_2 \pm 2f_1,$
$4f_1 \pm f_2, 4f_2 \pm f_1,$
$5f_1, 5f_2.$

The fifth-order term produces energy at f_1 and f_2, the third and fifth harmonics of f_1 and f_2, and several sum and difference combinations.

Observations

Figure 6-14 shows the output spectrum of a nonlinear amplifier when two sine waves are applied at slightly different frequencies, f_1 and f_2. The number associated with each spectral element is the order of the lowest term of the approximating polynomial that will generate the element. For example, if the lowest order that can produce a particular spectral element is a fifth-order combination of the input signals, then the spectral component is labeled with a "5."

564 | RADIO RECEIVER DESIGN

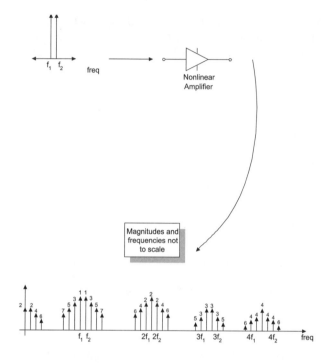

Figure 6-14 *Output spectrum of a nonlinear device when the input is two equal-amplitude sine waves of slightly different frequencies.*

Neighborhoods

The tones generated by the nonlinear process tend to concentrate in the general neighborhood of the harmonics of the two input signals. When the input signals are f_1 and f_2 and $f_1 \approx f_2$, the spurious output signals will cluster in the general area of

$$n \cdot \frac{f_1 + f_2}{2} \text{ where n = 0,1,2,3, ...} \qquad 6.23$$

Separation

The spurious signals tend to cluster in the neighborhoods given in Equation 6.23. If only the signals in a particular neighborhood are examined, we find the signals are separated by a frequency Δf where

$$\Delta f = |f_1 - f_2| \qquad 6.24$$

Power Decreases as n Increases

The power of a particular spurious signal is inversely proportional to its order. A 5th-order spurious signal will likely be stronger than a 7th-order spurious signal, and so on.

Spurious Frequencies

In equation form, the spurious output signal frequencies are given by

$$f_{out} = |\pm nf_1 \pm mf_2| \text{ where } \begin{cases} n = 0,1,2,3,\ldots \\ m = 0,1,2,3,\ldots \end{cases} \quad 6.25$$

where $m + n$ is the order of the spurious product.

6.7 Distortion Summary

A nonlinear device will generate distortion products whose frequencies are derived from the input frequencies. As a mathematical tool, the transfer function of a nonlinear device was modeled as a polynomial, then the terms of the polynomial under sinusoidal drive were analyzed.

The approximating polynomial holds only for moderately small power levels. As the input power increases, deviations between the device's actual transfer function and the approximation increase and eventually the model falls apart completely. Accordingly, the analysis is limited to weakly nonlinear systems, i.e., systems driven by just enough power to make the nonlinear aspects of the system noticeable. Because the model degenerates rather quickly with increasing power levels, two more approximations involving the transfer function of a nonlinear device will be used.

Small Signal Approximation (I)

For very small input signals (and ignoring the DC component), we can completely ignore the nonlinear aspects of most devices and simplify Equation 6.6 to

$$V_{out}(t) = k_1 V_{in}(t) \quad \text{for } V_{in}(t) \approx 0 \quad 6.26$$

This is the first small signal approximation.

Small Signal Approximation (II)

When the power of the input signals is relatively small yet large enough to generate noticeable distortion, we can make a second small-signal approximation. Instead of modeling the device's transfer function as linear, we will model it as a third-order polynomial with no dc term (i.e., $k_0 = 0$).

566 | RADIO RECEIVER DESIGN

$$V_{out}(t) = k_1 V_{in}(t) + k_2 V_{in}^2(t) + k_3 V_{in}^3(t) \qquad 6.27$$

This polynomial describes a weakly nonlinear device. The input power is still small and the device does exhibit slight nonlinearity.

Equation 6-27 is a third-order polynomial that will generate only second- and third-order distortion. The higher-order distortion products are ignored.

Summary

Table 6-1 summarizes the second- and third-order distortion components for single-tone and double-tone input signals.

Table 6-1 *Second- and third-order distortion generated by a nonlinear device under sinusoidal drive.*

Signal	2nd-Order Outputs	3rd-Order Outputs
$V_{in}(t) = A\cos(\omega t)$	$\dfrac{A^2}{2} + \dfrac{A^2}{2}\cos(2\omega t)$	$\dfrac{3A^3}{4}\cos(\omega t) + \dfrac{A^3}{4}\cos(3\omega t)$
$V_{in}(t) = A_1\cos(\omega_1 t)$ $+ A_2\cos(\omega_2 t)$	$\dfrac{A_1^2 + A_2^2}{2}$ $+\dfrac{A_1^2}{2}\cos(2\omega_1 t)$ $+\dfrac{A_2^2}{2}\cos(2\omega_2 t)$ $+ A_1 A_2 \cos[(\omega_1 + \omega_2)t]$ $+ A_1 A_2 \cos[(\omega_1 - \omega_2)t]$	$\left(\dfrac{3A_1^3}{4} + \dfrac{3A_1 A_2^2}{4}\right)\cos(\omega_1 t)$ $+\left(\dfrac{3A_2^3}{4} + \dfrac{3A_1^2 A_2}{4}\right)\cos(\omega_2 t)$ $+\dfrac{A_1^3}{4}\cos(3\omega_1 t) + \dfrac{A_2^3}{4}\cos(3\omega_2 t)$ $+\left(\dfrac{3A_1^2 A_2}{4}\right)\cos[(2\omega_1 + \omega_2)t]$ $+\left(\dfrac{3A_1^2 A_2}{4}\right)\cos[(2\omega_1 - \omega_2)t]$ $+\left(\dfrac{3A_1 A_2^2}{4}\right)\cos[(2\omega_2 + \omega_1)t]$ $+\left(\dfrac{3A_1 A_2^2}{4}\right)\cos[(2\omega_2 - \omega_1)t]$

Two Types of Distortion

From this point forward, two major types of distortion will be primarily discussed: second-order distortion and third-order distortion. Distortion caused by the second-order term of Equation 6.27 is second-order distortion; the distortion caused by the third-order term of Equation 6.27 is third-order distortion.

When the amplifier is operated in its weakly nonlinear input range, the amplitude of the nth harmonic tends to decrease as n increases. The second-order term of Equation 6.27 generates the strongest even-order spurious signals; the third-order term generates the strongest odd-order signal power. Higher-order distortion, however, can be a problem. The power of the second- and third-order spurious signals will be the largest of the unwanted products from a nonlinear device.

Also, the behavior of the second-order spurious signals is representative of all the even-order spurious signals; third-order spurious signals are representative of all the odd-order spurious signals. We will consider the characteristics of the higher-order spurious signals as deviations from the behavior of the second- and third-order behavior.

Finally, the various nonlinear distortion products can be filtered away if their frequencies are sufficiently removed from the desired signals. In some cases, we can design the systems to almost eliminate second-order distortion. This means that we can often ignore the stronger second-order distortion and concentrate on reducing the weaker third-order distortion.

6.8 Preselection

Figure 6-15 shows the receiving system model used in this discussion. In any receiver design, we now have to process some specific band of frequencies. The *band-pass filter* (BPF) between the antenna and the first amplifier is normally wide enough to accept this entire band of frequencies. For example, in the United States, the commercial FM radio band covers 87.8 to 108 MHz. In this case, the band-pass filter between the antenna and the first RF amplifier will allow 87.8 to 108 MHz to pass with relatively little attenuation. Alternately, the United States cellular telephone band inhabits the spectrum from 824 to 894 MHz. If we were building a cellular telephone receiver, the first band-pass filter in the system would pass 824 to 894 MHz.

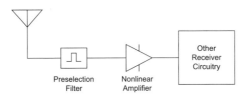

Figure 6-15 *Simple receiving system with a preselection filter.*

We almost universally place some type of filter between the antenna and the first amplifier because the filter limits the number of signals present in the receiver at one time. We are not interested in receiving these out-of-band signals and letting them into the receiver offers no benefit. Moreover, these unwanted signals can seriously degrade the receiver's performance.

Suppose you were listening to the FM radio in your car and you passed another driver who was using his cellular telephone. The car with the cellular phone is probably much closer to you than the FM transmitter. The undesired cellular telephone signal will be considerably stronger than the desired FM radio station. The band-pass filter between the antenna and first amplifier attenuates the unwanted signal (the cellular telephone frequency) while passing the desired signal (the FM radio station frequency). Without this filter, any large, undesired signal could cause the first amplifier to behave in a nonlinear fashion, producing distortion and perhaps completely masking the station's signal. High-performance receivers always have some form of filter between the antenna and the first amplifier. This is called *preselection,* and the first filter is a *preselection filter.*

6.9 Second-Order Distortion

Preselection and Second-Order Effects

The lower the order of the distortion, the larger the distortion power tends to be. Second-order distortion will therefore generate the largest spurious signals in the system. Anything we can do to minimize the nonlinear second-order effects will help us toward achieving a spurious-free system.

One Input Signal

For a single input signal whose frequency is f_0, Equation 6.13 indicates that second-order effects will generate distortion only at DC and at $2f_0$. For example, Figure 6-16 shows a wideband antenna connected to a nonlinear amplifier. A telephone radio receiver follows the amplifier. We will assume the receiver is tuned to 872.700 MHz.

Since the antenna is wideband, it will pick up the desired signal present at 872.700 MHz as well as signals above and below that frequency. Given we observe a signal at 872.700 MHz at the output of the nonlinear amplifier, how can we arrange things to be sure that the signal was not generated inside the amplifier? To answer this question, we need to apply knowledge of second-order effects.

One way we would observe a signal of frequency f_0 at the output of a second-order device is if we applied a signal at $f_0/2$ to the input. In the cellular telephone example, the $f_0/2$ frequency is 436.350 MHz. The single-tone case forces us to place a *high-pass filter* (HPF) between the antenna and the amplifier. The HPF must supply significant attenuation at $f_0/2$ or 436.350 MHz.

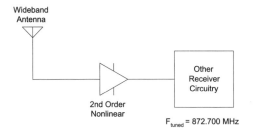

Figure 6-16 Wideband antenna connected to a nonlinear amplifier. No pre-selection filter.

In the more general case, if the receiver tunes from some lower frequency f_L to some upper frequency f_U, then the range of single-tone input signals which will generate in-band distortion due to second-order effects is

$$\frac{f_L}{2} \text{ to } \frac{f_U}{2} \qquad 6.28$$

In the example, cellular telephone receivers in the United States must receive 869 to 894 MHz. The troublesome $f_0/2$ frequencies are 434.5 to 447.0 MHz. To insure against single-tone second-order nonlinearities, the HPF of Figure 6-17 must provide high attenuation throughout this range.

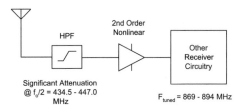

Figure 6-17 Wideband antenna connected to a nonlinear amplifier through a high-pass preselection filter.

Two Input Signals

Equation 6.21 describes the second-order outputs for two input signals. Any two signals whose sum or difference equals the tuned frequency might cause interference. In equation form, any two signals which satisfy

$$f_{tuned} = |f_1 \pm f_2| \qquad 6.29$$

can cause inband second-order distortion. Both signals must be present in the nonlinear amplifier at the same time and both must be at high levels to cause distortion.

Example 6.4 — Second-Order Distortion in the Cellular Telephone Band
The cellular telephone receiver is tuned to 872.700 MHz. What combination of the following signals can cause harmful second-order distortion?
a. 861.000 and 11.700 MHz,
b. 240.000 MHz and 1113.700 MHz,
c. 214.7 and 657.000 MHz.

Solution —
Using Equation 6.21, we find
a. The sum of 861.000 and 11.700 MHz is 872.700 MHz. This is a problematic combination.

b. Subtracting 240.000 from 1113.700 produces 872.7 MHz, which is again a problematic situation.

c. Adding 214.700 and 657.000 produces 871.700 MHz. This combination is not exactly at the tuned frequency but it is close. We are depending on some downstream filter to remove this distortion before it finds its way to the demodulator.

Problematic Frequencies

The system shown in Figure 6-18 consists of a wideband antenna followed by a preselection filter. The output of the filter goes into a nonlinear amplifier with a second-order characteristic. Finally, we feed the amplifier's output into a receiver with a tuning range of f_{LOW} to f_{HIGH}. The center frequency is f_0. We now derive the system constraints which will allow us to remove second-order distortion with filtering.

- The preselection filter must pass the signals between f_{LOW} and f_{HIGH}.

- We know the filter must suppress signals from $f_{LOW}/2$ to $f_{HIGH}/2$ so their second harmonics will not fall in the receiver's pass-band. Thus, the highest frequency f_{HIGH} should be kept less than twice the lowest frequency f_{LOW}. If this requirement is not enforced, we could apply a signal at f_{LOW} and its second harmonic would fall in the receiver pass-band. The receiver then cannot tell if a signal present at $2f_{LOW}$ is real or if it comes from second-order distortion in the amplifier. A system whose highest frequency is less than twice its lowest frequency is called a *suboctave system*.

- We must keep signals whose frequency sums lie between f_{LOW} and f_{HIGH} from combining in the amplifier. For example, a signal at $0.32f_0$ will combine with a signal at $0.68f_0$ to produce a signal at f_0.

Close examination of Equation 6.21 reveals that we have to suppress only one of the offending carriers. For example, if we apply $f_0/100$ and $99f_0/100$ to the system, we have to attenuate only the signal at $f_0/100$ to keep intermods from forming. A high-pass filter will solve this problem.

- Signals whose frequency differences lie between f_{LOW} and f_{HIGH} should be kept out of the amplifier. A signal at $3f_0$ will combine with a signal at $2f_0$ to produce a signal at f_0. A low-pass filter will solve this problem.

The frequency difference component also forces us to accept a suboctave system.

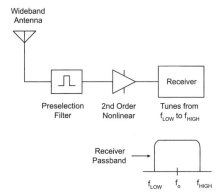

Figure 6-18 *Analysis of second-order distortion and the preselection filter bandwidth.*

The Second-Order Solution

Since we do not have any control over the signals present at the output of the antenna, we have to look at distortion almost from a statistical perspective. What are the most likely events that will cause harmful second-order distortion? We know a single tone at half the tuned frequency will cause problems. Since this is the most likely event, a high-pass filter has to be installed between the antenna and the first amplifier.

For two tones to combine to create in-band interference, both must have a large amplitude and the correct frequency relationship. Their sum or difference must be the tuned frequency. Combating this type of distortion requires both a high-pass and a low-pass filter. Ideally, we would like the preselection filter to be as narrow as possible, letting in only those frequencies the receiver is interested in seeing.

The sum and difference problem also forces us to keep the bandwidth of the system less than an octave. In a multioctave system, we cannot filter out the problem tones and have to rely on the nonlinear devices to hold up under strong signal conditions. Figure 6-19 shows a suboctave and a multioctave receiving system. Second-order distortion is a problem in systems that must process more than an octave of bandwidth at some point in the system

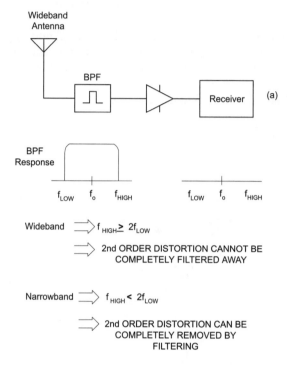

Figure 6-19 *Suboctave and multioctave receiving systems.*

6.10 Third-Order Distortion

Second-order distortion produces the strongest intermodulation products in a nonlinear device. In some situations second-order effects can be completely removed by using filters intelligently. However, there is no effective way to remove third-order intermodulation products from the systems, which is a fundamental limit on linearity in many common systems.

Preselection and Third-Order Effects

To diminish second-order distortion, a preselection band-pass filter is placed just after the antenna and before any nonlinear device. This filter was made as narrow as possible to pass desired signals while rejecting signals that might combine nonlinearly.

One Input Signal

Equation 6.15 describes the output of a third-order nonlinearity under sinusoidal drive. If the input frequency is f_1, the third-order output frequencies are f_1 and $3f_1$. The third-order nonlinearity generates a term at the fundamental frequency f_1. The fundamental component due to the nonlinearity will combine in some way with the fundamental component from the linear term of the amplifier to produce the fundamental output signal. In other words, some combination of the undistorted signal and the distortion makes up the output signal present at f_1.

Realization effects can change the relative phases between the undistorted signal (at f_1) and the third-order spurious signal (also at f_1). This third-order distortion can increase the apparent power gain of a nonlinear device (if the two signals are in phase) or it can decrease the power gain (if the two signals are 180° out of phase).

The spurious signal at $3f_1$ will fall outside of the preselection bandwidth in Figure 6-19 if $f_{HIGH} \leq 3 f_{LOW}$. In narrowband systems, when the bandwidth is less than an octave, the $3f_1$ component will always fall outside the preselection bandwidth. However, if the bandwidth is so large it can fit in a 3:1 frequency range, it can also fit in a 2:1 frequency range.

Two Input Signals

Equation 6.22 describes the output voltage of a third-order process when driven by two sinusoids of different amplitudes and frequencies. The third-order term produces energy at the fundamental frequencies (f_1 and f_2) and at three times the input frequencies ($3f_1$ and $3f_2$). We also find signals at ($2f_1 + f_2$), ($2f_2 + f_1$), ($2f_1 - f_1$) and ($2f_2 - f_1$). These new components demand examination, especially when the frequencies (f_1 and f_2) are restricted by the band-pass preselection filter.

In a simplified analysis, we can make the approximation that $f_1 \approx f_2$ since both of these signals must pass through the preselection filter. The third-order components are

- ($2f_1 + f_2$) and ($2f_2 + f_1$). When $f_1 \approx f_2$, then

$$2f_1 + f_2 \approx 3f_1 \approx 3f_2$$
$$\text{and} \qquad\qquad 6.30$$
$$2f_2 + f_1 \approx 3f_1 \approx 3f_2$$

These frequencies terms fall somewhere near the third harmonics of the input frequencies. In a suboctave system, these products can be filtered.

- ($2f_1 - f_2$) and ($2f_2 - f_1$). When $f_1 \approx f_2$, then

$$2f_1 - f_2 \approx f_1 \approx f_2$$
and
$$2f_2 - f_1 \approx f_1 \approx f_2$$
(6.31)

These distortion components will be problematic. When the input frequencies are about equal, these third-order terms occur very near the original input frequencies and are impossible to avoid. The only way to circumvent this problem is to depend upon the raw linearity of the system being designed.

Problematic Frequencies

Figure 6-20 illustrates the fundamental problem frequencies. We build a receiving system that is capable of receiving signals at f_1 through f_4. The input band-pass filter must be wide enough to pass this frequency range. If we apply two signals at f_2 and f_3, we may observe the output spectrum shown in Figure 6-20. The signals at f_2 and f_3 along with the nonlinear distortion at f_1 and f_4 can be observed.

It is impossible to determine whether the signals present at f_1 and f_4 come from the antenna or if they are internally generated spurious signals. If there are signals from the antenna at f_1 and f_4, nonlinear distortion could mask them, which constitutes a fundamental limitation.

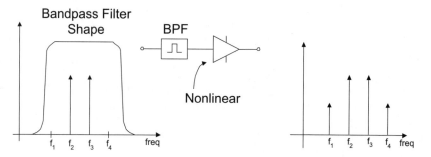

Figure 6-20 *The fundamental third-order problem. This distortion cannot be removed by filtering.*

The Third-Order Solution

The third-order solution is the same solution we used for the second-order problem. The preselection band-pass filter should be as narrow as possible, and the system should be kept suboctave to prevent second-order effects. However, the preselection filter will not prevent the third-order limitation caused by the $2f_1 - f_2$ and $2f_2 - f_1$ components.

6.11 Narrowband and Wideband Systems

Under a broad definition, a system is considered to be wideband if it contains more than an octave of bandwidth somewhere in its processing chain. A narrowband system never processes more than one octave. Narrowband systems are limited by third-order effects. Since second-order nonlinearities usually generate stronger distortion than third-order nonlinearities, it is useful to convert wideband systems into narrowband systems.

Generally, if a device is used in a narrowband system, we ignore its second-order performance and concentrate on its third-order performance. If we plan to use the same device in a wideband system, we are interested in its second-order distortion characteristics.

Wideband and Narrowband Examples

A system is *narrowband* if all of the filters in the processing path are less than an octave wide. A system is wideband if any of the filters in the RF path are more than an octave wide. Figure 6-21 shows an example of a wideband system. This is a 1 to 4 GHz receiver and the preselection filter covers the entire 1 to 4 GHz range. Since the receiver has a filter whose bandwidth must process more than one octave at a time, this is a *wideband* system.

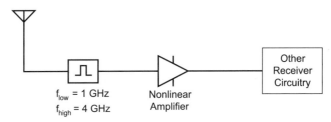

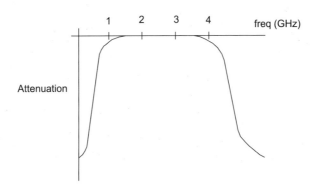

Figure 6-21 *A wideband receiving system. This system contains at least one filter with more than an octave of bandwidth.*

Figure 6-22 shows the same receiver with a slightly different filtering scheme. The receiver tunes over the same frequency range, but it now uses three switchable band-pass filters for preselection. When the receiver tunes in the 1 to 1.8 GHz band, we switch BPF_1 into the system. When the receiver tunes in the 1.8 to 2.7 GHz band, we switch in BPF_2 and, finally, when we tune the receiver anywhere in the 2.7 to 4.0 GHz range, we switch in BPF_3. Although the receiver still tunes over a 1 to 4 GHz range, this is now a narrowband system because we never have more than an octave's worth of frequency coverage present in the amplifier at one time.

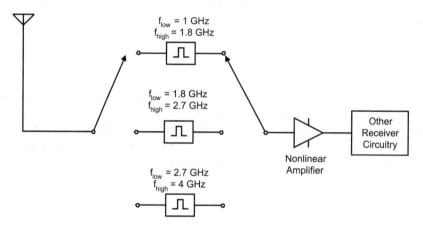

Figure 6-22 *A narrowband receiving system. The wideband system of Figure 6-21 has been converted into a narrowband system by the addition of switched filters. None of the filters in the processing chain is more than an octave wide.*

Another example: suppose we need a radar-warning receiver to tell us whether someone is shining radar on us. The radar signal is 120 MHz wide and might be centered anywhere from 1 to 4 GHz. We will start with the system in Figure 6-22 but will convert the signal to 160 MHz before processing (see Figure 6-23). The IF band-pass filter is centered at 160 MHz and has a 120 MHz bandwidth. The filter will pass 111 MHz to 231 MHz. The IF filter BPF_4 must pass more than an octave bandwidth. When the IF stage was added, the receiver was turned back into a wideband system.

LINEARITY | 577

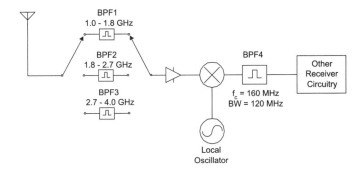

Figure 6-23 A wideband receiving system. The preselection structure is narrowband but the multi-octave IF filter defines this system as wideband.

In Figure 6-24 we have moved the IF center frequency to 700 MHz while keeping the 120 MHz bandwidth. The IF filter will pass 643 to 763 MHz. Since no part of this receiver has to process more than an octaves' worth of frequency at one time, this is a narrowband system. The definitions for wideband and narrowband systems that were presented here are somewhat loose. Depending upon the environment, the terms narrowband and wideband may have entirely different meanings.

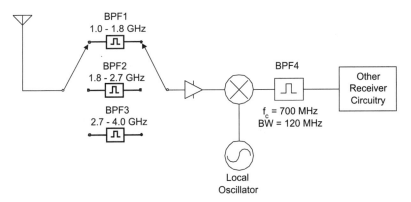

Figure 6-24 A narrowband receiving system. Moving the IF of Figure 6-23 up from 160 MHz to 700 MHz allows the IF filter to be suboctave.

6.12 Higher-Order Effects

We examined second- and third-order distortion in detail because they present fundamental limits on distortion in the cases that interest us. However, real devices exhibit higher-order nonlinearities.

Harmonics

If a signal of frequency f_0 is applied to a nonlinear device, the device will generate harmonics at

$$nf_0 \text{ where } n = 2,3,4, \ldots \qquad 6.32$$

Looking at the problem from a receiver perspective, if we are tuned to f_0, an input signal at

$$\frac{f_0}{n} \text{ where } n = 2,3,4, \ldots \qquad 6.33$$

can generate an interfering signal right on top of the tuned signal. The high-pass aspect of the preselection filter attenuates the troublesome input signals.

Multiple Input Signals

If multiple input signals are applied to a nonlinear device described by an nth order polynomial, the output frequencies are given by

$$f_{out} = \left| \pm nf_1 \pm of_2 \pm pf_3 \pm \cdots \right| \text{ where } \begin{cases} n = 0,1,2,\ldots \\ o = 0,1,2,\ldots \\ p = 0,1,2,\ldots \\ \vdots \end{cases} \qquad 6.34$$

Problems arise when one of the possible f_{out} frequencies lies on a signal of interest. It is therefore best to limit the number of signals a nonlinear device must process because it limits the number of signals that can combine and possibly cause interference.

6.13 Second-Order Intercept Point

Given an input signal power, how much second-order power will a particular nonlinear device generate? What will be the power at each of the second-order frequencies? The intercept concept quantifies the linearity of the device in a useful way. The linearity of different devices can be compared and the intercept information can be used to calculate the strength of the second-order distortion for a given fundamental input power.

Measuring Nonlinear Devices

To measure the second-order distortion generated by a nonlinear device, a tone at frequency f is applied to the input. The output power at f and at $2f$ is then measured. The input and output spectra are shown in Figure 6-25.

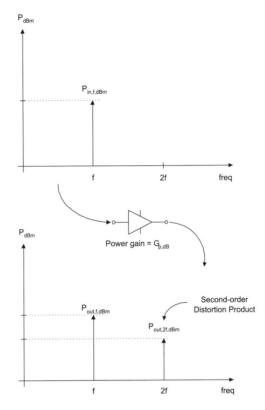

Figure 6-25 Measuring the second-order distortion generated by a nonlinear device.

After connecting a signal generator up to the *device under test* (DUT), we apply an input and watch the output on a spectrum analyzer. We must make certain that

- the signal generator output is free from second harmonic energy,
- the spectrum analyzer itself is not generating the second-order distortion observed on the screen.

Then the input port of the DUT is radiated with a single tone at f, the output power at f and the power of the second-order product at $2f$ is measured.

Definitions

The fundamental (or first harmonic) input power at frequency f is $P_{in,f,dBm}$ in dBm. The fundamental output power in dBm (still at frequency f) is $P_{out,f,dBm}$. We know

$$P_{out,f,dBm} = P_{in,f,dBm} + G_{p,dB} \qquad 6.35$$

where

$G_{p,dB}$ = the power gain of the amplifier in dB.

Let the output second-order distortion power at $2f$ be $P_{out,2f,dBm}$. Although the second-order signal does not exist on the input terminal of the nonlinear device, sometimes it is convenient to describe the equivalent input second-order power as

$$P_{out,2f,dBm} = P_{in,2f,dBm} + G_{p,dB} \qquad 6.36$$

where

$P_{in,2f,dBm}$ is the equivalent second-order input power in dBm.

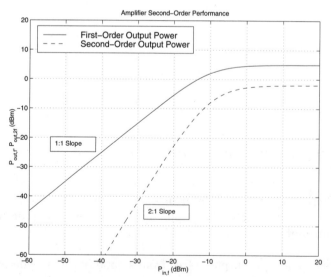

Figure 6-26 *Measured output power of a nonlinear device, at the fundamental frequency f and at the second-harmonic frequency $2f$.*

LINEARITY

There is no second-order signal present at the input of the amplifier because the amplifier generates it internally. We write Equation 6.36 only for mathematical manipulation. In the laboratory, we can easily measure $P_{out,f,dBm}$ and $P_{out,2f,dBm}$ for different values of $P_{in,f,dBm}$ and frequency f. Figure 6-26 shows the graph of these quantities for a particular device. All quantities are expressed in decibels. At low power levels, Figure 6-26 shows that the fundamental output power rises with a 1:1 slope with the fundamental input power (expressed in dB). This confirms the linear gain described by Equation 6.35.

The second order output power rises at a 2:1 slope with increasing input power (in dBm). The 2:1 slope is consistent with the device's second-order transfer function model of

$$V_{out} = k_2 V_{in}^2 \qquad 6.37$$

At high input power, the amplifier saturates. Both the fundamental and second-order output powers flatten out because the amplifier simply cannot supply any more signal. The flattening is not predicted by the simple third-order model.

Second-Order Intercept Point

If the fundamental and second-order output power curves were extended as shown in Figure 6-27, the two lines would intersect. The intersection is the *second-order intercept point* or *SOI* of that particular device. This is a direct measure of the amplifier's second-order performance.

Two possible *SOI* points, one for the input and one for the output, can be gathered from Figure 6-27. We will use *ISOI* for the input second-order intercept point and *OSOI* for the output second-order intercept point. The units for *ISOI* and *OSOI* are power.

Reading from Figure 6-27, the *ISOI* of this particular device is –3.1 dBm and the *OSOI* is +12.1 dBm. Note that

$$OSIO_{dBm} = ISOI_{dBm} + G_{p,dB} \qquad 6.38$$

This relationship is always true because of the way the terms were defined. Although the *SOI* point is a power level, a particular device should not be operated at its intercept point. This is a very high power level, which is well into saturation. The *SOI* point is simply a mathematical tool that allows us to describe the nonlinear behavior of a device. For example, the *ISOI* of the device in Figure 6-27 is –3.1 dBm. If we applied this power level to the input of the amplifier (i.e. $P_{in,f,dBm} = -3.1$ dBm), the fundamental output power level (reading $P_{out,f,dBm}$ from Figure 6-27) would only be 4.5 dBm. The output second-order power level ($P_{out,2f,dBm}$) would be –3.4 dBm.

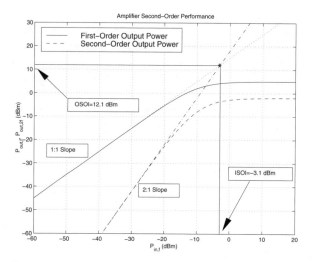

Figure 6-27 Definition of second-order intercept point from measured output powers at f and at 2f.

Quantifying Distortion Power

How does the *SOI* point relate to the fundamental input power level, the fundamental output power level and the second-order output power level? The geometry of Figure 6-28 provides the answer.

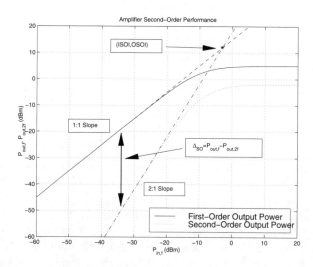

Figure 6-28 The geometry used to derive the relationships between fundamental input power and second-order distortion generated by a nonlinear device.

The fundamental and second-order lines both pass through the point $(ISOI_{dBm}, OSOI_{dBm})$. The slope of the fundamental line is unity, and the slope of the second-order line is two. The equation for a line with a known slope passing through a known point is

$$y - y_1 = m(x - x_1) \qquad 6.39$$

The equation of the line describing the fundamental input output power vs. the fundamental output power ($P_{in,f,dBm}$ vs. $P_{out,f,dBm}$) is

$$P_{out,f,dBm} - OSOI_{dBm} = 1 \cdot (P_{in,f,dBm} - ISOI_{dBm})$$
$$\Rightarrow P_{out,f,dBm} = P_{in,f,dBm} - ISOI_{dBm} + OSOI_{dBm} \qquad 6.40$$

Similarly, the equation for the line corresponding to the fundamental input power level vs. the second-order output power level (i.e., the $P_{in,f,dBm}$ vs. $P_{out,2f,dBm}$ line) is

$$P_{out,2f,dBm} - OSOI_{dBm} = 2 \cdot (P_{in,f,dBm} - ISOI_{dBm})$$
$$\Rightarrow P_{out,2f,dBm} = 2P_{in,f,dBm} - 2ISOI_{dBm} + OSOI_{dBm} \qquad 6.41$$

If we define $\Delta_{SO,dB}$ as the difference between the fundamental output power and the second-order output power (see Figure 6-28), we can write

$$\Delta_{SO,dB} = P_{out,f,dBm} - P_{out,2f,dBm}$$
$$= P_{in,f,dBm} - ISOI_{dBm} + OSOI_{dBm}$$
$$- (2P_{in,f,dBm} - 2ISOI_{dBm} + OSOI_{dBm}) \qquad 6.42$$
$$\Delta_{SO,dB} = ISOI_{dBm} - P_{in,f,dBm}$$

Similarly, we can show

$$\Delta_{SO,dB} = OSIO_{dBm} - P_{out,f,dBm} \qquad 6.43$$

Further manipulation of Equations 6.42 and 6.43 reveals

$$P_{in,2f,dBm} = 2P_{in,f,dBm} - ISOI_{dBm} \qquad 6.44$$

and

$$P_{out,2f,dBm} = 2P_{out,f,dBm} - OSOI_{dBm} \qquad 6.45$$

Example 6.5 — SOI and Spurious Power Levels

Given an amplifier with a 15 dB power gain, an $ISOI_{dBm}$ of –12 dBm and an input power level of –40 dBm, find

a. the amplifier's $OSOI_{dBm}$,
b. the levels of the signals at f and $2f$ at the output of the amplifier,
c. the equivalent input second order power.

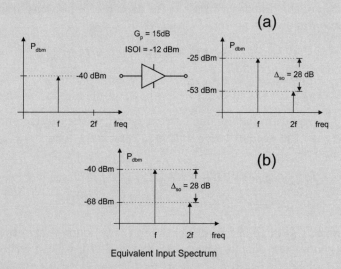

Figure 6-29 *Example 6.5. Find the amplifier's SOI and power gain.*

Solution —

Figure 6-29(a) is used as a reference.

a. Using Equation 6.38, we can write

$$OSOI_{dBm} = ISOI_{dBm} + G_{p,dB}$$
$$= -12 + 15 \qquad \qquad 6.46$$
$$= +3 \text{ dBm}$$

b. The fundamental output signal power is

$$P_{out,f,dBm} = P_{in,f,dBm} + G_{p,dB}$$
$$= -40 + 15 \qquad \qquad 6.47$$
$$= -25 \text{ dBm}$$

We use Equation 6.42 to find $\Delta_{SO,dB}$.

$$\Delta_{SO,dB} = ISOI_{dBm} - P_{in,f,dBm} \qquad 6.48$$
$$= -12 - (-40)$$
$$= 28 \text{ dBm}$$

The difference between the fundamental output signal power (at f) and the power of the second harmonic (at $2f$) is 28 dB. This is marked on Figure 6.29(a). The second-order output power is

$$P_{out,2f,dBm} = P_{out,f,dBm} - \Delta_{SO,dB}$$
$$= -25 - 28 \qquad 6.49$$
$$= -53 \text{ dBm}$$

c. The equivalent input second order power is

$$P_{in,2f,dBm} = P_{out,2f,dBm} - G_{p,dB}$$
$$= -53 - 15 \qquad 6.50$$
$$= -68 \text{ dBm}$$

Note that the value for $\Delta_{SO,dB}$ is the same on both the input and the output ports of the amplifier. Figure 6-29(b) shows the equivalent input spectrum.

Example 6.6 — Separating Spurious Signals from Real Signals

Figure 6-30(a) shows a wideband system covering 4 to 8 GHz and its output spectrum. How can we know whether the signal at 8 GHz is a second-order product or a valid input signal?

Solution —

Figure 6-30(a) shows the system and the output spectrum. Equation 6.44 and Figure 6-27 both show that a 1 dB decrease of the fundamental input power will result in a second-order output power drop of 2 dB. In other words, the slope of the $P_{in,f,dBm}$ vs. $P_{out,2f,dBm}$ line is two. Furthermore, since the input and output powers (in dBm) are related only by a constant, the fundamental output power vs. second-order output power curve also has a 2:1 slope.

If a 10-dB attenuator is placed on the input of the system [see Figure 6-30(b)], the fundamental input power presented to the amplifier drops by 10 dB. If the tone at 8 GHz drops by 20 dB, we know the 8-GHz tone is due to amplifier nonlinearity and not to a signal present on the antenna. If the 8 GHz tone drops by only 10 dB, we conclude the 8-GHz signal is real and originates from the antenna.

Figure 6-30(c) shows the output spectrum when the attenuator is placed in the system if the 8 GHz tone is due to amplifier nonlinearity. Note that the $\Delta_{SO,dBm}$ has changed. Figure 6-30(d) shows the output spectrum if the 8 GHz tone is due to a signal from the antenna. Note that the $\Delta_{SO,dBm}$ has not changed.

586 | RADIO RECEIVER DESIGN

This example illustrates a very useful method for identifying spurious responses. Usually, putting a pad of n dB on the input of a device will cause a $2n$ dB reduction of the amplifier's second-order distortion components. Occasionally, a particular spurious signal will have more than one source. In the example, there may be a signal present at the antenna port at 8 GHz, but the amplifier nonlinearity is hiding it. If the second-order distortion generated by the amplifier is only 5 dB above the signal from the antenna, the 8 GHz signal at the output will drop only 5 dB.

Moreover, in a typical receiving system, there are many places for an intermodulation product to be generated. Any or all of the devices in the receiver's cascade may be contributing power at the second-order frequency. The power of the spurious signals may not drop in a direct 2:1 ratio with the input power because of the way the distortion tones add up in the system. Anytime a $>n$ dB signal drops because an n dB attenuator is placed in a system, nonlinear behavior is the cause.

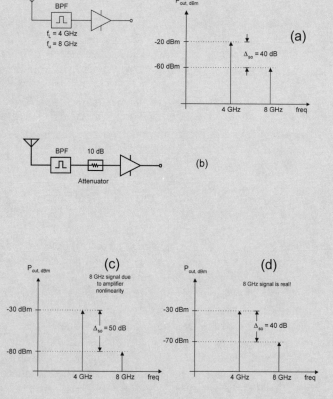

Figure 6-30 Example 6.6. *Identifying signals generated by second-order nonlinear distortion.*

LINEARITY

6.14 Third-Order Intercept Point

The third-order intercept point concept is used to compare the linearity of various devices in a useful way. It can also be used to calculate the amount of third-order distortion a device is likely to generate under a particular set of input conditions.

Measurement Technique

Figure 6-31 shows the third-order measurement procedure. We radiate the input terminal of a nonlinear device with two tones of equal power. The frequencies are f_1 and f_2. Then the output power and the power of the third-order products at $2f_1 - f_2$ and at $2f_2 - f_1$ are measured.

The two third-order tones at $2f_1 - f_2$ and $2f_2 - f_1$ will always exhibit the same power (see Equation 6.22) when A_1 equals A_2). We will assume both tones have equal power.

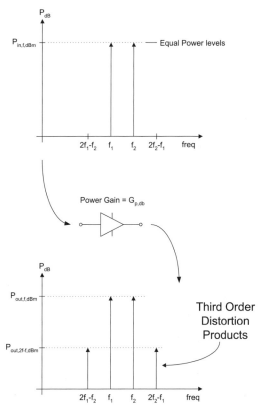

Figure 6-31 *Third-order intercept point measurement procedure.*

Measurement Requirements

A more detailed test schematic is shown in Figure 6-32. This measurement requires two signal sources and a summing network. We could perform an equivalent test by applying a single tone to the DUT and measuring the third harmonic at the device's output terminal. If the device under test operates over the entire f to $3f$ frequency range, this measurement would produce an equivalent description of the device's third-order performance. This characterization could be used to find the expected output levels of the components at $3f$ as well as the components at $2f_1 - f_2$ and $2f_2 - f_1$. In other words, we could make the characterization by generating only a single sine wave.

However, the single tone at $3f$ is not a problem. We are interested in the two tones at $2f_1 - f_2$ and $2f_2 - f_1$. These difference tones are the most troublesome because they usually cannot be filtered. The tones at the third harmonic are less difficult to filter.

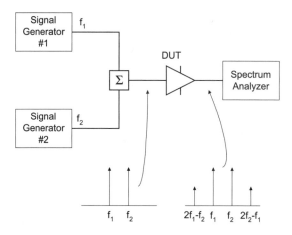

Figure 6-32 *Detailed test schematic for measuring the third-order intercept point.*

Definitions

Referring to Figure 6-31, the fundamental (or first harmonic) input power of each tone is $P_{in,f,dBm}$. The fundamental output power of each tone is $P_{out,f,dBm}$ and

$$P_{out,f,dBm} = P_{in,f,dBm} + G_{p,dB} \qquad 6.51$$

where $G_{p,dB}$ = the power gain of the amplifier in dB.

The power of each output tone at $2f_1 - f_2$ and $2f_2 - f_1$ will be equal. The third-order output power of a single tone is labeled $P_{out,2f\text{-}f,dBm}$. As in the second-order case, sometimes it is convenient to reference the third-order tones to the input.

$$P_{out,2f-f,dBm} = P_{in,2f-f,dBm} + G_{p,dB} \qquad 6.52$$

where $P_{in,2f\text{-}f,dBm}$ is the equivalent third-order input power. Again, this equation is for mathematical manipulation only. There are no signals at $2f_1 - f_2$ and $2f_2 - f_1$ present on the input of the nonlinear device. If we measure $P_{in,f,dBm}$, $P_{out,f,dBm}$ and $P_{out,2f\text{-}f,dBm}$ for many different values of $P_{in,f,dBm}$, we can graph these quantities as a function of $P_{in,f,dBm}$ (see Figure 6-33).

Figure 6-33 shows that the fundamental output power, expressed in dBm, rises at a 1:1 slope with increasing input power, also expressed in dBm. The third-order power rises at a 3:1 slope with increasing input power. At high levels of input power, the amplifier saturates. Both the fundamental and third-order output powers flatten out because the amplifier cannot supply any more output signal.

Before the amplifier saturates, we move from the weakly nonlinear region of the amplifier's transfer curve to the strongly nonlinear portion of the curve. The polynomial approximations do not apply for large input signals.

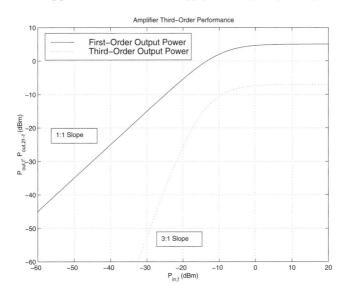

Figure 6-33 *Measured output power of a nonlinear device, at the fundamental frequency f and at the third-order distortion frequencies ($2f_2 - f_1$ or $2f_1 - f_2$).*

Third-Order Intercept Point

If the fundamental and third-order output power curves of Figure 6-34 are extended, the two lines intersect at the third-order intercept point or *TOI*. The device possesses both input (*ITOI*) and output (*OTOI*) third-order intercept points. The *TOI* point has units of power.

Figure 6-34 shows the characteristics of a device whose *ITOI* is –11.4 dBm and whose *OTOI* is 3.4 dBm. As in the second-order case, the *ITOI* and *OTOI* of a device are related by

$$OTOI_{dBm} = ITOI_{dBm} + G_{p,dB} \qquad 6.53$$

Applying an input power of $ITOI_{dBm}$ dBm would push the device into its strongly nonlinear region. The *TOI* point is only a mathematical tool that allows us to describe the nonlinear behavior of a device. For example, if we apply a fundamental power level of *ITOI* = –11.4 dBm (i.e., $P_{in,f,dBm}$ = –11.4 dBm) to the device described by Figure 6-33, the fundamental output power ($P_{out,f,dBm}$) level will be only 1.4 dBm. The output third-order power level ($P_{out,2f-f,dBm}$) would be –11.3 dBm.

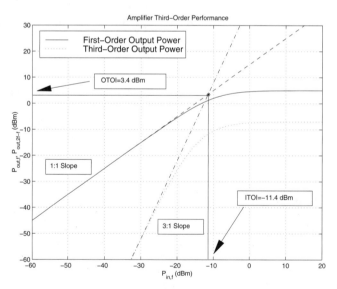

Figure 6-34 *The geometry used to derive the relationships between fundamental input power and third-order distortion generated by a nonlinear device*

Quantifying Distortion Power

The third-order intercept point describes the relationship between the fundamental input power and the third-order output powers. Figure 6-34 shows that the fundamental and third-order lines pass through the point ($ITOI_{dBm}$, $OTOI_{dBm}$). The slope of the fundamental line is unity, and the slope of the third-order line is three (recall the slope was two for the second-order case).

We can write the equation of the fundamental $P_{out,f,dBm}$ line as

$$P_{out,f,dBm} - OTOI_{dBm} = 1 \cdot (P_{in,f,dBm} - ITOI_{dBm})$$ 6.54
$$\Rightarrow P_{out,f,dBm} = P_{in,f,dBm} - ITOI_{dBm} + OTOI_{dBm}$$

The equation for the third-order $P_{out,2f\text{-}f,dBm}$ line is

$$P_{out,2f-f,dBm} - OTOI_{dBm} = 3P_{in,f,dBm} - 3ITOI_{dBm}$$ 6.55
$$\Rightarrow P_{out,2f-f,dBm} = 3P_{in,f,dBm} - 3ITOI_{dBm} + OTOI_{dBm}$$

$\Delta_{TO,dB}$ is defined as the difference between the fundamental output power and the third-order output power (see Figure 6-35).

$$\Delta_{TO,dB} = P_{out,f,dBm} - P_{out,2f-f,dBm}$$ 6.56
$$= P_{in,f,dBm} - ITOI_{dBm} + OTOI_{dBm}$$
$$\Delta_{TO,dB} = 2(ITOI_{dBm} - P_{in,f,dBm})$$

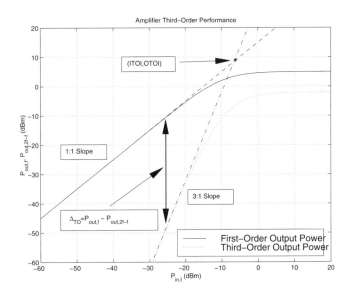

Figure 6-35 *Definition of third-order intercept point from measured output powers at f and at $2f_2 - f_1$ or $2f_1 - f_2$.*

We can also show

$$\Delta_{TO,dB} = 2(OTOI_{dBm} - P_{out,f,dBm}) \qquad 6.57$$

Combining these two equations produces

$$P_{in,2f-f,dBm} = 3P_{in,f,dBm} - 2ITOI_{dBm} \qquad 6.58$$

and

$$P_{out,2f-f,dBm} = 3P_{out,f,dBm} - 2OTOI_{dBm} \qquad 6.59$$

Example 6.7 — Measured Third-Order Intercept

A commercial RF amplifier has 16 dB of power gain, an $OTOI_{dBm}$ of +12 dBm and an input power level of –30 dBm, find
a. the amplifier's ITOI,
b. the levels of the signals at f_1 and f_2 along with the signal levels at $2f_1-f_2$ and $2f_2-f_1$ at the output of the amplifier,
c. the equivalent third-order input power.

Solution —

See Figure 6-36(a).

a. We can rewrite Equation 6.53 as

$$\begin{aligned}ITOI_{dBm} &= OTOI_{dBm} - G_{p,dB} \\ &= 12 - 16 \qquad\qquad 6.60 \\ &= -4 \text{ dBm}\end{aligned}$$

b. The fundamental output signal power is

$$\begin{aligned}P_{out,f,dBm} &= P_{in,f,dBm} + G_{p,dB} \\ &= -30 + 16 \qquad\qquad 6.61 \\ &= -14 \text{ dBm}\end{aligned}$$

We use Equation 6.56 to find $\Delta_{TO,dB}$.

$$\begin{aligned}\Delta_{TO,dB} &= 2(ITOI_{dBm} - P_{in,f,dbm}) \qquad 6.62\\ &= 2[-4 - (-30)] \\ &= 52 \text{ dB}\end{aligned}$$

The difference between the fundamental output signal power (at f) and the power of the third-order products (at $2f_1-f_2$ and $2f_2-f_1$) is 52 dB. This is marked on Figure 6-36(a). The third-order output power is

$$P_{out,2f-f,dBm} = P_{out,f,dBm} - \Delta_{TO,dB} \qquad 6.63$$
$$= -14 - 52$$
$$= -66 \text{ dBm}$$

c. The equivalent input third-order power is

$$P_{in,2f-f,dBm} = P_{out,2f-f} - G_{p,dB}$$
$$= -66 - 16 \qquad 6.64$$
$$= -82 \text{ dBm}$$

Figure 6-36(b) shows the equivalent input spectrum. $\Delta_{TO,dB}$ is the same on both the input and the output ports of the amplifier.

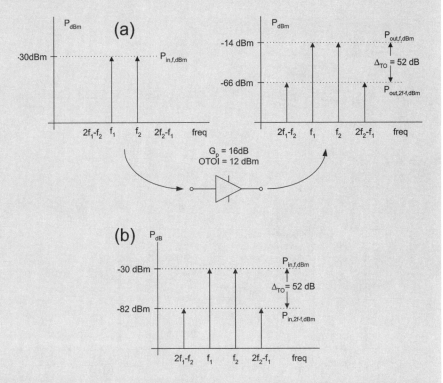

Figure 6-36 *Example 6.7. Input and output spectra of a nonlinear device generating third-order distortion.*

Example 6.8 — Spurious Signals and TOI

We are building a cellular telephone receiver for the US market where the RF channels are 30 kHz apart. Figure 6-37(a) shows the spectrum present at the output of a nonlinear amplifier. The spectrum shows signals every 30 kHz, in accordance with the specification. We suspect the two outer signals may be due to amplifier nonlinearity. How can this be determined?

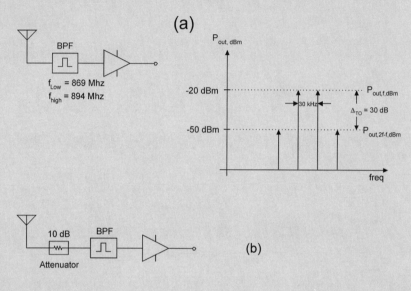

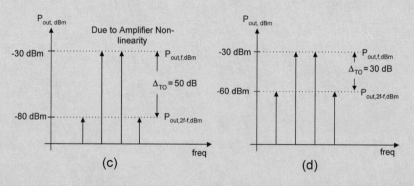

Figure 6-37 *Example 6.8. Identifying signals generated by third-order nonlinear distortion.*

Solution —

Equation 6.58 indicates that if we decrease the fundamental input power by 1 dB, the output third-order power will drop by 3 dB. Figure 6-37(b) shows the system with a 10 dB attenuator placed between the antenna and the amplifier. If the level of the two third-order tones at $2f_1 - f_2$ and $2f_2 - f_1$ drop by 30 dB when we insert the attenuator, the signals are due to amplifier nonlinearities. Figure 6-37(c) shows the output spectrum that results when the attenuator is placed in the system if the $2f_1 - f_2$ and $2f_2 - f_1$ tones are due to amplifier nonlinearity. Note that $\Delta_{TO,dBm}$ has changed. Figure 6-37(d) shows the output spectrum if two outer tones are due to signals from the antenna. Note that $\Delta_{TO,dBm}$ has not changed.

This method is used in practice to identify spurious responses. Placing an n dB attenuator on the input of a device will usually cause a $3n$ dB reduction of the device's third-order distortion. When a receiver is connected to an antenna, it is unlikely that two signals will arrive with the same power level. The example above is not very realistic. However, two signals entering a nonlinear device will produce third-order distortion at the $2f_1 - f_2$ and $2f_2 - f_1$ frequencies even if their power levels are different. The strength of nonlinear output signals depends upon the strength of both of the input signals causing the distortion.

6.15 Measuring Amplifier Nonlinearity

Equations 6.45 and 6.59 provide an easy way to measure the *SOI* and the *TOI* of an unknown device.

Second-Order Measurement

Rewriting Equation 6.45 produces

$$OSOI_{dBm} = 2P_{out,f,dBm} - P_{out,2f,dBm} \qquad 6.65$$

A test signal is applied to the nonlinear devices, then the fundamental output power and the second-order output power are measured. Equation 6.65 then yields the output second-order intercept for the amplifier. Figure 6-38 shows the test setup. We could have also written

$$OSOI_{dBm} = P_{out,f,dBm} + \Delta_{SO,dB}$$
$$\text{and} \qquad 6.66$$
$$ISOI_{dBm} = P_{in,f,dBm} + \Delta_{SO,dB}$$

We would measure the fundamental power and $\Delta_{SO,dB}$ (on either the input or the output), then use Equation 6.66 to find the SOI of the DUT.

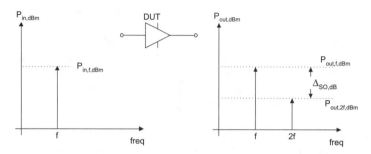

Figure 6-38 *Measuring the second-order intercept point of a nonlinear device.*

Example 6.9 — Amplifier Second-Order Measurements
Find the power gain, *ISOI* and *OSOI* of the DUT shown in Figure 6-39.

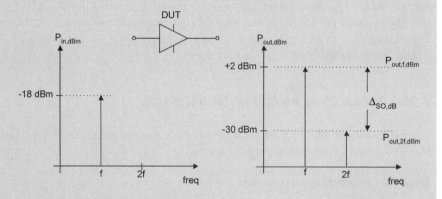

Figure 6-39 *Example 6.9. Measuring the second-order intercept point of an amplifier.*

Solution —
The power gain is

$$G_{p,dB} = P_{out,f,dBm} - P_{in,f,dBm}$$
$$= 2 - (-18) \qquad 6.67$$
$$= 20 \text{ dB}$$

Equation 6.65 yields the $OSOI_{dBm}$

$$OSOI_{dBm} = 2P_{out,f,dBm} - P_{out,2f,dBm} \qquad 6.68$$
$$= 2(2) - (-30)$$
$$= -34 \text{ dBm}$$

LINEARITY

The ISOI is

$$ISOI_{dBm} = OSOI_{dBm} - G_{p,dB}$$
$$= 34 - 20$$
$$= 14 \text{ dBm}$$
(6.69)

We could have used Equation 6.66 to find the OSOI.

$$OSOI_{dBm} = P_{out,f,dBm} + \Delta_{SO,dB}$$
$$= 2 + 32$$
$$= 34 \text{ dBm}$$
(6.70)

Third-Order Measurements

We can measure the *TOI* of any device using Equation 6.59. Rewriting this equation produces

$$OTOI_{dBm} = \frac{3}{2} P_{out,f,dBm} - \frac{1}{2} P_{out,2f-f,dBm}$$
(6.71)

Figure 6-40 shows the test setup.

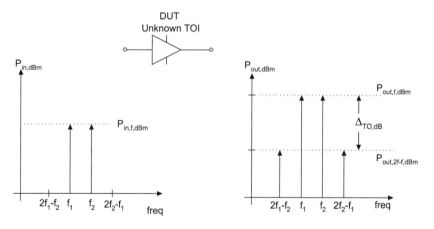

Figure 6-40 *Measuring the third-order intercept of a nonlinear device.*

We apply two signals to the DUT and measure the fundamental output power along with the output power at the intermod frequencies. Equation 6.71 yields the output third-order intercept of the DUT. Combining Equations 6.71 and 6.57 produces

$$OTOI_{dBm} = P_{out,f,dBm} + \frac{\Delta_{TO,dB}}{2}$$

and   6.72

$$ITOI_{dBm} = P_{in,f,dBm} + \frac{\Delta_{TO,dB}}{2}$$

These equations allow us to measure the fundamental power (either input or output) and the $\Delta_{TO,dB}$ in order to find the TOI of the DUT.

Example 6.10 — Amplifier Third-Order Measurements
Find the power gain, $ITOI$ and $OTOI$ of the DUT shown in Figure 6-41.

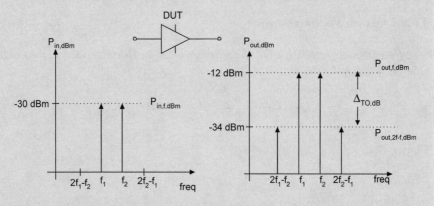

Figure 6-41 *Example 6.10. Find the amplifier's TOI and power gain.*

Solution —
The power gain is

$$G_{p,dB} = P_{out,f,dBm} - P_{in,f,dBm}$$
$$= -12 - (-30)$$ 6.73
$$= 18 \text{ dB}$$

Equation 6.71 gives us the $OTOI_{dBm}$

$$OTOI_{dBm} = \frac{3}{2} P_{out,f,dBm} - \frac{1}{2} P_{out,2f-f,dBm}$$ 6.74
$$= \frac{3}{2}(-12) - \frac{1}{2}(-34)$$
$$= -1 \text{ dBm}$$

The *ITOI* is

$$ITOI_{dBm} = OTOI_{dBm} - G_{p,dB}$$
$$= -1 - 18$$
$$= -19 \text{ dBm}$$

6.75

We could have used Equation 6.72 to find the *OTOI*.

$$OTOI_{dBm} = P_{out,f,dBm} + \frac{\Delta_{TO,dB}}{2}$$
$$= -12 + \frac{22}{2}$$
$$= -1 \text{ dBm}$$

6.76

6.16 Gain Compression and Output Saturation

Other concepts describing the linearity of a device are the 1 dB gain *compression point* and the *saturated output power*. Figure 6-42 shows output power vs. input power for a nonlinear device. The curve is linear at low input power levels. An n dB increase on the input port results in an n dB increase in output power.

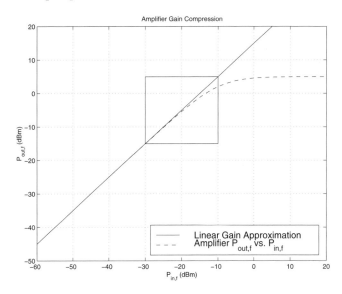

Figure 6-42 *Input power vs. output power plot of a nonlinear device. Note the output saturation.*

As the input power increases, the device becomes nonlinear and the amplifier is unable to supply the necessary output power to the load. This is a gradual process; the amplifier's gain usually drops off gently with increasing input power.

1 dB Compression Point

Figure 6-42 shows a linear approximation to the $P_{out,dBm}$ vs. $P_{in,dBm}$ curve. At low power levels (below approximately −30 dBm), the linear approximation follows the exact curve very well. However, the two curves depart at higher power levels.

Figure 6-43 shows the same two curves enlarged in the area where the two curves just begin to differ. The point where the two curves deviate by 1 dB is the amplifier's *1 dB compression point* or simply the *compression point*.

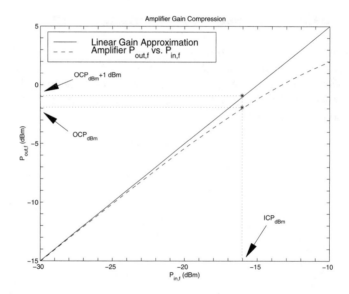

Figure 6-43 *Figure 6-42 enlarged around the device's 1 dB compression point.*

The device has both an *output compression point* (OCP_{dBm}) and an *input compression point* (ICP_{dBm}). The input compression point is the input power at which the amplifier's gain has dropped by 1 dB. The output compression point is the amplifier's output power when the input power is ICP_{dBm}. The ICP_{dBm} and the OCP_{dBm} are related.

$$OCP_{dBm} = ICP_{dBm} + (G_{p,dB} - 1) \qquad 6.77$$

This equation is a little different from the other equations that relate the input and output parameters of a device. Usually the equation takes the form of

$$\text{Output Spec}_{dBm} = \text{Input Spec}_{dBm} + G_{p,dB} \qquad 6.78$$

The *ICP* and the *OCP* of a device are related by the power gain minus one. Some of the literature defines the OCP_{dBm} as $ICP_{dBm} + G_{p,dBm}$. This is the point labeled "$OCP_{dBm} + 1$" on Figure 6-43. Here it is defined as stated in Equation 6.77.

The 1 dB compression point is an arbitrary boundary between the small signal operating range of a device and its large signal operating range. Before the concepts of *SOI* and *TOI*, the 1 dB compression point was the primary linearity specification. The intercept concepts allow us to calculate useful data when given the input power levels. The 1 dB compression point provides no such information.

Saturated Output Power

Saturated output power is the maximum fundamental power a device will produce under any input conditions. Figure 6-42 shows the P_{out} vs. P_{in} curve for a nonlinear device. The device shown in Figure 6-42 saturates at about 5.2 dBm of output power. Saturated output power is important in transmitter applications. Since we are generating the signal applied to the amplifier, we know there will be only one signal present in the power amplifier at one time. This limits the possible distortion products to the harmonics of the single input signal. Normally, a band-pass filter is placed after the power amplifier to suppress the harmonics of the fundamental output signal.

The saturated output power of an amplifier refers to the maximum amount of power the amplifier will produce at its output port. For example, a 10 watt amplifier can produce one sine wave at 10 watts, 2 sine waves at 5 watts each or 100 sine waves at 0.1 watts each. The amplifier can also produce a signal whose bandwidth is 20 MHz and whose power spectral density is 0.25 watts/MHz.

Example 6.11 — Satellite Transponders and Linearity

Figure 6-44 shows the simplified RF chain of a typical satellite transponder. The received signal, broadcast from the earth, passes through the receive antenna, a band-pass filter and into an amplifier. The mixer and oscillator combination then converts the received signal to a new frequency. After band-pass filtering, the signal is applied to a power amplifier, then to another band-pass filter and finally send into the transmit antenna.

This architecture is commonly called the *bent pipe*. To the user, the satellite appears to pull in the signals heading away from the earth, then appears to send the signals into a bent pipe to be rebroadcast to earth. Many cable television channels pass through these satellite systems. The signals are broadcast from the earth to the satellite at 5.925 GHz to 6.425 GHz. The signals are converted to 3700 to 4.200 GHz aboard the satellite and are rebroadcast back to the subscribers on the earth.

Note that the power gains of all the components are fixed. There is a one-for-one correspondence between the amount of power transmitted to the satellite and the strength of the signal we receive when the signal finally returns from the satellite. These satellite systems service different users who are physically separated on the earth. A user pays a fee based upon the bandwidth (or transponder space) he or she requires. The more spectrum the user requires, the larger the bill.

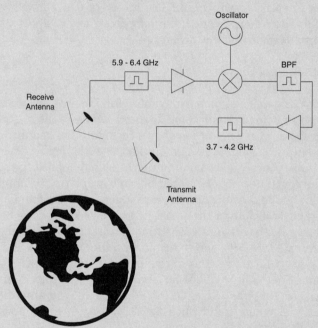

Figure 6-44 The RF chain of a typical "bent-pipe" satellite transponder.

There is another aspect to this problem. If the total transponder bandwidth is 500 MHz, for example, and the power amplifier produces 1000 watts, each user is limited to

$$\frac{1000 \text{ Watts}}{500 \text{ MHz}} = 2 \frac{\text{Watts}}{\text{MHz}} \qquad 6.79$$

unless he or she is willing to pay a premium.

Whenever several signals are present in a nonlinear device at the same time, the device will generate spurious signals. This applies directly to the satellite transponder. When the input power from each user is within the limits specified by the satellite operator, the spurious signals generated by the satellite are not strong enough to cause harm.

When the transmitted power is increased beyond the agreed-upon limits, it causes distortion in all the signals passing through the amplifier. In addition, due to other nonlinear effects in the receiver, the strongest signal will remove power from the other signals. In other words, all users must cooperate for the satellite transponder to work properly. The nonlinear distortion in the power amplifier limits the total amount of power that the satellite transponder can generate.

6.17 Comparison of Nonlinear Specifications

Figure 6-45 shows a summary of device nonlinearities. The nonlinearity of a particular device can be expressed in many ways: as second-order intercept, as third-order intercept, as compression point or as saturated output power. Several rules of thumb can be established when comparing the nonlinear characteristics of various devices.

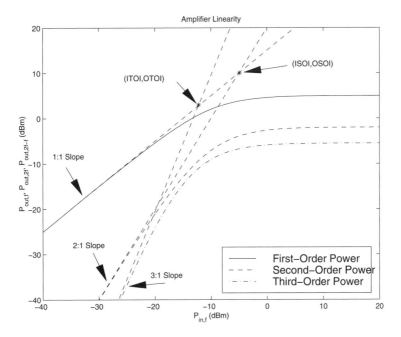

Figure 6-45 Summary of device nonlinearities.

SOI, TOI, then CP

For a common RF device, for example, an amplifier or mixer, the compression point is generally the smallest quantity, followed by the third-order intercept and the second-order intercept. The following relationships tend to be true:

$$SOI_{dBm} \approx TOI_{dBm} + (10 \text{ or } 15 \text{ dB})$$
$$TOI_{dBm} \approx CP_{dBm} + (10 \text{ or } 15 \text{ dB})$$
6.80

For a third-order polynomial approximation, the output compression point of a nonlinear device is 10.6 dB below the output third-order intercept or

$$OCP_{dBm} \approx OTOI_{dBm} - 10.6$$
6.81

which implies

$$ICP_{dBm} \approx ITOI_{dBm} - 9.6$$
6.82

Balanced Devices

Balanced devices are designed with a symmetrical output structure to reduce second-order distortion and increase the output power of the device. The balanced output increases the *SOI* about 15 dB. Unless the word "balanced" appears in the specifications or data sheet, the device probably is not balanced. If a device is balanced, then the following relationships should be true:

$$SOI_{dBm} \approx TOI_{dBm} + (30 \text{ or } 40 \text{ dB})$$
$$TOI_{dBm} \approx CP_{dBm} + (10 \text{ or } 15 \text{ dB})$$
6.83

Slopes

If we express the power levels in dB, we have seen that the second-order output power of a weakly nonlinear drops off at a 2:1 slope with decreasing input power. Third-order power device drops off at a 3:1 slope. This relationship holds as long as the input power is at least 20 dB below the ICP_{dBm}. Above that level, all bets are off, although the degradation is usually a gradual process.

Worst-Case Scenario

From a linearity perspective, the worst-case scenario is having to process many large undesired signals in a nonlinear device at the same time as a small, desired signal. The large, undesired signals may cause nonlinear distortion, which may even bury the small desired signal.

The worst-case scenario represents the world of practice. The antenna will supply an abundance of signals at a variety of frequencies and amplitudes. Since the possible intermodulation products fall at

$$f_{out} = |\pm nf_1 \pm of_2 \pm pf_3 \pm \cdots| \text{ where } \begin{cases} n = 0,1,2,\ldots \\ o = 0,1,2,\ldots \\ p = 0,1,2,\ldots \\ \vdots \end{cases} \quad 6.84$$

we can see why linearity is critical.

6.18 Nonlinearities in Cascade

Ultimately, when we build systems, we will string filters, attenuators, amplifiers and mixers together to perform receiving functions. We are interested in what happens to the system's linearity as we cascade nonlinear devices together. How does the linearity of the cascaded device compare with the linearity of the pieces used to build the cascade?

Nomenclature Refresher

Figure 6-46 shows the three-element cascade used in this analysis. We know the power gain and the output third-order intercept (OTOI) of each amplifier.

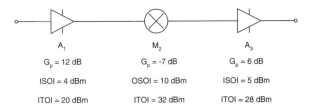

Figure 6-46 *Simple three-element cascade used to analyze the linearity of cascaded devices.*

In the following derivations, we define:

$G_{p,n}$ = the power gain of the nth element in the cascade,

$G_{p,cas}$ = the power gain of the cascade,
$ITOI_n$ = the input third-order intercept of the nth element in the cascade,
$OTOI_n$ = the output third-order intercept of the nth element in the cascade,
P_f = signal power at the fundamental frequency,
$P_{2f\text{-}f}$ = the third-order power at $2f_1 - f_2$ or at $2f_2 - f_1$.

Also, let

$P_{in,f,n}$ = the fundamental input power of each tone applied to element #n of the cascade. The two tones are at frequencies f_1 and f_2.

$P_{out,2f\text{-}f,n}$ = the third-order output power of one tone generated by cascade element #n. If the two input signals are at frequencies f_1 and f_2, then the third-order signals will be at $2f_2 - f_1$ and $2f_1 - f_2$.

$P_{out,2f\text{-}f,n,pt3}$ = the third-order output power generated by cascade element #n after it has been mathematically transferred to the end of the cascade (point 3 in Figure 6-46).

All terms are assumed linear if they lack a dB or dBm subscript. For example, G_p and $P_{out,f}$ are linear terms. $G_{p,dB}$ and $P_{out,f,dBm}$ are described in decibels. An *out* subscript ties a parameter to the output port, an *in* subscript to the input port.

Third-Order Intercept

Equation 6.59 expresses the relationship between the fundamental output power, the third-order output power and the OTOI of the nonlinear device. Equation 6.59 is repeated here.

$$P_{out,2f-f,dBm} = 3P_{out,f,dBm} - 2OTOI_{dBm} \qquad 6.85$$

We also know

$$P_{out,f,dBm} = P_{in,f,dBm} + G_{p,dB} \qquad 6.86$$

The linear equivalents for these two equations are

$$P_{out,2f-f} = \frac{P_{out,f}^3}{OTOI^2} \qquad 6.87$$

and

$$P_{out,f} = P_{in,f} G_p$$

Equation 6.59 and its linear equivalent, Equation 6.87, indicate that the third-order distortion power is proportional to the fundamental input power cubed and inversely proportional to the amplifier's OTOI squared.

Amplifier #1
Each tone we apply to the input of amplifier #1 has a power of $P_{in,1}$. The fundamental and third-order output powers from amplifier #1 are

$$P_{out,f,1} = G_{p,1} P_{in,f,1} \qquad 6.88$$

and

$$P_{out,2f-f,1} = \frac{P_{out,f,1}^3}{OTOI_1^2} \qquad 6.89$$

$$= P_{in,f,1}^3 \frac{G_{p,1}^3}{OTOI_1^2}$$

At the end of the cascade (point 3 in Figure 6-46), the third-order output power due to amplifier #1 is

$$P_{out,2f-f,1,pt3} = P_{out,2f-f,1} G_{p,2} G_{p,3}$$
$$= P_{in,f,1}^3 \frac{G_{p,1}^3 G_{p,2} G_{p,3}}{OTOI_1^2} \qquad 6.90$$

Amplifier #2
The fundamental signal power present at the input of amplifier #2 is

$$P_{in,f,2} = P_{out,f,1} = P_{in,f,1} G_{p,1} \qquad 6.91$$

The fundamental and third-order powers at the output of amplifier #2 are

$$P_{out,f,2} = G_{p,1} G_{p,2} P_{in,f,1} \qquad 6.92$$

and

$$P_{out,2f-f,2} = \frac{P_{out,f,2}^3}{OTOI_2^2} \qquad 6.93$$

$$= P_{in,f,1}^3 \frac{G_{p,1}^3 G_{p,2}^3}{OTOI_2^2}$$

The third-order power present at the output of the cascade due to amplifier #2 is

$$P_{out,2f-f,2,pt3} = P_{out,2f-f,2}G_{p,3}$$
$$= P_{in,f,1}^3 \frac{G_{p,1}^3 G_{p,2}^3 G_{p,3}}{OTOI_2^2} \quad 6.94$$

Amplifier #3
The fundamental power applied to amplifier #3 is

$$P_{in,f,3} = G_{p,1}G_{p,2}P_{in,f,1} \quad 6.95$$

The fundamental output power of amplifier #3 is

$$P_{out,f,3} = G_{p,1}G_{p,2}G_{p,3}P_{in,f,1} \quad 6.96$$

The third-order power present at the output port of the cascade due to amplifier #3 is

$$P_{out,2f-f,3} = P_{out,2f-f,3,pt3}$$
$$= \frac{P_{out,f,3}^3}{OTOI_3^2} \quad 6.97$$
$$= P_{in,f,1}^3 \frac{G_{p,1}^3 G_{p,2}^3 G_{p,3}^3}{OTOI_3^2}$$

Third-Order Power Summation
We now have third-order distortion power from three separate devices present at the output of the last amplifier in the cascade.

- $P_{out,2f-f,1,pt3}$ is the third-order power generated by amplifier #1. This power was multiplied by the gain of the last two stages to bring it to point 3, the output of the cascade.

- $P_{out,2f-f,2,pt3}$ is the distortion power generated by amplifier #2. This power was moved mathematically to the end of the cascade.

- $P_{out,2f-f,3,pt3}$ describes the third-order distortion power generated by amplifier #3 present at the end of the cascade.

Coherent vs. Noncoherent Summation
The third-order distortion products present at point 3 of Figure 6-46 are all at the same frequency. To combine the three signals accurately, we

have to know the relative phases of the signals as they exist at the end of the cascade. If all of the signals are in phase, the vector sum of these signals will be one quantity. If some of the signals are in phase while others have different phases, the vector sum of the signals is a different quantity.

The most nonlinear cascade results when we assume that the distortion voltages are all exactly in phase and add directly together, which is called *coherent summation*.

The term *noncoherent summation* means that the signal powers are added together. This does not produce a worst-case result but models the behavior of most systems. In practice, we usually observe results that suggest that signals are adding noncoherently. At some frequency, however, the distortion will add together in a coherent fashion, and the coherent numbers can be seen. The noncoherent summation results in a mostly right value. The coherent summation data represents a worst-case result.

Coherent Summation TOI Equation

First the nonlinear power components present at the output of the cascade are converted into voltages, then the addition is performed, and finally the sum is converted back into power. The coherent sum of several signals is

$$P_{total,coherent} = \left[P_1^{1/2} + P_2^{1/2} + P_3^{1/2} + \ldots + P_n^{1/2} \right]^2 \qquad 6.98$$

The total third-order power available at the output of the cascade is

$$P_{out,2f-f,cas} = P_{in,f,1}^3 \left[\sqrt{\frac{G_{p,1}^3 G_{p,2} G_{p,3}}{OTOI_1^2}} + \sqrt{\frac{G_{p,1}^3 G_{p,2}^3 G_{p,3}}{OTOI_2^2}} + \sqrt{\frac{G_{p,1}^3 G_{p,2}^3 G_{p,3}^3}{OTOI_3^2}} \right]^2 \qquad 6.99$$

The cascade gain is

$$G_{p,cas} = G_{p,1} G_{p,2} G_{p,3} \qquad 6.100$$

Combining these two equations produces

$$P_{out,2f-f,cas} = P_{in,f,1}^3 G_{p,cas}^3 \left[\sqrt{\frac{1}{G_{p,2}^2 G_{p,3}^2 OTOI_1^2}} + \sqrt{\frac{1}{G_{p,3}^2 OTOI_2^2}} + \sqrt{\frac{1}{OTOI_3^2}} \right]^2$$

$$= P_{out,f,1}^3 G_{p,cas}^3 \left[\frac{1}{G_{p,2} G_{p,3} OTOI_1} + \frac{1}{G_{p,3} OTOI_2} + \frac{1}{OTOI_3} \right]^2 \qquad 6.101$$

The third-order output power for the cascade is

$$P_{out,2f-f} = \frac{P_{out,f}^3}{OTOI^2} \qquad 6.102$$

These equations produce an expression for the cascade third-order intercept

$$\frac{P_{out,f,1}^3}{OTOI_{cas}^2} = P_{out,f,1}^3 \left[\frac{1}{G_{p,2}G_{p,3}OTOI_1} + \frac{1}{G_{p,3}OTOI_2} + \frac{1}{OTOI_3} \right]^2 \qquad 6.103$$

which can be simplified to

$$\frac{1}{OTOI_{cas}} = \frac{1}{G_{p,2}G_{p,3}OTOI_1} + \frac{1}{G_{p,3}OTOI_2} + \frac{1}{OTOI_3} \qquad 6.104$$

or, in terms of ITOI,

$$\frac{1}{ITOI_{cas}} = \frac{1}{ITOI_1} + \frac{1}{ITOI_2/G_{p,1}} + \frac{1}{ITOI_3/G_{p,1}G_{P,2}} \qquad 6.105$$

More algebra for an n-element cascade reveals

$$\frac{1}{OTOI_{cas}} = \frac{1}{G_{p,2}G_{p,3}G_{p,4}\cdots G_{p,n-1}OTOI_1}$$
$$+ \frac{1}{G_{p,3}G_{p,4}\cdots G_{p,n-1}OTOI_2}$$
$$+ \ldots$$
$$+ \frac{1}{OTOI_n} \qquad 6.106$$

and

$$\frac{1}{ITOI_{cas}} = \frac{1}{ITOI_1}$$
$$+ \frac{1}{ITOI_2/G_{p,1}}$$
$$+ \ldots$$
$$+ \frac{1}{ITOI_n/G_{p,1}G_{p,2}\cdots G_{p,n-1}} \qquad 6.107$$

Noncoherent Summation TOI Equation

Noncoherent summation assumes the phases of the third-order distortion from each cascade element are random. The distortion powers are simply added together.

$$P_{total,noncoherent} = P_1 + P_2 + P_3 + \ldots + P_n \qquad 6.108$$

The total third-order power available at the output of the cascade is

$$P_{out,2f-f,cas} = P_{in,f,1}^3 \left[\frac{G_{p,1}^3 G_{p,2} G_{p,3}}{OTOI_1^2} + \frac{G_{p,1}^3 G_{p,2}^3 G_{p,3}}{OTOI_2^2} + \frac{G_{p,1}^3 G_{p,2}^3 G_{p,3}^3}{OTOI_3^2} \right] \qquad 6.109$$

$$= P_{out,f,1}^3 \left[\frac{1}{G_{p,2}^2 G_{p,3}^2 OTOI_1^2} + \frac{1}{G_{p,3}^2 OTOI_2^2} + \frac{1}{OTOI_3^2} \right]$$

Again,

$$P_{out,2f-f} = \frac{P_{out,f}^3}{OTOI^2} \qquad 6.110$$

which combines to produce

$$\frac{P_{out,f,1}^3}{OTOI_{cas}^2} = P_{out,f,1}^3 \left[\frac{1}{G_{p,2}^2 G_{p,3}^2 OTOI_1^2} + \frac{1}{G_{p,3}^2 OTOI_2^2} + \frac{1}{OTOI_3^2} \right] \qquad 6.111$$

Simplifying reveals

$$\frac{1}{OTOI_{cas}^2} = \frac{1}{\left(G_{p,2} G_{p,3} OTOI_1\right)^2} + \frac{1}{\left(G_{p,3} OTOI_2\right)^2} + \frac{1}{\left(OTOI_3\right)^2} \qquad 6.112$$

and

$$\frac{1}{ITOI_{cas}^2} = \frac{1}{\left(ITOI_1\right)^2} + \frac{1}{\left(ITOI_2/G_{p,1}\right)^2} + \frac{1}{\left(ITOI_3/G_{p,1} G_{p,2}\right)^2} \qquad 6.113$$

The equations describing an n-element cascade are

and

$$\frac{1}{OTOI_{cas}^2} = \frac{1}{\left(G_{p,2}G_{p,3}G_{p,4}\cdots G_{p,n}OTOI_1\right)^2}$$
$$+ \frac{1}{\left(G_{p,3}G_{p,4}\cdots G_{p,n}OTOI_2\right)^2}$$
$$+ \cdots$$
$$+ \frac{1}{\left(OTOI_n\right)^2} \qquad 6.114$$

$$\frac{1}{ITOI_{cas}^2} = \frac{1}{\left(ITOI_1\right)^2}$$
$$+ \frac{1}{\left(ITOI_2/G_{p,1}\right)^2}$$
$$+ \cdots$$
$$+ \frac{1}{\left(ITOI_n/G_{p,1}G_{p,2}G_{p,3}\cdots G_{p,n-1}\right)^2} \qquad 6.115$$

Examining the TO Cascade Equation

The cascade Equations 6.104 through 6.107 and 6.112 through 6.115 provide insight into system design. To simplify the discussion, only Equations 6.106 and 6.107 are explained although the discussion applies to all the linearity equations.

Component Contribution

The linearity contribution of each component in the cascade is neatly tied up in a single term of the cascade equations. In a 3-element cascade, for example, the third component's contribution to the cascade's linearity is found only in the term

$$\frac{1}{ITOI_3/G_{p,1}G_{p,2}} \qquad 6.116$$

This is the only place in the cascade equation that contains the ITOI of the third element, which will make cascade analysis much easier.

Imaging

Concerning the linearity contribution of the third element in a cascade, the only term that needs to be analyzed is

$$\frac{1}{ITOI_3/G_{p,1}G_{p,2}} \qquad 6.117$$

If we think of the $ITOI_3$ as power in watts, for example, the term

$$ITOI_3/G_{p,1}G_{p,2} \qquad 6.118$$

appears as if we are mathematically moving that power level to the input of the cascade. Every term in the cascade equation can be thought of in this manner. When we perform cascade analysis in later chapters, we will move the TOI of each device in the cascade to a common port to perform comparisons and look for the weak link in the cascade's TOI chain.

The linearity of a cascade depends upon the linearity of each component and the power gain surrounding the component. Components preceded by a lot of power gain will experience higher signal levels and generate higher levels of distortion power. For example, if we apply a –80 dBm signal to the cascade of Figure 6-47, amplifier #1 will see a –80 dBm signal. Amplifier #2 will see a –60 dBm signal because of the 23 dB power gain of amplifier #1. Although amplifiers #1 and #2 have the same TOI, amplifier #2 will generate higher levels of distortion because it sees a larger signal. Likewise, amplifier #3 will see a –40 dBm signal and will generate the most distortion power of all the devices in the cascade. This is the reason that amplifier #3 dominates the TOI of the cascade.

When we translate the TOI of every component to a common port in a well-designed system, the translated TOI of each component should be about equal. If all the reflected TOIs are equal, then all of the elements in the cascade become nonlinear at the same input power level. If one element is too small, the TOI of the entire cascade will be dominated by that one weak link.

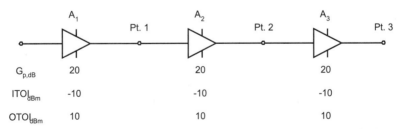

Figure 6-47 *Three-element cascade.*

614 | RADIO RECEIVER DESIGN

Resistors in Parallel

Equations 6.104 through 6.107 might look familiar to electrical engineers, because they resemble the equation that describes resistors in parallel.

$$\frac{1}{R_p} = \frac{1}{R_1} + \frac{1}{R_2} + \frac{1}{R_3} \qquad 6.119$$

Figure 6-48 shows the similarities between Equations 6.104 and 6.105 graphically. The value of each resistor represents the *TOI* of each component when it is referenced to a common port. Whenever a new device is added to a cascade, another parallel resistor is added to the equivalent circuits, lowering the *TOI* (resistance) of the cascade. Adding another component will never improve the linearity of the cascade.

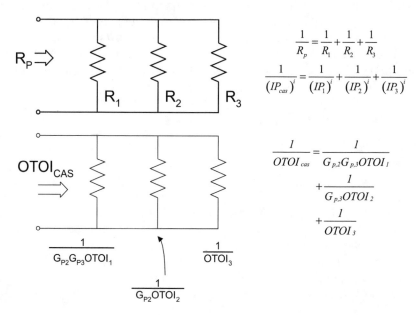

Figure 6-48 *The similarity of the TOI cascade equation to adding resistors in parallel.*

Second-Order Intercept

Figure 6-49 shows the system used in this analysis. Knowing the gain and *OSOI* of each amplifier, we want to find the *OSOI* of the cascade. Equation 6.45, repeated here, shows an important relationship between the fundamental and second-order power levels present in a nonlinear system.

$$P_{out,2f,dBm} = 2P_{out,f,dBm} - OSOI_{dBm} \qquad 6.120$$

We also know

$$P_{out,f,dBm} = P_{in,f,dBm} + G_{p,dB} \qquad 6.121$$

Converting these equations to linear expressions produces

$$P_{out,2f} = \frac{P_{out,f}^2}{OSOI} \qquad 6.122$$

and

$$P_{out,f} = P_{in,f} G_p \qquad 6.123$$

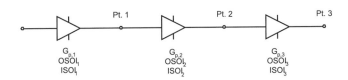

Figure 6-49 *Simple three-element cascade used to analyze the linearity of cascaded devices.*

More Definitions
- $P_{in,f,n}$ = the fundamental input power applied to amplifier #n.

- $P_{out,2f,n}$ = the second-order output power generated by amplifier #n. If the input signal is at a frequency of f, then the second-order output signal will be at $2f$.

- $P_{out,2f,n,pt3}$ = the second-order output power generated by amplifier #n present at the end of the cascade (point 3 in Figure 6-49). The second-order signals generated by an amplifier will experience the power gain of the succeeding stages.

Amplifier #1
First, we apply an input power of $P_{in,f,1}$ to amplifier #1 and calculate the fundamental and second-order output powers.

$$P_{out,f,1} = P_{in,f,1} G_{p,1} \qquad 6.124$$

and

$$P_{out,2f,1} = \frac{P_{out,f,1}^2}{OSOI_1} \qquad 6.125$$

$$= P_{in,f,1}^2 \frac{G_{p,1}^2}{OSOI_1}$$

At point 3 in Figure 6-49 (the end of the cascade), the second-order output power due to amplifier #1 is

$$P_{out,2f,1,pt3} = P_{out,2f,1} G_{p,2} G_{p,3} \qquad 6.126$$

$$= P_{in,f,1}^2 \frac{G_{p,1}^2 G_{p,2} G_{p,3}}{OSOI_1}$$

Amplifier #2

The input signal we applied to the input of amplifier #1 experiences the power gain of the first amplifier then arrives at the input to amplifier #2. The fundamental input power to amplifier #2 is

$$P_{in,f,2} = P_{out,f,1} = P_{in,f,1} G_{p,1} \qquad 6.127$$

The fundamental and second-order powers at the output of amplifier #2 are

$$P_{out,f,2} = P_{in,f,1} G_{p,1} G_{p,2} \qquad 6.128$$

and

$$P_{out,2f,2} = \frac{P_{out,f,2}^2}{OSOI_2} \qquad 6.129$$

$$= P_{in,f,1}^2 \frac{G_{p,1}^2 G_{p,2}^2}{OSOI_2}$$

The second-order output power present at the end of the cascade due to amplifier #2 is

$$P_{out,2f,2,pt3} = P_{out,2f,2} G_{p,3} \qquad 6.130$$

$$= P_{in,f,1}^2 \frac{G_{p,1}^2 G_{p,2}^2 G_{p,3}}{OSOI_2}$$

Amplifier #3

The fundamental input power to amplifier #3 is

$$P_{in,f,3} = P_{in,f,1} G_{p,1} G_{p,2} \qquad 6.131$$

The fundamental output power of amplifier #3 is

$$P_{out,f,3} = P_{in,f,1} G_{p,1} G_{p,2} G_{p,3} \qquad 6.132$$

The second-order power at the output port of the cascade generated in amplifier #3 is

$$\begin{aligned} P_{out,2f,3} &= P_{out,2f,3,pt3} \\ &= \frac{P_{out,f,3}^2}{OSOI_3} \\ &= P_{in,f,1}^2 \frac{G_{p,1}^2 G_{p,2}^2 G_{p,3}^2}{OSOI_3} \end{aligned} \qquad 6.133$$

Second-Order Power Summation

We now have three expressions that describe the second-order distortion power present at the output of the last amplifier in the cascade.

- $P_{out,2f,1,pt3}$ describes the distortion power generated by amplifier #1. This power was multiplied by the gain of the last two stages to bring it to point 3, the output of the cascade.

- $P_{out,2f,2,pt3}$ describes the distortion power generated by amplifier #2. This power was moved mathematically to the end of the cascade.

- $P_{out,2f,3,pt3}$ describes the second-order distortion power generated by amplifier #3 present at the end of the cascade.

Coherent Summation SOI Equation

The coherent sum of several signals is

$$P_{total,coherent} = \left[P_1^{1/2} + P_2^{1/2} + P_3^{1/2} + \ldots + P_n^{1/2} \right]^2 \qquad 6.134$$

The total second-order power available at the output of the cascade is

$$P_{out,2f,cas} = P_{in,f,1}^2 \left[\sqrt{\frac{G_{p,1}^2 G_{p,2} G_{p,3}}{OSOI_1}} + \sqrt{\frac{G_{p,1}^2 G_{p,2}^2 G_{p,3}}{OSOI_2}} + \sqrt{\frac{G_{p,1}^2 G_{p,2}^2 G_{p,3}^2}{OSOI_3}} \right]^2 \qquad 6.135$$

The cascade gain is

$$G_{p,cas} = G_{p,1}G_{p,2}G_{p,3} \qquad 6.136$$

Combining the cascade gain with Equation 6.135 produces

$$P_{out,2f,cas} = P_{in,f,cas}^2 G_{p,cas}^2 \left[\sqrt{\frac{1}{G_{p,2}G_{p,3}OSOI_1}} + \sqrt{\frac{1}{G_{p,3}OSOI_2}} + \sqrt{\frac{1}{OSOI_3}}\right]^2 \qquad 6.137$$

$$= P_{out,f,cas}^2 \left[\sqrt{\frac{1}{G_{p,2}G_{p,3}OSOI_1}} + \sqrt{\frac{1}{G_{p,3}OSOI_2}} + \sqrt{\frac{1}{OSOI_3}}\right]^2$$

Looking at the entire cascade as a single unit, the second-order output power of the cascade is

$$P_{out,2f,cas} = \frac{P_{out,f,cas}^2}{OSOI_{cas}} \qquad 6.138$$

Combining these two equations produces

$$\frac{P_{out,f,cas}^2}{OSOI_{cas}} = P_{out,f,cas}^2 \left[\sqrt{\frac{1}{G_{p,2}G_{p,3}OSOI_1}} + \sqrt{\frac{1}{G_{p,3}OSOI_2}} + \sqrt{\frac{1}{OSOI_3}}\right]^2 \qquad 6.139$$

which can be simplified to

$$\frac{1}{\sqrt{OSOI_{cas}}} = \frac{1}{\sqrt{G_{p,2}G_{p,3}OSOI_1}} + \frac{1}{\sqrt{G_{p,3}OSOI_2}} + \frac{1}{\sqrt{OSOI_3}} \qquad 6.140$$

The general expression for n devices in cascade is

$$\frac{1}{\sqrt{OSOI_{cas}}} = \frac{1}{\sqrt{G_{p,2}G_{p,3}G_{p,4}\ldots G_{p,n-1}OSOI_1}} + \frac{1}{\sqrt{G_{p,3}G_{p,4}\ldots G_{p,n-1}OSOI_2}} + \cdots + \frac{1}{\sqrt{OSOI_n}} \qquad 6.141$$

The input second-order intercept point for the 3-element cascade of Figure 6-49 is

$$\frac{1}{\sqrt{ISOI_{cas}}} = \frac{1}{\sqrt{ISOI_1}} + \frac{1}{\sqrt{ISOI_2/G_{p,1}}} + \frac{1}{\sqrt{ISOI_3/G_{p,1}G_{p,2}}} \qquad 6.142$$

For n devices in cascade, the expression is

$$\frac{1}{\sqrt{ISOI_{cas}}} = \frac{1}{\sqrt{ISOI_1}}$$
$$+ \frac{1}{\sqrt{ISOI_2/G_{p,1}}}$$
$$+ \cdots$$
$$+ \frac{1}{\sqrt{ISOI_n/G_{p,1}G_{p,2}\cdots G_{p,n-1}}} \qquad 6.143$$

Example 6.12 — SOI Cascade (Coherent Addition)

Figure 6-50 shows three devices in cascade. Find the ISOI of the cascade.

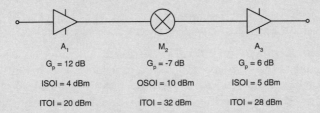

Figure 6-50 Example 6.12.

Solution —

Note the mixture of *ISOI*'s and *OSOI*'s. Rather than following Equation 6.141, first the *SOI* of each device is moved to the input of the cascade, then Equation 6.143 is applied.

Amplifier #1. We already know the $ISOI_{1,dBm}$ at the input of the cascade so

$$\text{ISOI}_{\text{amp1,input}} = 4 \text{ dBm} = 2.51 \text{ mW} \qquad 6.144$$

Mixer #2. Moving the *OSOI* of the mixer to the input of the cascade produces

$$\begin{aligned}\text{ISOI}_{\text{mix2,input}} &= OSOI_{dBm,2} - G_{P,dB,2} - G_{p,dB,1} \\ &= 10 + 7 - 12 \\ &= 5 \text{ dBm} \\ &= 3.16 \text{ mW}\end{aligned} \qquad 6.145$$

Amplifier #3. The equivalent input *SOI* of amplifier #3 is

$$\text{ISOI}_{\text{amp3,input}} = ISOI_{dBm,3} - G_{p,dB,2} - G_{P,dB,1}$$
$$= 5 + 7 - 12 \qquad \qquad 6.146$$
$$= 0 \text{ dBm}$$
$$= 1.00 \text{ mW}$$

Since all of the equivalent input *SOI*'s are about the same, each amplifier contributes about equally to the *SOI* of the cascade. Plugging the equivalent input *SOI* into Equation 6.143 produces

$$\frac{1}{\sqrt{ISOI_{cas}}} = \frac{1}{\sqrt{2.51}} \quad \text{(Amp1 contribution)}$$
$$+ \frac{1}{\sqrt{3.16}} \quad \text{(Mix1 contribution)} \qquad 6.147$$
$$+ \frac{1}{\sqrt{1.00}} \quad \text{(Amp3 contribution)}$$
$$ISOI_{cas} = 0.207 \text{ mW} = -6.8 \text{ dBm}$$

Noncoherent Summation SOI Equation

We calculated the equivalent *SOI* of a cascade by mathematically moving the distortion generated by the components in the cascade to a single point. We then summed up the distortion components assuming that they were all in phase (coherent addition). This produces the worst-case results described by Equations 6.141 and 6.143.

However, in practice, the distortion components are not always in phase, and we find that the results of Equations 6.141 and 6.143 are often too conservative. If we assume the phases of the distortion components are random by the time they arrive at a signal point, we can simply add the powers of all the distortion components together.

For a simple three-element cascade, noncoherent summation produces the following equations:

$$\frac{1}{OSOI_{cas}} = \frac{1}{G_{p,2}G_{p,3}OSOI_1} + \frac{1}{G_{p,3}OSOI_2} + \frac{1}{OSOI_3} \qquad 6.148$$

and

$$\frac{1}{ISOI_{cas}} = \frac{1}{ISOI_1} + \frac{1}{ISOI_2/G_{p,1}} + \frac{1}{ISOI_3/G_{p,1}G_{p,2}} \qquad 6.149$$

For the general *n*-element cascade, noncoherent summation produces

LINEARITY

$$\frac{1}{OSOI_{cas}} = \frac{1}{G_{p,2}G_{p,3}G_{p,4}\ldots G_{p,n-1}OSOI_1}$$
$$+ \frac{1}{G_{p,3}G_{p,4}\ldots G_{p,n-1}OSOI_2} \qquad 6.150$$

and

$$\frac{1}{ISOI_{cas}} = \frac{1}{ISOI_1}$$
$$+ \frac{1}{ISOI_2/G_{p,1}}$$
$$+ \ldots \qquad 6.151$$
$$+ \frac{1}{ISOI_n/G_{p,1}G_{p,2}\ldots G_{p,n-1}}$$

Example 6.13 — SOI Cascade (Noncoherent Addition)

Find the *ISOI* of the cascade of Figure 6-50 assuming noncoherent addition. This is the same cascade we examined earlier using coherent summation.

Solution —

Applying Equation 6.151 produces

$$\frac{1}{ISOI_{cas}} = \frac{1}{2.51} \quad \text{(Amp1 contribution)}$$
$$+ \frac{1}{3.16} \quad \text{(Mix1 contribution)} \qquad 6.152$$
$$+ \frac{1}{1.00} \quad \text{(Amp3 contribution)}$$

$ISOI_{cas} = 0.583$ mW $= -2.3$ dBm

The noncoherent assumption implies that the system's *ISOI* is −2.3 dBm. The coherent assumption produces a cascade *ISOI* of −6.8 dBm. As expected, noncoherent addition predicts that the cascade will have a higher *SOI*, i.e., it will be a more linear system than coherent addition.

Examining the SOI Cascade Equations

Equations 6.141 and 6.143 relate the cascade's second-order performance to the second-order performance of the components in the cascade. The component contributions, imaging and resistors in parallel observations we made for the third-order cascade equations apply equally well to the second-order equations.

Compression Point

The problem of compression points in cascade does not easily lend itself to analysis. Using the rule of thumb from Equation 6.80 it can be stated

$$CP_{dBm} \approx TOI_{dBm} - (10 \text{ or } 15 \text{ dB})\qquad 6.153$$

We calculate the cascade TOI then infer the cascade CP from the TOI.

6.19 Distortion Notes

Third-Order Measurement Difficulties

Occasionally, when we perform a third-order intercept test, we will observe the output spectrum shown in Figure 6-51.

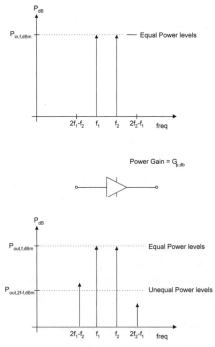

Figure 6-51 Third order intercept measurement difficulties. The distortion power levels are not identical.

Equation 6.72 indicates that we need to measure the difference between the fundamental output power and the third-order output power. Occasionally, when measuring the nonlinearity of a device, the two tones will not be at the same power level (as theory dictates).

What is the third-order intercept of this device? If the smaller Δ is measured, the result will be a smaller number for the device's *TOI*, i.e., the device will look less linear. Measuring the larger Δ yields a larger *TOI* number, and the device will appear to be more linear. The two Δs can also be averaged. It is wise to choose whichever method seems more reasonable for each particular application and be consistent.

Input vs. Output Specifications

Note the difference between input and output specifications.

$$OTOI_{dBm} = ITOI_{dBm} + G_{p,dB}$$
$$OSOI_{dBm} = ISOI_{dBm} + G_{p,dB} \qquad\qquad 6.154$$
$$OCP_{dBm} = ICP_{dBm} + G_{p,dB} - 1$$

If the reference port is not given, how can we differentiate the input specification and the output specification? The vendor will often specify the number on the port where it looks the best. Since a larger *SOI* (or *TOI*) is better than smaller *SOI* (*TOI*), the linearity of a device on the output port, if the device has gain, will be given. If the device is lossy (such as a mixer or a filter), the vendor will usually detail the input numbers. For example, an amplifier with a 10 dB power gain has an *OSOI* of +7 dBm and an *ISOI* of –3 dBm. The specifications would read "a +7 dBm third-order intercept." However, a mixer might have a gain of –10 dB (a loss of 10 dB), an *OTOI* of +7 dBm and an *ITOI* of +17 dBm. The vendor will usually specify "a +17 dBm third-order intercept."

Model Inadequacy

Occasionally, the results actually measured will deviate from the theory presented here because several simplifying assumptions were made during this analysis. When discussing the weakly nonlinear approximation and its implications, a fairly small input power level was assumed. If we apply too much power to any device, it will no longer behave in a weakly nonlinear fashion, and the analysis becomes invalid.

Another simplifying assumption was that the transfer function of any nonlinear device can be modeled as a cubic polynomial with a zero DC level or

$$V_{out}(t) = k_1 V_{in}(t) + k_2 V_{in}^2(t) + k_3 V_{in}^3(t) \qquad 6.155$$

If this assumption was always true, the output spectrum of any third-order nonlinear device would always show a specific symmetry. Figure 6-52(a) shows such a symmetrical spectrum. The power levels of symmetrically spaced signals are identical. However, we often observe spectra similar to the one shown in Figure 6-52 (b). The amplitude of the nonlinear products have lost their symmetry although the distortion products are still at exactly the same frequencies. The simple third-order relationship also predicts that the second-order products will rise at a 2:1 rate with increasing input power, and the third-order products will rise at a 3:1 rate.

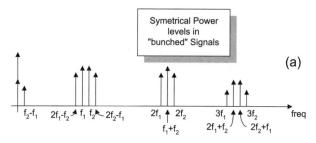

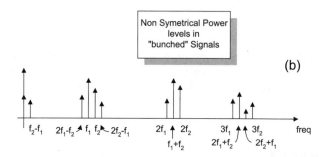

Figure 6-52 *Symmetrical and nonsymmetrical output spectra. This simple analysis predicts a symmetrical output spectrum (a). In practice, we often observe a nonsymmetrical output spectrum (b).*

The following mechanisms cause the model to break down and produce nonsymmetrical spectra.

- The model was developed assuming a weakly nonlinear system. If the input signals are too strong, the model rapidly breaks down.

- We have assumed that the gain and input impedance of the device stays constant over frequency and input power. This is not always the case.
- The third-order relationship between V_{in} and V_{out} describes a memory-less system. The output depends only upon the instantaneous input voltage. In reality, devices contain inductors and capacitors with memory. The present state depends on their instantaneous inputs and upon the past inputs.

However, the approximations inherent in the simple third-order equation hold under the most common conditions.

Linearity and Power Consumption

If linearity is such a problem, why not just build systems out of very linear components to avoid problematic *SOI*, *TOI* or any other nonlinear specification? Linearity in a particular device can only be reached at the expense of other factors. For example, an amplifier with a high *TOI* will tend to use more DC power than a device with a low *TOI*. The same is true for *SOI* and for CP. In other words, there is a strong correlation between the linearity of a device and its DC power consumption. A mixer with a high *TOI* will usually require more *local oscillator* (LO) drive power than a mixer with a low *TOI*. However, filters and simple resistive attenuators are usually very linear devices. The *TOI* of a band-pass filter, for example, can be several thousand watts. At high power levels, however, the inductors can saturate and the dielectrics of the capacitors can exhibit nonlinear properties.

Coherent vs. Noncoherent Addition

Coherent addition is more conservative than noncoherent addition, i.e., the coherent assumption indicates a less linear system than the noncoherent equations indicate. In a worst-case scenario, coherent addition should be used.

When designing low-noise receiving systems, it was found that well-designed cascades usually behave as though the distortion products are adding up noncoherently. For the most part, these systems have achieved the equivalent of noncoherent summation plus one or two dB. With wideband systems, the cascade *SOI* or *TOI* will stay at noncoherent levels over most of the frequency range of the system. However, over narrow frequency ranges, the *SOI* and *TOI* will increase to coherent summation levels.

In a well-designed system (where the equivalent intercept points of all the devices are equal), the difference between coherent and noncoherent summation is 4 to 5 dB. When designing a system, it is best to calculate the numbers for both the coherent and noncoherent cases to assess the variation likely to be expected over time and frequency.

Second-Order Distortion and Mixers

Second-order effects are extremely disadvantageous regarding amplifiers. They are useful, however, when a signal is moved from one center frequency to another with the help of a mixer. A mixer is a device built with enhanced second-order performance. We use the sum and difference products arising from the $k_2 V_{in}^2$ term to perform frequency translation, phase detection and other functions.

6.20 Nonlinearities and Modulated Signals

Nonlinearities are not just undesirable effects to be avoided; they affect modulated signals in both useful and harmful ways.

One Modulated Signal

First, we will explore what happens to a single modulated signal as it passes through a nonlinear device. The input waveform is

$$V_{in}(t) = [1 + A(t)] \cos[\omega t + \phi(t)] \qquad 6.156$$

where

$A(t)$ represents the AM modulation present on the signal and

$$0 \leq |A(t)| \leq 1 \qquad 6.157$$

$f(t)$ represents the PM or FM modulation present on the signal.

Second-Order Output

When the waveform of Equation 6.156 undergoes a second-order process, the output waveform is

$$\begin{aligned}
V_{out,2}(t) &= V_{in}^2(t) \\
&= [1 + A(t)]^2 \cos^2[\omega t + \phi(t)] \\
&= \frac{[1 + A(t)]^2}{2} + \frac{[1 + A(t)]^2}{2} \cos[2\omega t + 2\phi(t)] \\
&= \frac{1}{2} + A(t) + \frac{A^2(t)}{2} + \left[\frac{1}{2} + A(t) + \frac{A^2(t)}{2}\right] \cos[2\omega t + 2\phi(t)]
\end{aligned} \qquad 6.158$$

The second term on the last line of Equation 6.158 [the $A(t)$] is the amplitude modulation present on the input signal (see Equation 6.156). We can low-pass Equation 6.158 to remove the high-frequency components.

$$V_{out,2,LowPassed}(t) = \frac{1}{2} + A(t) + \frac{A^2(t)}{2} \qquad 6.159$$

The low-passed, second-order process returns DC, the AM signal and a distortion term involving $A_2(t)$. A device with second-order characteristics can be used to demodulate an AM waveform. We can recover the AM present on $V_{in}(t)$ even though the input waveform exhibits angle modulation [i.e., the FM or PM represented by $f(t)$].

The output waveform also contains the term $A_2(t)/2$ which is the AM signal squared. The demodulated waveform will exhibit second-order distortion. If the magnitude of $A(t)$ is a lot less than unity, the second-order distortion will be very small and can often be ignored.

Using a squaring operation to demodulate AM is a very common technique. The nonlinear element is usually a single diode or transistor. Equation 6.158 also contains a signal centered at the second harmonic of the input. The second-harmonic signal still possesses the original AM but the modulation now exhibits second-order distortion.

The angle modulation $f(t)$ has also changed. Let us assume that $f(t)$ represents frequency modulation and that the original frequency deviation was 100 kHz. The deviation at the second harmonic has doubled to 200 kHz via the $2f(t)$ term of Equation 6.158. This effect accounts for the increase in the phase noise of an oscillator when the frequency of an oscillator is doubled (see Chapter 4 for more details).

If the input signal were a quiet carrier, i.e., if both $A(t)$ and $f(t)$ were zero, then the output waveform would be

$$V_{out,2}^2(t) = \frac{1}{2} + \frac{1}{2}\cos(2\omega t) \qquad 6.160$$

The output frequency is twice the input frequency, which means a squaring device can be used to double the frequency of an oscillator.

Third-Order Outputs

If we apply the input waveform (see Equation 6.156) to a third-order nonlinear device, the output is

$$V_{out,3}(t) = V_{in}^3(t)$$
$$= [1 + A(t)]^3 \cos^3[\omega t + \phi(t)]$$
$$= [1 + 3A(t) + 3A^2(t) + A^3(t)]$$
$$\left\{ \frac{3}{4}\cos[\omega t + \phi(t)] + \frac{1}{4}\cos[3\omega t + 3\phi(t)] \right\}$$
6.161

The signal at the fundamental and at the third harmonic of the fundamental both still possess amplitude modulation. However, the AM envelope of the signal is distorted. The signal at the fundamental frequency still exhibits the undistorted angle modulation. The deviation of the signal present at the third harmonic has increased three-fold.

Frequency Deviation

Generally, the deviation of a signal will increase directly with the nonlinear order n. If we pass a signal with a 100 kHz bandwidth through a fourth-order device, the output signal will exhibit a 400 kHz deviation. This is one way to determine the order of a signal generated by a nonlinear effect. For example, a FM radio station has an FCC-defined bandwidth of about 150 kHz. If we find a FM radio station with a bandwidth of 3(150) = 450 kHz, we know it is most likely due to third-order distortion somewhere in your system.

Example 6.14 — Nonlinearities and Deviation

While tuning through the spectrum, you find a signal that sounds much like a commercial cellular telephone channel. It is a FM signal which sounds best when the receiver's IF bandwidth is 100 kHz. What is the order of the distortion term causing this signal?

Solution —

Since an undistorted cellular telephone signal has a Rf bandwidth of about 20 kHz, we infer that this signal was probably generated by a fifth-order nonlinearity.

One Modulated Signal, One Quiet Carrier

We now apply two signals simultaneously to the nonlinear device; one signal is modulated, and the other signal contains no modulation (quiet carrier). The modulated signal is

$$V_{\text{mod}}(t) = [1 + A(t)]\cos[\omega_1 t + \phi(t)]$$
6.162

We can describe the quiet carrier as

$$V_{quiet}(t) = \cos(\omega_2 t) \qquad 6.163$$

The input signal is the sum of these two waveforms or

$$\begin{aligned}V_{in}(t) &= V_{mod}(t) + V_{quiet}(t) \\ &= [1+A(t)]\cos[\omega_1 t + \phi(t)] + \cos(\omega_2 t)\end{aligned} \qquad 6.164$$

Second-Order Outputs

Applying a second-order process to the input waveform described by Equation 6.164 produces

$$\begin{aligned}V_{out,2}(t) &= V_{in}^2(t) \\ &= \{[1+A(t)]\cos[\omega_1 t + \phi(t)] + \cos(\omega_2 t)\}^2 \\ &= \frac{[1+A(t)]^2}{2} + \frac{1}{2} + \frac{[1+A(t)]^2}{2}\cos[2\omega_1 t + 2\phi(t)] \\ &\quad + \frac{1}{2}\cos(2\omega_2 t) \\ &\quad + [1+A(t)]\cos[(\omega_1 + \omega_2)t + \phi(t)] \\ &\quad + [1+A(t)]\cos[(\omega_1 - \omega_2)t + \phi(t)]\end{aligned} \qquad 6.165$$

The AM demodulation can be observed with the associated distortion and the signals generated at the second harmonics of the two input signals. The important components of the output waveform are centered approximately at $\omega_1 \pm \omega_2$.

$$[1+A(t)]\cos[(\omega_1 + \omega_2)t + \phi(t)]$$
$$\text{and} \qquad\qquad\qquad 6.166$$
$$[1+A(t)]\cos[(\omega_1 - \omega_2)t + \phi(t)]$$

As long as these two components do not overlap in the frequency domain, we have moved the modulated signal to two new frequencies without distortion. Both the AM and FM modulation is intact and undistorted.

Third-Order Components

The output signal produced when we apply Equation 6.164 to a device with a third-order characteristic is

$$\begin{aligned}
V_{out,3}(t) &= V_{in}^3(t) \\
&= \left\{[1+A(t)]\cos[\omega_1 t + \phi(t)] + \cos(\omega_2 t)\right\}^3 \\
&= \left\{\frac{3[1+A(t)]^3}{4} + \frac{3[1+A(t)]}{2}\right\}\cos[\omega_1 t + \phi(t)] \\
&+ \left\{\frac{3}{4} + \frac{3[1+A(t)]^2}{2}\right\}\cos(\omega_2 t) \\
&+ \frac{[1+A(t)]^3}{4}\cos[3\omega_1 t + 3\phi(t)] \\
&+ \frac{1}{4}\cos(3\omega_2 t) \\
&+ \frac{3[1+A(t)]^2}{4}\cos[(2\omega_1 + \omega_2)t + 2\phi(t)] \\
&+ \frac{3[1+A(t)]^2}{4}\cos[(2\omega_1 - \omega_2)t + 2\phi(t)] \\
&+ \frac{3[1+A(t)]}{4}\cos[(2\omega_2 + \omega_1)t + \phi(t)] \\
&+ \frac{3[1+A(t)]}{4}\cos[(2\omega_2 - \omega_1)t - \phi(t)]
\end{aligned}$$

6.167

Some tones are modulated, and some of those tones are modulated by distorted waveforms. The amount and type of modulation impressed on each output components are determined by the input signals that combined to form the particular component. It is interesting to note that the quiet carrier at ω_2 has acquired AM modulation from the signal at ω_1. This common effect is called *crossmodulation*.

Two Modulated Signals

Examining the effects of a nonlinear device on two modulated signals will provide insight into many common receiver problems. Each signal will have both AM and FM (or PM). The signals are

$$V_{mod,1}(t) = [1+A_1(t)]\cos[\omega_1 t + \phi_1(t)]$$

6.168

and

$$V_{\text{mod},2}(t) = [1 + A_2(t)]\cos[\omega_2 t + \phi_2(t)] \quad\quad 6.169$$

The input signal is the sum of these two waveforms or

$$\begin{aligned}V_{in}(t) &= V_{\text{mod},1}(t) + V_{\text{mod},2}(t) \\ &= [1+A_1(t)]\cos[\omega_1 t + \phi_1(t)] + [1+A_2(t)]\cos[\omega_2 t + \phi_2(t)]\end{aligned} \quad 6.170$$

Second-Order Outputs

Applying a of a second-order process to Equation 6.170 produces

$$\begin{aligned}V_{out,2}(t) &= V_{in}^2(t) \\ &= \{[1+A_1(t)]\cos[\omega_1 t+\phi_1(t)] + [1+A_2(t)]\cos[\omega_2 t+\phi_2(t)]\}^2 \\ &= [1+A_1(t)]^2 \cos^2[\omega_1 t + \phi_1(t)] \\ &\quad + 2[1+A_1(t)][1+A_2(t)]\cos[\omega_1 t+\phi_1(t)]\cos[\omega_2 t+\phi_2(t)] \\ &\quad + [1+A_2(t)]^2 \cos^2[\omega_2 t+\phi_2(t)]\end{aligned} \quad 6.171$$

Accordingly,

$$\begin{aligned}V_{out,2}(t) &= \frac{[1+A_1(t)]^2}{2} + \frac{[1+A_2(t)]^2}{2} \\ &\quad + \frac{[1+A_1(t)]^2}{2}\cos[2\omega_1 t + 2\phi_1(t)] \\ &\quad + \frac{[1+A_2(t)]^2}{2}\cos[2\omega_2 t + 2\phi_2(t)] \\ &\quad + [1 + A_1(t) + A_2(t) + A_1(t)A_2(t)]\cos[(\omega_1+\omega_2)t + \phi_1(t)+\phi_2(t)] \\ &\quad + [1 + A_1(t) + A_2(t) + A_1(t)A_2(t)]\cos[(\omega_1-\omega_2)t + \phi_1(t)-\phi_2(t)]\end{aligned} \quad 6.172$$

The modulation of the input signals is distributed to the second-order output terms in various ways. The nonlinear output signals will contain the modulation which was present on the input waveforms. However, the modulation will often be distorted. This is one useful way to determine if a signal is caused by nonlinearity. If the signal sounds distorted or requires an unusual bandwidth to demodulate, there is a good chance the signal is the result of nonlinearity.

In Chapter 3, the second-order nonlinearity was used to translate a signal from one center frequency to another. Referring to Equation 6.172, the two terms we will use to perform this translation are

$$[1 + A_1(t) + A_2(t) + A_1(t)A_2(t)] \cos[(\omega_1 + \omega_2)t + \phi_1(t) + \phi_2(t)]$$
$$\text{and} \qquad\qquad 6.173$$
$$[1 + A_1(t) + A_2(t) + A_1(t)A_2(t)] \cos[(\omega_1 - \omega_2)t + \phi_1(t) - \phi_2(t)]$$

We normally do everything possible to make certain that one signal is a quiet carrier, i.e., it has no amplitude or phase modulation. This is the local oscillator or LO. However, the LO will always possess some irreducible amplitude or phase modulation. Equation 6.172 states that any modulation present on the LO will be transferred to both of the output signals, i.e., any modulation present on the LO will contaminate the signal of interest.

In practice, the LO amplitude modulation can often be ignored, i.e., it can be assumed that $A_2(t) = 0$. When the local oscillator exhibits only phase noise, the last two terms of Equation 6.172 become

$$[1 + A_1(t)] \cos[(\omega_1 + \omega_2)t + \phi_1(t) + \phi_2(t)]$$
$$\text{and} \qquad\qquad 6.174$$
$$[1 + A_1(t)] \cos[(\omega_1 - \omega_2)t + \phi_1(t) - \phi_2(t)]$$

The oscillator phase noise is transferred directly to the two output signals. This effect often limits the processing of radar signals and can limit the ultimate bit error rate of digital phase-modulated signals.

Third-Order Outputs

Applying a third-order process to Equation 6.170 produces

$$\begin{aligned}
V_{out,3}(t) &= V_{in}^3(t) \\
&= \{[1 + A_1(t)]\cos[\omega_1 t + \phi_1(t)] + [1 + A_2(t)]\cos[\omega_2 t + \phi_2(t)]\}^3 \\
&= \left\{\frac{3[1 + A_1(t)]^3}{4} + \frac{3[1 + A_1(t)][1 + A_2(t)]^2}{2}\right\} \cos[\omega_1 t + \phi_1(t)] \\
&+ \left\{\frac{3[1 + A_2(t)]^3}{4} + \frac{3[1 + A_1(t)]^2[1 + A_2(t)]}{2}\right\} \cos[\omega_2 t + \phi_2(t)] \\
&+ \frac{[1 + A_1(t)]^3}{4} \cos[3\omega_1 t + 3\phi_1(t)] \\
&+ \frac{[1 + A_2(t)]^3}{4} \cos[3\omega_2 t + 3\phi_2(t)] \\
&+ \frac{3[1 + A_1(t)][1 + A_2(t)]^2}{4} \cos[(\omega_1 + 2\omega_2)t + \phi_1(t) + 2\phi_2(t)] \\
&+ \frac{3[1 + A_1(t)][1 + A_2(t)]^2}{4} \cos[(\omega_1 - 2\omega_2)t + \phi_1(t) - 2\phi_2(t)] \\
&+ \frac{3[1 + A_1(t)]^2[1 + A_2(t)]}{4} \cos[(2\omega_1 + \omega_2)t + 2\phi_1(t) + \phi_2(t)] \\
&+ \frac{3[1 + A_1(t)]^2[1 + A_2(t)]}{4} \cos[(2\omega_1 - \omega_2)t + 2\phi_1(t) - \phi_2(t)]
\end{aligned} \qquad 6.175$$

LINEARITY | 633

The modulation of both signals transfers to the third-order output terms in various combinations.

Crossmodulation

If two modulated signals (see Equation 6.170) are applied to to a nonlinear device whose transfer characteristic is

$$V_{out}(t) = k_1 V_{in}(t) + k_2 V_{in}^2(t) + k_3 V_{in}^3(t) \qquad 6.176$$

the output signal will be a combination of the linear, second-order and third-order components of the device. Let us examine all of the contributions to the output signal power present at one particular frequency, for example, f_1. Examination of Equations 6.170, 6.172, and 6.175 reveals that the total signal voltage present at f_1 will be

$$\left\{ k_1[1 + A_1(t)] + k_3 \frac{3[1 + A_1(t)]^3}{4} + k_3 \frac{3[1 + A_1(t)][1 + A_2(t)]^2}{2} \right\} \cos[\omega_1 t + \phi_1(t)] \qquad 6.177$$

where k_1 and k_3 refer to the amplifier's transfer polynomial. The constant k_1 represents the amplifier's linear power gain; k_3 refers to the amplifier's third-order distortion characteristic.

The signal present at f_1 contains the AM from the input signal at f_2. This is due to the third-order distortion of the amplifier. In general, if we examine the output present at one frequency, we will often find that the distortion of the amplifier has impressed the modulation of one signal onto a second signal (crossmodulation).

6.21 Linearity Design Summary

General Polynomial Model

We model the voltage transfer function of a nonlinear amplifier as a memoryless system whose input/output relationship is

$$V_{out}(t) = k_1 V_{in}(t) + k_2 V_{in}^2(t) + k_3 V_{in}^3(t)$$

Second-Order Intercept

If we define $\Delta_{SO,dB}$ as the difference between the fundamental output power and the second-order output power, we can write

$$\Delta_{SO,dB} = OSIO_{dBm} - P_{out,f,dBm}$$
$$\Delta_{SO,dB} = ISOI_{dBm} - P_{in,f,dBm}$$

The relationship between fundamental output power and second-order output power is

$$P_{out,2f,dBm} = 2P_{out,f,dBm} - OSOI_{dBm}$$

The relationship between fundamental input power and the equivalent second-order input power is

$$P_{in,2f,dBm} = 2P_{in,f,dBm} - ISOI_{dBm}$$

We can also write

$$OSOI_{dBm} = P_{out,f,dBm} + \Delta_{SO,dB}$$
and
$$ISOI_{dBm} = P_{in,f,dBm} + \Delta_{SO,dB}$$

Coherent Cascade Equation

The general expression for n devices in cascade is

$$\frac{1}{\sqrt{OSOI_{cas}}} = \frac{1}{\sqrt{G_{p,2}G_{p,3}G_{p,4}\ldots G_{p,n-1}OSOI_1}} + \frac{1}{\sqrt{G_{p,3}G_{p,4}\ldots G_{p,n-1}OSOI_2}} + \cdots + \frac{1}{\sqrt{OSOI_n}}$$

The input second-order intercept point for the 3-element cascade of Figure 6-49 is

$$\frac{1}{\sqrt{ISOI_{cas}}} = \frac{1}{\sqrt{ISOI_1}} + \frac{1}{\sqrt{ISOI_2/G_{p,1}}} + \frac{1}{\sqrt{ISOI_3/G_{p,1}G_{p,2}}}$$

For n devices in cascade, the expression is

$$\frac{1}{\sqrt{ISOI_{cas}}} = \frac{1}{\sqrt{ISOI_1}}$$
$$+ \frac{1}{\sqrt{ISOI_2/G_{p,1}}}$$
$$+ \cdots$$
$$+ \frac{1}{\sqrt{ISOI_n/G_{p,1}G_{p,2}\cdots G_{p,n-1}}}$$

Noncoherent Cascade Equation

For the general n-element cascade, noncoherent summation produces

$$\frac{1}{OSOI_{cas}} = \frac{1}{G_{p,2}G_{p,3}G_{p,4}\cdots G_{p,n-1}OSOI_1}$$
$$+ \frac{1}{G_{p,3}G_{p,4}\cdots G_{p,n-1}OSOI_2}$$
$$+ \cdots$$
$$+ \frac{1}{OSOI_n}$$

and

$$\frac{1}{ISOI_{cas}} = \frac{1}{ISOI_1}$$
$$+ \frac{1}{ISOI_2/G_{p,1}}$$
$$+ \cdots$$
$$+ \frac{1}{ISOI_n/G_{p,1}G_{p,2}\cdots G_{p,n-1}}$$

Third-Order Intercept

$\Delta_{TO,dB}$ is defined as the difference between the fundamental output power and the third-order output power.

$$\Delta_{TO,dB} = 2\left(OTOI_{dBm} - P_{out,f,dBm}\right)$$
$$\Delta_{TO,dB} = 2\left(ITOI_{dBm} - P_{in,f,dBm}\right)$$

The relationship between fundamental output power and third-order output power is

$$P_{in,2f-f,dBm} = 3P_{in,f,dBm} - 2ITOI_{dBm}$$

The relationship between fundamental input power and the equivalent third-order input power is

$$P_{out,2f-f,dBm} = 3P_{out,f,dBm} - 2OTOI_{dBm}$$

Coherent Cascade Equation

Using coherent addition, the *TOI* of a cascade given the element characteristics is

$$\frac{1}{OTOI_{cas}} = \frac{1}{G_{p,2}G_{p,3}G_{p,4}\cdots G_{p,n-1}OTOI_1}$$
$$+ \frac{1}{G_{p,3}G_{p,4}\cdots G_{p,n-1}OTOI_2}$$
$$+ \ldots$$
$$+ \frac{1}{OTOI_n}$$

and

$$\frac{1}{ITOI_{cas}} = \frac{1}{ITOI_1}$$
$$+ \frac{1}{ITOI_2/G_{p,1}}$$
$$+ \ldots$$
$$+ \frac{1}{ITOI_n/G_{p,1}G_{p,2}\cdots G_{p,n-1}}$$

Noncoherent Cascade Equation

Using non-coherent addition, the *TOI* of a cascade given the element characteristics is

$$\frac{1}{OTOI_{cas}^2} = \frac{1}{\left(G_{p,2}G_{p,3}G_{p,4}\ldots G_{p,n}OTOI_1\right)^2}$$
$$+ \frac{1}{\left(G_{p,3}G_{p,4}\ldots G_{p,n}OTOI_2\right)^2}$$
$$+ \cdots$$
$$+ \frac{1}{\left(OTOI_n\right)^2}$$

and

$$\frac{1}{ITOI_{cas}^2} = \frac{1}{\left(ITOI_1\right)^2}$$
$$+ \frac{1}{\left(ITOI_2/G_{p,1}\right)^2}$$
$$+ \cdots$$
$$+ \frac{1}{\left(ITOI_n/G_{p,1}G_{p,2}G_{p,3}\ldots G_{p,n-1}\right)^2}$$

Relationship Between Input and Output Parameters

Input and output parameters are related by the gain of the device.

$$OSIO_{dBm} = ISOI_{dBm} + G_{p,dB}$$
$$OTIO_{dBm} = ITOI_{dBm} + G_{p,dB}$$
$$OCP_{dBm} = ICP_{dBm} + \left(G_{p,dB} - 1\right)$$

We can also write

$$OTOI_{dBm} = P_{out,f,dBm} + \frac{\Delta_{TO,dB}}{2}$$

and

$$ITOI_{dBm} = P_{in,f,dBm} + \frac{\Delta_{TO,dB}}{2}$$

Relationships Between Linearity Specifications

To a rough approximation, we can write

$$SOI_{dBm} \approx TOI_{dBm} + (10 \text{ or } 15 \text{ dB})$$
$$TOI_{dBm} \approx CP_{dBm} + (10 \text{ or } 15 \text{ dB})$$
$$OCP_{dBm} \approx OTOI_{dBm} - 10.6$$
$$ICP_{dBm} \approx ITOI_{dBm} - 9.6$$

Nonlinearity Rules of Thumb

- A 1 dB increase (decrease) in fundamental input power level will cause a 2 dB increase (decrease) in signals caused by second-order distortion.

- A 1 dB increase (decrease) in fundamental input power level will cause a 3 dB increase (decrease) in signals caused by third-order distortion.

- Inserting an n dB attenuator in a signal path will cause a $2n$ dB decrease in second-order distortion and a $3n$ dB decrease in third-order distortion.

- The general form of the linearity cascade equations follows a resistors-in-parallel format. Adding a nonlinear device to a cascade will never improve the linearity of a cascade.

6.22 References

1. "Fundamentals of RF and Microwave Noise Figure Measurement." Hewlett-Packard Application Note 57-1 (July 1983).

2. Gross, Brian P., "Calculating the Cascade Intercept Point of Communications Receivers," *Ham Radio Magazine*, no. 8 (1980): 50.

3. Ha, Tri T. *Solid-State Microwave Amplifier Design*. New York: John Wiley and Sons, 1981.

4. Horowitz, Paul and Winfield Hill. *The Art of Electronics*. Cambridge: Cambridge UP, 1980.

5. Perlow, Stewart M. "Basic Facts about Distortion and Gain Saturation." *Applied Microwaves Magazine*, no. 5 (1989): 107.

6. Terman, Frederick E. *Electronic and Radio Engineering*. New York: McGraw-Hill, 1955.

7. Williams, Richard A. *Communications Systems Analysis and Design: A Systems Approach*. Englewood Cliffs, NJ: Prentice-Hall, 1987.

7

Cascade I — Gain Distribution

> What are you able to build with your blocks?
> Castles and palaces, temples and docks.
> —R.L.S.

7.1 Introduction

This chapter discusses how filters, amplifier noise, amplifier linearity, mixers, oscillators and transmission lines interact in cascade and how to distribute the gain along the cascade. Numerous examples are supplied to indicate the advantages and disadvantages with each system.

Description of the Problem

The receiving system of Figure 7-1 contains an antenna, a demodulator and the electronics between the two. The antenna port presents the receiver with different signals at different frequencies, with varying power levels and different types of modulation. The demodulator expects its input signal to always be centered at one frequency, the *intermediate frequency* (IF), ,and that its input signal to be at a particular power level.

The signal of interest is converted from its on-the-air frequency to the IF. Further, the gain of the receiver electronics is adjusted so the demodulator always sees a constant power level. This task is complicated by oscillator drift (both in the transmitter and receiver), phase noise, interfering signals, multipath and other difficulties.

640 | RADIO RECEIVER DESIGN

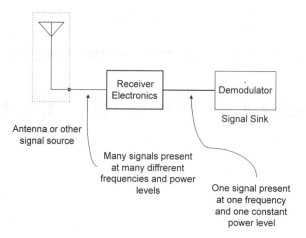

Figure 7-1 General receiving system.

Input and Output Requirements

Figure 7-2 shows some of the specifications of the antenna (signal source) and the signal sink (demodulator).

Signal Source
The signal source provides signals with the following characteristics:

- different center frequencies,
- different bandwidths,
- different signal power levels,
- a noise floor that changes with frequency,
- a nominal impedance environment.

Signal Sink
The signal sink may demand that the signal of interest,

- be centered at one particular frequency,
- exist in a filter large enough to pass the signal without distortion, yet narrow enough not to let excess noise through,
- be at a particular power level,
- exist in a specific impedance level.

System Specifications
The receiver converts signals from the relatively "messy" antenna port to the relatively clean and controlled environment required by the demod-

ulator. The receiver electronics of Figure 7-1 must perform this conversion under a variety of system specifications, including

- required frequency range,

- maximum permissible added noise (relates to the system's noise temperature or noise figure),

- maximum permissible signal distortion (relates to the system's linearity => SOI, TOI),

- allowed power range of the input signals that the system must process (very small to very large).

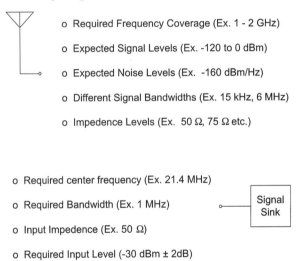

Figure 7-2 *Specifications of a signal source and a signal link.*

7.2 Minimum Detectable Signal (MDS)

The *minimum detectable signal* of a system is the minimum signal power we can apply to a receiver and still be able to detect that the signal is present at the output. We are not interested in demodulating the signal; we simply want to be able to determine whether it is present or not. Rather arbitrarily, we will define the MDS power as input signal whose signal power equals the equivalent input noise power or

$$S_{MDS} = N_{in} \qquad 7.1$$

where
N_{in} is the total noise power dissipated in the input resistor $R_{amp,in}$, N_{in} includes the noise coming in from the antenna as well as the receiver's own internally generated noise (see Figure 7-3).

When the input noise temperature is T_0, the total noise power dissipated in $R_{amp,in}$ is

$$N_{in} = F_{sys} k T_0 B_n \qquad 7.2$$
$$= S_{MDS}$$

so

$$S_{MDS} = F_{sys} k T_0 B_n \qquad 7.3$$

Converting to dBm

$$S_{MDS,dBm} = F_{sys,dB} - 174 \frac{\text{dBm}}{\text{Hz}} + 10 \log(B_n) \qquad 7.4$$

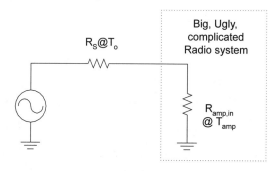

Figure 7-3 *Internal and external noise sources in a receiving system.*

Example 7.1 — Minimum Detectable Signal
A large parabolic antenna feeds satellite signals into a receiver with a 12 MHz bandwidth and a 1.5 dB noise figure.
a. What is the MDS?
b. Assuming the system needs about 20 dB of SNR to demodulate these signals, what is the minimum signal power the receiver can process?

Solution —

a. The input noise temperature of a satellite system is usually lower than T_0 or 290 K. Because we do not have any other information, we will assume that the input noise temperature is T_0. Using Equation 7.4, we find

$$S_{MDS,dBm} = 1.5 - 174 \frac{dBm}{Hz} + 10\log(12 \cdot 10^6)$$
$$= -101.7 \text{ dBm}$$

(7.5)

b. If the system requires a 20 dB SNR, the minimum input signal power is

$$S_{min} = S_{MDS,dBm} + SNR_{dB}$$
$$= -101.7 + 20$$
$$= -81.7 \text{ dBm}$$

(7.6)

7.3 Dynamic Range

The term *dynamic range* defines how well a system handles signals with varying power levels. System designers have developed various ways to specify this characteristic. Here we will discuss only the three most common specifications.

Linear Dynamic Range

Linear dynamic range is defined as the difference between the MDS and the *input compression point* (ICP) of a system.

$$DR_{lin,dB} = ICP_{dBm} - S_{MDS,dBm}$$

(7.7)

In practice, linear dynamic range is useful only when comparing systems. This specification does not provide mathematical insight into the amount of distortion power generated by a system.

Gain-Controlled Dynamic Range

Gain-controlled dynamic range (or noninstantaneous dynamic range) describes the ability of a system to process signals of different signal levels, one at a time. Only one signal is presented to the system at a time, then we let the system adjust itself to that one signal. For example, if you go outside on a dark, moonless night, at first you will not be able to see much. If you wait approximately ten minutes, your pupils will open and your eyes will become more sensitive. Similarly, you can also function in a bright, well-lit room after your eyes have had a chance to adjust.

Gain-controlled dynamic range specifies how well a system responds to input signals when they are applied one at a time (see Figure 7-4). The system is allowed to adjust in order to accommodate for the different signal strengths — similar to how the eye opens and closes to adjust for varying light levels. Gain-controlled dynamic range indicates what range of input signals the receiver is capable of processing. The lower end is set by the noise performance of the system. The upper end is usually set by some system nonlinearity.

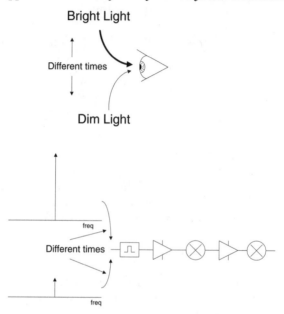

Figure 7-4 *Gain-controlled dynamic range in a receiver is similar to the behavior of the eye in dim and bright light conditions.*

7.4   Spur-Free Dynamic Range

Spur-free dynamic range (or instantaneous dynamic range) describes the ability of a receiver to process a small, desired signal in the presence of a large, unwanted signal. Another name for this quantity is *instantaneous dynamic range*. For example, imagine you are driving down the road late at night. A car in the other lane approaches you with its high beams on. Spur-free dynamic range describes your ability to see the relatively faint light coming off the lines on the road (the desired signal) in the presence of the other driver's high beams (the undesired signal). The equivalent receiver situation is processing a very small signal in the presence of a very large signal (see Figure 7-5).

We are interested in the range of input signals the system can process simultaneously without

- losing the desired signal in the noise. This is a noise problem principally controlled by the noise temperature of the receiver and the noise gathered by the receiving antenna.

- creating undesirable spurious products that can mask or distort the small, desired signal. This is a linearity problem described by the system's SOI and TOI.

Spur-free dynamic range is the most used and most useful dynamic range specification. Frequently, a major design goal is to maximize a system's spur-free dynamic range.

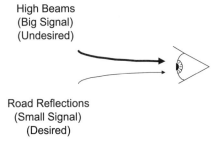

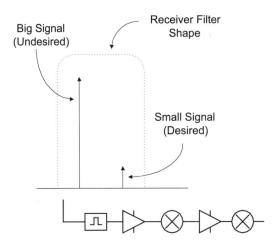

Figure 7-5 *Instantaneous dynamic range in a receiver is similar to the behavior of the eye when it views dim and bright objects at the same time.*

Derivation

Spur-free dynamic range is the difference between the MDS and the input power which will cause the third-order spurious products to be equal to the MDS. The low-level input condition is clear: the input signal equals the MDS. There are two high-level input conditions, one for second-order limited systems and one for systems limited by third-order distortion.

Second Order

Figure 7-6 illustrates the high-level input conditions when second-order distortion is a concern. The input consists of a signal tone and a noise floor. We increase the signal power until the second-order spurious product just peaks out of the noise. At this point, the power in the spurious tone equals the system MDS. The power in the second-order tone is

$$P_{in,2f} = S_{MDS} \quad\quad 7.8$$

We know

$$P_{in,2f,dBm} = 2P_{in,f,dBm} - ISOI_{dBm} \quad\quad 7.9$$

When the second-order distortion power just equals the system's MDS, the fundamental input power is

$$P_{in,f,dBm} = \frac{S_{MDS,dBm} + ISOI_{dBm}}{2} \quad\quad 7.10$$

Second-order spur-free dynamic range is almost universally ignored because it creates only a problem in wideband systems (e.g., RADARs and ELINT receivers). Most of the world is interested in communications systems which are inherently narrowband and are limited by third-order distortion.

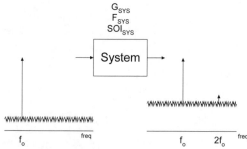

Figure 7-6 *High-level input conditions with second-order distortion limiting the system dynamic range.*

Third Order

Figure 7-7 illustrates the high-level input conditions when third-order distortion is the limiting distortion. The signal power of the two input tones is increased until the third-order spurious products are just discernable in the output noise. At this point, the power in each spurious tone equals the system MDS. This is the highest input signal we will allow.

Given a system with a minimum detectable signal of S_{MDS}, a noise figure of F_{sys} and an input third-order intercept of ITOI, we want to find the fundamental input power when the third-order power equals the S_{MDS} or

$$P_{in,2f-f} = S_{MDS} \qquad 7.11$$

When discussing linearity, we found that

$$P_{out,2f-f,dBm} = 3P_{out,f,dBm} - 2OTOI_{dBm} \qquad 7.12$$

Combining Equation 7.12 with Equation 7.11 produces

$$P_{in,f,dBm} = \frac{1}{3}S_{MDS,dBm} + \frac{2}{3}ITOI_{dBm} \qquad 7.13$$

This is the maximum input signal power that the system can process without producing detectable third-order spurious signals.

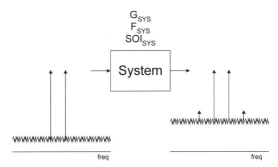

Figure 7-7 *High-level input conditions with second-order distortion limiting the system dynamic range.*

Dynamic Range

Using the third-order criteria, the spur-free dynamic range is the difference between the maximum allowable input signal ($P_{in,f,dBm}$ of Equation 7.13) and the system's minimum detectable signal, or

$$SFDR_{dB} = P_{in,f,dBm} - S_{MDS,dBm}$$
$$= \frac{1}{3}S_{MDS,dBm} + \frac{2}{3}ITOI_{dBm} - S_{MDS,dBm}$$
$$= \frac{2}{3}\left[ITOI_{dBm} - S_{MDS,dBm}\right] \qquad 7.14$$
$$= \frac{2}{3}\left[ITOI_{dBm} - F_{sys}kT_0B_n\right]$$

A large spur-free dynamic range is desirable. The SFDR will be improved (increased) if we increase the ITOI, decrease the noise figure or decrease the noise bandwidth. The spur-free dynamic range applies only to spurious signals produced by the nonlinearities in the receiver's cascade chain. Spurious signals can arise in a receiver through other mechanisms such as poor LO rejection or electromagnetic interference generated within the receiver itself.

Example 7.2 — Spur-Free Dynamic Range

Given the cascade shown in Figure 7-8, find the MDS and the SFDR in a 30 kHz bandwidth.

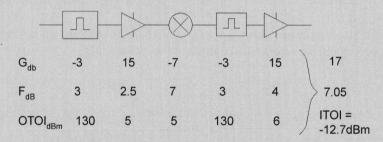

Figure 7-8 Cascade of Example 7.2. Find the MDS and the SFDR.

Solution —

The cascade power gain $G_{p,cas}$ is 17 dB, the cascade noise figure F_{sys} is 7.05 dB and the cascade's input TOI is –12.7 dBm. Using Equation 7.4, the MDS is

$$S_{MDS,dBm} = F_{sys,dB} - 174\frac{dBm}{Hz} + 10\log(B_n) \qquad 7.15$$
$$= 7.05 - 174\frac{dBm}{Hz} + 10\log(30{,}000)$$
$$= -122.2 \text{ dBm}$$

Equation 7.14 yields the SFDR.

$$SFDR_{dB} = \frac{2}{3}[ITOI_{dBm} - S_{MDS.dBm}]$$
$$= \frac{2}{3}[-12.7 - (-122.2)]$$
$$= 73 \text{ dB}$$

7.16

Figure 7-9 shows the results of these calculations graphically. The output noise floor is

$$N_{out,dBm} = G_{p,dB} + F_{sys,dB} - 174\frac{\text{dBm}}{\text{Hz}} + 10\log(B_n)$$
$$= 17 + 7 - 174 + 10\log(30{,}000)$$
$$= -105.2 \text{ dBm}$$

7.17

Completing the math, we can detect the third-order distortion products when the fundamental input power is –49 dBm.

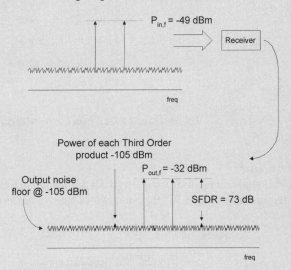

Figure 7-9 *Solution for Example 7.2. The system MDS and SFDR are shown graphically.*

Front-End Attenuators and Spur-Free Dynamic Range

The design goal is often to maximize the SFDR of a system. We can "move" this dynamic range to any input power level using an attenuator on the front end of the receiver. Let us look at an example (see Figure 7-10).

650 | RADIO RECEIVER DESIGN

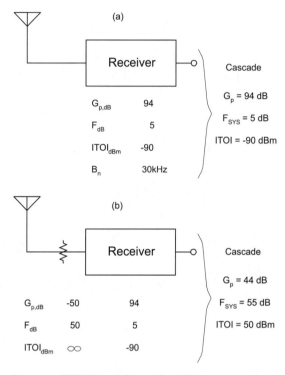

Figure 7-10 *System SFDR with and without an attenuator on the receiver's input.*

Without Attenuator

Figure 7-10(a) shows a receiving system with a 5 dB noise figure, a 30 kHz noise bandwidth and a –90 dBm ITOI. The system has an MDS of

$$S_{MDS,dBm} = F_{sys,dB} - 174\frac{\text{dBm}}{\text{Hz}} + 10\log(B_n) \qquad 7.18$$
$$= 5 - 174\frac{\text{dBm}}{\text{Hz}} + 10\log(30{,}000)$$
$$= -124 \text{ dBm}$$

The SFDR is

$$SFDR_{dB} = \frac{2}{3}\left[ITOI_{dBm} - S_{MDS.dBm}\right] \qquad 7.19$$
$$= \frac{2}{3}\left[-90 - (-124)\right]$$
$$= 22.7 \text{ dB}$$

This is a fairly poor SFDR.

With Attenuator

The system with the attenuator is shown in Figure 7-10(b). When we add the attenuator, the gain of the system drops to 44 dB, the noise figure rises to 55 dB and the system's ITOI rises to –40 dBm. The MDS is

$$S_{MDS,dBm} = 55 - 174\frac{dBm}{Hz} + 10\log(30{,}000) \qquad 7.20$$
$$= -74 \text{ dBm}$$

The attenuator has made the system nearly deaf. However, the SFDR is

$$SFDR_{dB} = \frac{2}{3}\left[ITOI_{dBm} - S_{MDS.dBm}\right]$$
$$= \frac{2}{3}\left[-40 - (-74)\right] \qquad 7.21$$
$$= 22.7 \text{ dB}$$

The SFDR is the same as before. The receiver simply performs over a different range of input power levels. In general, a receiver can process signal power levels in the range of

$$MDS_{dBm} \text{ to } (MDS_{dBm} + SFDR_{dB}) \qquad 7.22$$

Without the attenuator, the system can process signals over the input power range of

$$-124 \text{ to } (-124 + 83) \qquad 7.23$$
$$= -124 \text{ dBm to } -41 \text{ dBm}$$

With the attenuator, the system can process signals over the input power range of

$$-74 \text{ to } (-74 + 83) \qquad 7.24$$
$$= -74 \text{ dBm to } +9 \text{ dBm}$$

An input attenuator does not affect the SFDR of a receiver. It affects only the range of input signals. These calculations assume we are using a resistive or other highly linear attenuator.

7.5 Dynamic Range Notes

Specifications

Since there are two or three separate dynamic range specifications, we have to be clear about which dynamic range is meant. Engineers designing fixed-gain systems usually specify the *spur-free dynamic range* (SFDR); designers who are speaking about receiving systems often quote the gain-controlled dynamic range. Moreover, there is the second-order/third-order ambiguity.

Other Sources of Spurious Signals

There are many sources of spurious signals in a receiver. The dynamic range quantities specified here only address the nonlinearity of the system (i.e., the second- and third-order intercepts of the components making up the system). Other effects can cause spurious signals in a receiver.

Dual Conversion Systems

Figure 7-11(a) shows one source of these spurious products. In a dual conversion receiver, the signal is converted from the RF to the first IF. After filtering, the signal is then converted from the first IF to a different IF, which requires two separate local oscillators to be running in the receiver simultaneously. Due to the finite attenuation of the band-pass filter between the mixers and port-to-port leakage, LO_1 can find its way into mixer M_2 and LO_2 can find its way into mixer M_1. These two signals will mix and produce spurious products at

$$|mf_{LO1} \pm nf_{LO2}| \quad for \quad \begin{cases} n = 0,1,2,3... \\ m = 0,1,2,3... \end{cases} \qquad 7.25$$

When one of the products falls inside any one of the IF filters in the radio, the result is a spurious signal. This type of spurious response (internally generated spurious) is not related to the linearity of the amplifiers and mixers that make up the system. The amount of spurious signal power we will see is purely a function of how well the two local oscillators were kept apart.

Other Nonlinearities

Figure 7-11(b) shows an example architecture which will generate spurious products. A 350 MHz RF signal will pass through the RF BPF and generate second-order distortion in the mixer. This effect produces a 700 Mhz signal at the mixer's output. This is a second-order effect which is not addressed by Equation 7.14; hence a spurious product generated in this way will not be predicted by the equation. This spurious product will be very large and will almost certainly reduce the measured spur-free dynamic range of the receiver. If the second-order spurious product is a problem, the conversion scheme of the receiver can be redesigned to solve this problem (see Chapter 8).

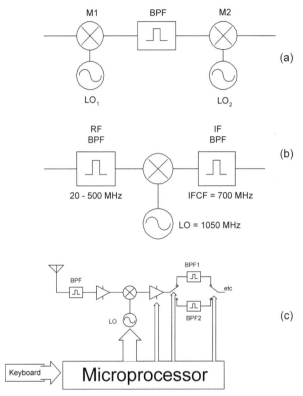

Figure 7-11 Sources of spurious signals in a receiving system.

Digital Logic

Another serious source of spurious products in a receiving system is the always present *digital logic*. Figure 7-11(c) shows a typical microprocessor-controlled receiving system. Although designers take great pains to isolate the sensitive receiving circuits from the digital logic, digital signals occasionally find their way onto the receiver's signal path. Consider a receiver with a 60 MHz IF with a 10 MHz crystal for its microprocessor clock. Since the microprocessor is probably a 0 to 5 volt square wave, it is rich in strong harmonics. The 6th harmonic of a 10 MHz microprocessor clock will fall directly on top of the 60 MHz IF center frequency.

The receiver's control logic often contains counters, gating and other functions that produce many frequency components that are not obvious. For example, a microprocessor using a 10 MHz system clock might produce components at 45.6 kHz or 22.3 MHz. Digital control lines that must pass into sensitive RF compartments are especially dangerous. These lines are usually filtered heavily. Although digital logic and microprocessors can be a serious problem for a receiving system, they are just too useful not to include in a modern system.

Measuring the SFDR

Receiver *users*, as opposed to receiver *designers*, are not interested in the source of the spurious signal. They are interested only in whether the signal being observed is real or an internally generated phantom. Testing should reflect the concerns of the receiver user.

The SFDR is measured by tuning the receiver with the antenna port terminated in a matched load. Any signals that show up are generated internally in the receiver. The equivalent input power level of the largest of these internally generated spurious products is then used in place of S_{MDS} in Equation 7.14. Internally generated spurious signals limit the receiver's dynamic range.

7.6 Gain Distribution

The goal is to appropriately convert signals present at the output of the antenna into a format compatible with some downstream signal processor, which involves changing both the power level of the signal and the frequency of the signal and removing other signals that are close to the signal of interest. In this chapter, we focus on the gain aspect of the problem. Frequency conversion and filtering aspects will be discussed in Chapter 8.

Required Gain

Figure 7-12 shows a receiving system in its initial design stages. Let us assume the noise figure of the completed receiver must be 5 dB and its ITOI will be 0 dBm. The receiver must support frequencies of 500 to 1000 MHz. The demodulator requires a signal with a 30 kHz bandwidth. The signal must be centered at 21.4 MHz and must be at least –30 dBm. The minimum detectable signal of this system will be

$$S_{MDS,dBm} = F_{sys,dB} - 174\frac{\text{dBm}}{\text{Hz}} + 10\log(B_n)$$
$$= 5 - 174\frac{\text{dBm}}{\text{Hz}} + 10\log(30{,}000) \qquad 7.26$$
$$= -124 \text{ dBm}$$

Since we need −30 dBm at the output of the receiver even under MDS input conditions, the receiver must have a power gain of

$$G_{p,dB} = -30 - (-124) \qquad 7.27$$
$$= 94 \text{ dB}$$

The system needs 94 dB of power gain. We must also translate any signal in the 500 to 1000 MHz range to a 21.4 MHz IF center frequency. What is the best way to distribute the gains and losses of the system to achieve a maximum SFDR?

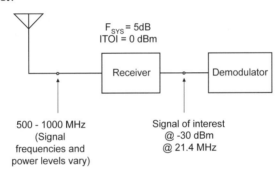

Figure 7-12 *Receiving system in its initial design stages. Only high-level specifications are shown.*

Three Gain Distributions

Figure 7-13 shows three ways to distribute the 94 dB of power gain.

- In Figure 7-13 (a), we have placed all of the gain at the front of the cascade. After the signal passes through the lossy components, there is a total of 94 dB through the cascade.

- In Figure 7-13(b), we have placed all of the power gain at the very end of the cascade. Again, after accounting for the lossy elements, we achieve a total cascade gain of 94 dB.

- Figure 7-13(c) shows a distributed system. We apply gain in various places throughout the cascade and in concert with the losses. The total cascade gain remains 94 dB.

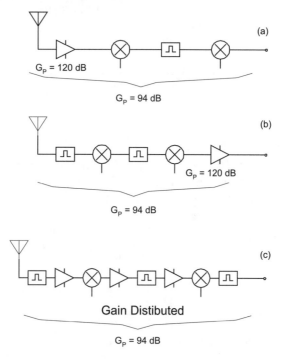

Figure 7-13 Three ways to distribute the power gain in a receiving system. (a) Place all of the gain at the front end of the receiver. (b) Place all of the gain at the end of the receiver cascade. (c) Distribute the gain throughout the receiver according to some rule.

Which method will produce the system with the largest dynamic range? We can find a solution to this problem by examining several of the equations developed over the last several chapters. For the n-element cascade of Figure 7-14, we can write

1. The noise temperature of a cascade given the noise temperature and gain of its components is

$$T_{cas} = T_1 + \frac{T_2}{G_{p,1}} + \frac{T_3}{G_{p,1}G_{p,2}} + \ldots + \frac{T_n}{G_{p,1}G_{p,2}G_{p,3}\cdots G_{p,n-1}} \qquad 7.28$$

CASCADE I — GAIN DISTRIBUTION | 657

Similarly, the noise figure of a cascade given the noise figure and gain of its components is

$$F_{cas} = F_1 + \frac{F_2 - 1}{G_{p,1}} + \frac{F_3 - 1}{G_{p,1}G_{p,2}} + \ldots + \frac{F_n - 1}{G_{p,1}G_{p,2}G_{p,3}\cdots G_{p,n-1}} \qquad 7.29$$

2. The ITOI of a cascade, given the ITOI and gain of its components, assuming coherent addition, is

$$\frac{1}{ITOI_{cas}} = \frac{1}{ITOI_1}$$
$$+ \frac{1}{ITOI_2/G_{p,1}}$$
$$+ \frac{1}{ITOI_3/G_{p,1}G_{p,2}} \qquad 7.30$$
$$+ \ldots$$
$$+ \frac{1}{ITOI_n/G_{p,1}G_{p,2}\cdots G_{p,n-1}}$$

We will discuss only the ITOI equation in this chapter but the arguments are valid for OTOI and the SOI equations as well. We will begin by examining some common themes to understand Equations 7.28, 7.29 and 7.30.

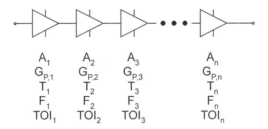

Figure 7-14 A general n-element cascade.

Excess Gain

The three cascade equations contain expressions for *excess gain*. Excess gain describes the amount of power gain between the cascade input and a particular component. Figure 7-15 shows an example of excess gain calculation. The first row of numbers, placed under each component, is the power gain of the component. The row of numbers centered between components is the excess gain of the cascade up to that point.

658 | RADIO RECEIVER DESIGN

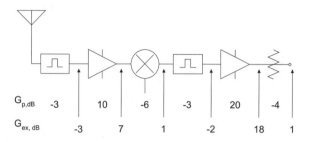

Figure 7-15 *Excess gain calculation. In a receiver, the excess gain is the amount of gain between the antenna and any given point in the receiver.*

To find the excess gain of any point in a cascade, simply add up the power gains (in dB) of all the components before the point of interest. Remember, gains are associated with components; excess gain is associated with the points between components. The excess gain appears in Equations 7.28, 7.29 and 7.30 as

$$G_{excess} = G_{p,1} G_{p,2} G_{p,3} \cdots$$
$$G_{excess,dB} = G_{p,1,dB} + G_{p,2,db} + G_{p,3,dB} + \cdots \quad \quad 7.31$$

Translation

On closer inspection, we find that the individual terms of the cascade equations mathematically translate the component specifications to either the input or output port.

Linearity

To determine the linearity contribution of the third element in a cascade, for example, the third term of Equation 7.30 is analyzed.

$$\frac{1}{ITOI_3/G_{p,1}G_{p,2}} \quad \quad 7.32$$

The term $ITOI_3$ represents a power in watts. The term

$$ITOI_3/G_{p,1}G_{p,2} \quad \quad 7.33$$

represents mathematically translating that power level to the input of the cascade. Every term in the TOI cascade equation can be viewed in this manner. For example, Figure 7-16 shows three amplifiers in cascade. It is not obvious which amplifier is limiting the cascade's ITOI until we translate the TOI of each amplifier to the input of the cascade.

CASCADE I — GAIN DISTRIBUTION | 659

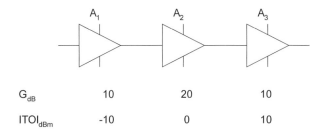

G_{dB}	10	20	10
$ITOI_{dBm}$	-10	0	10

Figure 7-16 *Three amplifiers in cascade. Determine which amplifier limits the cascade TOI.*

Amplifier A_1. This amplifier's TOI is already referenced to the cascade's input.

$$A_1\text{'s TOI referenced to cascade input} = ITOI_1 \qquad 7.34$$
$$= -10 \text{ dBm}$$

Amplifier A_2. The relevant part of Equation 7.30 is

$$ITOI_2/G_{p,1} \qquad 7.35$$

Accordingly, we find that

$$A_2\text{'s TOI referenced to cascade input} = ITOI_2/G_{p,1} \qquad 7.36$$
$$= 0 - 10 \text{ dBm}$$
$$= -10 \text{ dBm}$$

Amplifier A_3. The relevant part of Equation 7.30 is

$$ITOI_3/G_{p,1}G_{p,2} \qquad 7.37$$

Accordingly, we find that

$$A_3\text{'s TOI referenced to cascade input} = ITOI_3/G_{p,1}G_{p,2}$$
$$= 10 - 30 \text{ dBm} \qquad 7.38$$
$$= -20 \text{ dBm}$$

After moving the ITOI of all the amplifiers to a common point, we find two amplifiers both produce a –10 dBm equivalent input TOI while the third amplifier produces a –20 dBm value. Since the ITOI of a device is linearly related to the amount of power the device can handle before it becomes nonlinear, the third amplifier is limiting the TOI of the cascade. Equation 7.30 shows the ITOI for this cascade is −20.8 dBm.

Noise Temperature

Translation also works for the noise temperature. Figure 7-17 shows three amplifiers in cascade. Which one limits the noise performance of the cascade? Equation 7.28 contains the answer.

Amplifier A_1. The noise temperature contribution of this amplifier is already referenced to the cascade's input.

$$A_1\text{'s noise temperature at the cascade input} = T_1 \qquad 7.39$$
$$= 290 \text{ K}$$

Amplifier A_2. Equation 7.28 indicates

$$A_2\text{'s noise temperature at the cascade input} = T_1/G_{p,1} \qquad 7.40$$
$$= 8881/10$$
$$= 888.1 \text{ K}$$

Amplifier A_3. Equation 7.28 reveals

$$A_3\text{'s noise temperature at the cascade input} = T_3/G_{p,1}G_{p,2} \qquad 7.41$$
$$= 8881/1000$$
$$= 8.9 \text{ K}$$

Translating the noise to a common port reveals that amplifier A_2 limits the noise performance.

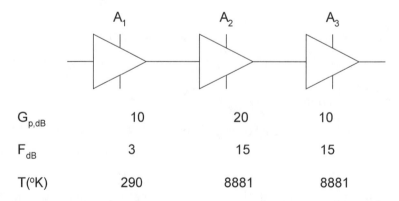

$G_{p,dB}$	10	20	10
F_{dB}	3	15	15
T(°K)	290	8881	8881

Figure 7-17 *Three amplifiers in cascade. Determine which amplifier limits the cascade's noise performance.*

7.7 Gain and Noise

Noise Temperature

Figure 7-14 and Equation 7.28 describe an n-element. We can write

$$\begin{aligned}
T_{cas} = &\; T_1 & \text{[Component \#1]} \\
&+ \frac{T_2}{G_{p,1}} & \text{[Component \#2]} \\
&+ \frac{T_3}{G_{p,1} G_{p,2}} & \text{[Component \#3]} \\
&+ \ldots & \\
&+ \frac{T_n}{G_{p,1} G_{p,2} G_{p,3} \cdots G_{p,n-1}} & \text{[Component \#n]}
\end{aligned} \qquad 7.42$$

- *Component #1.* This term represents the noise temperature contributed to the cascade by amplifier A_1. Note that this is the only place T_1 appears in the equation. This piece of the equation indicates that, if we want a cascade with a low noise temperature, the first amplifier should have a low noise temperature. Any change in the noise temperature of the first element directly affects the cascade's noise temperature. We also note that the cascade noise temperature will never be any lower than the noise temperature of the first component.

- *Component #2*: This term represents the noise temperature contribution of amplifier A_2. This is the only place where the noise temperature of amplifier A_2 exists in the cascade equation. In effect, the noise temperature of amplifier A_2 is reduced by the gain of amplifier A_1. The noise temperature contribution of amplifier A_2 will be small either when A_1 has a large gain or when A_2 has a low noise temperature (or both). For example, if A_1 has 20 dB of gain and A_2 has a 290 K noise temperature, the noise temperature contribution of A_2 is

$$\frac{T_2}{G_{p,1}} = \frac{290}{100} = 2.9 \, \text{K} \qquad 7.43$$

The gain of A_1 has greatly reduced the noise contribution of A_2.

- *Component #3.* This term represents the noise contribution of A_3 to the cascade. In this case, the noise temperature of A_3 is reduced by $G_{p,1} G_{p,2}$, the gains of A_1 and A_2. Because A_3 should contribute only an insignificant amount of noise to the cascade, the power gain of amplifiers A_1 and A_2 should be as large as practical.

- *Component #n.* We can see that the noise contribution of the nth term is reduced by the gain of the $(n - 1)$ components before the nth component. The noise temperature of the nth component is reduced by the excess gain of the cascade at the input to the nth component.

Relative Levels

Figure 7-18 shows the noise levels present in a cascade when the excess gain of the cascade is large. Most of the cascade's noise power comes from the antenna and the noise generated by the first amplifier. This is true because the large excess gain reduces the contribution of the subsequent amplifiers. After the input noise experiences the gain of A_1, the noise power from A_1's output is much larger than the noise added by A_2.

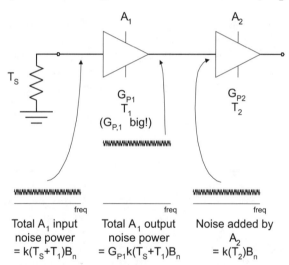

Figure 7-18 *Relative noise levels in a cascade shown when the excess gain is large. The output noise of amplifier A_1 is much greater than the noise generated by amplifier A_2's input resistor. Thus, most of the system noise originates from T_S and T_1.*

Figure 7-19 shows the relative noise levels when the excess gain of the cascade is small. When the excess gain preceding a particular amplifier is small, the noise contribution of that amplifier is large. For example, in Figure 7-19, the noise added by amplifier A_2 is significant because of the small excess gain prior to A_2.

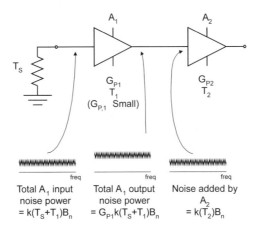

Figure 7-19 *Relative noise levels in a cascade when the excess gain is small. The noise generated by amplifier A_2's input resistor is significant with respect to the output noise of amplifier A_1. The system noise originates from T_S, T_1 and T_2.*

If we want amplifier A_2 to add only a small amount of noise to the cascade, Figure 7-18 and Equation 7.28 show

$$\text{Noise at } A_1\text{'s output} \gg \text{Noise added by } A_2$$

$$G_{p,1}k(T_s + T_1)B_n \gg kT_2 B_n \qquad 7.44$$

$$G_{p,1}(T_s + T_1) \gg T_2$$

If $T_S = T_0$, we can speak in terms of noise figure and the equation becomes

$$G_{p,1}(T_0 + T_1) \gg T_2$$

$$G_{p,1}[T_0 + T_0(F_1 - 1)] \gg T_0(F_2 - 1) \qquad 7.45$$

$$\Rightarrow G_{p,1}F_1 \gg F_2 - 1$$

$$\Rightarrow G_{p,1}F_1 + 1 \gg F_2$$

If the $G_{p,1}F_1$ product is large, we can make one final approximation.

$$G_{p,1}F_1 \gg 1$$

$$\Rightarrow G_{p,1}F_1 \gg F_2 \qquad 7.46$$

Using decibels, we can write

$$G_{p,1,dB} + F_{1,dB} \gg F_{2,dB} \qquad 7.47$$

In other words, add the gain of the first amplifier to its noise figure (all in dB). If that number is much greater than the noise figure of the second amplifier, the noise added to the cascade by the second amplifier will be insignificant.

A Rule of Thumb

A successful strategy is to keep the excess gain plus the noise figure of the first amplifier at least 15 dB greater than the noise figure of the second amplifier.

$$G_{p,1,dB} + F_{1,dB} > F_{2,dB} + 15 \qquad 7.48$$

The approximation is experimental but, under the worst conditions, this rule insures that the second component will contribute less than 0.1 dB of noise figure to the cascade. If the difference is greater than 15 dB, the second device will add even less noise to the cascade. However, although high excess gains are very good for low-noise cascades, they degrade linearity. Remember that the goal of cascade design is to maximize the SFDR, which depends on both low cascade noise temperature and on high cascade linearity.

Example 7.3 — Noise Cascades

Consider a system that has been specified to have a maximum noise figure of 12 dB. The designers are considering adding a new component at the end of the cascade to increase the gain. The new component has 13 dB of gain and a 7.5 dB noise figure. How much excess gain do we need to be sure this new component will not significantly alter the noise figure of the current cascade?

Solution —

Figure 7-20 shows the situation. Equation 7.48 reveals that the most gain is required from the partial cascade when the cascade's noise figure is very low. For a worst-case analysis, we will assume the noise figure of the partial cascade is 1 dB. Equation 7.48 produces

$$G_{p,1,dB} + F_{1,dB} > F_{2,dB} + 15$$
$$G_{p,1,dB} + 1 > 7.5 + 15 \qquad 7.49$$
$$\Rightarrow G_{p,1,dB} > 21.5 \text{ dB}$$

If we assume the worst case, we need about 21.5 dB of excess gain prior to the last amplifier. We can reduce the amount of required excess gain by decreasing the noise figure of the last amplifier.

CASCADE I — GAIN DISTRIBUTION | 665

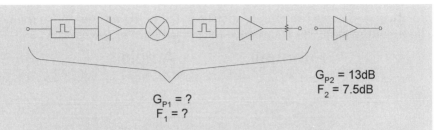

Figure 7-20 *Cascade for Example 7.3. What is the effect of adding a component to the end of the cascade?*

Front-End Attenuators

In Chapter 5, we demonstrated that the noise figure of a resistive attenuator equals its attenuation if the attenuator is at room temperature. In equation form,

$$\text{Attenuator Noise Figure} = \text{Attenuator Loss} \qquad 7.50$$

for attenuators at T_0

Figure 7-21 shows a simple two-element cascade consisting of a resistive attenuator followed by an amplifier. The noise figure of the attenuator equals the attenuator value or

$$F_{Attn} = \frac{1}{G_{Attn}} \qquad 7.51$$

or

$$F_{Attn,dB} = -G_{Attn,dB}$$

The noise figure of the cascade is

$$F_{cas} = F_1 + \frac{F_2 - 1}{G_1} \qquad 7.52$$

Under the special conditions where the noise figure of the first device is inversely related to its gain, the cascade noise figure becomes

$$F_{cas} = F_1 + \frac{F_2 - 1}{1/F_1}$$
$$= F_1 + F_1(F_2 - 1) \qquad 7.53$$
$$= F_1 F_2$$
$$F_{cas,dB} = F_{1,dB} + F_{2,dB}$$

In other words, we can simply add the value of the attenuator to the noise figure of the amplifier to produce the noise figure of the cascade.

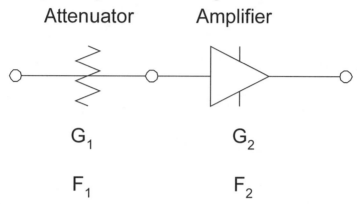

Figure 7-21 *Two-element cascade consisting of a resistive attenuator and an amplifier.*

Example 7.4 — TV Antenna Amplifier

Figure 7-22 shows a typical household television receiving system. An outdoor TV antenna sits on a mast on the roof of the house. The antenna is connected to the television by 50 feet of RG-59 coaxial cable, which has a nominal loss of 6.5 dB at 500 MHz. The noise figure of the television is 15 dB. We would like to add a TV amplifier to the system to improve the snowy channels. The amplifier has a 4 dB noise figure and 20 dB of gain. We have three choices.

1. We can attach the amplifier directly to the television. This is the least trouble because the amplifier is inside the house and we can plug it directly into the wall socket by the TV.

2. We can place the amplifier inside, right where the cable enters the house. This places the amplifier about equidistant between the antenna and TV, which means that it is more difficult to provide power to the amplifier than in solution 1.

3. Finally, we can place the amplifier right on the mast, next to the antenna. This is a lot of trouble because we have to protect the amplifier against the weather and provide power to the amplifier.

Find the noise figure of the system with no amplifier. Also, find the noise figure reduction in each of the three situations described above.

CASCADE I — GAIN DISTRIBUTION | 667

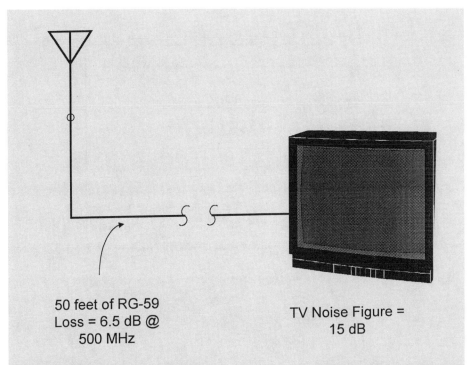

Figure 7-22 Typical household television system with antenna, cable and television set.

Solution —

Cable Alone. Figure 7-23 shows the system cascade without the amplifier. Table 7-1 shows the relevant information about the cascade.

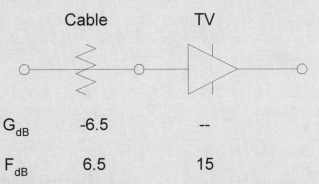

Figure 7-23 Cascade equivalent of a household television system before adding an amplifier.

Table 7-1 *Television antenna cascade.*

	Cable	TV
G_{dB}	-6.5	--
G	0.224	--
F_{dB}	6.5	15
F	4.47	31.6
T (°K)	1005	8881
F (referenced to cascade input)	4.47	$\dfrac{F_2-1}{G_1} = \dfrac{31.6-1}{0.224} = 136.8$
T (referenced to cascade input - °K)	1005	$\dfrac{T_2}{G_1} = \dfrac{8881}{0.224} = 39{,}688\,°K$

The cascade noise figure and temperature are

$$F_{cas} = F_1 + \frac{F_2-1}{G_1} \qquad T_{cas} = T_1 + \frac{T_2}{G_1}$$
$$= 4.47 + 136.8 \qquad\quad = 1005 + 39{,}688 \qquad 7.54$$
$$= 141.3 \qquad\qquad\quad\; = 40{,}675\,K$$
$$= 21.5\,dB$$

Table 7-1 shows that the television is still producing the bulk of the noise in the cascade. In this simple case, we could use Equation 7.53 to find the noise figure of the cascade as 6.5 + 15 = 21.5 dB.

1. *Amplifier next to the TV.* Figure 7-24 shows the system cascade when the amplifier is placed immediately before the television.

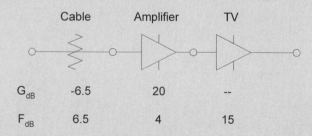

Figure 7-24 *Cascade equivalent of a household television system after adding an amplifier directly to the TV.*

Table 7-2 *Television antenna cascade with amplifier at television.*

	Cable	Amplifier	TV
G_{dB}	-6.5	20	--
G	0.224	100	--
F_{dB}	6.5	4	15
F	4.47	2.51	31.6
$T\ (°K)$	1005	438	8881
F (referenced to cascade input)	4.47	$\dfrac{F_2-1}{G_1} = \dfrac{2.51-1}{0.224} = 6.74$	$\dfrac{F_3-1}{G_1 G_2} = \dfrac{31.6-1}{(0.224)(100)} = 1.37$
T (referenced to cascade input °K)	1005	$\dfrac{T_2}{G_1} = \dfrac{438}{0.224} = 1955\,°K$	$\dfrac{T_3}{G_1 G_2} = \dfrac{8881}{(0.224)(100)} = 397\,°K$

In this case, the cascade noise figure and temperature are

$$F_{cas} = F_1 + \frac{F_2-1}{G_1} + \frac{F_3-1}{G_1 G_2} \qquad T_{cas} = T_1 + \frac{T_2}{G_1} + \frac{T_3}{G_1 G_2} \qquad 7.55$$

$$= 4.47 + 6.74 + 1.37 \qquad\qquad = 1005 + 1955 + 397$$

$$= 12.6 \qquad\qquad\qquad\qquad\quad = 3{,}357\,K$$

$$= 11.0\,\text{dB}$$

Simply adding the amplifier before the TV has decreased the system noise figure to 11.0 dB. The amplifier adds about as much noise as the cable. The noise from the TV is now insignificant.

2. *Amplifier in the middle of cable.* Figure 7-25 shows the system when the amplifier is about halfway between the antenna and the television.

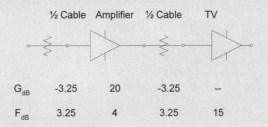

Figure 7-25 *Cascade equivalent of a household television system with an amplifier placed in the middle of the cable.*

Table 7-3 Television antenna cascade with amplifier at cable mid-point.

	½ Cable	Amplifier	½ Cable	TV
G_{dB}	-3.25	20	-3.25	--
G	0.473	100	0.473	--
F_{dB}	3.25	4	3.25	15
F	2.11	2.51	2.11	31.6
T (°K)	323	438	323	8881
F (referenced to cascade input)	2.11	$\frac{F_2-1}{G_1} = \frac{2.51-1}{0.473}$ $= 3.19$	$\frac{F_3-1}{G_1 G_2} = \frac{2.11-1}{(0.473)(100)}$ $= 0.02$	$\frac{F_4-1}{G_1 G_2 G_3} = \frac{31.6-1}{(0.473)(100)(0.473)}$ $= 1.37$
T (referenced to cascade input - °K)	323	$\frac{T_2}{G_1} = \frac{290}{0.473}$ $= 926\,°K$	$\frac{T_3}{G_1 G_2} = \frac{323}{(0.473)(100)}$ $= 7\,°K$	$\frac{T_3}{G_1 G_2 G_3} = \frac{8881}{(0.473)(100)(0.473)}$ $= 397\,°K$

The noise figure and noise temperature cascade equations are

$$F_{cas} = F_1 + \frac{F_2-1}{G_1} + \frac{F_3-1}{G_1 G_2} + \frac{F_4-1}{G_1 G_2 G_3} \quad T_{cas} = T_1 + \frac{T_2}{G_1} + \frac{T_3}{G_1 G_2} + \frac{T_4}{G_1 G_2 G_3} \qquad 7.56$$

$$= 2.11 + 3.19 + 0.03 + 1.37 \qquad = 323 + 926 + 7 + 397$$
$$= 6.7 \qquad = 1{,}653\,K$$
$$= 8.3\,dB$$

The system noise figure is down to 8.3 dB, which means there is a 13.2 dB decrease over the original system. The second term of the cascade equation contributes the lion's share of the noise to the system.

3. *Amplifier on the antenna mast.* Figure 7-26 shows the system when we place the amplifier on the antenna mast as close to the antenna as possible.

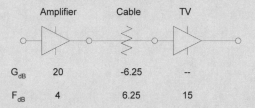

Figure 7-26 *Cascade equivalent of a household television system with an amplifier placed directly at the antenna.*

CASCADE I — GAIN DISTRIBUTION | 671

Table 7-4 Television antenna cascade with amplifier at antenna.

	Amplifier	Cable	TV
G_{dB}	20	-6.5	--
G	100	0.224	--
F_{dB}	4	6.5	15
F	2.51	4.47	31.6
T (°K)	438	1005	8881
F (referenced to cascade input)	2.51	$\frac{F_2-1}{G_1} = \frac{4.47-1}{100} = 0.03$	$\frac{F_3-1}{G_1 G_2} = \frac{31.6-1}{(100)(0.224)} = 1.37$
T (referenced to cascade input - °K)	438	$\frac{T_2}{G_1} = \frac{1005}{100} = 10.1 °K$	$\frac{T_3}{G_1 G_2} = \frac{8881}{(100)(0.224)} = 397 °K$

In this case, the cascade noise figure and temperature are

$$F_{cas} = F_1 + \frac{F_2-1}{G_1} + \frac{F_3-1}{G_1 G_2} \quad T_{cas} = T_1 + \frac{T_2}{G_1} + \frac{T_3}{G_1 G_2}$$
$$= 2.51 + 0.03 + 1.37 \qquad\qquad = 438 + 10.1 + 397$$
$$= 3.91 \qquad\qquad\qquad\qquad = 845 \text{ K}$$
$$= 5.9 \text{ dB}$$

7.57

Now the cable contributes only 10 K of noise temperature to the cascade, the TV supplies 397 K, and the front-end amplifier supplies 438 K. The cascade noise figure is down to 5.9 dB. The system's noise figure can be further decreased by increasing the gain of the first amplifier or decreasing its noise figure. However, high excess gain causes linearity problems in a cascade. Table 7-5 presents a summary.

Table 7-5 Noise performance summary.

Amplifier Position	System Noise Figure	System Noise Temperature
None	21.5 dB	40,675 °K
At TV	11.0 dB	3,357 °K
Halfway	8.3 dB	1,653 °K
At Antenna	5.9 dB	845 °K

The best noise figure occurs when we place the amplifier on the mast, next to the antenna. However, we still get a significant improvement when we attach the amplifier right to the TV input terminals. In more complex systems we have many parameters to juggle. We deal with filters, mixers and amplifiers with different specifications. Remember, these tables address only noise problems, not linearity concerns. For a complete analysis, we need to run a computer program.

Conclusions

For a low noise cascade, we want to use an amplifier with a low noise temperature as the first amplifier in the cascade. Equation 7.28 tells us that, if all we care about is noise temperature, we want to keep the excess gain in front of every component in the cascade as large as possible. Noise figure constitutes only one part of the concerns. We also need to consider linearity, which suffers when we build a cascade with high excess gain.

7.8 Gain and Linearity

This section examines the effects of gain distribution on the linearity of a cascade. Although we will chiefly discuss TOI, the arguments apply to SOI as well. Figure 7-27 shows a general n-element cascade that we will use for the analysis.

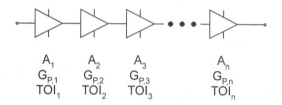

Figure 7-27 *General n-element cascade.*

OTOI vs. ITOI

Most design engineers are interested in the port that interfaces to the outside world, i.e., the antenna port. In transmitter design, the levels of the unwanted signals the system radiates into free space are important. Since the output of the cascade is attached to the antenna, transmitter designers pay more attention to output specifications, i.e., the OTOI or OSOI rather than input specifications.

In transmitter design, the smallest and largest signals the system will be able to reliably process are significant. Since the input of the cascade is attached to the antenna, receiver designers are more interested in the cascade's input specifications, i.e., the ITOI or ISOI.

This book concentrates on the input specifications of the cascade although the arguments will be valid for the output specifications as well. We will assume that all the coherent distortion products add coherently (see Chapter 6 for details).

Linearity Cascade Equation

Equation 7.30 describes the linearity behavior of the cascade in Figure 7-27. The equation is reproduced here.

$$\begin{aligned}\frac{1}{ITOI_{cas}} = &\frac{1}{ITOI_1} && \text{[Component \#1]} \\ &+ \frac{1}{ITOI_2/G_{p,1}} && \text{[Component \#2]} \\ &+ \frac{1}{ITOI_3/G_{p,1}\,G_{p,2}} && \text{[Component \#3]} \\ &+ \ldots \\ &+ \frac{1}{ITOI_n/G_{p,1}\,G_{p,2}\ldots G_{p,n-1}} && \text{[Component \#n]} \end{aligned} \qquad 7.58$$

The concepts of excess gain and translation apply. Each term of Equation 7.58 expresses the linearity contribution of each component in the cascade.

Resistors in Parallel

Equation 7.58 resembles the equation that describes adding resistors in parallel.

$$\frac{1}{R_p} = \frac{1}{R_1} + \frac{1}{R_2} + \frac{1}{R_3} \qquad 7.59$$

Figure 7-28 shows the similarities. The value of each resistor represents the TOI of each component when it is referenced to the input port. Whenever we add a new device to a cascade, we effectively add another parallel resistor to the equivalent circuits and lower the TOI (resistance) of the cascade. We cannot improve the linearity of the cascade by adding another component; we can only make the linearity worse.

674 | RADIO RECEIVER DESIGN

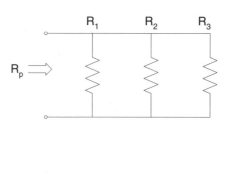

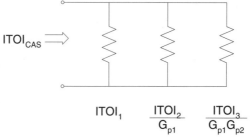

Figure 7-28 *The equation describing resistors in parallel is similar to the equation describing the linearity of a cascade given the linearity of the components which make up the cascade. Adding a component to a cascade will always decrease the linearity of the cascade.*

Pieces of the Linearity Cascade Equation

Referring to the analogy of resistors in parallel, each term of Equation 7.58 signifies

- *Component #1.*

$$\frac{1}{ITOI_1} \qquad 7.60$$

The first term represents the distortion the first component contributes to the cascade. Any change in the ITOI of the first element directly affects the cascade's linearity one-for-one. This is the only term in Equation 7.58 where the ITOI of amplifier A_1 appears.

- *Component #2.*

$$\frac{1}{ITOI_2/G_{p,1}} \qquad 7.61$$

This term represents the ITOI contribution of amplifier A_2 to the cascade. This is the only place where the ITOI of amplifier A_2 exists in the general equation. The ITOI contribution of amplifier A_2 is reduced by the gain of amplifier A_1. In this case, excess gain is harmful.

Remember the parallel resistor analogy of Figure 7-28. If we add a low-value resistor, i.e., a low ITOI to the cascade in parallel, we limit the parallel resistance to a low value (a low cascade ITOI). The cascade linearity will always be limited by the component whose ITOI is the smallest after it has been moved to a common point.

- *Component #3*. This term represents the linearity contribution that amplifier A_3 makes to the cascade. In this case, the ITOI of A_3 is reduced by $G_{p,1}G_{p,}$ (the excess gain up to the input of A_3). If A_3 is to contribute only a small amount of distortion to the cascade, the combination of $ITOI_3/G_{p,1}G_{p,2}$ has to be as small as practical.

- *Component #n*. We can see that the ITOI contribution of the nth term is reduced by the excess gain from the $(n-1)$ components preceding the nth component. In other words, the linearity of the nth component is reduced by the excess gain of the cascade at the input to the nth component.

Example 7.5 — Three-Element Cascade
Figure 7-29 shows a three-element cascade. Examine the cascade.

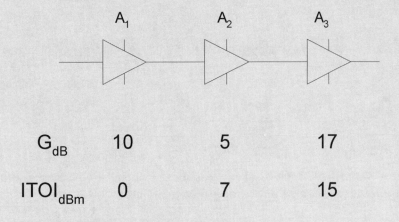

Figure 7-29 Simple three-element cascade (Example 7.5).

Solution —

Table 7-6 describes the cascade.

Table 7-6 *Example 7.5. Linearity contributions in the three-element cascade of Figure 7-29. Reference point is the cascade's input port.*

	A_1	A_2	A_3
G_{dB}	10	5.0	17
G	10	3.16	50.1
$ITOI_{dBm}$	0.0	7.0	15.0
$ITOI_{mW}$	1.0	5.0	31.6
$ITOI_{mW}$ (referenced to cascade input)	1.0	$\dfrac{ITOI_2}{G_1} = \dfrac{5.0}{10} = 0.50$	$\dfrac{ITOI_3}{G_1 G_2} = \dfrac{31.6}{(3.16)(10)} = 1.0$
$ITOI_{dBm}$ (referenced to cascade input)	0.0	-3.0	0.0

Table 7-6 reveals that the first and third amplifiers contribute equally to the cascade ITOI. The second component contributes the most nonlinearity to the cascade because it exhibits the lowest linearity compared to a common point. Complete evaluation of Equation 7.58 indicates that the ITOI of the cascade is

$$\frac{1}{ITOI_{cas}} = \frac{1}{ITOI_1} + \frac{1}{ITOI_2/G_{p,1}} + \frac{1}{ITOI_3/G_{p,1}G_{p,2}}$$

$$= \frac{1}{1.0} + \frac{1}{0.5} + \frac{1}{1.0} \quad\quad 7.62$$

$$= 4.0$$

$$ITOI_{cas} = 0.25 \text{ mW}$$

$$= -6.0 \text{ dBm}$$

We can perform this identical process on the output of the cascade. The equation for cascade OTOI in terms of component OTOI is

$$\frac{1}{OTOI_{cas}} = \frac{1}{G_{p,2}G_{p,3}OTOI_1} + \frac{1}{G_{p,3}OTOI_2} + \frac{1}{OTOI_3} \quad\quad 7.63$$

Component ITOI is related to component OTOI via

$$OTOI_n = ITOI_n G_{p,n} \quad\quad 7.64$$

Equation 7.63 can be written as

$$\frac{1}{OTOI_{cas}} = \frac{1}{G_{p,1}G_{p,2}G_{p,3}ITOI_1} + \frac{1}{G_{p,2}G_{p,3}ITOI_2} + \frac{1}{G_{p,3}ITOI_3} \quad 7.65$$

Again, this equation demonstrates the concept of moving the TOI specification of a component to a particular port, then adding the equivalent TOIs together. Table 7-7 was built applicable to Equation 7.65.

Table 7-7 *Example 7.5. Linearity contributions in the three-element cascade of Figure 7-29. Reference point is the cascade's output port.*

	A_1	A_2	A_3
G_{dB}	10	5.0	17
G	10	3.16	50.1
$ITOI_{dBm}$	0.0	7.0	15.0
$ITOI_{mW}$	1.0	5.0	31.6
$ITOI_{mW}$ (referenced to cascade output)	1580*	792**	1580***
$ITOI_{dBm}$ (referenced to cascade output)	32.0	29.0	32.0

*
$$G_{p,1}G_{p,2}G_{p,3}ITOI_1 = (10)(3.16)(50.1)(1.0)$$
$$= 1580$$

**
$$G_2 G_3 ITOI_2 = (3.16)(50.1)(5.0)$$
$$= 792$$

***
$$G_3 ITOI_3 = (50.1)(31.6)$$
$$= 1580$$

Components #1 and #3 contribute equally to the cascade TOI; component #2 is the weak link, contributing the most nonlinearity to the cascade. If the component TOIs and gains are expressed in dBM, we can move the linearity specification to any port by simple addition or subtraction.

Strength of Distortion Products

Equation 7.12, repeated here, indicates that the strength of the distortion products produced by a nonlinear device is a strong function of the fundamental input power.

$$P_{out,2f-f,dBm} = 3P_{out,f,dBm} - 2OTOI_{dBm} \qquad 7.66$$

In a cascade, the strength of the fundamental signal into a particular device is determined by the gain preceding the device. To keep the distortion products low (and the linearity high), the excess gain through the cascade should be kept as low as possible.

For example, if a –80 dBm signal is applied to the cascade of Figure 7-30, amplifier #1 will see a –80 dBm signal. Amplifier #2 will see a –60 dBm signal because of the 20 dB power gain of amplifier #1. Although amplifiers #1 and #2 have the same TOI, amplifier #2 will generate higher levels of distortion because it sees a larger signal. Likewise, amplifier #3 will see a –40 dBm signal and will generate the most distortion power of all the devices in the cascade, thus dominating the TOI of the cascade.

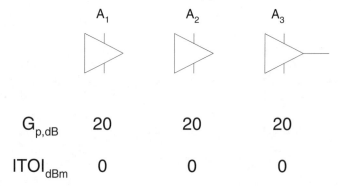

Figure 7-30 *Three-element cascade. Amplifier A_3 will always see a larger signal than either A_1 or A_2, so A_3 will limit the cascade's linearity.*

Conflict

Good noise performance dictates that we should keep the excess gain as high as possible. Good linearity performance forces us to keep the excess gain as low as possible. We have a fundamental compromise between low noise temperature and high linearity.

Rule of Thumb

When the TOI of every component is translated to a common port, the translated TOI of each component should be about equal. If all the translated TOIs are equal, then all of the elements in the cascade will become nonlinear at the same input power level. If one translated TOI is much smaller than all the rest, that component will set the linearity of the cascade.

CASCADE I — GAIN DISTRIBUTION | 679

Example 7.6 — Component TOI and Excess Power Gain

Figure 7-31 contains three identical amplifiers in cascade. Each amplifier has a power gain of 20 dB (or 100x), an OTOI of +10 dBm (or 10 mW) and an ITOI of −10 dBm (or 0.10 mW). Analyze the cascade. Which component dominates the linearity?

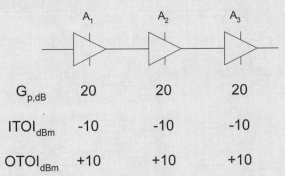

Figure 7-31 Three-element cascade of Example 7.6. Component TOIs are poorly matched to the excess gains.

Solution —

First, let us mathematically move the ITOI of each amplifier to the cascade's input. We adjust the TOI of each amplifier according to the gain preceding the amplifier.

Table 7-7 Example 7.6. Linearity contributions in the three-element cascade of Figure 7-31. Amplifier A_3 limits the cascade linearity.

	A_1	A_2	A_3
G_{dB}	20	20	20
G	100	100	100
$ITOI_{dBm}$	-10.0	-10.0	-10.0
$ITOI_{dBm}$ (referenced to cascade output)	$ITOI_1 = -10$	$ITOI_2 - G_{p,1} = -10 - 20$ $= -30 \, dBm$	$ITOI_3 - G_{p,1} - G_{p,1} = -10 - 20 - 20$ $= -50 \, dBm$

Clearly, amplifier #3 dominates the cascade TOI (which is −50.04 dBm). The third amplifier becomes nonlinear at a much lower power level than the other two amplifiers. The 40 dB power gain preceding amplifier #3 reduces its effective linearity when we translate its TOI to the cascade's input.

Example 7.8 — Matching Component TOI to Excess Power Gain

Figure 7-32 shows another three-component cascade. This time, we have selected the TOI of each component based upon the gain preceding the component. Analyze the cascade. Which component dominates the cascade TOI?

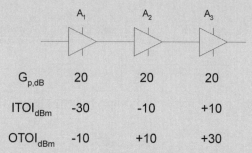

	A_1	A_2	A_3
$G_{p,dB}$	20	20	20
$ITOI_{dBm}$	-30	-10	+10
$OTOI_{dBm}$	-10	+10	+30

Figure 7-32 Three-element cascade of Example 7.8. Determine coherent and noncoherent cascade linearities.

Solution —

Table 7-9 Example 7.7. Linearity contributions in the three-element cascade of Figure 7-33. Each amplifier contributes equally to the cascade's TOI.

	A_1	A_2	A_3
G_{dB}	20	20	20
G	100	100	100
$ITOI_{dBm}$	-30.0	-10.0	+10.0
$ITOI_{dBm}$ (referenced to cascade output)	$ITOI_1 = -30$	$ITOI_2 - G_{p,1} = -10 - 20$ $= -30$ dBm	$ITOI_3 - G_{p,1} - G_{p,1} = 10 - 20 - 20$ $= -30$ dBm

The translated TOI of each component is the same. None of the components dominates the linearity of the cascade (-34.7 dBm); hence none of the amplifiers is the weak link in the chain.

Coherent vs. Noncoherent Addition

In Chapter 6, two different equations describing cascade linearity were derived. One equation assumed that the phase of the distortion products were coherent and in-phase, so the distortion voltages added. This is the worst-case assumption. The second equation assumed that the phases of the distortion products were noncoherent and random, so the distortion powers added. This is the noncoherent assumption.

When designing low-noise receiving systems, we have found that well-designed cascades usually behave as though the distortion products are adding up noncoherently. For the most part, these systems followed the noncoherent equations plus one or two dB. With wideband systems, the cascade SOI or TOI will stay at noncoherent levels over most of the frequency range of the system. However, over narrow frequency ranges, the SOI and TOI will increase to coherent summation levels. In a well-designed system (where the equivalent intercept points of all the devices are equal), the difference between coherent and noncoherent summation is 4 to 5 dB. When designing a system, we usually calculate the numbers for both the coherent and noncoherent cases. This gives us an idea of the variation we are likely to expect over time and frequency. For reference, the coherent and noncoherent cascade equations are

Third-Order Intercept (Coherent)

$$\frac{1}{ITOI_{cas}} = \frac{1}{ITOI_1}$$
$$+ \frac{1}{ITOI_2/G_{p,1}}$$
$$+ \ldots$$
$$+ \frac{1}{ITOI_n/G_{p,1}G_{p,2}\cdots G_{p,n-1}}$$

7.67

Third-Order Intercept (Noncoherent)

$$\frac{1}{ITOI_{cas}^2} = \frac{1}{(ITOI_1)^2}$$
$$+ \frac{1}{(ITOI_2/G_{p,1})^2}$$
$$+ \ldots$$
$$+ \frac{1}{(ITOI_n/G_{p,1}G_{p,2}G_{p,3}\cdots G_{p,n-1})^2}$$

7.68

Second-Order Intercept (Coherent)

$$\frac{1}{\sqrt{ISOI_{cas}}} = \frac{1}{\sqrt{ISOI_1}}$$
$$+ \frac{1}{\sqrt{ISOI_2/G_{p,1}}}$$
$$+ \cdots$$
$$+ \frac{1}{\sqrt{ISOI_n/G_{p,1}G_{p,2}\cdots G_{p,n-1}}}$$

7.69

Second-Order Intercept (Noncoherent)

$$\frac{1}{ISOI_{cas}} = \frac{1}{ISOI_1}$$
$$+ \frac{1}{ISOI_2/G_{p,1}}$$
$$+ \cdots$$
$$+ \frac{1}{ISOI_n/G_{p,1}G_{p,2}\cdots G_{p,n-1}}$$

7.70

Example 7.9 — TOI Cascade (Coherent and Noncoherent Addition)

Figure 7-33 shows three devices in cascade. Find the ITOI of the cascade using both coherent and noncoherent summation.

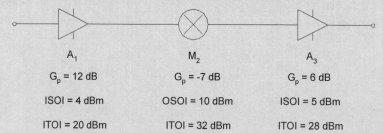

A_1	M_2	A_3
G_p = 12 dB	G_p = -7 dB	G_p = 6 dB
ISOI = 4 dBm	OSOI = 10 dBm	ISOI = 5 dBm
ITOI = 20 dBm	ITOI = 32 dBm	ITOI = 28 dBm

Figure 7-33 Three-element cascade of Example 7.9. Determine the coherent and noncoherent cascade TOI.

Solution —

We have a mixture of input and output specifications. First, the TOI of each device is moved to the input of the cascade, then the appropriate equations are applied.

Amplifier #1. We know the $ITOI_{1,dBm}$ at the input of the cascade thus

$$ITOI_{amp1,input} = 20 \text{ dBm} = 100 \text{ mW} \qquad 7.71$$

Mixer #2. Moving the OTOI of the mixer to the input of the cascade produces

$$\begin{aligned}ITOI_{mix2,input} &= 32 + 7 - 12 \\ &= 27 \text{ dBm} \\ &= 501 \text{ mW}\end{aligned} \qquad 7.72$$

Amplifier #3. The equivalent input TOI of amplifier #3 is

$$\begin{aligned}ITOI_{amp3,input} &= 28 + 7 - 12 \\ &= 23 \text{ dBm} \\ &= 200 \text{ mW}\end{aligned} \qquad 7.73$$

Amplifier #1 is the lowest ITOI and will dominate the result; amplifier #3 is the next lowest. The coherent TOI equation above reveals that the coherent summation ITOI is

$$\begin{aligned}\frac{1}{ITOI_{cas}} &= \frac{1}{100 \text{ mW}} + \frac{1}{501 \text{ mW}} + \frac{1}{200 \text{ mW}} \\ ITOI_{cas} &= 58.8 \text{ mW} \\ &= 17.7 \text{ dBm}\end{aligned} \qquad 7.74$$

The noncoherent assumption produces

$$\begin{aligned}\frac{1}{ITOI_{cas}^2} &= \frac{1}{(100 \text{ mW})^2} + \frac{1}{(501 \text{ mW})^2} + \frac{1}{(200 \text{ mW})^2} \\ ITOI_{cas} &= 88.1 \text{ mW} \\ &= 19.4 \text{ dBm}\end{aligned} \qquad 7.75$$

In this particular case, the noncoherent assumption produces a result that is 1.7 dB better than the coherent assumption.

Example 7.10 — Cascade TOI (Coherent and Noncoherent Cases)

Find the ITOI of the cascade shown as Figure 7-34 for both the coherent and noncoherent cases.

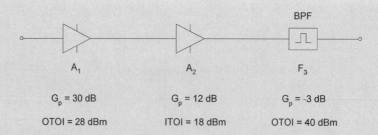

Figure 7-34 Three-element cascade of Example 7.9. Determine the coherent and noncoherent cascade TOI.

Solution —
First, we will move the TOI specification for each component to the input of the cascade.
Amplifier #1. Moving the OTOI of amplifier #1 to the cascade's input produces

$$ITOI_{amp1,input} = 28 - 30$$
$$= -2 \text{ dBm}$$
$$= 0.631 \text{ mW}$$
(7.76)

Amplifier #2. The equivalent input TOI of this amplifier referenced to the cascade's input is

$$ITOI_{amp2,input} = 18 - 30$$
$$= -12 \text{ dBm}$$
$$= 63.1 \cdot 10^{-3} \text{ mW}$$
(7.77)

Filter #3. The filter's equivalent cascade input TOI is

$$ITOI_{filter3,input} = 40 + 3 - 12 - 30$$
$$= 1 \text{ dBm}$$
$$= 1.26 \text{ mW}$$
(7.78)

Amplifier #2 is the weak link in the cascade. For coherent addition, we find

$$\frac{1}{ITOI_{cas}} = \frac{1}{0.631} \quad (\text{Amp}_1 \text{ contribution})$$
$$+ \frac{1}{63.1\text{E}-3} \quad (\text{Amp}_2 \text{ contribution})$$
$$+ \frac{1}{1.26} \quad (\text{Filter}_3 \text{ contribution})$$
$$ITOI = 54.9 \cdot 10^{-3} \text{ mW} = -12.6 \text{ dBm}$$
(7.79)

The noncoherent power summation produces

$$\frac{1}{ITOI_{cas}^2} = \frac{1}{(0.631)^2} \quad \text{(Amp}_1 \text{ contribution)}$$
$$+ \frac{1}{(63.1\text{E}-3)^2} \quad \text{(Amp}_2 \text{ contribution)} \quad\quad 7.80$$
$$+ \frac{1}{(1.26)^2} \quad \text{(Filter}_3 \text{ contribution)}$$
$$ITOI = 62.7 \cdot 10^{-3} \text{ mW} = -12.0 \text{ dBm}$$

In this case, the difference between coherent and noncoherent addition is only 0.6 dBm. Since we have one component contributing most of the distortion to the cascade, it does not matter much how we add up the distortion products. If all of the components in the cascade contributed equally to the cascade TOI, there would be a greater difference between the coherent and noncoherent summations.

7.9 System Nonlinearities

Detection

Given a common wideband receiving system connected to an antenna, how can we determine if it is suffering from linearity problems? Several clues may help to determine nonlinearities.

- Signals are not where they are supposed to be. For example, we observe FM radio stations at 500 MHz when they are supposed to be at 100 MHz.

- Signals appear and disappear abruptly at the same time. This effect can be caused by constant signals combining with push-to-talk signals ("walkie-talkies"). If either signal is very strong, all of the intermodulation products developed by the combination will also exhibit the on/off rate of the push-to-talk.

- Signals seem to have more than one type of modulation present. In other words, a signal with both AM and FM characteristics (with different modulating signals) can be a result of intermodulation.

- The noise floor of the receiver bounces up and down, especially at push-to-talk rates.

- A signal is found with the same information on it at several different frequencies. Although this is occasionally done on purpose, it can also result from intermodulation.

- If distortion in a tunable receiver is suspected, view the IF port on a spectrum analyzer as the receiver is tuned. Intermodulation products will move through the IF at different rates than signals from the antenna. The distortion products will often be observed moving in opposite directions as the receiver is tuned.

Example 7.11 — Receiving Problems

When tuning through the spectrum, you can hear an FM radio station (which normally broadcasts at 101.1 MHz) at 247 MHz. This is a result of the FM broadcast at 101.1 MHz combining with some other signal in a nonlinear device according to

$$f_{out} = |\pm n \cdot f_1 \pm m \cdot f_2| \text{ where } \begin{cases} n = 0,1,2,3, ... \\ m = 0,1,2,3, ... \end{cases} \qquad 7.81$$

where $m + n$ is the order of the spurious product. If f_1 is the FM broadcast station at 101.1 MHz, find all of the possible values of f_2 resulting in an output frequency of 247 MHz. Assume $m + n \leq 5$.

Solution —

For this particular problem, we can write

$$247 = n \cdot 101.1 \pm m \cdot f_2 \qquad 7.82$$

or

$$f_2 = \left| \frac{247 \pm n \cdot 101.1}{m} \right| \qquad 7.83$$

Table 7-10 *Possible spurious products generated in the receiver of Example 7.11.*

n,m	f_2 (MHz) (+, -)	n,m	f_2 (MHz) (+, -)	n,m	f_2 (MHz) (+, -)
0,0	*,*	1,1	348.1, 145.9	2,3	149.7, 14.9
0,1	247, 247	1,2	174.1, 73.0	3,0	*,*
0,2	123.5, 123.5	1,3	116, 48.6	3,1	550.3, 56.3
0,3	82.3, 82.3	1,4	87, 36.5	3,2	275.2, 28.2
0,4	61.8, 61.8	2,0	*,*	4,0	*,*
0,5	49.4, 49.4	2,1	449, 44.8	4,1	651.4, 157.4
1,0	*,*	2,2	224.6, 22.4	5,0	*,*

Table 7-10 shows all of the frequencies that can combine with 101.1 MHz to produce a signal at 247 MHz (up to order 5). The "*" indicates that there is no solution for these particular values of n and m.

Correction

Two major corrective actions can be taken when a system exhibits non-linear behavior: the linearity of the system can be increased or the signals that are causing the system to distort can be removed. The system's linearity can be improved by using more linear components. Limiting the input signal power often requires some tradeoffs.

First, the complete spectrum that is being applied to the system should be examined. Both in-band and out-of-band signals have to be included in the search. Very strong signals that are far removed from the center frequency of the system can still cause problems. Ideally, all of the signals present within the bandwidth of every component in the system should be considered. *Push-to-talk* (PTT) systems often cause intermittent nonlinear behavior in receiving systems because the transmitters can be very close to the receiving antenna. The intermittent nature of the problem can also make this type of interference difficult to track down and solve. Airport communications, taxi cab transmitters and FM radio stations can all be sources of strong signals which are likely to cause intermodulation problems. Once the offending signals are located, they can be removed with a low-pass, high-pass or notch filter if the filter will not also remove the desired signals. Sometimes the receive antenna just has to be re-oriented to reduce the amplitude of the interferer.

7.10 TOI Tone Placement

Figure 7-35 shows a typical receiving system consisting of amplifiers, filters, mixers and attenuators. The last filter in the system (BPF_4) is typically the narrowest filter in the system, i.e., it is just large enough to pass the entire signal of interest. If it were any wider, it would pass more noise than necessary and degrade the system performance. This filter also sets the system noise bandwidth. By definition, there is only one signal present in the system after it passes through BPF_4. This single signal is the signal to be collected and demodulated.

Third-order distortion results from unwanted signals combining to form phantom signals that might obscure the wanted signal. Since BPF_4 passes only the one signal of interest, we can safely state that there are no unwanted signals present in the system beyond BPF_4. In other words, the components following BPF_4 do not contribute to the nonlinearity of the cascade. The TOI of these components can be considered infinite.

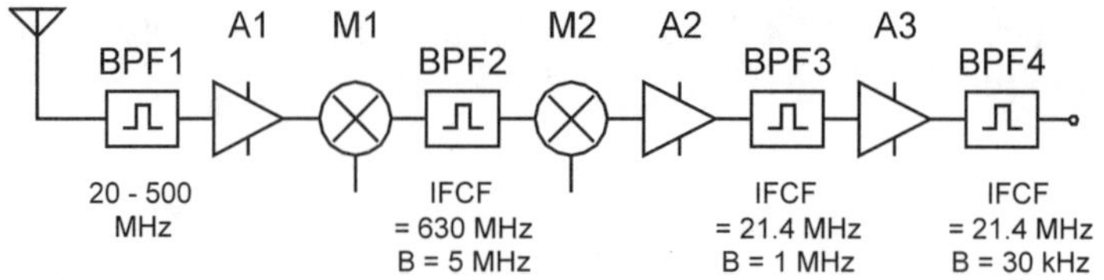

688 | RADIO RECEIVER DESIGN

Figure 7-35 Practical receiving system consisting of amplifiers, filters, mixers and attenuators.

TOI Input Tones

Suppose we wanted to measure the TOI of the receiver in Figure 7-35. The third-order test requires us to apply two tones to the receiver and look for third-order distortion products. We measure the power of all four tones present at the output of the receiver, then we can calculate the receiver's TOI. We usually set the frequency of the two tones so that the nonlinear distortion will appear at the receiver's tuned frequency.

100 kHz Tone Spacing

By way of example, the TOI test will be examined when the two input tones are spaced 100 kHz apart. Figure 7-36(a) shows the relationship between the two tones, the receiver's tuned frequency and the filter bandwidths of Figure 7-35. The receiver is tuned to the upper distortion product. Note that the two input signals pass through the entire receiver until they are finally stopped by BPF_4. Every amplifier and mixer in the cascade has an opportunity to generate distortion power since every component must process the large input signals. However, the components beyond BPF_4 do not contribute to the cascade's distortion power because BPF_4 severely attenuates the large input signals. Components which are downstream from BPF_4 do not see the large input signals.

1 MHz Tone Spacing

Figure 7-36(b) shows two large input signals spaced out to 1 MHz. Since the receiver is tuned to the upper third-order distortion product, the two input signals are 1 and 2 MHz below the tuned frequency. Both input signals will experience some attenuation due to BPF_3 (because of its 1 MHz bandwidth).

The components before BPF_3 contribute to the cascade's distortion (and its TOI) but the components after BPF_3 will contribute less than they did when the tones were only 100 kHz apart. When we increase the input tone spacing, the cascade appears to have a higher TOI because fewer components "see" the large input signals and fewer components can contribute to the output distortion power.

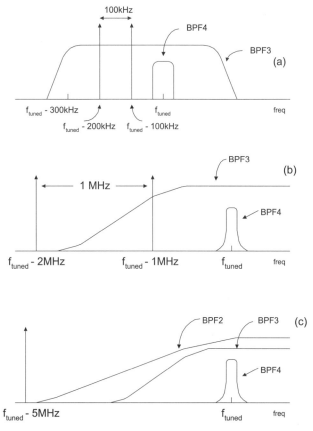

Figure 7-36 *Testing receiver TOI with two input signals. The relationship between the spacing of the two test tones, the receiver's tuned frequency and the bandwidths of the filters affects the TOI measurement.*

5 MHz Tone Spacing

Figure 7-36(c) shows the results when the input tones are 5 MHz apart. Only components between the antenna and the input of BPF_2 contribute distortion to the cascade. Increasing the tone spacing again has made the cascade look even more linear.

Figure 7-37 shows how the measured TOI of a receiver changes with the measurement tone spacing. The actual tone spacing has no significance in itself. The tone spacing relative to the bandwidth of the band-pass filters in the system is the important metric.

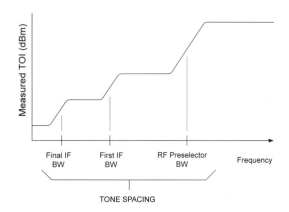

Figure 7-37 *When measuring the TOI of a receiver, the linearity will change with the spacing of the test tones used to perform the test.*

The measured linearity of the cascade changed depending upon how the test was performed. The cascade did not change — its linearity was simply measured under different conditions. When a linearity test is performed on a receiver, how far apart should the input tones be placed? If it is possible to specify, consider the receiver's operational environment. Generally, the tones are specified in a way that they can be stopped only by the final IF filter. This lets every component in the cascade up to the final IF filter contribute to the system's nonlinearity.

In a channelized system, such as a commercial FM radio receiver, the tones are placed on the two adjacent channels. For example, if we tune the receiver to 99.1 MHz, we will place the two tones at 98.7 and 98.9 MHz (or 99.3 and 99.5 MHz). When the TOI of a receive is specified, the test conditions should also be specified including

- the IF bandwidths used,
- the tuned frequency of the receiver,
- frequencies of the two tones.

Application to Gain Distribution

Given that every device prior to the final IF filter is allowed to contribute to the nonlinearity of the cascade, the excess gain should be kept as low as possible prior to the final IF filter. After the final IF filter, only one signal is present in the receiver. Since all of the undesired signals have been suppressed at this point, gain is available without a linearity penalty. In other words, gain can be applied freely after the final IF filter without incurring a degradation in the cascade's third-order intercept.

7.11 Automatic Gain Control (AGC)

So far it was assumed that the signal present at the input port of the receiver is very weak. In the example, we applied the MDS to the system and calculated the amount of power gain needed to bring this signal up to the level required by the downstream processor and found that the MDS was –124 dBm. We needed 94 dB of gain to make this signal compatible with the demodulator. If we design such a system and supply a signal at the MDS power, the receiver will process it. However, the signals arriving at the input of a typical receiver will not be that weak. Some could be very strong (upwards of 0 dBm, in many cases), and strong signals will cause the high-gain system to distort, which will influence the processing of signals negatively.

Example 7.12 — SNR of Large Signals
What is the equivalent input SNR of a –30 dBm signal presented to a receiver with a 5 dB noise figure and a 30 kHz noise bandwidth?

Solution —
We know

$$S_{in,dBm} = -30 \text{ dBm} \qquad 7.84$$

If we assume a 290 K antenna noise temperature, the receiver's input noise power is

$$\begin{aligned} N_{in,dBm} &= F_{dB} - 174 \frac{\text{dBm}}{\text{Hz}} + 10\log(B_n) \\ &= 5 - 174 \frac{\text{dBm}}{\text{Hz}} + 10\log(30{,}000) \\ &= -124 \text{ dBm} \end{aligned} \qquad 7.85$$

The SNR is

$$\begin{aligned} \left(\frac{S}{N}\right)_{dB} &= -30 - (-124) \\ &= 94 \text{ dB} \end{aligned} \qquad 7.86$$

AGC Bandwidth
When designing a receiver with an AGC function, an AGC bandwidth is implied (or equivalently, AGC attack and decay time). For example, suppose we wanted to design an AM receiver to process human speech that nominally covers 300 to 3000 hertz. The information to be processed is present in the amplitude variation of the signal that is received. If the AGC is

fast enough, it will change the gain of the receiver as the amplitude of the AM waveform changes. The fast AGC will remove the AM modulation from the signal to be received. In this case, the bandwidth of the AGC should be much less than 300 hertz and perhaps as small as 3 hertz.

Attack and Decay Times

AGC bandwidth is often specified in terms of attack and decay times rather than bandwidth. Attack time describes the time it takes for a receiver to adjust to a signal whose amplitude suddenly increases. Decay time refers to how quickly the receiver adjusts to a signal whose amplitude suddenly decreases.

Noise and Linearity Tradeoffs

For strong signals, the SNR is usually very large. It is helpful to exchange some of the abundant signal power (and hence, some of the signal-to-noise ratio) for some linearity. Earlier we found that an attenuator placed in front of a receiver performs the exact function we require. It preserves the receiver's dynamic range and allows us to operate at higher signal powers.

Figure 7-38 shows an example. We design a receiver with enough gain to process an input signal at –124 dBm. If the signal arrives with –30 dBm of power, the receiver will go into compression and the demodulator will see a severely distorted signal [Figure 7-38(a)]. Let us assume we want to keep a 50 dB SNR on the signal present at the receiver's input (this is more than enough SNR for even the most finicky demodulator). The equivalent receiver input noise is –124 dBm. For an input SNR of 50 dB, we can write

$$\left(\frac{S}{N}\right)_{dB} = 50 \text{ dB}$$
$$= S_{in,dBm} - (-124) \qquad 7.87$$
$$S_{in,dBm} = -74 \text{ dBm}$$

This equation indicates that the output SNR will be 50 dB if the receiver sees an input signal of –74 dBm. An attenuator at the front end of the receiver can be employed directly between the antenna and the receiver to reduce the –30 dBm antenna signal to –74 dBm at the receiver input [Figure 7-38(b)]. The 50 dB of SNR at the receiver's output will remain and the receiver will not have to process the strong –30 dBm signal.

The value of the attenuator is

$$\text{Attn}_{dB} = -30 - (-74) \qquad 7.88$$
$$= 44 \text{ dB}$$

The problem can be solved by placing a 44-dB attenuator between the antenna and the receiver.

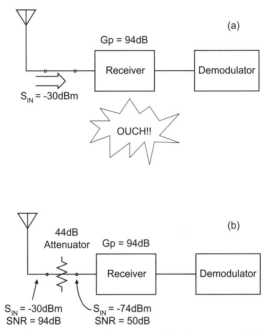

Figure 7-38 Trading noise performance for linearity. A large signal can drive a high-gain system into distortion. We adjust the receiver's gain based upon the input power to process a wide range of signal powers.

Receiver Automatic Gain Control

In an actual receiver, the process of adjusting the receiver's gain based upon the power level of the input signal is mechanized. At some point in the cascade chain, the power of the desired signal has to be measured. We then adjust the gain of the cascade to force the signal of interest to be at one set power level at the input to the demodulator.

The gain variation is built in the receiver using voltage-variable attenuators (devices whose attenuation depends upon some input voltage) or voltage variable gain stages (amplifiers whose power gain depends upon some externally applied voltage). Figure 7-49 shows a diagram.

For practical reasons, the gain reduction is applied at several places in the cascade. As the input signal power increases, the gain reduction is first applied at places in the cascade where the excess gain is large. Then, as the input signal power increases, the power gain at other stages in the receiver are decreased.

Most commonly, the gain reduction occurs in two stages. As the input signal level increases, the gain reduction is performed first at the sections of the receiver that are farthest from the antenna. When the gain reduction has reached its maximum value, a second gain reduction (or "delayed AGC") is applied to the sections of the receiver that are closest to the antenna.

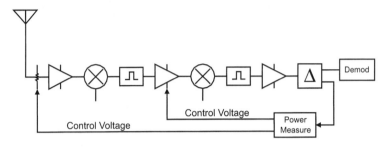

Figure 7-39 *Automatic gain control (AGC) applied to separate stages in the receiver. This technique is commonly referred to as "delayed AGC."*

7.12 Cascade Gain Distribution Rules

We are finally in a position to answer the question we posed earlier in this chapter. Given that we need 94 dB of gain, which configuration of Figure 7-13 will produce the system with the largest dynamic range? The answer is that none of the candidate systems is really adequate. From a noise perspective, Figure 7-13(a) is ideal because it applies a great deal of gain at the very beginning of the cascade. This system will result in a minimum noise figure.

From a linearity perspective, Figure 7-13(a) is a very poor choice. Since every component beyond the first amplifier is supposed to process a very large signal, each component will produce a lot of distortion and the linearity of the cascade will suffer. Figure 7-13(b) presents a very linear solution because all of the components process very small signals. However, Figure 7-13(b) is a poor performer in terms of noise because each lossy component takes precious signal power away from the receiver while contributing noise to the system. Figure 7-13(c) is a reasonable tradeoff between noise and linearity. Taking both noise and linearity into account, we arrive at the following design rules of thumb.

Excess Gain

Excess gain should be kept between 15 and 25 dB. At low levels of excess gain, individual components contribute too much noise to the cascade. At high levels of excess gain, individual components add distortion because they are forced to process signals that are too large.

Follow the 15 and 25 dB excess gain rule until you pass through the narrowest filter in the system. This filter determines your noise bandwidth. By definition, we have only one signal present in this filter. Distortion, which results from unwanted signals combining to form phantom signals, is no longer a problem since we are processing only one signal. After passing through this filter, enough gain has to be added to make certain that the system's MDS will experience enough power gain to bring it up to the levels required by further processing.

IF Bandwidth

To avoid second-order nonlinearities, all the bandwidths in the system should be kept to much less than an octave. This might mean choosing a high center frequency for one of the IFs if processing signals with large bandwidths. Building a system containing a multioctave bandwidth should be avoided because second-order distortion is typically much stronger than third-order distortion.

Stability

An added benefit to placing most of the gain after the final IF filter is stability. Whenever a system has power gain, oscillation is possible. Because IF frequencies are usually lower than the received RF signals, oscillation is usually less of a problem in the IF stages of a receiver.

AGC

Supply enough gain so the system can process its MDS. Rely on the receiver's AGC to reduce the gain so that the system can process larger signals. Remember, when dealing with a large signal, i.e., one with a large SNR, noise performance can be traded off for linearity.

Limit Bandwidth

Almost without exception, it is a good idea to limit bandwidth when possible because it will

- decrease the possibility of producing spurious signals in nonlinear devices. A device processing a large number of unwanted signals is more likely to generate an in-band spurious response than a device processing a small number of unwanted signals.

- limit stability problems. Oscillation occurs because a signal finds its way from a system's output back to its input. Limiting the bandwidth of a system limits the number of signals that can pass through the system with gain, which reduces the chance of oscillation.

- limit external pickup of unwanted signals. As discussed earlier, there are many sources of spurious responses in an actual receiver (particularly the internal digital logic). Limiting the bandwidth reduces the chance that the wideband spectrum generated by the digital logic will experience gain through the RF system.

Matching

Most components used to build receivers are designed, built and tested assuming they will be operated in a wideband Z_0 impedance environment. If the ports are not terminated, the devices will likely misbehave and not perform as expected. This is especially the case with mixers and filters.

Occasionally amplifiers can be counted on to provide a wideband match to the outside world, although this is a poor design practice. If misterminated, a poorly designed amplifier can oscillate or change its gain and noise characteristics. However, there are many examples in functioning receivers which have poorly terminated amplifiers providing a match to external devices. Attenuators are useful matching tools, because the maximum return loss of a resistive attenuator is twice its attenuation value (i.e., a 5 dB attenuator will always present at least a 10 dB return loss to anything connected to it). Although some signal loss will occur, the matching provided by the attenuator is often worth it.

War Story – Amplifier Stability and Terminal Loading

The author was working with a system that contained a parabolic dish antenna. Using good design practices, the antenna feed was followed immediately by a wideband amplifier. After working for about an hour, the author noted that the RF environment was behaving strangely. The noise floor, as viewed on a spectrum analyzer, would rise about 40 dB as the antenna was moved past one particular azimuth. Finally, the author found a huge signal present at the same azimuth and reasoned that the large signal was causing the front-end amplifier to go nonlinear, which can raise the system's noise floor and cause other impractical effects.

On closer examination, the large signal disappeared when the antenna was moved just a few degrees. This was unusual considering the power in the signal and the sidelobe levels of the antenna. Closer investigation revealed that the frequency of the signal changed slightly as the author changed the antenna's azimuth! In this particular facility, the antenna was some distance from the receiver and the control functions occurred by remote control. When the author finally looked at the antenna, he found that it faced a piece of steel supporting structure when it "received" the large, troublesome signal.

When the antenna was pointing at the metal, the front-end amplifier was exposed to a poor impedance match on its input and the amplifier broke into self-oscillation. Since the reflection coefficient of the poor match changed slightly as the author moved the antenna, the frequency of oscillation changed the antenna moved. Changing the defective front-end amplifier solved the problem.

Linearity and Power Consumption

An amplifier with a high TOI will usually require more DC power than an amplifier with a low TOI. High DC power usually means the amplifier will dissipate much heat. If power consumption and heat dissipation can be ignored, we can build a very linear system. Similarly, a mixer with a high TOI will usually require a higher LO drive level than a mixer with a low TOI. This increases the complexity of the LO and complicates the receiver's internal isolation problems.

Balancing TOI with Excess Gain

When we translate the linearity specifications of each component in the cascade to a common port, the translated specs of every component should be about equal. For example, if all the translated TOI's are equal, all of the elements in the cascade become nonlinear at the same input power level. If one element is too small, the TOI of the entire cascade will be dominated by that one weak link.

7.13 Cascades, Bandwidth and Cable Runs

In some situations, we are forced to place an antenna a long distance from the receiving system. In these cases, the attenuation of the cable connecting the antenna and the receiver is a significant factor in a system's gain distribution. The cable attenuation is even more problematic when we are handling wideband signals because the cable's attenuation changes with frequency. Table 7-11 lists the attenuation of three types of coaxial cable over frequency.

Table 7-11 *Attenuation of three types of cable (100-foot run).*

Freq (MHz)	Cable Attenuation (100 foot run) - dB		
	RG-58A	RG-223	1/4-inch Heliax
10	1.4	1.3	0.5
100	4.9	4.0	2.0
200	7.3	5.7	2.8
300	9.1	7.1	3.3
400	11.0	8.3	4.0
500	13.5	9.4	4.3
600	15.0	10.4	5.0
700	17.0	11.3	5.2
800	18.0	12.2	6.0
900	19.0	12.9	6.1
1000	20.0	13.3	6.8
3000	41.0	36.0	12.0
5000	-	51.0	18.0
8000	-	74.0	23.0
10000	-	85.0	29.0

Figure 7-40 shows this data graphically; Examples 7.12 and 7.13 illustrate the problem.

Example 7.13 — Narrowband Cable Runs

Suppose we want to move a signal which is centered at 8 GHz signal and is 30 MHz wide. Although 30 MHz sounds like a large number, the percentage bandwidth is very small. This signal will fall between 7.85 GHz to 8.15 GHz; the percentage bandwidth is less than 0.4%.

Figure 7-40 shows that the attenuation of 100 feet of 1/4-inch Heliax cable is about 25 dB. Since the percentage bandwidth is so small, the cable's attenuation over the signal bandwidth is approximately constant. In other words, the transmission line can be modeled as a frequency-independent attenuator, and amplifiers are simply inserted at the proper places in the cable to keep the excess gain between 15 and 25 dB.

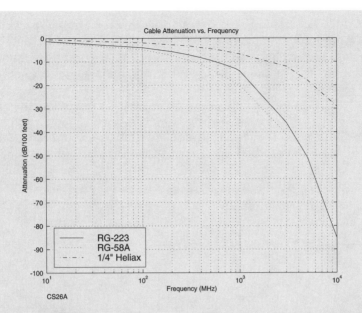

Figure 7-40 Cable attenuation vs. frequency for three cable types (450-foot run).

Example 7.14 — Wideband Cable Runs

Suppose we have to transport signals in the range of 10 to 1000 MHz from an antenna to a receiver. Further, let us suppose the distance is 450 feet. Figure 7-41 shows the cable attenuation for a 450-foot run over a 10 to 1000 MHz frequency range. Over this frequency range, the loss of the RG-58A cable varies from 6 dB at 10 MHz to about 90 dB at 1000 MHz, a difference of 84 dB. Low-frequency signals will suffer little attenuation through the cable, and high-frequency signals will suffer a lot of attenuation. The net result is that, at the end of the cable, we can see a lot of strong, low-frequency signals and few high-frequency signals.

It will be difficult for the amplifiers in the signal path to remain linear in this situation because the gain of the system changes with frequency. For example, if an amplifier is placed in the cable to increase the strength of the high-frequency signals, the amplifier will have to process the strong, low-frequency signals that did not experience much attenuation. Linearity could be a problem. Figure 7-41 shows that the differential attenuation for the RG-223 cable is 58 dB; the attenuation difference for the 1/4-inch Heliax is about 28 dB. The low-loss cable eases the wideband problem not because of its low attenuation but because a low-loss cable usually exhibits lower differential attenuation with frequency. We can also design an amplitude-compensating network to flatten out the attenuation before gain can be applied.

Usually the hardest cable run to realize is a wideband (several octaves) link between an antenna and a remote receiver. In this case, the primary design factor is the cable's attenuation characteristics over frequency. The other factors quickly become unimportant when a low-level set of signals has to be moved from the antenna to the receiver.

In mobile or test situations, we often place too much emphasis on the ease of handling and on the ease of connector attachment, violating good engineering practices. As a direct result of poor cable selection we may miss many low-level signals (which can result in insufficient signal strength at the receiver).

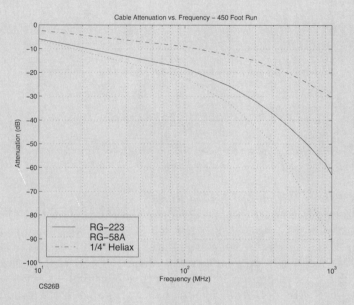

Figure 7-41 Example 7.14. Cable attenuation vs. frequency for three cable types (450-foot run).

7.14 Cascade I Design Summary

Resistive Attenuators

Return Loss of an Attenuator

The worst-case return loss of a resistive attenuator is twice the value of the attenuation. For example, when looking at any component through a 10 dB attenuator, the power that reflects off of the attenuator and back into the source will always be at least 20 dB less than the incident power.

Front-End Attenuators
A resistive attenuator placed between the antenna and a receiver will preserve the dynamic range of the receiver but will allow the system to process stronger signals. Noise performance is being traded for linearity.

Noise Figure and Attenuators
If a resistive attenuator (whose noise temperature is T_0) is placed in front of a device, the noise figure of the device will increase by the value of the attenuator. In other words, the value of the attenuator can be added to the noise figure of the amplifier to produce the noise figure of the two-element cascade.

Cascade Evaluation

Noise/Linearity Translation
Given a cascade, the best analysis approach is to move the noise or linearity specification of each component either to the cascade's input or to its output and then perform the summation or comparison at that port. When the intercept of each component is translated to a common port, we would like the translated intercept of each component to be about equal. If all the translated intercepts are equal, all of the elements in the cascade will become nonlinear at the same input power level. If one translated intercept is much smaller than all the rest, it will set the linearity of the cascade. From a noise perspective, the first component in the cascade should set the cascade's noise performance.

Cascade Design Rules of Thumb
1. Keep the excess gain between 15 and 25 dB. At low levels of excess gain, individual components contribute too much noise to the cascade. At high levels of excess gain, individual components add distortion because they are forced to process signals that are too large.

2. Take every opportunity to limit bandwidth. To avoid second-order nonlinearities, keep all the bandwidths in the system much less than an octave. This might mean choosing a high center frequency for one of the IFs if signals with large bandwidths are to be processed . Limiting bandwidth restricts the number of strong signals that can attack your receiver and cause it to intermod.

3. Place most of the gain after the final IF filter for stability and linearity. Whenever a system has power gain, oscillation is possible. Because IF frequencies are usually lower than the received RF signals, oscillation is usually less of a problem in the IF stages of a receiver. Also, high gain builds strong signals and strong signals cause intermod products. By

definition, only the desired signal is present in the receiver's IF bandwidth after the final IF filter, which means that so multisignal intermodulation products are no longer an issue.

4. Supply enough gain so the system can process its MDS. Rely on the receiver's AGC to reduce the gain so that the system can process larger signals. Remember, when dealing with a large signal (i.e., one with a large SNR), noise can be traded for linearity.

5. Use resistive attenuators to trim gain and improve interstage matching.

6. To increase the SFDR of a system, either increase the system's ITOI or decrease the system's noise figure, or decrease the noise bandwidth.

7.15 References

1. Gross, Brian P. "Calculating the Cascade Intercept Point of Communications Receivers." *Ham Radio Magazine*, no. 8 (1980): 50.

2. Ha, Tri T. *Solid-State Microwave Amplifier Design*. New York: John Wiley and Sons, 1981.

3. Perlow, Stewart M. "Basic Facts About Distortion and Gain Saturation." *Applied Microwaves Magazine* 1, no. 5 (1989): 107.

4. Terman, Frederick E. *Electronic and Radio Engineering*. New York: McGraw-Hill, 1955.

5. Williams, Richard A. *Communications Systems Analysis and Design: A Systems Approach*. Englewood Cliffs, NJ: Prentice-Hall, 1987.

8

Cascade II — IF Selection

Suppose one of you wants to build a tower. Will he first sit down and estimate the cost to see if he has enough money to complete it? For if he lays the foundation and is not able to finish it, everyone who sees it will ridicule him.
—Luke 14:18-29

8.1   Introduction

This chapter discusses design strategies and considerations for frequency conversion schemes.

Cost

It is desirable to build an inexpensive system that includes the cost of mixers, the filter, amplifiers, LOs and their support circuitry, the demodulator, microprocessor and other components. For example, cellular telephones use a first IF of 45 MHz. Every television in the world uses this IF and plenty of cheap components are available that work in that frequency range. Common IFs include 10.7 MHz, 21.4 MHz, 455 kHz, 70 MHz and 160 MHz.

Inertia

Designing a conversion scheme for a high-performance receiver can be quite involved once you consider all the trade-offs, so there is some validity to designing a receiver using a previously fielded conversion.

Physical Size

Physical size, commercial components, low power and common IFs are usually interdependent.

Power Consumption

Small physical size makes it difficult to remove heat. Small systems require that the power supply be very efficient and that the receiver itself require little power. Often battery life has to be considered as well.

Further, the design of the power supply can influence a system in unexpected ways. For example, the switching frequencies present in a switch-mode power supply can appear as sidebands on the LOs of a receiver. The switching frequency affects the step size and noise performance of PLL-type frequency synthesizers. It can also limit the receiver's spur-free dynamic range.

Spurious Considerations

A receiving system should generate as few spurious signals as possible. Spurious signals usually arise from two events. Unwanted signals allowed to pass through nonlinear elements in the receiver can generate new signals that are at the same frequency as a desired signal. These intermodulation products can cover up or distort the desired signals.

The local, digital logic and power supply oscillators present in a receiver can find their way onto the desired signals present in a receiver and distort them. These internal signal sources can also seep into the receiver's nonlinear elements and combine to form new signals.

Narrowband vs. Wideband Design

A *narrowband* system, in contrast to a wideband system, does not process more than one octave of signal bandwidth at one time. In a *wideband* design, the second-order and third-order distortion has to be taken into account. This allows more undesired signals to be present at one time and increases the spurious problems. In short, a narrowband design is easier than a wideband design.

8.2 Review

Figure 8-1 shows a simple conversion scheme. The goal is to convert a swatch of spectrum centered at the RF to some IF.

CASCADE II — IF SELECTION | 705

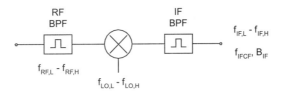

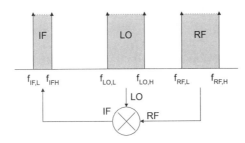

Figure 8-1 *A simple frequency conversion scheme defining variables for analysis.*

Observations

The following analysis is based upon three assumptions.

- At any time, every signal in the range of $f_{RF,L}$ to $f_{RF,H}$ will pass through the RF BPF to the RF port of the mixer.

- At any time, every signal in the range $f_{LO,L}$ to $f_{LO,H}$ will be present on the mixer's LO port.

- We are interested in signals that pass through the IF BPF. We want to determine the sum or difference component of the mixing process. Any other mixing component that passes through the IF BPF is undesired and constitutes an error.

Conversion Equation

The conversion equation is

$$f_{IF} = |mf_{LO} \pm nf_{RF}| \qquad 8.1$$

where

f_{LO} = any frequency from $f_{LO,L}$ to $f_{LO,H}$,
f_{RF} = any frequency from $f_{RF,L}$ to $f_{RF,H}$,
$m = 0, 1, 2, 3 ...,$
$n = 0, 1, 2, 3 ... .$

For each value of m and n, the band of frequencies from $f_{LO,L}$ to $f_{LO,H}$ will mix with the band of frequencies from $f_{RF,L}$ to $f_{RF,H}$ to produce two frequency bands on the mixer's IF port. The two bands fall in the ranges of $f_{IF,-,L}$ to $f_{IF,-,H}$ for the lower sideband and from $f_{IF,+,L}$ to $f_{IF,+,H}$ for the upper sideband. Figure 8-1 represents the operation for only one value of m and one value of n. If we change either m or n, the upper and lower IF sidebands will shift in frequency. To calculate $f_{IF,-,L}$, $f_{IF,-,H}$, $f_{IF,+,L}$ and $f_{IF,+,H}$, we evaluate the conversion equations at the band edges of the RF and LO ports. For the lower sideband (using $f_{IF} = |f_{LO} - f_{RF}|$), we find

$$f_{IF,-,L} = |mf_{LO,L} - nf_{RF,H}|$$
$$f_{IF,-,H} = |mf_{LO,H} - nf_{RF,L}|$$

8.2

For the upper sideband (using $f_{IF} = f_{LO} + f_{RF}|$), we can write

$$f_{IF,+,L} = |mf_{LO,L} + nf_{RF,L}|$$
$$f_{IF,+,H} = |mf_{LO,H} + nf_{RF,H}|$$

8.3

Example 8.1 — Cellular Telephone
Figure 8-2 shows a conversion scheme for a cellular telephone receiver.

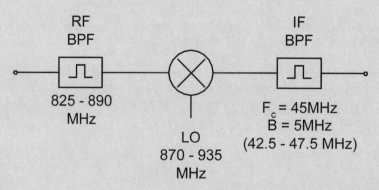

Figure 8-2 *Example 8.1. Simplifier conversion scheme for a cellular telephone receiver.*

We can write
- $f_{RF,L} = 825$ MHz,
- $f_{RF,H} = 890$ MHz,
- $f_{LO,L} = 870$ MHz,
- $f_{LO,H} = 935$ MHz,
- $f_{IF,L} = 42.5$ MHz,
- $f_{IF,H} = 47.5$ MHz,
- $f_{IFCF} = 45$ MHz,
- $BIF = 5$ MHz.

Draw the output spectrums for the $m \cdot f_{LO} \times n \cdot f_{RF}$ combinations of 1×1, 1×2, 2×1, 2×2, 3×1 and 3×3 products.

Solution —
Using Equations 8.2 and 8.3, we can draw Figure 8-3. The frequencies are 1×1.

$$f_{IF,-,L} = (1)(870) - (1)(890) = -20 \text{ MHz}$$
$$f_{IF,-,H} = (1)(935) - (1)(825) = 110 \text{ MHz}$$
$$f_{IF,+,L} = (1)(870) + (1)(825) = 1695 \text{ MHz}$$
$$f_{IF,+,H} = (1)(935) + (1)(890) = 1825 \text{ MHz}$$

8.4

The 1x1 output contains frequencies which overlap the 45 MHz IF.
1×2.

$$f_{IF,-,L} = (1)(870) - (2)(890) = -910 \text{ MHz}$$
$$f_{IF,-,H} = (1)(935) - (2)(825) = -715 \text{ MHz}$$
$$f_{IF,+,L} = (1)(870) + (2)(825) = 2520 \text{ MHz}$$
$$f_{IF,+,H} = (1)(935) + (2)(890) = 2715 \text{ MHz}$$

8.5

2×1.

$$f_{IF,-,L} = (2)(870) - (1)(890) = 850 \text{ MHz}$$
$$f_{IF,-,H} = (2)(935) - (1)(825) = 1045 \text{ MHz}$$
$$f_{IF,+,L} = (2)(870) + (1)(825) = 2565 \text{ MHz}$$
$$f_{IF,+,H} = (2)(935) + (1)(890) = 2760 \text{ MHz}$$

8.6

2×2.

$$f_{IF,-,L} = (2)(870) - (2)(890) = -40 \text{ MHz}$$
$$f_{IF,-,H} = (2)(935) - (2)(825) = 220 \text{ MHz}$$
$$f_{IF,+,L} = (2)(870) + (2)(825) = 3390 \text{ MHz}$$
$$f_{IF,+,H} = (2)(935) + (2)(890) = 3650 \text{ MHz}$$

8.7

The 2×2 products also contain frequencies that overlap the 45 MHz IF. Since these spurious signals appear in-band, they might be problematic.

3×1.

$$f_{IF,-,L} = (3)(870) - (1)(890) = 1720 \text{ MHz}$$
$$f_{IF,-,H} = (3)(935) - (1)(825) = 1980 \text{ MHz}$$
$$f_{IF,+,L} = (3)(870) + (1)(825) = 3435 \text{ MHz}$$
$$f_{IF,+,H} = (3)(935) + (1)(890) = 3695 \text{ MHz}$$

8.8

3×3.

$$f_{IF,-,L} = (3)(870) - (3)(890) = -60 \text{ MHz}$$
$$f_{IF,-,H} = (3)(935) - (3)(825) = 330 \text{ MHz}$$
$$f_{IF,+,L} = (3)(870) + (3)(825) = 5085 \text{ MHz}$$
$$f_{IF,+,H} = (3)(935) + (3)(890) = 5475 \text{ MHz}$$

8.9

The 3×3 component also contains signals that overlap the 45 MHz IF. These accidental in-band spurious signals are potentially a problem. The final graph of Figure 8-3 shows the sum of the outputs we analyzed above. Note that the 1×1, the 2×2 and the 3×3 products all can produce signals that fall within the 45 MHz IF band-pass filter.

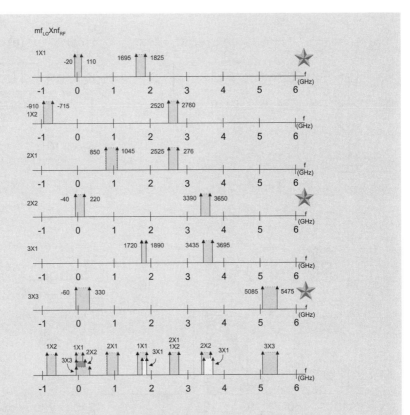

Figure 8-3 *Example 8.1. A partial list of the mxn mixer products generated by the conversion scheme of Figure 8-2.*

8.3   Image Noise

In a poorly designed system, the conversion process will move out-of-band noise to the IF along with the desired signal. Figure 8-4(a) shows the front end of a receiving system. The antenna is modeled as a voltage source in series with a noisy resistor. The noisy resistor models the antenna noise. We will assume the temperature of R_{ant} is 290 K (or T_0).

The RF BPF passes 10 to 50 MHz. We follow the RF BPF with a low noise figure amplifier ($F = 3$ dB; $T_{amp,in} = 290$ K) whose gain is 25 dB. A mixer converts a signal from the RF up to 145 MHz using a low-side LO (LSLO). Figure 8-4(b) shows where the LO, RF and image frequencies fall.

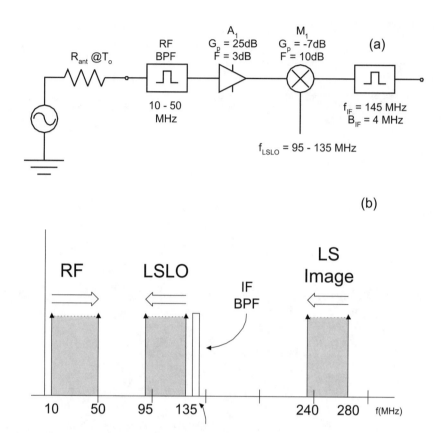

Figure 8-4 A conversion scheme exhibiting image noise problems. (a) The conversion scheme and (b) a graphical view of the conversion scheme showing the RF, LSLO, IF and image frequencies.

It is necessary to make various simplifying assumptions.

- The noise performance of the amplifier does not change when a poor match is presented to its input terminal. In other words, the value of $T_{amp,in}$ remains constant, even when the RF BPF presents the amplifier with a poor match above 500 MHz.

- Amplifier A_1 provides constant gain up to approximately 300 MHz.

- For the sake of calculation, we will assume a 1-hertz noise bandwidth.

- A RF signal is applied at 30 MHz and −161 dBm to the front end of the receiver.

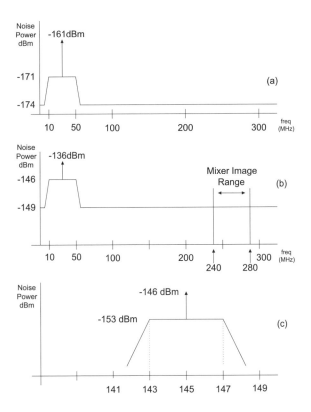

Figure 8-5 *The noise and signal powers at various points of Figure 8-4. (a) The noise power present at the input of amplifier A_1. (b) The spectrum present at the input to mixer M_1. (c) The spectrum present at the output of the 145 MHz band-pass filter.*

Figure 8-5(a) shows the noise and signal power dissipated in the input resistor of amplifier A_1 over frequency. Inside the RF passband, the noise from the antenna (at T_0) adds to the internally generated noise of amplifier A_1 to produce the noise increase from 10 to 50 MHz. In the RF passband, the noise power dissipated in A_1's input resistor is

$$\begin{aligned} N_{in,in\text{-}band} &= k(T_{amp} + T_{ant})B_n \\ &= k(290 + 290)(1) \\ &= 8.00 \cdot 10^{-21} \text{ W} \\ &= -171 \text{ dBm} \end{aligned}$$

8.10

The input SNR is 10 dB.

In the stopband of the RF BPF, the noise from the antenna is severely attenuated by the RF BPF. This drives the antenna noise temperature to 0 K outside the passband of the RF BPF. The out-of-band noise dissipated in A_1's input resistor is

$$\begin{aligned} N_{in,out-of-band} &= k\big(T_{amp} + T_{ant}\big)B_n \\ &= k(290 + 0)(1) \\ &= 4.00 \cdot 10^{-21} \text{ W} \\ &= -174 \text{ dBm} \end{aligned} \qquad 8.11$$

Figure 8-5(b) shows the noise and signal power at the output of amplifier A_1. A significant amount of noise up to and beyond 300 MHz due to the amplifier's gain and noise figure is noticeable. However, the SNR (measured in a 1-hertz bandwidth around the signal) remains 10 dB.

The noise figure of mixer M_1 is 10 dB (or 10 in linear terms). This means that M_1's noise temperature (i.e., the temperature of the mixer's imaginary input resistor) is

$$\begin{aligned} T_{mix,in} &= T_0\big(F_{mix} - 1\big) \\ &= 290(10 - 1) \\ &= 2610 \text{ K} \end{aligned} \qquad 8.12$$

The mixer's input resistance dissipates because of its own internally generated noise.

$$\begin{aligned} N_{mix} &= kT_{mix}B_n \\ &= k(2610)(1) \\ &= 6.0 \cdot 10^{-21} \text{ W} \\ &= -164.4 \text{ dBm} \end{aligned} \qquad 8.13$$

At the mixer's image frequency band (240 to 280 MHz), we find two sources of input noise: the noise from A_1 (at −149 dBm) and the noise resulting from the mixer's own internal noise (at −164 dBm). The external noise is 15 dB (or 32 times) bigger than mixer's internal noise. The major source of noise at this node is the amplified noise supplied by A_1.

At the mixer's input port, signals present at both the RF and at the image frequencies are both converted to the IF center frequency with equal efficiency. The excess noise present at the image frequency of mixer M_1 is similar to any other signal.

Figure 8-5(c) shows the spectrum on the output of M_1. The SNR has changed from 10 dB to only 7 dB. We lost 3 dB in SNR because the noise at the image frequency of the mixer (240 to 280 MHz), along with the signal and noise present at the desired frequency (10 to 50 MHz), was converted to the IF center frequency. Since the noise powers at the RF and its image were about equal, the net result is a loss of 3 dB of SNR (or an increase in the effective noise temperature of the mixer).

Figure 8-6 shows the solution to this problem. Adding a filter between amplifier A_1 and mixer M_1 attenuates the noise generated by the amplifier at the image frequency and preserves the received SNR. Figure 8-6(b) shows the spectrum applied to the RF port of the mixer. Note that we did not have to use a BPF for the filter between A_1 and M_1. We have to attenuate the noise at 240 to 280 MHz. Figure 8-6(c) shows the output spectrum at 145 MHz. The signal's SNR is preserved.

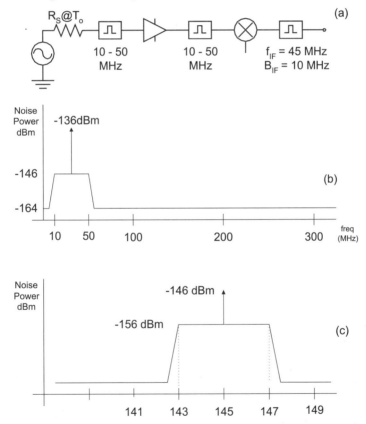

Figure 8-6 *Reducing the image noise problem of Figures 8-4 and 8-5. (a) The cascade. Note the 10 to 50 MHz BPF between the amplifier and mixer. (b) The spectrum present at the input to the mixer. (c) The spectrum present at the output of the 145 MHz band-pass filter. Note the improved SNR.*

8.4 Upconversion vs. Downconversion

When a signal is moved from one frequency to another [see Figure 8-7(a-d)], we can

- upconvert with a HSLO,
- upconvert with a LSLO,
- downconvert with a HSLO,
- downconvert with a LSLO.

Upconvert means to move a signal from its original frequency to a higher frequency; *downconvert* means to move a signal from its original frequency to a lower frequency. *Low-side LO* (LSLO) indicates that the LO frequency is less than the RF frequency. *High-side LO* (HSLO) means that the LO frequency is greater than the RF frequency. In equation form, we can write

$$\text{Low-Side LO} \Rightarrow f_{LO} < f_{RF}$$
$$\text{and} \qquad \qquad 8.14$$
$$\text{High-Side LO} \Rightarrow f_{LO} > f_{RF}$$

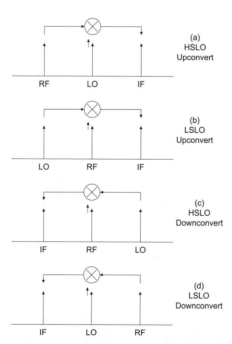

Figure 8-7 *The four possible types of conversion schemes (a) HSLO upconverter, (b) LSLO upconverter, (c) HSLO downconverter and (d) LSLO downconverter.*

Upconversion with HSLO

To upconvert, the signal is moved to a higher center frequency. Whenever a HSLO is used and the lower sideband is selected, frequency inversion will occur. The receiver architecture of Figure 8-8 shows a system that will convert a signal from anywhere in the 20 to 500 MHz range to 700 MHz using a HSLO.

Note the directions of the horizontal arrows on the spectrum plot of Figure 8-8. As we tune the receiver from 20 MHz upwards, the LO starts at 720 MHz and moves upwards also. The HS image frequency starts at 1420 and moves upward in frequency.

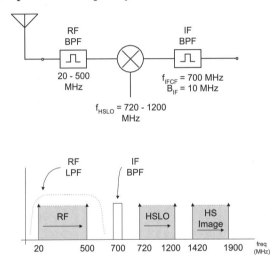

Figure 8-8 Architecture and frequency plan for a HSLO upconverter.

Strengths

- The IF and image rejection, as well as the LO radiation at the antenna port, can all be effectively serviced using a simple low-pass filter as the RF BPF. Very high rejection can be achieved with low complexity and small physical size.

- It is a very cheap and simple solution to the many spurious problems present in a typical receiver.

- The simplicity of this system makes it very easy to realize a complex, yet precise measurement system. The architecture of Figure 8-8 is typical of many spectrum analyzers sold today.

Weaknesses

The major weakness of this system is that its dynamic range is limited due to harmonics of the RF falling within the IF passband. For example, if the receiver is tuned to 80 MHz, the microprocessor tunes the LO to 780 MHz to convert the 80 MHz RF signal to a signal centered at 700 MHz. A 350 MHz signal is applied to the front end of the system (while it is still tuned to 80 MHz). The second harmonic of 350 MHz is 700 MHz. The second-order nonlinearity of the mixer will generate a 700 MHz signal from both the 350-hertz signal and from the 80 MHz signal. The signal powers will compete with each other for control of the IF.

If the 700 MHz energy generated from the 350 MHz signal is greater than the 700 MHz energy generated from the 80 MHz signal, the wanted signal energy from 80 MHz will be covered up by the distortion energy. The receiver is still tuned to 80 MHz, yet it is demodulating a signal coming from the antenna at 350 MHz.

This system is also vulnerable to higher-order distortion products. For example, if we apply 700/3 = 233.33 MHz to the receiver, the third harmonic of this RF signal falls within the IF passband. The same argument holds for 175 MHz (=700/4), 140 MHz (=700/5), 116.67 MHz (=700/6), and so on. This design can be improved by selecting an IF which is higher in frequency. For example, if we set the first IF to be 1100 MHz, second-order distortion is no longer a problem because the highest second-order distortion product produced by the system will be 1000 MHz. Remember that the lower-order products tend to exhibit stronger responses than higher-order products. With the IF at 1100 MHz, there is no second-order distortion, and the higher-order products remain problematic. As a general rule of thumb for a design consisting of an upconverter with a HSLO, it is desirable to place an IF at least 2.1 times higher than the highest RF signal or

$$f_{IFCF} \geq (2.1)f_{RF,H} \qquad 8.15$$

Upconversion with LSLO

Figure 8-9 shows an upconverting receiver using a LSLO. The receiver tunes from 20 to 500 MHz and uses an IF of 700 MHz. Note the directions of the arrows on the spectrum plot of Figure 8-9. As we tune the receiver from 20 MHz upwards, the LO starts at 680 MHz and moves downward. The HS image frequency starts at 1380 MHz and moves downward in frequency.

CASCADE II — IF SELECTION | 717

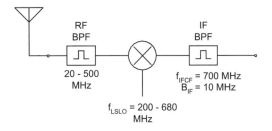

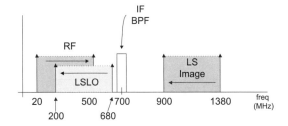

Figure 8-9 Architecture and frequency plan for a LSLO upconverter.

Strengths

As with the HSLO upconverter, we can use a LPF as the RF BPF and achieve very good IF and image rejection.

Weaknesses

- Since the LO overlaps the RF bandwidth, the front-end preselection LPF does not always provide attenuation for a LO signal that may be leaking out of the antenna port.

- As with the HSLO upconverter, this architecture is subject to harmonics of the RF appearing in the IF. The IF should be places as high in frequency as possible to mitigate this effect.

LO Effects of HSLO vs. LSLO Upconverters

The LO of these two upconverting schemes must tune over the same number of hertz. However, since the HSLO is at a higher center frequency, its LO tunes over a smaller percentage bandwidth than the LSLO converter does. In the first HSLO upconverter (see Figure 8-8), the HSLO tuned from 720 to 1200 MHz. This is a 480 MHz bandwidth centered at 960 MHz, or a 50% bandwidth. In the LSLO upconverter (see Figure 8-9), the LO must tune from 200 to 680 MHz. This is also a 480 MHz tuning range but it is centered at 440 MHz. The LSLO oscillator must tune over a 110% range.

As discussed in Chapter 4, if the tuning range of an oscillator (in percentage of center frequency) can be reduced, its phase noise can be improved at the same time. In terms of phase noise, the HSLO system will tend to be quieter.

Downconverting with HSLO

Figure 8-10 shows a system used to convert signals in the 100 to 500 MHz range to 60 MHz. Note that as we increase the tuned frequency, the HSLO and HS image frequencies also increase.

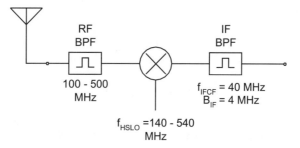

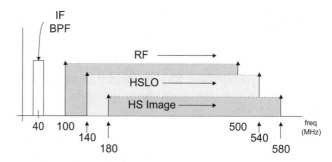

Figure 8-10 *Architecture and frequency plan for a HSLO downconverter.*

Since the RF and LO are both above the IF, their harmonics will not fall into the IF. This increases the dynamic range of the system considerably.

Weaknesses
The IF, image and LO rejection problems have increased dramatically over the upconversion cases. The most common solution to these rejection problems is to use a tracking RF BPF as the preselection filter. This will provide the IF and image rejection as well as suppress the LO leakage to the antenna port. It will, however, also increase the cost of the receiver significantly.

Downconverting with LSLO

Figure 8-11 shows a system used to convert signals in the 100 to 500 MHz range to 40 MHz using a LSLO. Note that as we increase the tuned frequency, the HSLO and HS image frequencies also increase.

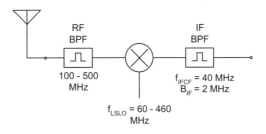

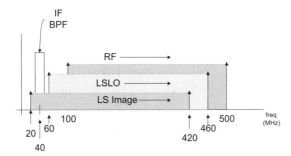

Figure 8-11 Architecture and frequency plan for a LSLO downconverter.

Strengths
The RF and LO are still above the IF, and there is no possibility of harmonics of either signal fall into the IF. This increases the dynamic range of the system considerably.

Weaknesses
The IF, image and LO rejection problems remain difficult issues in regards to the upconversion cases. To provide adequate IF and image rejection as well as to suppress the LO leakage to the antenna port, we have to use a tracking RF BPF as the preselection filter. Again, this will add significant cost and complexity to the receiver.

Conclusions

Simplicity
Upconverters are generally simpler than downconverters. Upconverters provide LO, IF, and image rejection all with a simple low-pass front-end filter.

Downconverters usually require tracking band-pass filters or switched filters on the RF port to accomplish these tasks. Upconverters have limited dynamic range because harmonics of RF signals can fall into the IF passband.

LO Rejection

The HSLO upconverter usually has a smaller amount of LO power present on the antenna port than the LSLO upconverter because the HSLO always falls outside of the RF passband. In a LSLO upconverter, the LO can fall inside the RF passband.

Oscillator Tuning Range

Whereas the HSLO upconverter relies on a higher LO frequency (which may be difficult to realize), the synthesizer has to tune over a smaller fractional bandwidth. This usually reduces the phase noise of the oscillator and makes for a cleaner demodulation.

When to Upconvert and When to Downconvert

Eventually, we must bring the signal to some common frequency to demodulate it and, more often than not, the final IF is at a relatively low frequency such as 10.7 MHz. Despite inherent problems, we have to use a downconverter in some part of the receiver. In a narrowband environment, the difference between upconversion and downconversion schemes is small for the following reasons:

- In both cases, the LO must tune over only a very small percentage bandwidth, which means that the LO design considerations even out.

- A careful design can achieve all of the LO, image and IF rejection needed in a downconverter using band-pass filters.

In conclusion, it is desirable to upconvert in a wideband environment. Once the signal has passed through the first IF filter, we have considerably simplified the problem. We can then downconvert from there to move the signal to the final IF for demodulation, digitization and so on.

War Story — Unintended Effects in Conversion Schemes

The effects of conversion schemes can cause unexpected problems. Once the author designed a voltage-controlled oscillator (VCO). Since these devices are nonlinear and not easy to characterize theoretically, designers spend much time on the bench verifying their designs over variations in power supply voltages, temperature, load impedances, and such.

One of the characteristics of a poor VCO design is that, at some frequency, the oscillator's output signal will widen. The oscillator will exhibit modulation and the noise floor in the vicinity of the oscillator frequency will increase. Your author spent several days debugging one particular VCO. It worked well but always misbehaved at one single frequency. This was a fairly benign design: he had built similar oscillators before and was fairly confident. However, nothing he could do would solve the problem.

Desperate, the author recovered an old, reliable oscillator and placed it in his test jig. This oscillator exhibited the same problem in the exact same frequency range although its design had been fielded. Eventually, he connected a commercial signal generator into his test setup and the problem reappeared.

Figure 8-12 shows the block diagram of the spectrum analyzer that was used. The center frequency of the band-pass filter is significant. The problem occurred when his VCO was tuned to 405 MHz ($f_{IF}/5$). The strong signal of the oscillator was producing harmonics in the mixer that fell directly into the passband of the IF. The problem was caused by the spectrum analyzer.

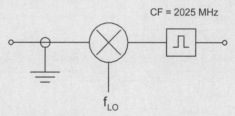

Figure 8-12 Simplified block diagram of the input stage of a spectrum analyzer.

8.5 LOs, Tuning Range and Phase Noise

In Chapter 4, we found that a local oscillator tuning over a large percentage bandwidth will tend to be noisy. One of the major problems in building a wideband receiver is local oscillator phase noise. Since the LO must tune over a large percentage bandwidth to cover the entire tuning range of the receiver, the oscillator is likely to be noisy.

Large Tuning Range, Small Step Size

Figure 8-13 shows one possible architecture for a wideband receiver. The receiver must tune from 20 to 500 MHz in 1 kHz steps (i.e., the receiver must tune to 20.000 MHz, 20.001 MHz, 20.002 MHz ... 499.999 MHz, 500.000 MHz.) An IF of 700 MHz and a HSLO will be used first. LO_1 must tune over a 720 to 1200 MHz range (a percentage bandwidth of 52%) in 1 kHz steps.

722 | RADIO RECEIVER DESIGN

A phase-locked loop oscillator is ideal for this receiver because it can easily realize the step size and the tuning range. Phase noise problems due to the large tuning range can be expected. We know that a PLL will exhibit improved phase noise when we are within a loop bandwidth of the carrier. However, practical considerations limit the loop bandwidth of a typical PLL synthesizer to less than 10% of the step size or 100 hertz in this case.

Two Synthesizers

Figure 8-13(b) shows that the first synthesizer tunes from 720 to 1200 MHz in 1 MHz steps. This allows for the loop bandwidth of the first synthesizer to be about 100 kHz, quieting the phase noise of this oscillator significantly. The signal of interest now lies within ±500 kHz of the center frequency of the first IF filter (we have to make the first IF filter wide enough to accommodate this variation).

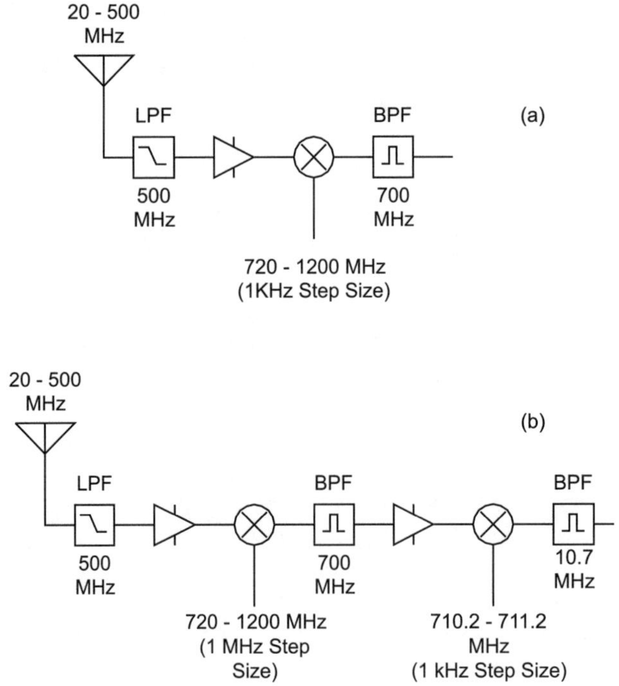

Figure 8-13 *Two possible architectures for a wideband receiver with a small tuning resolution. (a) One synthesizer tunes over the entire LO range with a small step size. (b) Two synthesizers: one tunes over a wide range with a large step size, the second synthesizer tunes over a small range with a small step size.*

We will use the second conversion stage to achieve the 1 kHz step size. The second LO is tunable with a 1 kHz step size, but it tunes only over 1 MHz. Since the tuning range of this oscillator is relatively small, the phase noise of the basic oscillator will be small, and the small loop bandwidth will be appropriate. This scheme allows us to achieve a wide tuning range with a small step size.

Direct Digital Synthesizer

The second small-step-size synthesizer can be realized in several ways. Although DDS technology is capable of a large percentage tuning range and a very small step size, the center frequencies are limited. However, Figures 8-14(a) and 8-14(b) show two methods using a direct digital synthesizer.

Figure 8-14(a) uses a multiplier to achieve the 1 kHz step size at the center frequency we require. Note that the multiplier multiplies the step size as well as the center frequency of the oscillator.

Figure 8-14(b) shows a DDS with an offset oscillator. A second fixed oscillator (either a PLL or a crystal) is used in combination with the DDS to translate the DDS output to the frequency we require for the second conversion. The band-pass filter following the mixer removes the various spurious products generated in the mixer.

The DDS output can contain spurious signals at relatively high levels. This conversion scheme can reduce the dynamic range of the receiver. Adding another oscillator and mixer to a receiver can cause isolation and spurious responses in the receiver's output.

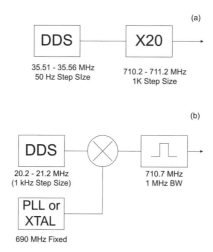

Figure 8-14 Using a direct digital synthesizer (DDS) to produce a high-frequency LO with a 1 kHz step size.

8.6 IF Selection Guidelines

Designing a conversion scheme can be difficult because the final performance of the receiver is often determined by quantities that cannot be readily measured or calculated. Sometimes, it is impossible to know how well the design is going until a long way into the design process at which time changes are difficult and expensive. Here are a few useful rules of thumb that can be used in the initial stages of designing a conversion scheme.

Suboctave Preselection

Use suboctave preselection filters to reduce the effects of second-order distortion. Second-order effects are usually the strongest of the nonlinear responses because the order is so low. Figure 8-15(a) shows a system with a single 100 to 500 MHz RF preselection filter. Instead of a single filter, it is sometimes wise to use several filters as shown in Figure 8-15(b). Note that each filter is less than an octave in bandwidth.

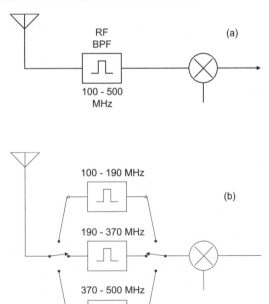

Figure 8-15 Suboctave preselection reduces the effects of second-order distortion. (a) The wideband RF BPF is greater than an octave wide and can cause second-order problems to occur. (b) Replacing the wideband RF BPF with several suboctave filters prevents second-order distortion from becoming a problem.

Image Noise

Image noise can seriously reduce the sensitivity of a system. We have to be sure to provide filtering to eliminate it.

Upconverting

When upconverting, the harmonics of the RF should be kept out of the IF filter. At best, the IF placed in a way that at least the 2nd harmonic of the RF falls below the passband of the IF filter. A rough rule of thumb is to place the center frequency of the IF passband at least 2.1 times the highest RF frequency we want to process.

Butterworth Approximations

We use the Butterworth filter approximations to find the approximate attenuation of the filters in the system. This will allow us a rough guess of the attenuation each filter will provide against unwanted signals although the Butterworth approximations will not predict the filter's ultimate attenuation.

Downconverting

The insertion loss of a band-pass filter tends to increase as its percentage bandwidth (or its B_{IF}/f_{IFCF} ratio) decreases. In other words, a filter with a small percentage bandwidth will tend to be lossier than a filter with a large percentage bandwidth. To realize a narrow channel spacing (or a narrow IF bandwidth), we must eventually downconvert if we want to perform the filtering with a reasonable insertion loss. For a narrowband signal of 30 kHz, for example, the signal has to be moved to a low frequency (i.e., 10.7 MHz or 455 kHz) to filter and demodulate it properly.

Filter Technologies

Some filter technologies, such as crystal filters and SAW filters, will provide very small percentage bandwidths at high frequencies, but can be bulky, expensive or suffer from high insertion loss. However, technology moves on, and it is useful to survey the current vendor catalogs before committing to a design.

Oscillator Center Frequency

High frequency oscillators tend to be noisier than low frequency oscillators. Oscillators that must tune over a wide percentage bandwidth tend to be noisier than oscillators that tune over a small percentage bandwidth.

Common IF Frequencies

When possible, common IF frequencies should be used to take advantage of the plethora of cheap, small commercial parts available. Some common IFs are

- 21.4 MHz (high-end receivers),
- 10.7 MHz (commercial receivers),
- 45 MHz (television, cellular telephones),
- 70 MHz (satellite television, military gear),
- 455 kHz (commercial equipment),
- 160 MHz (commercial satellite equipment).

Bad IF Frequencies

Avoid placing IFs in sections of the spectrum where there are many strong signals. The receiver's IF rejection has to be that much better in order to function well. For example, in the United States, it would be inefficient to design a system whose first IF falls within the commercial FM band. The receiver is likely to encounter very strong signals in this frequency range, which puts a heavy burden on the rejection of the RF BPF at the IF center frequency. Also, it is wise to avoid the commercial aircraft bands, the cellular telephone band and common amateur radio frequencies.

LO Harmonics

Since the local oscillator is one of the strongest signals present in a receiver, it is unwise to design a conversion scheme that allows harmonics of the LO to overlap the IF passband. The harmonics of the very strong LO will likely be larger than most of the RF signals the receiver will process. This is true even for high harmonic orders.

8.7 Practical Design Considerations

Separate Compartments

If possible, provide separate, isolated compartments for each subsystem of the receiver. A separate compartment for each of the following receiver subsystems is ideal.

- *Switching power supply.* The switching transients will get everywhere, especially the voltage-controlled oscillators used in the frequency synthesizers.

- *Microprocessor.* The high-level digital signals present in clocked logic are damaging to any high gain system. The harmonics of the digital waveforms can be a problem to several hundred MHz.

- *Voltage-controlled oscillators (VCOs).* The VCOs used in PLL frequency synthesizers are especially sensitive to power supply, radiated and conducted noise.

- *Digital PLL Components.* In the ideal case, a separate compartment for the PLL circuitry other than the VCOs is provided. This circuitry is a unique combination of digital and analog systems working together, which can be affected by other components. However, these components are often packaged in the same compartment as the VCOs with only minor problems.

- Each compartment should house only one LO. In other words, we do not want two separate oscillators present in the same compartment at the same time. The interface between compartments provides a unique opportunity to increase the absolute attenuation of the filter that separates the two mixers.

Single Printed Circuit Board Systems

Cost and manufacturability are important considerations which may force us to make design choices that are not ideal. Although placing all of the sensitive circuitry in a compartment all by itself would be desirable, we are often forced to place them on the same physical printed circuit board as the noisy circuitry. With careful layout techniques, we can achieve reasonably high levels of isolation on a single circuit board [1].

Ground Planes

A ground plane provides a low impedance return path for power supply and signal currents (the key term is low impedance). We can also call on a properly designed ground plane to provide isolation between components (to increase the ultimate attenuation of a filter, for example) and to form printed-circuit transmission lines (i.e., microstrip) directly on the circuit board.

Power Supplies

Many of the subsystems of a receiver are sensitive to power supply variations. Commercial voltage-controlled oscillators, for example, usually come with a frequency-pushing specification. This specification tells us how much the frequency of the oscillator will change if its power supply voltage is varied. If the power supply of the oscillator contains a component at 60 hertz, for example, the oscillator will exhibit sidebands at ±60 hertz from the carrier. Amplifiers are also affected by power supply variations. The power supply of a receiver is usually one of the most neglected subsystems although it is crucial to a functioning system.

Switching Power Supplies

A switching power supply is common in battery-powered systems. Switching power supplies will accommodate the changing battery voltage and generate all of the voltages required by the various receiver subsections. However, they are a notorious source of high-level transients that are hard to filter.

Switching power supplies operate using fast, high-current switches to gate the DC battery voltage into AC waveforms. Then, by means of switching diodes and other means, the AC waveforms are converted back into noisy DC voltages. The noisy DC voltages are then filtered to produce the *still* noisy DC voltages used throughout the receiver. Switching power supplies currently operate from 20 kHz to beyond 200 kHz. These relatively low frequencies require large inductors and large capacitors to filter adequately. This is a difficult problem in a miniature or cost-sensitive design.

The switcher is frequently used to generate a voltage that is three or four volts higher than actually needed. Then a linear regulator is used to subregulate. Linear regulators are fairly quiet (although specifications vary) and, if properly designed and used, they will remove most of the switching transients from the switcher's output voltage.

Much of the noise from a switching power supply comes about from the high-speed transients present in the circuitry. These switching transients can have peak values of several amps, and their magnitudes are directly related to the DC current supplied by the power supply. A system that draws as little power as possible will help suppress the noise of the switcher.

Finally, the large transient currents developed by switching power supplies usually end up travelling in the system's ground plane. These large transient currents will cause transient voltage spikes on the ground plane in the neighborhood of the switching supply circuitry. (V = IR; since the ground plane has a nonzero resistance, the current transients will cause voltage transients.). In short, the sensitive and low-level circuitry should be kept as physically far away from the switching supply as practical. Imagine the path that the transient currents from the switcher must take and place the sensitive circuitry accordingly.

Linear Regulators

Without exception, linear regulators are always quieter than switching regulators even though they generate some noise. Linear regulators are simple control systems, which contain a pass transistor (or FET) and an internal voltage reference. It is necessary to monitor the output of the regulator and change the base (or gate) conditions on the pass transistor so that the output voltage equals the reference voltage (after scaling).

There are two potential problems here. The first is that the control process generates noise. Also, some regulators are quieter than others. It pays to read the regulator's specifications. A noisy linear regulator used to supply a voltage-controlled oscillator, for example, can cause excess phase noise to appear on the oscillator's output.

The second potential obstacle involves the control bandwidth of the regulator. Figure 8-16 shows an illustration of this concept. The first voltage source (VDC), represents the pure DC applied to the regulator. The input voltage is normally several volts higher than the regulator's output voltage (V_{out}). The second voltage source (V_{noise}) represents noise present on the unfiltered power supply. For this analysis, we assume a simple sinusoid whose frequency can be changed at will.

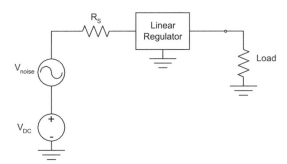

Figure 8-16 Using a linear regulator to reduce power supply noise. The regulator will reduce power supply noise if the frequency of the noise is lower than the control bandwidth of the regulator. Noise outside the regulator's control bandwidth may not be suppressed.

When V_{noise} is at a very low frequency (1 hertz, for example), the control loop inside the regulator may have no trouble following this variation and suppressing it at the output. If we increase the frequency of V_{noise}, we will eventually reach a frequency where the loop will have trouble following the variation on the regulator's input. Some component of V_{noise} will be present on the regulator's output. As the frequency of V_{noise} increases, the regulator will eventually become ineffective against V_{noise} and any component of V_{noise} will pass unaffected through the regulator and onto its output.

In short, linear regulators contain a control bandwidth. Inside this bandwidth, they will apply some measure of attenuation to noise at the regulator's input. However, little or no attenuation of V_{noise} components beyond the regulator's control bandwidth will occur. If a regulator is used to subregulate some noisy DC voltage, it is necessary to ensure the regulator is capable of suppressing the particular frequencies. Low dropout regulators tend to have particularly small bandwidths.

Digital Logic

Every receiver built today contains a microprocessor or utilizes some form of digital control. This advancement, although necessary and proper, appears problematic. The high-level digital signals derived from the microprocessor's clock present isolation problems. The harmonics of the digital waveforms can extend to several hundred MHz, well into the passband of most radio receivers.

However, to be able to transmit information from the digital portion of the receiver to the rest of the system, it is necessary to run conductors from the microprocessor portion of the receiver to the frequency synthesizers, the demodulator and to the various IF and RF subsections.

Continuously Clocked vs. Unclocked Logic

From a noise perspective, we can place digital logic into two categories: *continuously clocked* and *unclocked*. *Continuously clocked logic* is defined as logic that requires a continuous clock to operate. For example, most microprocessors require a continuous clock or they will malfunction. *Unclocked logic* does not require a continuous clock to operate. Flip-flops, shift registers and static memory are examples.

Receiver Design Considerations

Continuously clocked logic poses a threat to the performance of a receiver because the digital waveforms are always present as long as the receiver has power. Unclocked logic is not a source of noise because both its inputs and outputs are not changing (unless we specifically clock the logic and command it to change). Unclocked logic can be placed in close proximity to sensitive electronic components without a noise penalty.

Number of Conductors

Designers are very concerned about the number of wires which transverse the boundaries between the separate subsections of a receiver. There are good reasons to minimize these interconnections:

- *Electromagnetic Interference.* Each conductor passing through the boundary between a noise-generating piece of the system and a noise-sensitive part of the system represents an opportunity for an interfering signal to escape from the noisy subsection and contaminate the quiet subsection. Even though a conductor may appear to show only a quiet DC level on an oscilloscope, receivers can easily detect signals far below the observational threshold of oscilloscope.

- *Ease of Manufacture.* A smaller number of interconnections minimizes the number of operations required to build the receiver. This can reduce the complexity and cost of the receiver.

- *Cost.* Decreasing the interconnections reduces the assembly cost as well as the cost of associated components such as wire and EMI filters.

- *Reliability.* Every solder connection has a non-zero probability of failure. Minimizing the number of connections will increase the reliability of the unit.

- *Miniaturization.* In a very small system, the interconnections between separate subsystems can take a surprising amount of physical space. The fewer connections, the more space will be available for other functions.

Digital Control of the Receiver Subsystems

Figure 8-17 shows the information that must be passed from the controller to the rest of the receiver. It is necessary to switch different preselection filters into the system depending upon the tuned frequency of the receiver. There are six sets of suboctave preselection filters: 20 to 35, 35 to 60, 60 to 110, 110 to 200, 200 to 350 and 350 to 500 MHz. We have to choose a first IF filter and provide digital control to both PLL synthesizers.

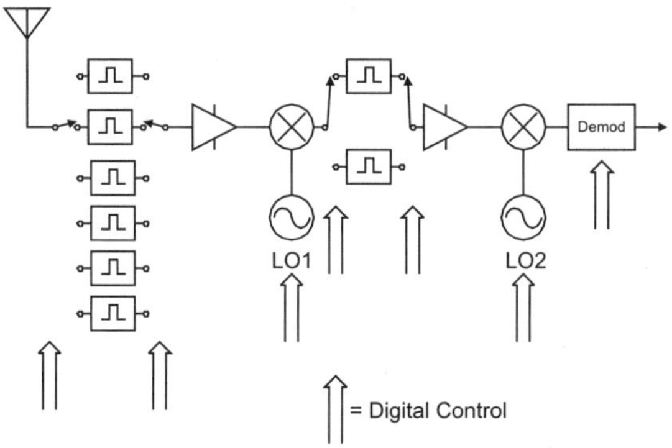

Figure 8-17 *Portions of a receiver which require digital control.*

Figure 8-18 shows one way to move the control data from module to module. Three lines are brought from the microprocessor to the noise-sensitive compartment (filtered heavily, if necessary). The three lines run into a shift register — one line carries the data, one line is the clock and one line is the latch enable. The latch enable line transfers data from the internal shift register flip-flops to an output latch, where the data is held until the next latch enable toggle.

When the microprocessor changes the state of the digital control lines in the noise-sensitive compartment, it shifts the new data into the shift register using the clock and data lines. The outputs of the shift register IC do not change until the microprocessor toggles the latch enable line. Finally, when all the data has been shifted in, the microprocessor toggles the latch enable line and the states of the control lines in the module change to their new values. This method is easily expandable. Figure 8-18 shows two 8-bit shift registers in series are connected to produce sixteen control outputs. Some makers of integrated circuits have adapted this technique in their chip sets, requiring only two or three digital lines to perform multibit control over the IC's function.

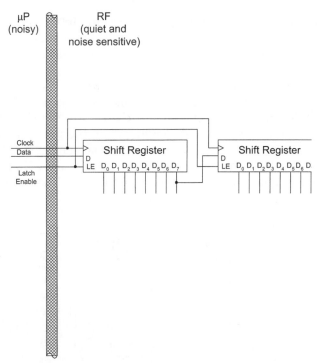

Figure 8-18 One method used to move digital data from module to module within a receiver. Only three wires are required.

8.8　A Typical System

Figure 8-19 shows the architecture of an ideal receiving system. Most receivers will not contain all components because factors such as cost or size ultimately affect the performance of the receiver.

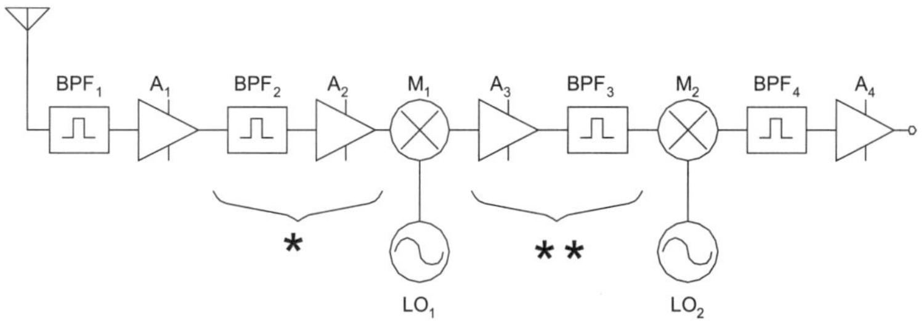

Figure 8-19 *Architecture of an ideal receiving system. Many systems will omit parts of the architecture shown here due to cost, physical size constraints or other trade-offs.*

Filter BPF$_1$

A BPF$_1$ filter is the RF preselection filter. It immediately limits the bandwidth of the system to the range of frequencies processed. Allowing the receiver to process signals outside of the range of interest will not help the receiver's performance. Large out-of-band signals will diminish the receiver's performance. This filter should be as narrow as possible without excessive loss. Any loss at this point in the receiver adds directly to the receiver's noise figure and limits the receiver's MDS. BPF$_1$ and BPF$_2$ (if present) are responsible for the image and IF rejection of the receiver.

In high-performance wideband receivers, this BPF usually consists of a voltage-tunable band-pass filter (see Figure 8-20). The complete receiver tunes over a 500 to 1000 MHz frequency range. BPF$_1$ and BPF$_2$ are voltage-tuned band-pass filters with a 5% bandwidth (i.e., the bandwidth is 5% of the center frequency). A microprocessor inside the receiver changes the center frequency of the band-pass filter to the receiver's tuned frequency. The function of BPF$_1$ and BPF$_2$ is to limit the bandwidth to the maximum practical extent as early in the receiver chain as possible. This discourages unwanted out-of-band signals from creating in-band intermodulation products.

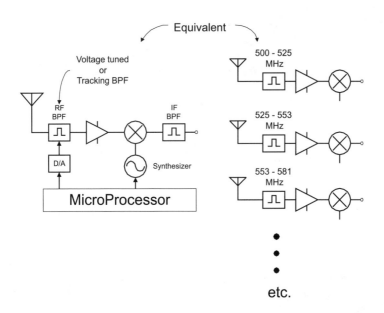

Figure 8-20 *Voltage-tuned preselection filters and an equivalent filter configuration.*

Amplifier A_1

A_1, as the first amplifier in the system, should have a low noise figure. The amplifiers should possess enough gain to move the signal to the next amplifier without the excess gain dropping below 15 dB.

The ITOI of this amplifier is not a major concern. Since there is no gain before this amplifier, it will always see relatively small signals. The reverse isolation of amplifier A_1 keeps internally generated signals such as LO_1, LO_2 and any unwanted signals generated by the microprocessor from leaving the receiver.

Filter BPF_2

Filter BPF_2 is an optional component and may be used depending on consideration such as cost and space. If amplifier A_1 has significant gain at mixer M_1's image frequency, the wideband noise from A_1 present at the image frequency will be converted to the IF center frequency by M_1. This will degrade the effective noise figure of mixer M_1.

Amplifier A_2 is also an optional component. If A_2 is omitted, mixer M_1 does not see a wideband match on its RF port. As a result, the mixer's linearity may suffer. This filter is partially responsible for the mixer's image and IF rejection as well as the suppression of internally generated signals exiting the receiver.

Amplifier A_2

Amplifier A_2, another optional component, provides gain and helps produce a wideband match to M_1's RF port.

Mixer M_1

Mixer M_1 converts the RF to the first IF. The finite LO:RF isolation of this mixer is partially responsible for the presence of LO_1 at the receiver's antenna port. The finite LO:IF isolation of M_1 makes it possible for LO_1 to find its way to M_2. Two large signals present in a mixer at the same time cause intermodulation products. If the sum or difference of any of the harmonics of LO_1 and LO_2 fall inside any IF filter, a spur occurs. The greater the LO:IF isolation of mixer M_1, the less LO power leaks through the mixer into the receiver's first IF and the lower the danger of a detectable spur. The RF:IF rejection of mixer M_1 helps suppress the IF rejection of the receiver.

Amplifier A_3

Amplifier A_3 provides a wideband match to M_1 and gain in the receiver chain if needed. If there is no filter between M_1 and A_3, then A_3 must be able to process the LO_1 leakage from M_1. Since this is a fairly strong signal ($\simeq -30$ dBm), amplifier A_3 must have good linearity. The reverse isolation of A_3 helps keep the local oscillator from mixer M_2 out of mixer M_1. If BPF_3 is placed between M_1 and A_3, then amplifier A_3 does not have to be strictly linear. Since linearity usually comes at the cost of more DC power, it is sometimes possible to save power with this filter.

Filter BPF_3

The first IF should be selected with the characteristics of BPF_3 in mind. The rolloff, ultimate attenuation and bandwidth of this band-pass filter are all important in determining the spurious responses of the receiver. BPF_3 rejects all of the out-of-band spurious products generated by mixer M_1, including the image, all of the $|m \cdot LO \pm n \cdot RF|$ products and LO_1.

BPF_3 should be as narrow as possible so that it will reject as many of the unwanted signals as possible. However, BPF_3 must be wide enough to pass the widest signal likely to be encountered. The ultimate rejection of BPF_3 is important because this characteristic of BPF_3 sets the lower limit to the strength of the unwanted signals passing between M_1 and M_2. Since

LO_1 will experience gain through amplifier A_3, we would like BPF_3 to be as lossy as possible at LO_1. Sometimes we arrange to place a notch in BPF_3 at the frequency of LO_1. Amplifier A_3 and band-pass filter BPF_3 are often reversed in the cascade.

Mixer M_2

Mixer M_2 is responsible for the second conversion in the receiver. It uses LO_2 to convert signals present at the first IF to the second IF. Mixer M_2 is less important to the operation of the receiver than mixer M_1. Due to its physical position in the cascade, LO_2 is much less likely to find its way to the antenna port than LO_1. Also, at this point in the cascade, the band-pass filters will be considerably narrowed and spurious problems caused by unwanted signals have been significantly reduced.

Filter BPF_4

Filter BPF_4 is the narrowest band-pass filter in the system. It sets the noise bandwidth and the sensitivity of the receiver. This filter must be wide enough to pass the signal of interest without significant distortion. If it is wider, it will pass more noise than is absolutely necessary (thus decreasing the SNR of the signal presented to the demodulator). If BPF_4 is too wide, it may also pass strong, adjacent signals that can ruin the performance of the demodulator. This filter is responsible for the adjacent channel rejection of the receiver.

Amplifier A_4

Up to this point in the cascade, we have been very careful about applying gain to the signal — too little and we bury the signal in noise; too much and we create distortion. However, since there is now only one signal present in the system beyond BPF_4 and since we normally make the bandwidth of BPF_4 less than an octave, linearity is much less of an issue than it was prior to passing the signal through BPF_4. Amplifier A_4 provides the bulk of the gain in the receiver that is applied at this point.

8.9 Design Examples

Example #1

This example is a fairly expensive, high-performance receiver. The most important design parameters are performance, low power and relatively small size.

CASCADE II — IF SELECTION | 737

- *Tuning Range*: 20 to 500 MHz.
- *Tuning Plan*: Frequency synthesized local oscillators locked to an internal reference.
- *Noise Figure*: 8.5 dB maximum.
- *Input Third-Order Intercept*: –10 dBm, 10 kHz IF bandwidth. Tones placed at f_{tuned} + 100 kHz and at f_{tuned} + 200 kHz.
- *Reference Accuracy*: ±3.5 ppm over –10 to +55 °C temperature range.
- *Tuning Step Size/Tuning Speed*: 2.0 kHz/25 ms
- *Front-End RF Preselectors*: The bandwidth of the RF preselection filters will be less than 15% of the tuned frequency.
- *Detector Modes*: AM, FM.
- *IF Bandwidths*: 10, 50, 200 and 1000 kHz (selectable). IF filter shape factors: 3:1 nominal (6 to 60 dB).
- *AM Sensitivity*: For an input signal level of –110 dBm, a 1 kHz, 60% modulated AM tone, the audio SINAD will be greater than 10 dB. Measured in a 10 kHz IF bandwidth and a 3 kHz audio bandwidth.
- *FM Sensitivity*: For an input signal level of –110 dBm, a 1 kHz FM tone which deviates over 80% of the 50 kHz IF bandwidth, the SINAD ratio will be greater than 10 dB. Measured with a 3 kHz audio BW.
- *Third-Order Intercept*: The ITOI of the receiver will be greater than –10 dBm with test signals at 0.5 and 1.0 MHz from the tuned frequency with the 50 kHz IF bandwidth selected.
- *Image Rejection*: The power difference between a test signal at the image frequency and at the tuned frequency will be greater then 65 dB.
- *IF Rejection*: The power difference between a test signal at either of the first IF frequencies and at the tuned frequency will be greater than 65 dB.
- *Local Oscillator Radiation*: The power of any local oscillator present at the antenna port of the receiver should be less than –100 dBm.
- *Local Oscillator Phase Noise*: The combined incidental FM of all the local oscillators is less than 100 hertz$_{RMS}$ when measured in an audio bandwidth of 3 kHz.
- *Internally Generated Spurious Signals*: All spurious signals should be less than –110 dBm relative to the antenna input. No more than six spurious responses should be greater than –120 dBm.
- *Automatic Gain Control*: The AGC shall act to keep less than 2 dB of variation in the IF signal for an input signal variation of 80 dB.
- *Computer Control* via RS-232 with hardware reset.
- *Operating/Storage Temperature*: –10° to +55° C, operating –55° to 70° C, storage.
- *Power Requirement*: 6 to16 VDC, 1.25 W maximum.

Analysis of the Specifications

This is a digitally controlled receiver, capable of autonomous action. It requires a microprocessor, which can perform fairly sophisticated controlling functions, such as switching of preselectors, demodulator selection and user interfaces. The disadvantage is that the microprocessor is also a huge source of noise. Large isolation problems can be anticipated that have to be addressed when the design process begins.

The front-end preselection filters enable the receiver to operate in high-interference (urban) environments. These filters severely limit the number of signals that can pass into the receiver and reduce possible intermodulation products. The preselectors also provide high rejection at the local oscillator frequencies to avoid LO radiations from the receiver's antenna terminal.

Tuned preselection has historically been used in communication receivers to reduce interference of desired weak signals by strong interfering signals. The interference falls into two general types: *single tone* and *double tone*. Single-tone spurious signals are produced when the harmonics of the interfering signal combine with the receiver's local oscillator in the first mixer. RF preselectors reduce the level of the interfering signals between the antenna input and the mixer.

A second single-tone spurious response occurs when the image signal (which is separated from the desired signal by twice the IF) mixes with the first LO. The preselectors provide the only discrimination to these image frequencies. The preselectors also protect the receiver from a signal whose frequency is the same as the first IF (i.e., direct IF feedthrough).

Third-order intermodulation products are the most troublesome because band-pass filtering alone cannot suppress these distortion products. Two signals can still produce a false signal in the passband regardless of the narrowness of the preselector filter. However, the probability of two signals producing third-order distortion products is directly proportional to the preselector bandwidth. The narrower the preselector filter, the better the third order performance for a given system TOI.

The final advantage of RF preselection is the reduction of the first LO leakage power from the antenna connector. The minimization of the LO leakage is particularly important in multiple receiver configurations, where the LO of one receiver can appear as a valid signal to another receiver. In summary, the advantages of preselection are

- Reduction of higher-order spurious products.
- Improvement of second- and third-order intermodulation performance.
- Improvement of image and IF feedthrough rejection.
- Reduction of the conducted LO leakage.

Overall Block Diagram

Figure 8-21 shows the gross block diagram of the overall receiver. The antenna output is first passed into a set of preselection filters. These filters perform initial filtering and contain enough gain to set the noise figure of the receiver. Then the RF is converted to an IF of either 60 MHz or 145 MHz, depending upon the tuned frequency. With this conversion scheme, the receiver must generate an LO of 160 to 560 MHz for the first conversion but this LOs step size is limited to 100 kHz. The 60/145 MHz signal of interest is then converted to the final IF of 21.4 MHz using either a 81.4 MHz LO or a 166.4 MHz LO. Finally, the demodulation at 21.4 MHz is performed.

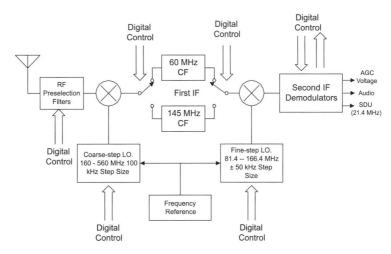

Figure 8-21 Block diagram of a candidate receiving system.

RF Preselection Filters

Figure 8-22 shows a block diagram of the voltage-tuned 20 to 500 MHz preselector filters. The microprocessor switches the 20 to 500 MHz antenna input to the appropriate set of voltage-tuned band-pass filters. The preselector tuning voltage comes from a 12-bit DAC, which is driven by the system microprocessor. The receiver contains a ROM-based calibration table to look up the proper DAC value when given the tuned frequency. The ROM table accommodates the nonlinear tuning characteristics of the preselectors. The processor also switches power to the individual amplifiers (VCC_1 through VCC_4).

The losses in the input switch and the first two-pole band-pass filter add directly to the receiver's noise figure. The amplifier isolates the two filters and establishes the receiver noise figure. The gain of the first amplifier is fairly low (approximately 12 dB). High-power signals may be present in the RF path.

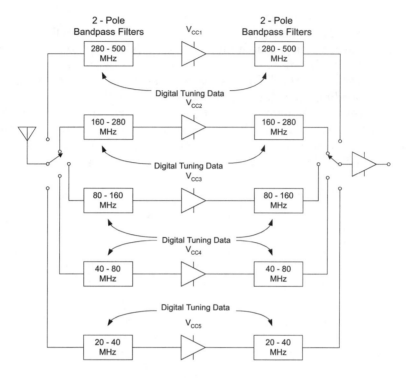

Figure 8-22 *Structure of the voltage-tuned preselection filters (20 to 500 MHz) at the receiver's front end.*

The microprocessor supplies power to an amplifier only when its band is selected. The amplifiers in the bands that are not selected are powered off and provide isolation from out-of-band signals. Table 8-1 lists the five sub-bands that tune from 20 to 500 MHz.

Table 8-1 *RF Preselection filter band breaks.*

Preselector	Frequency Range
Band 1	20- 40 MHz
Band 2	40 – 80 MHz
Band 3	80 – 160 MHz
Band 4	160 – 280 MHz
Band 5	280 – 500 MHz

The output of the preselector module is routed into the first mixer in the 60/145 MHz IF converter module.

Coarse-Step Synthesizer/First LO

The conversion scheme requires that the first LO tune from 160 to 560 MHz. Oscillator phase noise is a strong function of tuning range (in percent), so five separate, power-switched oscillators are used (see Figure 8-23). for a block diagram. Since each oscillator tunes less than an octave, the system phase noise can be reduced while maintaining the required tuning range.

The microprocessor provides power control for each separate VCO. Only one oscillator is powered up at any one time, greatly reducing the likelihood of internally generated spurious signals.

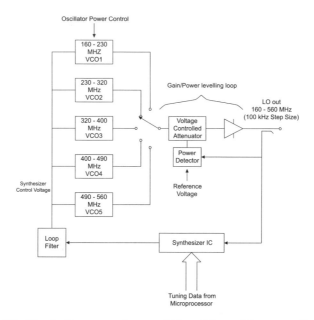

Figure 8-23 *Block diagram of a candidate LO architecture. Several power-switched VCOs are used to reduce phase noise.*

The breakdown of the VCO frequencies is shown in Table 8-2.

Table 8-2 *Frequency synthesizer VCO band breaks.*

VCO 1	160 – 230 MHz
VCO 2	230 – 320 MHz
VCO 3	320 – 400 MHz
VCO 4	400 – 490 MHz
VCO 5	490 – 560 MHz

The step size of the first synthesizer is 100 kHz. The large step size allows for a large synthesizer loop bandwidth, which again helps reduce the phase noise. The large loop bandwidth also enables fast tuning speed.

One effect of the 100 kHz step size is that the first mix will not exactly center the signal of interest in the first IF band-pass filter. The signal will be ±50 kHz removed from the first IF center frequency and will be accounted for in the second mix. The LO leveling loop insures that the LO power remains constant as we tune the receiver. Constant LO power preserves the system's performance over the receiver tuning range.

First IF

The first IF section contains the first mixer and the first IF filters (Figure 8-24). The two first IF center frequencies are 60 and 145 MHz. Filter selection for the tuned frequency is based on an analysis of the spurious signals that might be generated. The microprocessor again switches the signal and controls the power of each amplifier.

The first synthesizer tunes the 20 to 100 MHz band by tuning the first LO from 165 to 245 MHz, upconverting the RF to 145 MHz. The 100 to 500 MHz band is downconverted to 60 MHz using an LO of 160 to 560. This conversion scheme allows "reusing" some of the first LO tuning range. Table 8-3 describes the IF selection and local oscillator frequencies with respect to tuned frequency.

Table 8-3 *IF and local oscillator frequencies.*

Tuned Frequency	First IF Frequency	First LO Frequency (100 kHz Step Size)
20 – 85 MHz	145 MHz	165 – 230 MHz (VCO 1)
85 – 100 MHz	145 MHz	230 – 245 MHz (VCO 2)
100 – 170 MHz	60 MHz	160 – 230 MHz (VCO 1)
170 – 260 MHz	60 MHz	230 – 320 MHz (VCO 2)
260 – 340 MHz	60 MHz	320 – 400 MHz (VCO 3)
340 – 430 MHz	60 MHz	400 – 490 MHz (VCO 4)
430 – 500 MHz	60 MHz	490 – 560 MHz (VCO 5)

The first IF also contains the first instance of gain control in the receiver. One of the AGC voltages developed in the demodulator is fed to a voltage-controlled attenuator. When the signal of interest is large, the AGC acts to reduce the receiver gain through VAGC.

CASCADE II — IF SELECTION | 743

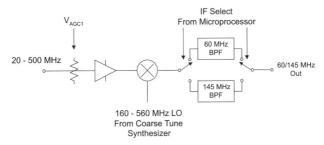

Figure 8-24 Block diagram of the first conversion and the first IF stages. The IF is switched to reduce intermodulation distortion and to allow the receiver to "re-use" some of its LO range. The first LO has a 100 kHz step size.

Fine Step Synthesizer/Second LO

The fine step synthesizer/second LO converts the first IF (at either 60 or 145 MHz) down to 21.4 MHz. Figure 8-25 shows the architecture of the second frequency synthesizer tuning in 2 kHz steps. These LOs are centered about 81.4 and 166.4 MHz, depending upon which first IF is selected. When the first IF is 60 MHz, the second LO tunes 81.4 MHz ± 50 kHz. When the first IF is 145 MHz, the second LO tunes from 166.4 MHz ± 50 kHz. The second LO tunes only ±50 kHz because only the first IF needs to be centered in the final IF bandwidth. The narrow tuning range coupled with the fine step size preserves the tuning speed of the receiver.

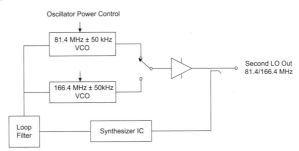

Figure 8-25 Block diagram of the second LO. Two separate oscillators are required because the receiver contains two radically different first IFs. The second LO has a 2 kHz step size.

Second IF

Figure 8-26 shows the second IF. The fine-step frequency synthesizer converts the 60/145 MHz first IF to the 21.4 MHz second IF. The microprocessor selects the proper band-pass filter based on the user's choice of IF bandwidth. We use a second AGC voltage (V_{AGC2}), to control the receiver gain based on the received signal strength.

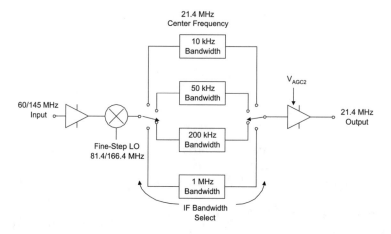

Figure 8-26 *Block diagram of the second IF at 21.4 MHz. Several final IF filters can be selected.*

Demodulators

The demodulator block diagram is shown in Figure 8-27. Demodulators develop much of the information used by the receiver.

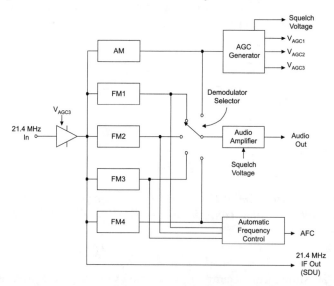

Figure 8-27 *Block diagram of the demodulator section. FM demodulation requires a separate detector for each final IF bandwidth. We generate the AGC control voltages and the automatic frequency control (AFC) signals in the demodulator.*

The AM detector is used to generate several *automatic gain control* (AGC) voltages for the receiver. These voltages are designed to reduce the receiver's gain in a carefully controlled manner. The objective is to trade off noise performance for linearity without giving up noise performance entirely. Often these AGC voltages are delivered to the system's microprocessor for further processing. The microprocessor then generates the signals that control the receiver's gain via a DAC.

The AM detector also generates a squelch signal. The receiver audio is turned off when there is no signal present. When a signal arrives, the squelch is released, the audio amplifier is enabled and the user can hear the signal.

The FM detectors generate a signal used for *automatic frequency control* (AFC). When the receiver is slightly off-tuned, we can measure the frequency error with the FM detectors and center the signal up in the selected IF bandwidth. Again, this signal often goes into the microprocessor and the microprocessor makes the tuning decisions.

Digital Interfaces

Most of the receiver blocks require microprocessor control. We have to carefully control the physical interfaces between the very noisy microprocessor and the sensitive RF components.

Aggregate Plan

Table 8-4 describes the operation of the receiver over its tuned frequency range.

Table 8-4 *Aggregate receiver control.*

Tuned Frequency (MHz)	LO Frequency (MHz)	IF (MHz)	Preselector (MHz)
20 – 40	165 – 185 (VCO 1)	145	20 – 40 (RF 1)
40 – 80	185 – 225 (VCO 1)	145	40 – 80 (RF 2)
80 – 85	225 – 230 (VCO 1)	145	80 – 160 (RF 3)
85 – 100	230 – 245 (VCO 2)	145	80 – 160 (RF 3)
100 – 160	160 – 220 (VCO 1)	60	80 – 160 (RF 3)
160 – 170	220 – 230 (VCO 1)	60	160 – 280 (RF 4)
170 – 260	230 – 320 (VCO 2)	60	160 – 280 (RF 4)
260 – 280	320 – 340 (VCO 3)	60	160 – 280 (RF 4)
280 – 340	340 – 400 (VCO 3)	60	280 – 500 (RF 5)
340 – 430	400 – 490 (VCO 4)	60	280 – 500 (RF 5)
430 – 500	490 – 560 (VCO 5)	60	280 – 500 (RF 5)

Example #2

At the time of its manufacture, this physically small and low-power receiver achieved a state-of-the-art size/power/performance capability.

- *Tuned Frequency*: 20 to 500 MHz
- *Tuning Step Size*: 5 kHz
- *Noise Figure*: 8 dB (maximum).
- *IF Bandwidths*: 30 kHz, 200 kHz, and 1 MHz.
- *Input Third-Order Intercept Point, Out-of-Band: Two Tones*: F_0 + 3.5 MHz and f_{tuned} +7.0 MHz with 1 MHz bandwidth selected: –11 dBm (minimum).
- *Input Third-Order Intercept Point, In-Band Two Tones*: F_0 + 0.5 MHz and F_0 + 1.0 MHz with 30 kHz IF bandwidth selected: +20 dBm minimum.
- *Internally Generated Spurious Signals*: –120 dBm maximum, referenced to the antenna input.
- *Image Rejection*: > 60 dB.
- *IF Rejection*: > 60 dB.
- *Reference Accuracy*: ±5 ppm, maximum, over temperature for three years.
- *Incidental Frequency Modulation*: < 300 hertzRMS, measured over a 100-hertz to 3.0 kHz bandwidth.
- *Demodulation* FM, AM.
- *FM Sensitivity*: Signal deviation of IFBW/2, audio SINAD of 10 dB (in a 3 kHz audio bandwidth). The FM sensitivities are: –108 dBm (30 kHz IFBW), –100 dBm (200 kHz IFBW) and –93 dBm (1 MHz IFBW).
- *AM Sensitivity*: Modulation index of 60%. Audio SINAD of 10 dB (in a 3 kHz audio bandwidth). The AM sensitivities are: –104 dBm (30 kHz IFBW), –96 dBm (200 kHz IFBW) and –89 dBm (1 MHz IFBW).
- *IF Bandwidths*: 30 KHz, 200 kHz and 1 MHz.
- *IF Filter Shape Factors* (60dB/3dB): 30 kHz bandwidth: 2:1, 200 kHz: 2.5:1, 1 MHz: 2.0:1.
- *AGC Characteristics*: 10 ms attack time, 80 ms decay time.
- *IF Output* (10.7 MHz): –30 dBm ± 2.5 dB, 80 dB AGC range.
- *Maximum Power Dissipation* less than 1.2 watts (with prime power at 6 Vdc).

For this receiver, the prioritized characteristics are:
1. size,
2. power consumption,
3. performance.

Analysis of the Specifications

This receiver requires a digital control system that does not have to be very sophisticated. The design does not include a microprocessor. After the receiver is tuned, there is no active digital circuitry. This quiets the inside of the receiver and greatly alleviates the isolation demands.

The small physical size dictates a simple conversion scheme. We choose to upconvert the entire range to 700 MHz and then to filter with a SAW filter. SAW filters provide reasonable temperature stability and very narrow bandwidths. Upconverting to the first IF provides good spurious and image rejection and low LO radiation. However, second-order problems are present.

We also used a SAW-based second LO for small size. Although the SAW oscillators are not very stable, this design uses a compensation scheme to partially compensate for this shortcoming.

Figure 8-28 shows a top-level diagram of this receiver. The receiver has four basic functions: frequency converter, LO generator, demodulator and digital controller. In the course of the design, we will develop candidate architectures for each of these functions. We will discuss the merits and problems of each architecture before converging on the final design. The small physical size of this receiver dictates that we use the smallest number of frequency conversions possible.

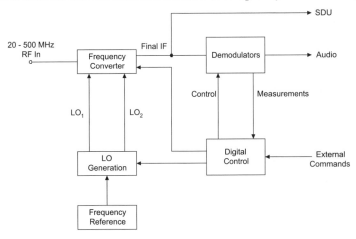

Figure 8-28 High-level block diagram of a candidate receiving system.

Selection of the First IF

Figure 8-29 shows a possible single-conversion architecture. The desired signal is immediately moved down to 21.4 MHz and the demodulation is performed. However, this simple architecture does not provide any image rejection and low LO radiation will be difficult to achieve. We readily abandon this approach.

The RF tuning range sets most of the constraints for the first IF. Simplicity and modest performance requirements force us to upconvert to the first IF. Filter roll-off considerations place the minimum center of this filter at approximately 600 MHz. Ideally, we would like the first IF to be at least 2.1 times the highest RF, which places the first IF above 1050 MHz.

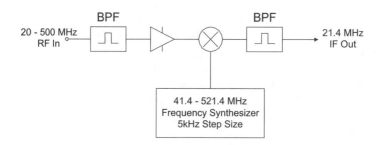

Figure 8-29 High-level block diagram of a single conversion architecture. This approach has no hope of achieving the required specifications.

To protect the receiver against strong undesired signals, the bandwidth has to be limited as quickly as possible. The RF preselectors will limit the RF bandwidth to about 15% of the tuned frequency. Ideally, we would like the first IF filter to be as narrow as the maximum final IF bandwidth. The maximum bandwidth of the final IF is 1 MHz, and a 2 MHz bandwidth for the first IF is ideal.

At the time of this design, the technology most suitable for the first IF filter (center frequency > 600 MHz with a 2 MHz bandwidth) was a *surface acoustic wave* (SAW) filter. SAW filters with center frequencies from 500 to 725 MHz were available with 2 MHz bandwidths.

Image rejection is determined by the RF preselection filters, the first IF and the LO selection. High image rejection results when we maximize the difference between the first IF and the highest RF. The IF should be close to 725 MHz rather than close to 500 MHz.

With a high-side LO it is easier to keep the LO from the antenna port. It will also be easier to filter the LO-IF leakage using the first IF filter. We have identified a possible first IF of 700 MHz. Let us look at the effect on the rest of the system.

Selection of the Second IF

Physical size is crucial: we want to use standard, commercially available and physically small parts in the demodulator. Second IFs of 10.7 MHz or 21.4 both meet these requirements. Can we build or buy the three IF filters we need at either of these frequencies? Which IF produces the smallest volume?

We can build or buy the 30 kHz IF filter at either 10.7 or 21.4 MHz in both discrete and monolithic crystal filters. The smallest physical volume for the 30 kHz filter is the 21.4 MHz version (monolithic crystal filters). The 200 kHz filter is smallest at 10.7 MHz if we use a common FM radio ceramic filter. We can build the 1 MHz filter from discrete components at either 10.7 or 21.4 MHz.

A 10.7 MHz second IF seems the most feasible. We decided on a first IF of 710.7 MHz, using high-side LO injection to convert the RF to the first IF. The second IF was 10.7 MHz, and we decided to use a low-side LO to convert the 710.7 MHz to 10.7 MHz.

The Design of the First LO

The characteristics of the LOs are:

- *First LO Frequency Coverage*: 730.7 to 1210.7 MHz.
- *Second LO Frequency*: 700 MHz.

Figure 8-30 shows a possible LO architecture using a frequency-synthesized first LO. A crystal oscillator and multiplier generate the second LO. This architecture will work but the multiply-and-filter scheme used to generate the second LO requires much filtering and shielding to develop a clean, spurious-free second LO.

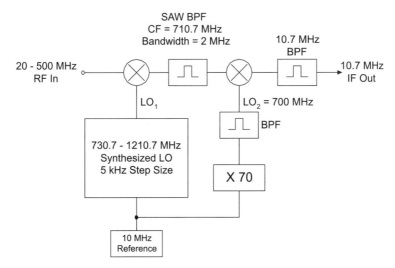

Figure 8-30 LO architecture using a single synthesized first LO and a x70 multiplier to generate the fixed second LO.

Figure 8-31 shows a different architecture using two frequency synthesizers. The first LO tunes in 5 kHz step sizes and the second LO is fixed at 700 MHz. This scheme achieves good frequency stability because both local oscillators are referenced to one crystal reference but two complete synthesizers will require more surface area and power. After considering several other architectures, the two synthesizer designs of Figure 8-31 were accepted and built.

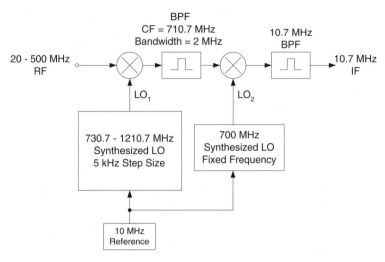

Figure 8-31 LO architecture using two synthesized LOs.

8.10 Cascade II Design Summary

System Considerations

Good receiver design requires careful consideration of the entire system — not only the RF components. The designer must consider the digital requirements of the receiver and the noise/isolation problems the digital circuits will introduce.

Frequency Conversion Schemes

Preselection filters are the only protection against the first mixer's image response. The noise present at the image frequency of every conversion has to be considered. The choices between HSLO/LSLO and upconverting/downconverting involve VCO design, filter issues, signal bandwidths, spurious signal specifications and physical size. Conversion scheme design requires much analysis and iteration. Component realization plays an important part in the design. It is helpful to refer to many different catalogs.

Narrowband and Wideband Systems

Keep the design narrowband, i.e., all signal paths less than an octave of bandwidth, if at all possible. This converts a second-order problem into a more manageable third-order problem.

8.11 References

1. Dexter, Charles E. "Design Considerations for Miniaturized Receivers." *Watkins-Johnson Technical Notes* 16, no.5 (1989).

2. "Fundamentals of RF and Microwave Noise Figure Measurement." *Hewlett-Packard Application Note* 57-1 (July 1983).

3. Gross, Brian P. "Calculating the Cascade Intercept Point of Communications Receivers." *Ham Radio Magazine*, 8 (1980): 50.

4. Ha, Tri T. *Solid-State Microwave Amplifier Design*. New York: John Wiley and Sons, 1981.

5. Hayward, W. H. *Introduction to Radio Frequency Design*. Englewood Cliffs, NJ: Prentice-Hall, 1982.

6. Horowitz, Paul and Winfield Hill. *The Art of Electronics*. Cambridge: Cambridge UP, 1980.

7. Perlow, Stewart M. "Basic Facts About Distortion and Gain Saturation." *Applied Microwave,* vol. 1, no. 5 (1989): 107.

8. Rohde, Ulrich L. and T. T. N. Bucher. *Communications Receivers: Principles and Design*. New York: McGraw-Hill, 1988.

9. Sabin, William E. and Edgar O. Schoenike. *Single-Sideband Circuits and Systems*. New York: McGraw-Hill, 1987.

10. Watson, Robert. "Receiver Dynamic Range." Parts 1 and 2. *Microwaves and RF*, no.12 (1986): 113; no. 1 (1987): 99.

11. Williams, Richard A. *Communications Systems Analysis and Design: A Systems Approach*. Englewood Cliffs, NJ: Prentice-Hall, 1987.

9

Appendix

9.1   Gaussian or Normal Statistics

Gaussian Probability Density Function

The equation for the general *Gaussian probability density function* (GPDF) is

$$GPDF(x) = \frac{1}{\sigma\sqrt{2\pi}} e^{\left[-\frac{(x-\mu)^2}{2\sigma^2}\right]} \qquad 9.1$$

where
μ = the mean or average,
σ = the standard deviation,
x = the variable under consideration.

The Gaussian distribution is also referred to as a *normal distribution*. It is common to speak of *a Gaussian or normal distribution with mean μ and standard deviations*. Thermal noise is almost universally assumed to be a zero mean process. The mean value of the noise is zero, i.e., the noise has no DC component because it has been passed through a capacitor. The GPDF for thermal noise is

$$GPDF_{\mu=0}(x) = \frac{1}{\sigma\sqrt{2\pi}} e^{\left[-\frac{x^2}{2\sigma^2}\right]} \qquad 9.2$$

Normalized Gaussian Probability Density Function

For the sake of calculation, we sometimes assume that the mean of the GPDF in Equation 9.1 equals zero and that the standard deviation equals 1. These assumptions produce the *normalized Gaussian probability density function* (NGPDF). The equation for the NGPDF is

$$NGPDF(x) = \frac{1}{\sqrt{2\pi}} e^{\left[-\frac{z^2}{2}\right]} \qquad 9.3$$

where z = the normalized value of x.

Figure 9-1 shows the normalized Gaussian probability density function. Figure 9-2 illustrates the area A(z) and tail functions T(s) of the Gaussian PDF. Figure 9-3 shows the area and tail functions.

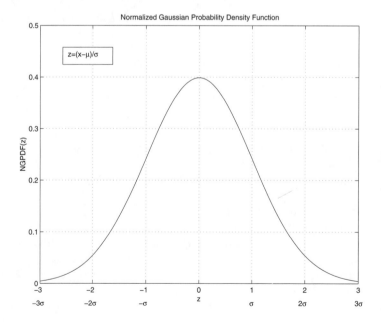

Figure 9-1 *The normalized Gaussian probability density function (NGPDF).*

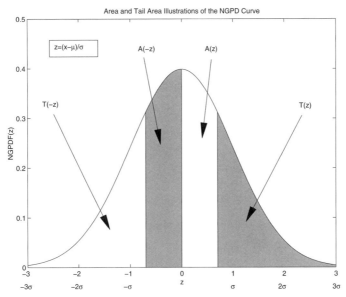

Figure 9-2 *The NGPDF and associated areas. The area under the curve from 0 to z is denoted by A(z). The area under the curve from z to infinity is denoted as T(z).*

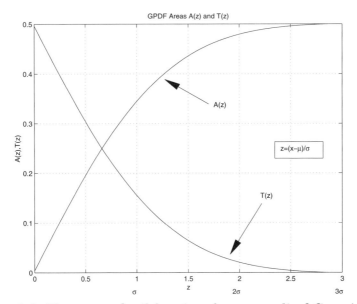

Figure 9-3 *The area and tail functions for a normalized Gaussian probability density function.*

Normalizing Data

Given a Gaussian population with a mean μ and a standard deviation σ, the normalized variable z is related to the unnormalized variables μ, σ, and x by

$$z = \frac{x - \mu}{\sigma} \qquad 9.4$$

This normalization translates $\mu - \sigma$ to -1, μ to 0, and $\mu + \sigma$ to $+1$.

Properties of the NGPDF

The NGPDF has several useful properties.

1. If a random process follows a Gaussian distribution, the area under the Gaussian curve between two values of z is the expected percentage of events that will lie between the two values of z. For example, the probability that a Gaussian random variable will fall between z_1 and z_2 is

$$\Pr[z_1 \leq z \leq z_2] \qquad 9.5$$

This probability is given by the shaded area shown in Figure 9-4.

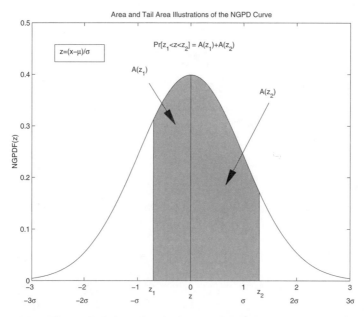

Figure 9-4 The probability that a Gaussian random value will fall between z_1 and z_2 is given by the area under the NGPDF curve between z_1 and z_2.

2. The entire area under the NGPDF curve is unity or

$$\Pr[z_1 \leq z \leq z_2] = \Pr[-\infty \leq z \leq \infty] \qquad 9.6$$
$$= 1$$

The NGPDF is symmetrical about the $z = 0$ line. Note that the NGPDF will not equal zero no matter how large the value of z.

Example 9.1 — Expected Values of Resistors
Suppose we have measured a sufficiently large number of 5100 ohm resistors from one manufacturer and found that the sample has a mean value of 5180 ohms and a standard deviation of 215 ohms.
a. What percentage of resistors will have a value in the range of 5000 to 5200 ohms?
b. What percentage of resistors will be greater than 5500 ohms?
c. What percentage of the resistor population will be less than 4500 ohms?

Solution —
First, we normalize the given data using Equation 9.4. We know $\mu = 5180\ \Omega$ and $\sigma = 215\ \Omega$.
a. Using Equation 9.4 to normalize the data, we find

$$z_{5000} = \frac{5000 - 5180}{215} \qquad z_{5200} = \frac{5200 - 5180}{215} \qquad 9.7$$
$$= -0.837 \qquad \qquad = 0.0930$$

Using Figure 9-3, the area between $z_{5000} = -0.837$ and $z = 0$ is 0.2995. Similarly, the area between $z_{5200} = 0.093$ and $z = 0$ is 0.0359. The total area between z_{5000} and z_{5200} is $0.2995 + 0.0359 = 0.335$. This means that 33.5% of the resistors from the population will have values between 5000 ohms and 5200 ohms.

b. Using Equation 9.4, we will normalize $x = 5500$ ohms to

$$z_{5500} = \frac{5500 - 5180}{215} \qquad 9.8$$
$$= 1.49$$

Figure 9-3 tells us that the area under the Gaussian curve above $z_{5500} = 1.49$ is 0.0681. This means 6.8% of the resistors from the population will have values greater than 5500 ohms.

c. Using Equation 9.4, we will normalize $x = 4500 \, \Omega$ and find

$$z_{4500} = \frac{4500 - 5180}{215}$$
$$= -3.16 \qquad 9.9$$

Figure 9-3 does not show data above $z = 3.0$. However, calculation reveals that the area above $z = 3.16$ is 0.0008. Since the Gaussian curve is symmetrical about the mean, we know the area below $z_{4500} = -3.16$ is the same as the area above $z = 3.16$. 0.08% of the resistors from the population will have values of less than 4500 ohms.

Example 9.2 — Thermal Resistor Noise

Assume we have measured a noisy signal and found that its mean is 0 volts and its standard deviation is 890 nV$_{RMS}$. How often will the magnitude of this noise voltage be greater than 2220 nV?

Solution —

We use Equation 9.4 to normalize z_{2220} nV to a Gaussian distribution with zero mean and a standard deviation of 890 nV. This operation produces

$$z_{2220nV} = \frac{2200 - 0}{890}$$
$$= 2.49 \qquad 9.10$$

Figure 9-3 indicates that the area of the Gaussian curve lying above $z = 2.49$ is 0.0064, which indicates that the noise voltage will be greater than 2220 nV 0.64% of the time. Using symmetry, we know that the noise voltage will be less than –2220 nV 0.64% of the time. Accordingly, the magnitude of the noise voltage will be greater than 2220 nV 1.28% of the time.

Complimentary Error Function

Calculations dealing with noise and demodulation often involve the area under the NGPDF curve between some point z and $\pm$ infinity, which is the *complimentary error function* $Q(z)$ or, mathematically, the integral of the NGPDF.

$$Q(z) = \frac{1}{\sqrt{2\pi}} \int_{z}^{\infty} e^{\left[-\frac{z^2}{2}\right]} dz \qquad 9.11$$

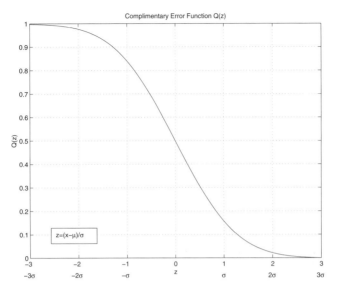

Figure 9-5 *The complimentary error function Q(z). This function is also called the cumulative normalized Gaussian probability density function, or the co-error function.*

The complimentary error function is shown in Figure 9-5. *Q(z)* is also called the *cumulative normalized Gaussian probability density function*, or the *co-error function*.

Approximations for *Q(z)*

Mathematically, we find that Equation 9.11 is not directly integrable. There are many situations, however, where it is convenient to express *Q(z)* as a simple equation. In these cases, we can form several approximations. For z >= 3, we can approximate *Q(z)* by

$$Q(z) \approx \frac{1}{z\sqrt{2\pi}} e^{\left[-\frac{z^2}{2}\right]} \quad \text{for } z \geq 3 \qquad 9.12$$

This approximation, which becomes more accurate as z increases, denotes that, for large z, the function *Q(z)* decreases at an exponential rate. This exponential fall-off is responsible for the "waterfall" bit error rate curves of detection theory. A more accurate approximation for *Q(z)* is

$$Q(z) \approx \frac{1}{(1-a)z + a\sqrt{z^2+b}} \frac{1}{\sqrt{2\pi}} e^{\left[-\frac{z^2}{2}\right]} \quad \text{for } z \geq 0 \qquad 9.13$$

where
 $a = 0.344$,
 $b = 5.344$.

9.2 Statistics and Noise

Gaussian statistical functions are used to describe the random thermal noise that is present in a receiving systems. This noise exhibits a Gaussian amplitude distribution. A Gaussian-distributed random process is completely described by two parameters: the mean, (or average) and the standard deviation.

Mean (μ)

The mean marks the center of symmetry of the Gaussian curve, as shown in Figure 9-1. We universally assume that thermal noise has a mean of zero.

Standard Deviation (σ)

Graphically, the standard deviation σ is the distance from the center of symmetry of the curve to the two points at which the curve changes from curving downward to curving upward (see Figure 9-1). The standard deviation is a measure of the variability of the data; a change in σ changes the outline of the curve. A small standard deviation indicates a population that is tightly concentrated about the mean and that the curve is high and narrow. A large standard deviation indicates that the population is spread out and that the Gaussian curve is low and wide. A change in value of μ does not change the outline of the curve. It moves the whole curve farther to the right or left depending upon whether μ increases or decreases. The standard deviation of thermal noise is its RMS value.

$$\sigma_{Noise} = V_{Noise,RMS} \qquad 9.14$$

or

$$\sigma_{Noise} = I_{Noise,RMS} \qquad 9.15$$

Variance

The variance of a Gaussian process is the square of the standard deviation, or

$$\text{Variance} = \sigma^2 \qquad 9.16$$

For thermal noise, the standard deviation is the RMS value of the noise voltage or current. Since the variance is the square of the voltage or current, the variance is a measure of the noise power.

Example 9.3 — Signals and Noise
If we have 0.43 volts$_{RMS}$ of noise voltage riding on a 1 volt DC level, how often will the noise voltage cause the sum to go below zero volts?

Solution —
Figure 9-6 shows the problem graphically.

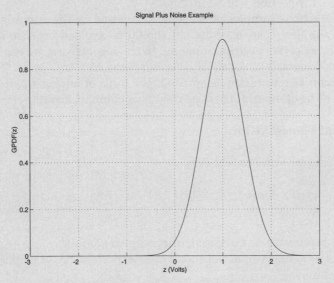

Figure 9-6 *Example 9.3. The probability density function of 0.43 volts$_{RMS}$ riding on 1 volt$_{DC}$.*

We know the mean value of Figure 9-6 is 1 volt and the standard deviation is 0.43 volts$_{RMS}$. Using Equation 9.4, we first normalize the data.

$$z = \frac{x - \mu}{\sigma} = \frac{0 - 1}{0.43}$$
$$= -2.33 \qquad 9.17$$

Figure 9-3 shows that this z corresponds to a tail probability, $T(z)$, of about 1%. On the average, noisy DC voltage will be negative about 1% of the time.

9.3 Cancellation and Balance of Cosine Waves

Frequency-selective multipath fading is caused by a single signal taking more than one path from the transmit antenna to the receive antenna. For the sake of calculation, if we assume there were two paths, one signal will experience a phase and an amplitude change with respect to the other before they both arrive at the receive antenna. If the two copies of the signal are equal in amplitude and differ in phase by 180°, the received signal will be completely cancelled.

In a double-balanced mixer, we strive to increase the LO:IF isolation by arranging for one copy of the LO to arrive at the IF port along with a copy of the LO signal multiplied by –1 (or equivalently, $1\angle 180°$). If we are successful, the LO signal will be completely cancelled at the mixer's IF port, and the LO-IF rejection will be infinite.

These are two examples of cancellation caused by the summing of two signals. In the first case, cancellation should be avoided because it results in frequency-selective multipath fading. In the second case, we can use cancellation to achieve high levels of isolation between mixer ports. This type of cancellation also provides insight into the nulls of antenna patterns and the effects of limiters and filters as they pass complex signals.

Two-Signal Cancellation

The model we will analyze is

$$V_{cancel}(t) = \cos(\omega t) + \rho \cos(\omega t + \phi) \qquad 9.18$$

where
$\cos(\omega t)$ represents the original, uncorrupted signal,
$\rho \cdot \cos(\omega t + \phi)$ represents the signal after experiencing an amplitude and phase change.

The sum of these two signals is the cancelled signal. Exact cancellation occurs when $\rho = 1$ and $\phi = 180°$. To determine the power change between the original cosine wave $\cos(\omega t)$ and $V_{cancel}(t)$, we can use vector analysis. Equation 9.18 can be rewritten as a single cosine with an amplitude and phase change.

$$\begin{aligned} V_{cancel}(t) &= \cos(\omega t) + \rho \cos(\omega t + \phi) \\ &= \xi \cos(\omega t + \theta) \end{aligned} \qquad 9.19$$

where

$$\xi^2 = 1 + 2\rho\cos(\phi) + \rho^2$$

and

$$\theta = \tan^{-1}\left[\frac{\rho\sin(\phi)}{1+\rho\cos(\phi)}\right]$$

9.20

The ratio of the power in $V_{cancel}(t)$ to the power in the original cosine wave is ξ^2. The change in phase between the original cosine wave and $V_{cancel}(t)$ is θ.

Characteristics of Cancellation

Figures 9-7 and 9-8 show the signal attenuation ξ^2 with respect to the phase difference between the two signals. As expected, the worst signal attenuation occurs when ρ is unity and ϕ is 180°. Under these conditions complete cancellation occurs.

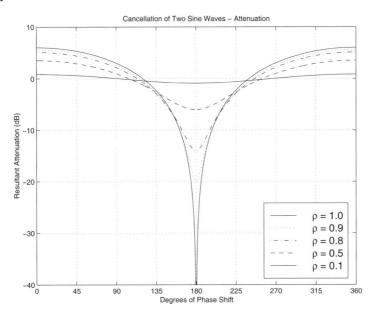

Figure 9-7 *Equation 9.20. The resultant power when two sine waves of different phase and amplitude are additively combined. The most attenuation occurs when the amplitudes of the two sine waves are equal and the phase difference is 180°.*

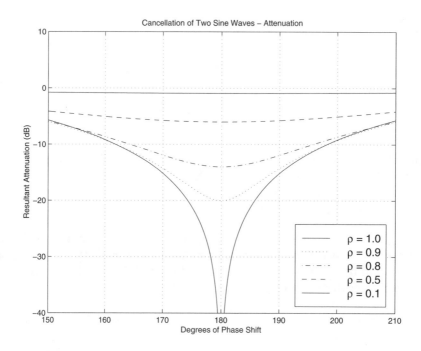

Figure 9-8 *Close-up view of Figure 9-7, emphasizing the behavior of Equation 9.20 at 180°.*

A small change in ϕ results in a large change in the combined signal strength. For example, the $\rho = 1$ curve of Figure 9-8 shows the signal cancellation will change from completely cancelled at $\phi = 180°$ to -15 dB with a phase change of 10°. An *increase* in the cancelled signal can be observed when

$$-90° \leq \phi \leq 90° \qquad 9.21$$

The maximum increase is 6 dB (when ρ is unity and ϕ is 0°). The same ρ that produces the maximum increase in signal power also produces the worst attenuation. The difference lies in the phase ϕ. Figures 9-9 and 9-10 show the phase resultant signal. When ρ is in the neighborhood of unity, the received phase changes abruptly when ϕ passes through 180°. This sudden phase change plays havoc with phase-encoded signals.

APPENDIX | 765

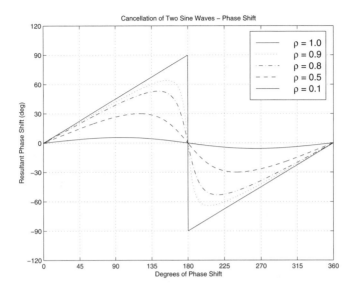

Figure 9-9 *Equation 9.20. The resultant phase when two sine waves of different phase and amplitude are additively combined. Note the abrupt phase change when the amplitudes of the two sine waves are equal and the phase difference is 180°.*

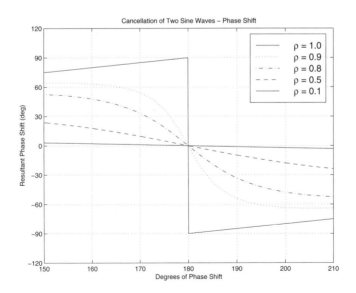

Figure 9-10 *Close-up view of Figure 9-9 emphasizing the behavior of Equation 9.20 at 180°.*

Example 9.4 – Multipath Fading (I)
When you pull your car up to a stop light and the FM radio fades out, you may be able to bring the signal back in by scootching your car forward a few feet. When the FM signal disappeared, two (or more) signals from the transmitter were arriving at the car exactly out-of-phase and cancelled. Moving the car changes the phase difference between the arriving signals by a small amount and may result in a large change in the received signal strength.

Example 9.5 – Multipath Fading (II)
One of the authors used to live in a townhouse that was along the flight path to an airport. When an airplane flew along one particular flight path, his television reception would fade in and out as the airplane moved. The rate of change was slow at first, taking perhaps three seconds to change from a perfect picture to a completely unwatchable picture. As the airplane continued along its approach toward the airport, the rate of change of the signal strength would increase until the picture would fade in and out several times a second. Finally, as the geometry continued to change, the television reception cycle time gradually decreased back to its three-second fade in/fade out behavior. This entire interference event lasted perhaps 20 seconds from start to finish.

A commercial airliner is an excellent signal reflector. The signal the author received from the airplane was about as strong as the signal he received directly from the television transmitter (in other words, ρ was about equal to unity). As the airplane moved along its flight path, the changing geometry caused the signals arriving at his receive antenna to add up alternately in phase and out-of-phase.

9.4 References

1. Borjesson, Per Ola and Carl-Erik W. Sundberg. "Simple Approximations of the Error Function $Q(x)$ for Communications Applications." *IEEE Transactions on Communications* COM-27, no. 3 (1979): 639-643.

2. Noether, Gottfried E. *Introduction to Statistics: A Nonparametric Approach*. 2nd ed. Boston: Houghton Mifflin, 1976.

Index

A
Adjacent Channel Masking, 356
Amplifier
    Coherence Time, 455
    Eqiuvalent Model, 444
    Gain Compressions, 599
    Large Signal Conditions, 549
    Small Signal Conditions, 548
    Small Signal Model, 546
    Voltage Transfer Curve, 543
        Polynomials, 550
Amplifier Noise Model, 481
Amplifiers, Radio Frequency, 78
Amplitude Modulation (AM), 105
Amplitude Noise, 354
Amplitude Ripple, 171
Antenna Noise Model, 474
Antenna Noise Temperature, 474
Arithmtic Center Frequency, 216
Attenuation
    Ultimate, 171, 172
Attenuator
    Front-End, 665
Automatic Frequency Control (AFC), 407
Automatic Gain Control (AGC), 691

B
Balance
    Cosine Waves, 762
Baluns, 320
    Dot Convention, 321
Band-Stop Filter, 168
Bandwidth
    Band-Pass Filter, 217
    Digital Signals, 28
Bel, 4
Bessel Filter
    Poles and Zeros, 189
Bessel Functions, 115, 365
Block Frequency Converters, 284
Butterworth Filter
    Low-Pass, 179
    Poles and Zeros, 180

C
c. See Speed of Light
Cable Runs, 697
Cancellation (Cosine Waves), 762

Cascade
  Attenuators, Front-End, 665
  Coherent vs. Noncoherent Addition, 680
  Design Rules for Gain Distribution, 694
  Equations, 507
  Excess Gain, 657
  Gain and Linearity, 672
  Gain Distribution, 654
    Linearity, 657
    Noise, 656
  Gain Distribution Rules, 694
  Gain Performance, 498
  Noise
    Component VSWR, 500
    Equation Analysis, 507
    Noise Temperature Analysis, 501
    Practical Effects, 500, 506
  Noise Factor, 504
  Noise Figure Analysis, 497
  Noise Temperature, 501
  OTOI vs. ITOI, 672
  Output Specifications vs. Input Specifications, 672
  Relative Levels of Noise, 662
  Resistors in Parallel Analogy for Linearity, 673
  Translation of Component Specifications, 658
  Translation of Linearity Specifications, 658
  Translation of Noise Specifications, 660
Cavity Filters, 252
Ceramic Filters, 252
Ceramic Resonator Filters, 253
Channelized Frequency Converters, 284
Chebychev Filter
  Equi-Ripple Frequency vs. 3-dB Frequency, 184
  Low-Pass, 182
  Poles and Zeros, 185
Coherent Addition, 680
Coherent Summation, 609
Complimentary Error Function, 758
Conversion Loss, 276
Conversion Scheme Design, 285
  HSLO vs. LSLO, 714
  IF Selection Guidelines, 724
  Local Oscillators, 721
  Phase Noise, 721
  Tuning Range, 721
  Typical System, 733
  Upconversion vs. Downconversion, 714
Crossmodulation, 633
Crystal Filters, 251
Crystal Reference Oscillators, 433

D

Data
  Normalizing, 756
dBHz, 11
dBK, 11
dBm, 7
dBmV, 9
dBV, 8
dBW, 7
Decibels, 14
  Gains and Losses, 15
  Meaning of Specific Digits, 14
  Significant Figures, 17
Digital Logic, 730
Digital Signals, Bandwidth, 28
Distortion
  Coherent vs. Noncoherent Addition, 625
  Filtering, 567
  Harmonic, 543
  Higher-Order Effects, 577
  Input vs. Output Specifications, 623
  Linearity and Power Consumption, 625

INDEX | 769

Model Inadequacy, 623
Notes, 622
Second-Order, 568
  Filtering, 568
  Problematic Frequencies, 570
  Small Signal Approximation (I), 565
  Small Signal Approximation (II), 565
  Summary, 565
Third-Order, 572
  Filtering, 572
  Problematic Frequencies, 574
  Third-Order Measurement Difficulties, 622
Dividers and Phase Noise, 381
Dot Convention, 321
Double-Balanced Mixer (DBM), 329
Downconversion, 714
Drift, 399
Dual Tone Analysis. See Two-Tone analysis
Dynamic Range, 643
  Attenuators, 649
  Conversion Schemes, 652
  Digital Logic, 653
  Gain-Controlled, 643
  Linear, 643
  Measurement, 654
  Specification Ambiguity, 652
  Spur-Free, 644
    Second Order Limited, 646
    Third Order Limited, 647
  Spurious Signals, 652

E

Elliptic Filter
  Poles and Zeros, 193
Equal-Ripple Bandwidth, 217
Examples
  Amplifiers and Noise
    174 dBm/Hz, 468
    Antenna Noise Temperature, 475
    Cascade Gain Calculation, 499

Combining Noise Sources, 462
ENR, 512
Equivalency of the Voltage and Current Noise Models, 451
Expected Values of Resistor Noise, 456
Gains and Losses in Cascades, 520
Input Noise Power, 485
Lossy Devices in Cascade, 530
Lossy Devices in Cascade II, 531
Measuring Noise Temperature, 513
Noise Energy, 468
Noise Factor of a Cascade, 505
Noise Factor of Attenuators at 0 K, 531
Noise Factor of Attenuators at Room Temperature, 528
Noise Figure Measurement Error, 519
Noise Figure/Noise Factor/Noise Temperature Conversions, 490
Noise Power, 467
Noise Temperature Measurement, 514
Noise Temperature of a Cascade, 504
Noise Temperature of a Resistive Attenuator, 534
Power Transfer in Resistors of Unequal Temperatures, 472
Resistor Thermal Noise Current, 451
Resistor Thermal Noise Voltage, 450
S/N Degradation, 495
Satellite Antenna Noise Temperature, 477
Shot Noise, 480
Signal-to-Noise Objectives, 487
System Noise Floor (General Case), 482
System Noise Floor with Input Noise Power, 492

Table of Noise Figure/Noise Temperature, 490
Appendix
  Expected Values of Resistors, 757
  Multipath Fading (I), 766
  Multipath Fading (II), 766
  Signals and Noise, 761
  Thermal Resistor Noise, 758
Cascade I
  Cascade TOI, 683
  Component TOI and Excess Power Gain, 679
  Matching Component TOI to Excess Power Gain, 680
  Minumum Detectable Signal, 642
  Noise Cascades, 664
  Receiving Problems, 686
  SNR of Large Signals, 691
  Spur-Free Dynamic Range, 648
  Three-Element Cascade, 675
  TOI Cascade, 682
  TV Antenna Amplifier, 666
  Wideband Cable Runs, 699
Cascade II
  Cellular Telephone, 706
Filters
  Band-Pass Filters, 207, 218, 219
  Butterworth BPF Attenuation, 241
  Butterworth Filter Equations, 243
  Butterworth High-Pass Filter, 247
  Butterworth LPF Attenuation, 239
  Butterworth Noise Bandwidth, 246
  Butterworth Notch Filter, 249
  Butterworth Pole Positions, 237
  Diplexers, 204
  Dominant Poles and Zeroes in Filter Responses, 159
  Elliptic Filter I, 205
  Elliptic Filter II, 206
  Lossy Butterworth Band-Pass Filter, 245
  Noise Bandwidth Calculation, 233
  Shape Factor of a Band-Pass Filter, 221
  Shape Factor of a Butterworth Band-Pass Filter, 242
  Transfer Function Ultimate Phase, 152
  Transfer Function Ultimate Roll-Off Rate, 149
Introduction
  Bessel Functions and FM, 117
  Bessel Functions and PM, 125
  Characteristic Impedance, 33
  dB Power Conversions, 4
  dBHz, 11
  dBK, 11
  dBm and dBW, 7
  FM Bandwidth, 121
  FM Bandwidth Under Small beta Conditions, 130
  Gains and Losses in Cascades, 16
  Measuring the Propagation Velocity, 34
  Measuring Velocity Factor, 35
  Mismatch Loss, 47
  Mismatched Transmission Line, 40
  Parasitic Components, 26
  Power Amplifiers and Decibels, 14
  Power Measurements, 9, 10
  Reflection Coefficient, 40
  Reflection Coefficient and Reactive Terminations, 44
  Satellite Delay Time, 23
  Small beta and FM, 129
  Small beta and Spectrum Plots, 132
  Transmission Line Voltages (High Z-Load), 52
  Transmission Line Voltages (Low Z-Load), 52
  Transmission Lines, 28
  Wavelength and Frequency, 22
Linearity
  Amplifier Power Gain, 546
  Amplifier Second-Order

INDEX | 771

Measurements, 596
Amplifier Third-Order
  Measurements, 598
Measured Third-Order Intercept,
  592
Nonlinearities and Deviation, 628
Satellite Transponders and
  Linearity, 601
Second-Order Distortion and the
  FM Broadcast Band, 558
Second-Order Distortion in the
  Cellular Telephone Band, 570
Separating Spurious Signals from
  Real Signals, 585
SOI and Spurious Power Levels,
  584
SOI Cascade (Coherent Addition),
  619
SOI Cascade (Noncoherent
  Addition), 621
Spurious Signals and TOI, 593
Third-Order Distortion and the
  FM Broadcast Band, 561
Mixers
  Cellular Telephone, 310, 311
  Cellular Telephone and Spectrum
    Inversion, 294
  Commercial FM Radio, 299
  Conversion Equations, 274
  DC and RF:IF Rejection, 315
  High-Side and Low-Side LO, 291
  Mixer Conversion Loss, 277
  Mixer Conversion Loss and
    Isolation, 279
  Mixer Spurious Products, 305
  Mixers in Cascade, 283
Oscillators
  Cellular Telephones and Drift, 402
  Frequency Division and Phase
    Noise, 381
  Frequency Multiplication and
    Drift, 404
  Incidental Frequency Modulation,
    390
  Incidental Frequency Modulation
    and SNR, 393
  Incidental Phase Modulation, 385
  Incidental Phase Modulation and
    SNR, 388
  Multiplied Oscillator Phase Noise,
    380
  NCO Rollover
  Oscillator Drift in %, 400
  Oscillator Drift in ppm, 400
  Stability in ppm, hertz and %, 405
  VCO Tuning Constant or VCO
    Gain, 412
Receiving System Design
  Example #1, 736
  Example #2, 746
Excess Gain, 657
Excess Noise Ratio (ENR), 511

F
Filter Terminology, 165
  Group Delay, 173
  Passband, 165
  Stopband, 165
  Transition Band, 165
Filter Types, 166
  Band-Stop Filters, 168
  High-Pass Filters, 166
  Low-Pass Filters, 166
Filters
  Amplitude Equalizer, 228
  Band-Pass, 215, 222
    Comparison of Amplitude
      Responses, 212
    Comparison of Band-Pass Filter
      Types, 222
    Terminology, 215
  Bessel Low-Pass, 187
    Comparison with Butterworth
      Low-Pass, 189
    Phase and Group Delay, 188
  Butterworth Band-Pass, 240

Magnitude Response, 240
Noise Bandwidth, 245
Butterworth Band-Stop, 248
   Magnitude Response Equation, 248
Butterworth High-Pass, 247
   Magnitude Response, 247
Butterworth Low-Pass, 179
   Detailed Description, 236
   Magnitude Response Equation, 238
   Phase and Group Delay, 181
   Pole Positions, 237
Chebychev Low-Pass, 182
   Comparison with Butterworth Low-Pass, 185
   Phase and Group Delay, 186
Element Impedances, 201
Elliptic Low-Pass, 192
   Comparison with Butterworth Low-Pass, 192
   Phase and Group Delay, 194
   Specifications, 194
Equal-Ripple Group Delay, 190
Finite-Impulse Response (FIR), 191
Gaussian, 190
Input Impedance, 199
Isolators, 253
Low-Pass Filter Comparison, 195
   Group Delay Plots, 197
   Passband Amplitude Responses, 198
   Phase Responses, 197
   Pole/Zero Plots, 196
   Stopband Amplitude Responses, 198
   Transient Responses, 207
Noise Bandwidth, 229
Nonlinearities, 254
Power-Handling Capacity, 254
Realizations, 250
   Cavity Filters, 252
   Ceramic Filters, 252
   Ceramic Resonator Filters, 253
   Crystal Filters, 251
   LC Filters, 251
   Surface Acoustic Wave Filters, 253
Terminal Impedances, 199
Time Domain, 187
Transient Response, 207
   Band-Pass Filters, 210
Transitional, 191
Transmitting and Receiving, 225
Vibration Sensitivity, 255
Flicker Noise, 361
FM Radio, 556
   Third-Order Distortion, 561
Frequencies
   Band-Pass Filter, 216
Frequency Accuracy, 399
   Timing Accuracy, 406
Frequency and Wavelength, 21
Frequency Conversion Equations, 705
Frequency Drift. See Frequency Accuracy
Frequency Inversion, 292
Frequency Modulation, 111
Frequency Synthesizer, 348, 408

G

Gain Compression, 543, 599
   Compression Point, 600
Gaussian Noise, 95
Gaussian Probability Density Function (GPDF), 753
Gaussian Statistics, 753
Geometric Center Frequency, 216
Group Delay, 173

H

Harmonic Distortion, 543
Harmonics
   Definitions, 541
High-Pass Filter, 166
High-Side LO (HSLO), 288
H-Parameters, 65
HSLO vs. LSLO, 714

INDEX | 773

I
Image Frequencies, 296
Image Noise, 709
Impedance
    Filters, 199
Incidental Frequency Modulation, 389, 393
Incidental Phase Modulation, 382, 393
Initial Roll-Off, 171, 172
INMARSAT Satellite Link, 3
Input Impedance
Insertion Loss
    Band-Pass Filter, 215
Intercept Point
    Second-Order, 578
    Third-Order, 587
Intermediate Frequency Port, 268
ISOI, 581
Isolators, 253

L
LC Filters, 251
Leeson Model, 360
Limiting
    Oscillator, 362
Linear Systems, 144
Linearity
    Gain Distribution, 657
    Translation in a Cascade, 658
LO Frequency Calculation, 288
Local Oscillator (LO) Port, 269
Lossy Devices, 519
    Attenuation, 522
    Characteristic Impedance, 521
    Gain Performance, 519
    Noise Figures, 530
    Noise Performance, 521
    Receiving Systems, 524
Low-Pass Filters, 166
    Comparison of Transient Responses, 209
    Transient Response Rules of Thumb, 210
Low-Side LO (LSLO), 288

M
Matching, 72
    Complex Loads, 74
    Maximum Power Transfer, 72
    Reasons to Match/Reasons not to Match, 75
Mathematics, decibel, 12
Maximum Power Transfer, 72
Microphonics, 255
Minimum Detectable Signal (MDS), 508, 641
Mismatch Loss, 45
Mixers
    Attenuator, Variable, 338
    Baluns, 320
    Block vs. Channelized Systems, 284
    BPSK Modulator, 339
    Cascade Equation, 283
    Conversion Loss, 276
    Conversion Scheme Design, 285
    Double-Balanced (DBM), 329
    Port Swapping, 333
    Filtering, 274
    Frequency Inversion, 292
    Frequency Translation Equations, 269, 272
    Frequency Translation Mechanisms, 263
    Gain, 519
    High-Side LO (HSLO), 288, 291
    Image Frequencies, 296
    Image Noise, 300
    Impedance Mismatch, 335
    Linearity, 336
    LO Frequency Calculation, 288
    LO Noise, 335
    LO Power, 334
    Low-Side LO (LSLO), 288, 290
    Noise Performance, 519
    Phase Detector, 339

774 | INDEX

Port Interchangeability, 269
Port Nomenclature, 267
   Intermediate Frequency (IF) Port, 268
   Local Oscillator (LO) Port, 269
   Radio Frequency (RF) Port, 268
Port-to-Port Isolation, 278
   LO:IF Isolation, 279
   LO:RF Isolation, 278
   RF:IF Isolation, 279
RF Switch, 337
Second-Order Distortion, 626
Second-Order Response, 264
Single-Balanced Mixers (SBM), 320
Single-Ended Mixers (SEM), 313
Spurious Calculations, 307
Spurious Frequencies, 301
Television Receive-Only (TVRO) Conversion Scheme, 286, 296, 302
Time-Domain Multiply, 267
Zero Hertz, 270

Model
RF Device, 443

Modulation
Bessel Functions, 365
General, 371
Phase Modulation Review, 364
Small beta Approximations, 367, 375, 378

N

Narrowband Systems, 575
Noise
Amplifier Model, 481
Antenna Model, 474
Antenna Temperature, 474, 478
Bandwidth, 464
Coherence Time, 455
Combining Independent Sources, 461
Current Source Model, 450
Equivalency of the Voltage and Current Models, 452
Equivalent Input Power, 481, 490
Equivalent Output Power, 484, 494
Excess Noise Ratio (ENR), 511
Frequency Spectrum, 454
Fundamentals, 446
Gain Distribution, 656
Internally Generated, 479
Mean, 760
Measurement of Component Noise Temperature, 509
   Practical Effects, 515
Noise Factor, 487
Noise Figure, 487
Partition, 480
Physical Temperature, 479
Power Spectral Density, 468
Properties, 454
Recombination, 479
Relative Levels in a Cascade, 662
Response of Voltmeters, 464
RMS Value, 448
Shot, 480
Signal-to-Noise Ratio, 494
Signal-to-Noise Ratio (SNR), 484
Sources, 510
Standard Deviation, 760
Standard Temperature, 467
Statistical Description, 448, 455
Statistics, 760
System Temperature, 484
Variance, 760
Voltage Source Model, 449

Noise Bandwidth of a Filter, 229
Noise Factor, 487
Noise Figure, 487
Cascade Analysis, 497
Definition, 488
Fundamentals, 487
Noise Factor, 489
Noise Power, 488
Noise Temperature, 487
Signal-to-Noise Ratio, 494

Noise, Gaussian, 95
Noisy Resistors in a System, 465, 469

INDEX | 775

Nomenclature, 1
  Decibels vs. Linear, 1
Noncoherent Addition, 680
Noncoherent Summation, 611
Nonlinear Systems
  Cascades, 605
    Coherent vs. Noncoherent
      Addition, 625
    Coherent vs. Noncoherent
      Summation, 608
    Resistors in Parallel, 614
    Second-Order Intercept, 614
    Third-Order Intercept, 606
    Third-Order Intercept Point, 606
  Comparison of Specifications, 603
  Measurement, 595
  Single-Tone Analysis, 552
  Two-Tone Analysis, 556
Nonlinearities
  Detection, 685
  Filters, 254
Normal Statistics. See Gaussian Statistics
Normalized Gaussian Probability Density Function (NGPDF), 754
Numerically Controlled Oscillator (NCO)

O

Oscillator
  Ideal, 345
  Limiting, 362
  Practical, 346
  Spurious Signals, 395
OSOI, 581
OTOI, 590

P

Passband, 165
Phase Detector, 339
Phase Modulation (PM), 121
Phase Noise, 352, 394
  Adjacent Channel Masking, 356

  Amplitude Noise, 354
  Dividers, 379
  Flicker Noise, 361
  Incidental Frequency Modulation, 389
  Incidental Phase Modulation, 382
  Leeson Model, 360
  Mixers, 355
  Multipliers, 379
  Representations, 352
  Resonator Q, 363
Phase-Locked Loop (PLL)
Frequency Synthesizer, 408
Physical Size, 23
  Electrical Effects, 23
  Rules of Thumb, 24
Pole/Zero Thought Experiment, 160
Poles and Zeros
  Bessel Low-Pass Filter, 189
  Butterworth Low-Pass Filter, 179
  Chebychev Low-Pass Filter, 185, 186
  Dominant, 152
    Magnitude Characteristics, 155
    Phase Characteristics, 158
  Elliptic Low-Pass Filter, 193
  Evaluating Pole/Zero Plots, 160
  Group Delay Response, 173
  Magnitude Response, 146, 168
    Complex Frequency, 148
    Ultimate Roll-Off, 149
  Phase Response, 150
    Complex Frequency, 151
    Ultimate Phase, 151
  Thought Experiment, 160
Polynomial Approximations, 550
Port Swapping, 333
Power Supplies, 728
Preselection, 567

Q

$Q(z)$, 758

R

Radio Frequency (RF) Port, 268
Receiving Systems
  Design Example #1, 736
  Design Example #2, 746
Reference Oscillators, 433
Reflection Coefficient, 37
Response
  Re-entrant, 171, 172
Return Loss, 57
RF Switch, 337
Ripple
  Amplitude, 171
RLC, Pole/Zero Review, 144
Roll-Off
  Initial, 171, 172
  Ultimate, 171, 172
Rules of Thumb
  Cascade Linearity, 678
  Distortion, 563
  Gain Distribution Rules, 694
  Harmonic Distortion, 555
  IF Selection Guidelines, 724
  Noise, 664
  Noise Bandwidth of Various Band-Pass Filters, 236
  Noise Properties, 458
  PLL Frequency Synthesizer
  Practical Design of Receiving Systems, 726
  TOI, 678
  Transinient Response of Low-Pass Filters, 210

S

Saturated Output Power, 601
Second-Order Intercept Point, 578
Series RLC, Pole/Zero Review, 144
Shape Factor, 220
Signal Standards, 6
Signals, 86
  Comparison of FM and PM Waveforms, 122
  Double Sideband Amplitude Modulation, 105
    Effects of Group Delay, 176
    Phasor Representation, 176
  Frequency Modulation, 111
    Bandwidth (Carson's Rule), 119
    Bessel Functions, 115
    Small beta Approximations, 127
  Gaussian Noise, 95
    Zero Crossings, 98
  Nonlinear Systems, 626
  Phase Modulation, 121
    RF Bandwidth, 126
  Sine Wave and Gaussian Noise, 99
    Equivalent Phase and Amplitude Deviation, 103
  Sine Waves (Ideal, Noiseless), 86
  Sine Waves (Two, Ideal, Noiseless), 89
    Equivalent Phase and Amplitude Deviation, 94
Signal-to-Noise Ratio (SNR), 484, 494
Sine Waves (Ideal, Noiseless), 86
Sine Waves (Two, Ideal, Noiseless), 89
Single-Balanced Mixer (SBM), 320
Single-Ended Mixer (SEM), 313
Single-Tone Analysis, 552
Small beta Approximations, 127, 367
  Frequency and Phase Modulation, 127
SOI, 578
S-parameters, 66
Speed of Light, 23
  Approximations, 23
  Numerical Value, 23
Spurious Calculations, 307
Spurious Frequencies in Mixers, 301
Standard Temperature, 467
Standards, Signal

INDEX | 777

dBf, 8
dBHZ, 11
dBK, 11
dBm, 7
dBmV, 9
dBV, 8
dBW, 7
Volume Units, 8
Statistics
Complimentary Error Function, 758
Approximations for Q(z), 759
Gaussian, 753
Gaussian Probability Density Function, 753
Noise, 760
Normalized Gaussian Probability Density Function (NGPDF), 754
Stopband, 165
Summaries, Chapter
Amplifier and Noise Data, 536
Cascade I Design, 700
Cascade II Design, 750
Filter Design, 255
Introduction Data, 133
Linearity Design, 633
Mixer Design, 340
Oscillator Design, 435
Superposition, 542
Surface Acoustic Wave (SAW) Filters, 253
System Noise Temperature, 484
Systems
Linear, 542
Linear and Nonlinear Systems, 542
Narrowband and Wideband, 575
Nonlinear, 549
Voltage Transfer Curve, 543
Weakly Nonlinear, 543

T
Tables
Absolute Power Ratios Expressed in dBm, 13

Aggregate Receiver Control, 745
Attenuation of Three Types of Cable, 698
Component Contributions to Cascade Linearity, 676, 677, 679, 680
Decibel Equivalents of Linear Power Ratios, 13
ENR vs. Frequency, 511
Frequency and Wavelength, 22
Frequency Synthesizer VCO Band Breaks, 741
IF and Local Oscillator Frequencies, 742
Matching Component TOI to Excess Power Gain, 677
Measured and Calculated Noise Figures of a Resistive Attenuator at Various Physical Temperatures, 534
Measured Noise Figure of an Attenuator at Various Physical Temperatures, 533
Mixer Spurious Components, 303
NCO Phase Accumulator Rollover, 420
NCO Phase Trunction, 423
Noise Bandwidth of Butterworth Filters, 246
Noise Figures and the Equivalent Noise Temperatures, 490
Noise Performance Summary, 671
Nonlinear Distortion Equations, 566
Numerical Integration of Filter Passband, 235
Oscillator Aging and Timing Inaccuracies, 407
Peak Values of Gaussian Noise, 458
Physical Size and Electrical Characteristics, 24
Possible Spurious Products, 686
Power in Mixer Spurious Components, 305
Resistor Noise Voltages, 525

RF Preselection Filter Band Breaks, 740
Rise Time Related to Bandwidth and Physical Dimensions, 29
Television Antenna Cascade, 668
Television Antenna Cascade with Amplifier at Antenna, 671
Television Antenna Cascade with Amplifier at Cable Mid-Point, 670
Television Antenna Cascade with Amplifier at Television, 669
Television Receive-Only (TVRO) Conversion Scheme, 286
Third-Order Intercept Point
  Measurement, 687
TOI, 587, 590
Tone Placement, 687
Transient Response of Filters, 207
Transition Band, 165
Transmission Lines, 25
  Characteristic Impedance, 32
  Clock Skew, 26
  Complex Reflection Coefficient, 43
  Electrical Models, 30
  Identifying, 27
  Input Impedance with Arbitrary Load, 59
  Mismatch Loss, 45
  Propagation Time, 25
  Propagation Velocity, 33
  Pulsed Input Signals, 36
  Reflection Coefficient, 37
  Return Loss, 57
  Sine-Wave Input Signals, 48
  Velocity Factor, 35
  Voltage Minimums and Maximums, 50
  Voltage Standing Wave Ratio (VSWR), 53
  Waves Propagating, 32
Tuning Speed, 407
Two Ports, 62
  H-Parameters, 65
  Relationships, 70
  S-Parameters, 66
  Y-Parameters, 64
  Z-Parameters, 63
Two-Tone Analysis, 556

U

Ultimate Attenuation, 172
Ultimate Bit Error Rate, 358
Ultimate Phase, 151
Ultimate Roll-Off, 149, 172
Ultimate Signal-to-Noise Ratio, 358
Upconversion, 714

V

Voltage Standing Wave Ratio (VSWR), 53
Volume Units, 8

W

War Stories
  Amplifier Stability (I), 85
  Amplifier Stability (II), 85
  Amplifier Stability and Terminal Loading, 696
  Bandwidth of Digital Signals, 29
  Filter Vibration, 255
  Frequency Synthesizer Design, 83
  Reflection Coefficients and Speech Synthesis, 59
  Specialized Amplifiers, 85
  Unintended Effects in Conversion Schemes, 720
  VSWR and Microwave Ovens, 55
  VSWR and Screen Room Testing, 55
Wavelength and Frequency, 21
Wideband Systems, 575

Y

Y-Parameters, 64

Z

Z-Parameters, 63